Results and Problems in Cell Differentiation

Volume 74

Results and Problems in Cell Differentiation is an up-to-date book series that presents and explores selected questions of cell and developmental biology. Each volume focuses on a single, well-defined topic. Reviews address basic questions and phenomena, but also provide concise information on the most recent advances. Together, the volumes provide a valuable overview of this exciting and dynamically expanding field. This series is indexed in PubMed.

Malgorzata Kloc • Jacek Z. Kubiak •
Marta Halasa
Editors

Monocytes and Macrophages in Development, Regeneration, and Disease

Springer

Editors
Malgorzata Kloc
Houston Methodist Research Institute
and The Houston Methodist Hospital
Houston, TX, USA

Jacek Z. Kubiak
Institute of Genetics and Development
of Rennes, CNRS
University of Rennes
Rennes, France

Marta Halasa
Houston Methodist Research Institute
and The Houston Methodist Hospital
Houston, TX, USA

ISSN 0080-1844 ISSN 1861-0412 (electronic)
Results and Problems in Cell Differentiation
ISBN 978-3-031-65943-0 ISBN 978-3-031-65944-7 (eBook)
https://doi.org/10.1007/978-3-031-65944-7

This Springer imprint is published by the registered company Springer Nature Switzerland AG
The registered company address is: Gewerbestrasse 11, 6330 Cham, Switzerland

Preface

This volume consists of three parts covering various aspects of monocyte/macrophage involvement in the organ and tissue structure and functioning under healthy and diseased conditions. The first part describes the role of resident macrophages in the homeostatic maintenance of the tissue and how macrophages interact with the extracellular matrix in healthy tissues and during tissue repair and remodeling. It reviews how dietary carbohydrates, fats, and amino acids, by induction of cellular metabolic shifts, affect the immunomodulatory functions of macrophages in healthy and inflamed tissues. It also describes how the macrophage cell cycle and circadian rhythm affect the immune and non-immune functions of macrophages. The second part of the volume focuses on tissue/organ-specific subpopulations of macrophages. It describes the origin and immune- and non-immune roles of testicular, adipose tissue, and liver macrophages. In adult mammals, testicular macrophages, besides their immunological functions, regulate male hormone synthesis by Leydig cells, spermatogonia proliferation and differentiation, sperm formation, and testis development. In the fetal testes, macrophages regulate testis vasculature. Adipose tissue macrophages (ATM) are life-long constituents of adipose tissue. They regulate lipolysis, mitobiogenesis, and thermogenesis in adipocytes. In response to lipid overload in obesity, the ATMs switch on the pro-inflammatory processes. During homeostasis, liver resident macrophages, the Kupffer cells, perform phagocytosis, present antigens, and regulate metabolism. During liver injury, Kupffer cell numbers decrease, and monocyte-derived macrophages take over, driving and resolving injury and regulating inflammation, fibrosis, and regeneration. The third part of the volume covers the subjects of macrophage involvement in the development and progression of respiratory dysfunction, lung fibrosis, osteoporosis, and breast cancer. It also explores how histone deacetylase activities in macrophages affect chronic rejection of transplanted organs, how monocytes/macrophages interact with helminth parasites, how mitochondria, metabolism, and circadian rhythmicity affect macrophage interactions with tuberculosis bacteria, and how, by being a viral reservoir, macrophages participate in HIV infection of the nervous system. We hope that this book covering a variety of macrophage features and functions will inform and inspire the experts and newcomers to the macrophage field.

Houston, TX, USA Malgorzata Kloc
Rennes, France Jacek Z. Kubiak
Houston, TX, USA Marta Halasa

Contents

Part I
Nonimmunologic and Immunologic Functions of Monocytes/Macrophages

Chapter 1
Homeostatic Functions of Tissue-Resident Macrophages and Their Role in Tissue Maintenance

Luís Crisóstomo, Ae Mäkinen, Besmir Hyseni, Hans Bergman, and Alexander Mildner

Abstract Tissue-resident macrophages are best known for their indispensable role in immunological reactions, where they contribute to immune defense and resolution of inflammation. However, recent studies have also uncovered that they provide crucial tissue-specific functions that support organ homeostasis and maintenance. Accordingly, defects in macrophage function or development can disrupt the delicate balance of organ homeostasis, leading to pathological conditions. Therefore, understanding the functions and development of macrophages within a tissue is critical for comprehending the interplay between immune and stromal cells, which together maintain organ physiology. This knowledge has clinical implications, such as in organ transplantation or irradiation, where monocyte-derived cells with different functions may replace the original macrophage population. In this chapter, we aim to provide an overview of the tissue-specific homeostatic functions of various macrophage populations, emphasizing that macrophages are essential components of each organ and play a vital role in ensuring the organism's survival, beyond their role in immunity.

Luís Crisóstomo and Ae Mäkinen contributed equally.

L. Crisóstomo · B. Hyseni · A. Mildner (✉)
MediCity Research Laboratory, University of Turku, Turku, Finland

InFLAMES Research Flagship, University of Turku, Turku, Finland
e-mail: alexander.mildner@utu.fi

A. Mäkinen · H. Bergman
MediCity Research Laboratory, University of Turku, Turku, Finland

M. Kloc et al. (eds.), *Monocytes and Macrophages in Development, Regeneration, and Disease*, Results and Problems in Cell Differentiation 74,
https://doi.org/10.1007/978-3-031-65944-7_1

1.1 Introduction

Phagocytosis is a cellular process by which cells engulf and digest foreign particles such as pathogens or cellular debris to remove them from the body. The process of phagocytosis was first observed by Alexander Ecker in 1847 when he described the presence of numerous red blood cells within vacuole-rich splenic cells (Ecker 1847). That a specialized cell type, so-called phagocytes and especially macrophages ('large eaters', from the Greek word *μακρός* (*makrós*) = large, *φαγεῖν* (*phagein*) = to eat), are mainly involved in the cellular uptake and immune response was discovered by Ilya Mechnikoff's ground-breaking research (Metchnikoff 1905). Mechnikoff conducted his first experiments on starfish larvae and observed, after inserting a rose thorn into the transparent larval body, the accumulation of cells that were involved in phagocytosis at the site of the wound, suggesting their critical function in the innate immune response. From today's point of view, two things are remarkable about this observation: of course, the identification of phagocytes as immune sentinels by Mechnikoff and the existence of a functioning innate immune system in starfish larvae, a conclusion that could be easily overlooked. In fact, all eukaryotic organisms, starting from the single-cell amoeba *Dictyostelium discoideum*, rely on phagocytes as the main innate immune component. Under unfavorable conditions, *D. discoideum* transforms into a social organism and creates multicellular aggregates, known as the slug state, which can travel greater distances in search of better conditions. In this state, the newly assembled multicellular "organism" initiates the differentiation of a specialized cell type called sentinel cells (Chen et al. 2007), which exhibit phagocytic, extracellular trap-forming, and reactive oxygen species-producing functions (Zhang et al. 2016). Sentinel cells share these functions with macrophages and granulocytes in humans. Similar effector cell types can be found in other eukaryotes such as sponges (amebocytes in their mesoglea), cnidarians (interstitial cells), invertebrates (hemolymph-borne hemocytes), and coelomate animals (coelomocytes), showing that they and their functions have been conserved throughout the entire phylogeny of eukaryotes up to macrophages and granulocytes in more highly developed species such as humans. Consequently, the innate immune system, with phagocytes as the primary effector cell population, is present in all eukaryotes. In contrast, an advanced immune system with an additional adaptive arm is only found in vertebrates, beginning with early-jawed fish (emergence of the RAG gene) or jawless fish (the variable lymphocyte receptor (VLR) family). To put it another way, more than 95% of eukaryotes (all invertebrates) rely exclusively on innate immunity, while the remaining 5% of eukaryotes (vertebrates) additionally possess an adaptive immune system.

The primary function and evolutionary force behind phagocytes may be pathogen elimination and immune defense, as exemplified by the above-mentioned case of multicellular *D. discoideum,* but phagocytes and especially macrophages also play key homeostatic roles during development, aging, and wound healing. Consequently, macrophages are besides red blood cells and megakaryocytes, one of the first hematopoietic cell types that appear during embryogenesis and that seed every organ at these early developmental stages. Their main function during development is the removal of apoptotic cells, which accumulate during organ expansion and organization. The importance of macrophages for proper embryonic development can be observed in hemocyte-deficient *Drosophila melanogaster*. When hemocytes are depleted during embryogenesis by the genetic induction of apoptosis specifically in this cell lineage, the animals show embryonic lethality with strong developmental defects in the central nervous system (CNS) (Defaye et al. 2009). However, no morphological changes were detected when hemocyte depletion was induced during the later larval stage, highlighting the important role of phagocytes during embryogenesis. The coordinated and complex functions of macrophages in development can also be observed during epimorphic regeneration. The amphibian model organism axolotl, an aquatic salamander, utilizes macrophages to promote limb regeneration after amputation. The depletion of macrophages in this model leads to excessive fibrosis and collagen accumulation, which permanently hinders limb regeneration (Godwin et al. 2013). This demonstrates that macrophages are crucially involved in tissue organization, remodeling, and outgrowth.

From these examples, it is easy to infer that all macrophage populations are similar, and that phagocytosis is a uniform process, but this is not actually the case. Therefore, an important question is: what determines the versatile identities and functions of macrophages in tissues? Within the past decade, a growing number of studies have contributed to providing an answer to this question. Within their tissue of residence, macrophages are exposed to local environmental factors and signals that vary substantially depending on the specific tissue and its cellular composition. These tissue-specific signals and cell contacts/interactions induce signaling pathways, influence chromatin accessibility of genes and gene regulatory sequences, and drive tissue-specific gene programs in macrophages that ultimately determine their organ-specific cellular functions. For instance, the CNS is composed of specialized glial cells, which clearly differ from the cells found in the lungs or spleen and form a unique environment. Microglia, the macrophages of the CNS, receive factors from neurons that imprint microglial functions that reciprocally allow them to participate in neuronal maintenance and synaptic pruning. Alveolar macrophages in the lungs are *per se* not able to perform the same functions as microglia. Instead, they are exposed to high concentrations of surfactant lipids that need to be degraded. Accordingly, lung epithelial cells provide specific factors that enable alveolar macrophages to catabolize and degrade these lipids. Red pulp macrophages in the spleen on the other hand are exposed to senescent red blood cells. Phagocytosis and degradation of these iron-containing cells requires a specific gene program that is induced by red blood cell-derived heme in macrophages. These examples indicate that the surrounding tissue dictates the identity and function of tissue-resident macrophages. This further shows that macrophages and tissue-specific cells engage in bidirectional communication and mutually influence each other to secure the functionality of an organ.

Given the emerging evidence that macrophages form an integral part of every tissue by fulfilling organ-specific functions and developmental cues, this chapter concentrates on the homeostatic role of various tissue-resident macrophages, including alveolar macrophages in the lungs, splenic macrophages, Kupffer cells in the liver, cardiac macrophages, microglia in the CNS, macrophages in the kidneys, and adipose tissue macrophages. This chapter summarizes how disruption of macrophage homeostasis in these selected organs can affect organ development and function and consequently lead to pathological manifestations. For a detailed description of macrophage origin or their role in pathology, the reader is referred to other recent, excellent reviews (Hoeffel and Ginhoux 2018; Park et al. 2022; Lazarov et al. 2023; Mass et al. 2023).

1.2 Precursors of Tissue-Resident Macrophages

To better comprehend and contextualize the functions of tissue-resident macrophages, it is important to understand their ontogeny and developmental processes. For decades, tissue-resident macrophages in humans and mice were believed to be derived from circulating monocytes. Even though monocytes can differentiate into tissue-resident macrophages, the necessity of macrophages to perform apoptotic cell clearance during organ development precedes the appearance of monocytes during embryogenesis, at least in the case of the CNS (Alliot et al. 1999; Hoeffel et al. 2015). This indicates that another origin of macrophages must exist and indeed it was shown that tissue-resident macrophages can derive from two sources: from precursors in the yolk sac (YS) and monocytes, either fetal or adult. Currently, two scenarios are being discussed:

The hematopoietic potential of the YS was first observed in mammals during two distinct phases, also referred to as "waves". The earliest hematopoietic precursor cells in the mouse embryo derive from the posterior plate mesoderm of the extra-embryonic YS around embryonic days E7–E7.5. These "early" or primitive erythro-myeloid progenitors (EMPs) can differentiate into megakaryocytes, (nucleated) red blood cells, and primitive macrophages (Fig. 1.1; Scenario 1). Using a fate-mapping model based on the expression of *Runx1*, Ginhoux and colleagues observed that labeling YS hematopoietic cells on day E7 led to labeled tissue-resident macrophages on days E10–E13, which in some tissues such as the brain persisted into adulthood (Ginhoux et al. 2010). This indicates that "early" EMPs are the precursors of tissue-resident macrophages, at least of microglia in the CNS. Starting from around E8.0, hematopoiesis expands to the YS hemogenic endothelium, initiating the second transient definitive wave of "late" EMPs. "Late" EMPs can also give rise to the erythroid and megakaryocytic lineages, but in addition, they can also produce macrophages, mast cells, and granulocytes (Palis et al. 1999; Bertrand et al. 2005). These *Myb*-dependent EMPs can seed other tissues, especially the fetal liver, where they differentiate into fetal monocytes that subsequently colonize other tissues and further differentiate into tissue-resident macrophages (Hoeffel et al. 2015) (Fig. 1.1). When "late" EMPs are targeted using the *Runx1*-dependent fate-mapping approach after E8, the number of labeled tissue-resident macrophages in embryos increases significantly, whereas labeled microglia are nearly absent in the adult brain (Ginhoux

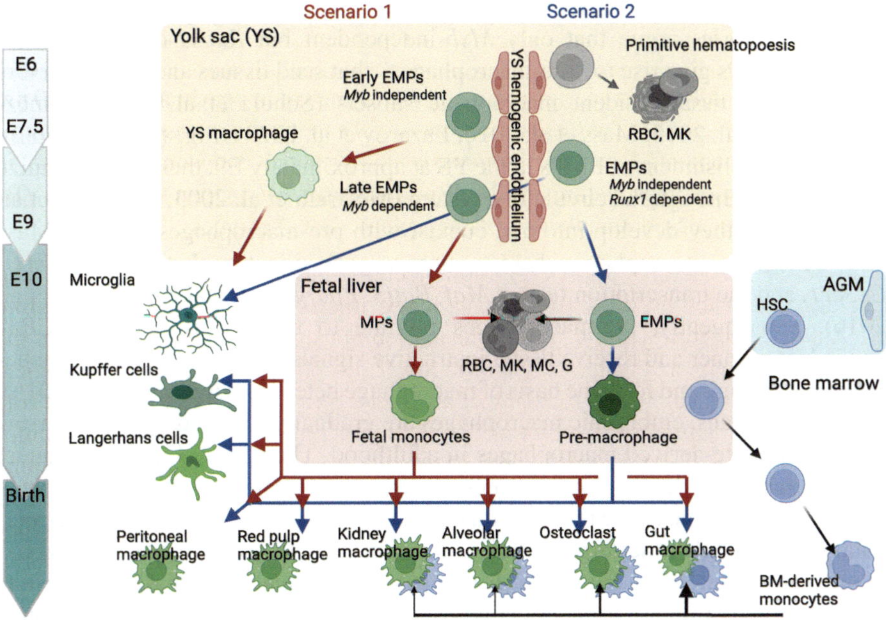

Fig. 1.1 Development of tissue-resident macrophages in mice. Two main differentiation routes are currently being discussed based on (Lazarov et al. 2023; Hoeffel and Ginhoux 2018). Scenario 1 (left; in red) postulates that early *Myb*-independent erythromyeloid precursors (EMPs) give rise to red blood cells (RBC), megakaryocytes (MK), and yolk sac (YS) macrophages, which then infiltrate the developing central nervous system and differentiate into microglia. Late *Myb*-dependent EMPs colonize the liver, where these myeloid precursors (MPs) can differentiate into RBC, MK, mast cells (MC), granulocytes (G), and fetal monocytes. Fetal monocytes are the main progenitors of tissue-resident macrophages. Some macrophage pools can then be progressively replaced or supplemented with bone marrow-derived monocytes during adulthood. In scenario 2 (right; in blue), primitive hematopoiesis results in the production of RBC and MK, but not EMPs. EMPs arise later and are *Myb*-independent and *Runx1*-dependent. These cells can expand to the fetal liver and can differentiate into RBC, MK, MC, G, and pre-macrophages. From there, pre-macrophages infiltrate in a CX3CR1-dependent manner all organs and their tissue-specific identity is regulated by specific sets of transcription factors. Definitive hematopoiesis begins in the aorta-gonad-mesonephros (AGM), from which hematopoietic stem cells (HSC) migrate to the liver and subsequently to the bone marrow. HSCs give rise to monocytes that infiltrate certain tissues and differentiate into tissue-resident macrophages. Created with BioRender.com

et al. 2010). This suggests that most tissue-resident macrophages originate from "late" EMPs, while microglia derive from "early" EMPs, a hypothesis supported by subsequent studies (Hoeffel et al. 2015). However, the classification of EMPs into "early" and "late" stages has also been met with skepticism. *Runx1* is not specific for "early" and "late" EMPs and so far, no marker or fate-mapping system could be established that definitely distinguishes these functionally overlapping cell types. It was also questioned if *Runx1* is expressed by "early" EMPs (Gomez-Perdiguero and Geissmann 2013; Lazarov et al. 2023), and therefore the contribution of this subset to macrophage establishment is controversially discussed in the field

(Schneider and Kopf 2015; Hoeffel and Ginhoux 2018; Lazarov et al. 2023). Instead, some laboratories argue that only *Myb*-independent but *Runx1*-dependent non-primitive EMPs give rise to "pre-macrophages" that seed tissues and are precursors of almost all tissue-resident macrophage subsets (Schulz et al. 2012; Gomez-Perdiguero et al. 2015; Mass et al. 2016; Lazarov et al. 2023) (Fig. 1.1; Scenario 2). After the establishment of EMPs in the YS at approximately E9, these cells colonize the fetal liver through the circulatory system (McGrath et al. 2003; Rantakari et al. 2016), where they develop into and coexist with pre-macrophages at E9.5 (Mass et al. 2016). Pre-macrophages share a common gene signature that includes *Csf1r*, *Cx3cr1*, and the transcription factors *Maf*, *Batf3*, *Pparg*, *Irf8*, and *Zeb2* (Mass et al. 2016). Subsequently, pre-macrophages migrate to the organs in a CX3CR1-dependent manner and receive tissue-instructive signals, which initialize the tissue-specific signature and form the basis of macrophage heterogeneity (Mass et al. 2016).

In most organs, embryonic macrophages are gradually replaced by bone marrow (BM) monocyte-derived macrophages in adulthood. The speed of the replacement process varies between organs. Tissue-resident macrophages of some organs such as the intestine or heart are replenished quickly by classical monocytes during post-natal phases, while other organs such as the skin or CNS are replenished slowly, if at all (Patel et al. 2021) (Fig. 1.1). It is important to mention that all these results were obtained from experiments conducted in laboratory inbred mice that were housed under specific pathogen-free conditions. It is essential to determine whether the repopulation of tissue-resident macrophages by monocytes occurs more frequently in humans who are constantly exposed to changing environmental influences that cause stress, infections, and sleep deprivation. As these factors can affect macrophage competition, further research is needed to translate the findings obtained in mice with the situation in humans (Hume 2023).

One reason for the different replacement rates between organs observed in mice may be related to the inflammatory or stress state of a particular organ. For example, intestinal macrophages are exposed to microbial products that can cause immune activation and oppose their homeostatic self-renewal capacity. Moreover, shear stress, as observed in the heart due to muscle contraction, might contribute to a compromised macrophage niche that is replenished by monocytes. Under specific conditions, such as irradiation or transplantation (Ajami et al. 2007; Mildner et al. 2007), infection (Li et al. 2022), or macrophage niche liberation by depletion models, such as diphtheria toxin injections (Lund et al. 2018), monocytes can rapidly colonize all affected organs and differentiate into tissue-resident macrophage subsets. However, it remains a matter of debate whether monocyte-derived cells that infiltrate a compromised niche completely adapt to the new tissue environment and are equivalent to and transcriptionally indistinguishable from their embryonic counterparts in terms of self-renewal capacity and function (Lund et al. 2018; Shemer et al. 2018). Nevertheless, the experimental findings that monocytes can replenish macrophages under specific conditions have led to the "macrophage niche" concept (Guilliams et al. 2020), which considers and incorporates factors like tissue accessibility, availability, and macrophage precursor plasticity, thereby representing an excellent theoretical model for the understanding of macrophage replenishment by monocytes.

1.3 Development and Homeostatic Function of Alveolar Macrophages

The primary function of the lungs is to facilitate gas exchange, i.e., mediate the diffusion of oxygen from inhaled air into the bloodstream within the lung alveoli while carbon dioxide is expelled during exhalation. Since the lungs are constantly exposed to external factors that are part of the inhaled air, the lungs also play a vital role in immune defense. Alveolar macrophages (AMs) represent primary tissue-resident immune cells of the lungs and serve as the first line of defense against pathogens such as bacteria, viruses, dust, and man-made pollutants. Located in the alveolar lumen and therefore outside of the body, these cells play a crucial role in maintaining effective immune surveillance of the lung tissue. However, the alveolar space undergoes significant changes in oxygen pressure and microflora, and AMs must be able to adapt to these changes to maintain homeostatic tissue integrity. Therefore, the development and function of AMs are closely linked to the development and function of the lung tissue and are regulated by tissue-specific signals that induce a unique transcriptomic signature in AMs.

During mouse embryogenesis, the lungs are colonized by various macrophage lineages. Fetal lung macrophages derived from EMPs in the yolk sac are the first macrophages to inhabit the pulmonary tissue, preceding E12.0 (Guilliams et al. 2013). Subsequently, liver-derived fetal monocytes accumulate in the interstitial lung tissue starting at approximately E14.0. At the same time, the number of fetal lung macrophages begins to decrease before they almost completely disappear by the first postnatal week (Guilliams et al. 2013; Tan and Krasnow 2016). A small fraction of F4/80^{high} Mac2^{low} fetal-derived cells remained detectable in the lung interstitium along the mesothelium as well as perivascularly in adult *Runx1*$^{Cre/wt}$ *Rosa26*$^{TdTomato/TdTomato}$ reporter mice (Tan and Krasnow 2016). Although the precise function of fetal lung macrophages is not yet clear, they are likely involved in the elimination of apoptotic pulmonary cells during organ development.

Once in the lung tissue, fetal monocytes are exposed to local environmental factors that induce their differentiation into immature alveolar macrophages (preAMs) (Guilliams et al. 2013). One important factor involved in this process is the granulocyte/macrophage colony-stimulating factor (GM-CSF, encoded by *Csf2*), which is expressed by type 2 alveolar epithelial cells and influences the development and maintenance of preAMs and mature AMs throughout their lifetime (Gschwend et al. 2021). In addition, TGF-β was shown to regulate early AM differentiation in an autocrine manner and reduce AM frequency when deleted (Yu et al. 2017).

Fate-mapping and parabiosis experiments in mice have indicated that monocytes rarely infiltrate the alveolar niche and do not develop into AMs during adulthood

(Hashimoto et al. 2013), although they can differentiate into AMs when transferred to an empty alveolar niche (van de Laar et al. 2016). Accordingly, 5–10% of the AM population undergoes homeostatic proliferation, as measured by Bromodeoxyuridine incorporation (Guilliams et al. 2013). However, experiments with newly developed mouse models that allow the permanent labeling of the entire monocyte pool indicated that monocytes do develop into AMs during aging and account for 50–80% of all tissue-resident macrophages in the aged lung (Liu et al. 2019b; Zhang et al. 2024). In addition, *Ccr2*-deficient humans are characterized by reduced numbers of AMs, suggesting that BM-derived monocytes might differentiate into AMs and are recruited to the lung in a CCR2-dependent manner (Neehus et al. 2024).

After the differentiation of preAMs, the external environment and high oxygen levels after birth appear to be crucial for the further maturation process of AMs. Mice lacking the L-plastin (encoded by *Lcp1*) protein, which helps with accurate AM colonization in the alveolar lumina, show compromised development from preAMs to mature AMs (Todd et al. 2016). Additionally, a deficiency in the Von Hippel-Lindau tumor suppressor protein (encoded by *Vhl*) desensitizes AMs to oxygen after birth and inhibits the formation of mature AMs (Izquierdo et al. 2018).

Because the number of alveoli exceeds the number of AMs in the lungs (Westphalen et al. 2014), AMs need to be motile cells to survey all alveoli for pathogen colonization. Indeed, confocal intravital microscopy has indicated that AMs constantly patrol multiple alveoli through the pores of Kohn (Neupane et al. 2020). While this migration might secure immune surveillance, AMs are also involved in homeostatic functions, of which removal of pulmonary surfactant lipoproteins is the most important. Surfactant components are constantly secreted by type II alveolar epithelial cells and form a thin fluid film on the air-liquid interface that covers the entire alveolar wall (Fig. 1.2). Surfactants play an immunological role as opsonizing agents of pathogens but are mainly involved in reducing the surface tension of the alveoli, thereby preventing alveoli collapse during respiration. However, constant production requires constant degradation to prevent accumulation and to maintain physiological balance. Therefore, the turnover of these surfactant components is tightly regulated: while approximately half of the surfactant is recycled and catabolized by type II alveolar epithelial cells, the other half is cleared by AMs. The regular consumption of substantial quantities of phospholipids can result in lipotoxicity, necessitating specialized metabolic machinery in AMs to facilitate the complete breakdown of lipids and the recycling of cholesterol. The induction of these lung-specific catabolic programs is ensured by GM-CSF, which is secreted by type II alveolar epithelial cells and acts as an instructive cytokine. Defects in surfactant degradation can cause accumulation of surfactant in the alveoli, a syndrome termed pulmonary alveolar proteinosis (PAP). Primary PAP is a condition characterized by impaired GM-CSF signaling, which can either be caused by hereditary factors, such as mutations in *Csf2* or its receptor *Csf2r*, or by autoimmunity resulting from auto-antibodies against GM-CSF. Therefore, reduced numbers or dysfunctional AMs can only insufficiently clear surfactants, which accumulate, hinder proper gas exchange, and cause PAP. These findings highlight the essential constitutive role of GM-CSF in AM and surfactant homeostasis. Secondary PAP, in contrast, is defined by reduced

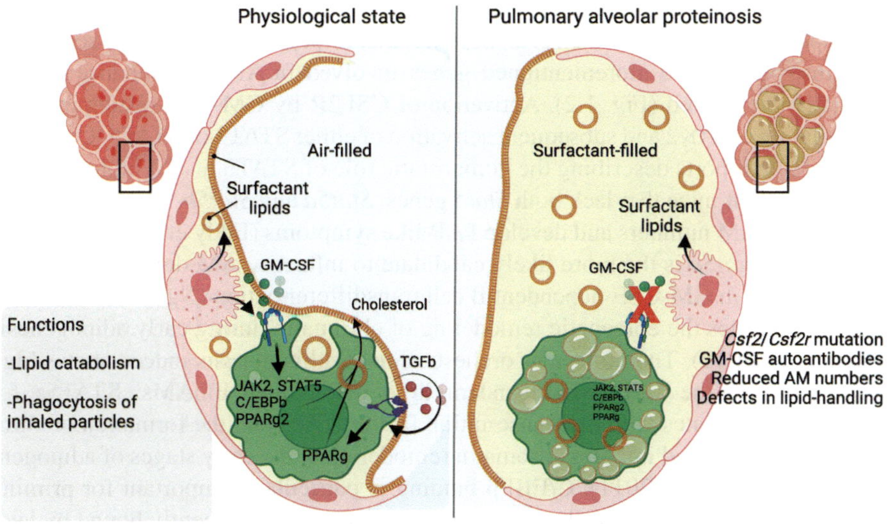

Fig. 1.2 Representation of alveolar macrophage function. Left: Under physiological conditions, alveolar macrophages are dependent on GM-CSF, which is secreted by lung type II epithelial cells. GM-CSF induces JAK2 activation in alveolar macrophages, which activates *Pparg* isoform 2 via C/EBPβ, and possibly STAT5. *Pparg* expression can also be induced by autocrine TGF-β signaling. PPARγ is crucial for the catabolism of surfactant lipids into cholesterol. Right: When alveolar macrophage functions are disturbed either because of autoantibodies against GM-CSF, mutations in the GM-CSF pathway, or genes relevant for alveolar macrophage maintenance and function such as lipid handling capacity, alveolar macrophages are not able to catabolize surfactant lipids, which subsequently accumulate, reduce air-filled space, and cause pulmonary alveolar proteinosis. Created with BioRender.com

numbers or impaired functions of AMs due to hematological disorders or other gene mutations. This is for instance the case in *Ccr2*-deficient patients who suffer from secondary PAP due to recruitment defects of monocytes into the alveoli (Neehus et al. 2024).

Reconstitution of AM numbers and function, for instance, by subcutaneous or inhaled administration of recombinant GM-CSF, is an effective treatment option for patients with *Csf2* deficiency and restores the expression of lipid handling genes such as PPARγ and CD36 in AMs (Seymour and Presneill 2002). Deficiency of several genes involved in lipid handling have been identified to induce PAP-like symptoms in mice, including *Pparg* (Schneider et al. 2014), *Cebpb* (Cain et al. 2013; Dörr et al. 2022), *Abcg1* (de Aguiar Vallim et al. 2017), *Abca1* (Bates et al. 2005), *Stat5* (Eddy et al. 2017), *Tgfb1* (Yu et al. 2017), and *Bach2* (Ebina-Shibuya et al. 2017). Among these genes, the transcription factor PPARγ appears to be essential for lipid metabolism. Myeloid-restricted *Pparg* deficiency in mice leads to the accumulation of surfactants in the lung alveoli and neutral lipids and cholesterol within AMs (Schneider et al. 2014). AM cell numbers were also reduced in these mice, indicating that PPARγ is important for the survival of AMs, or that lipid-laden AMs undergo cell death, leading to a long-term compromised AM pool. Therefore,

PPARγ might prevent lipotoxicity in macrophages, similar to the situation observed in adipocytes (Medina-Gomez et al. 2007). However, the exact molecular sequences and interplay of the aforementioned genes involved in AM maintenance is only partially deciphered (Fig. 1.2). Activation of CSF2R by GM-CSF results in phosphorylation of JAK2 and subsequent activation of either STAT3 or STAT5. Currently, there are no reports describing the homeostatic role of STAT3 in AM biology, while *Stat5*-deficient mice that lack both *Stat5* genes, *Stat5a* and *Stat5b*, are characterized by reduced AM numbers and develop PAP-like symptoms (Eddy et al. 2017). Of the two genes, *Stat5a* is the more likely candidate to influence AM development since it is involved in GM-CSF-dependent B cell transdifferentiation (Nguyen et al. 2024) and orchestrates the epigenetic remodeling of chromatin during early adipogenesis (Jung et al. 2012). The molecular orchestration of adipogenesis indeed shows high similarities to the GM-CSF-dependent lipid metabolism in AMs. STAT5a for instance participates with the transcription factor C/EBPβ in the formation of chromatin 'hotspots' and initiates chromatin remodeling at the early stages of adipogenesis (Siersbæk et al. 2011). C/EBPβ binding is particularly important for priming chromatin regions, because the open chromatin can be subsequently bound by late-acting adipogenic factors, including PPARγ (Siersbæk et al. 2011). Accordingly, a high binding overlap between PPARγ and C/EBPβ is evident during adipogenesis (Siersbæk et al. 2011). C/EBPβ is also involved in AM maintenance as its depletion reduces the number of AMs (Cain et al. 2013). Subsequent research showed that C/EBPβ specifically induces in a GM-CSF-dependent manner the expression of *Pparg* isoform 2 (PPARγ2) (Dörr et al. 2022), which is the main isoform involved in lipotoxicity prevention in adipocytes (Medina-Gomez et al. 2007). Under homeostatic conditions, *Pparg*2 is specifically expressed by AMs, while kidney, spleen, peritoneal, and adipose tissue-resident macrophages only express *Pparg* isoform 1 (PPARγ1) (Dörr et al. 2022). Although PPARγ1 and PPARγ2 are identical except for an additional 30 amino acids at the N-terminus of PPARγ2, both isoforms seem to regulate distinct sets of target genes with genes involved in the PPAR signaling pathway mainly controlled by PPARγ2, at least in white adipose tissue (Hu et al. 2022). Further research is needed to dissect the distinct functions of PPARγ1 and PPARγ2 in macrophage biology, but the high molecular similarity between adipogenesis and AMs in terms of C/EBPβ and PPARγ2 usage might indicate a conserved gene regulatory program beyond these cell lineages.

Not only GM-CSF-induced signaling can drive the expression of *Pparg*. TGFβ was also shown to induce *Pparg*. Myeloid-restricted deficiency of either *Tgfbr2* or *Tgfb1* led to reduced AM development and blocked their differentiation during mouse embryogenesis (Yu et al. 2017). These results indicate that TGFβ functions in an autocrine manner.

The conserved natural killer cell receptor *Nkrp1b* also affects AM maintenance through the control of metabolic genes responsible for lipid turnover (Scur et al. 2022). However, this regulation appeared to be independent of GM-CSF and PPARγ. Instead, increased lipid uptake, storage, and processing pathways can be detected in *Nkrp1b*-deficient AMs, which is accompanied by the downregulation of cholesterol efflux pumps (Scur et al. 2022). Therefore, increased uptake and reduced efflux account for the foamy phenotype of *Nkrp1b*$^{-/-}$ AMs.

In addition to tissue-specific gene expression that characterizes AMs, the metabolic state can influence the biology of macrophages. It was recently shown that a myeloid-restricted deficiency of mitochondrial transcription factor A (encoded by *Tfam*) disrupted oxidative phosphorylation (Oxphos) in macrophages and influenced the development of lipid handling macrophages, particularly AMs (Wculek et al. 2023). *Tfam*-deficiency leads to increased cholesterol accumulation in AMs, cellular stress, and cell-cycle arrest, and culminates in PAP-like syndrome in mice (Wculek et al. 2023). This emphasizes that the ability to perform Oxphos might be a deterministic requirement for proper lipid handling and potentially applicable to other diseases in which macrophages encounter high lipid concentrations (Wculek et al. 2022). However, further research is required to link the metabolic state of macrophage subsets with tissue-specific gene signatures. In particular, it needs to be shown how tissue-specific niche instruction influences the metabolism of macrophages, and *vice versa*.

Of note, the lung interstitium also contains two or three other macrophage subsets, which develop in a GM-CSF independent but macrophage colony-stimulating factor (M-CSF; encoded by *Csf1*) dependent manner. The two main subsets are phenotypically characterized as Lyve1^{low}MHCII^{hi} and Lyve1^{hi}MHCII^{low}, respectively (Chakarov et al. 2019). Lyve1^{low}MHCII^{hi} interstitial macrophages are associated with nerve bundles, whereas Lyve1^{hi}MHCII^{low} macrophages are in proximity to blood vessels. Depletion of Lyve1^{hi}MHCII^{low} cells influences bleomycin-induced fibrosis, but the physiological role of both subsets in lung tissue maintenance is unknown.

In conclusion, AMs adapted to a lipid-rich environment by expressing a unique set of genes involved in lipid catabolism, which enables them to participate in surfactant clearance. Defects in this crucial function can cause pathological manifestations, which highlight the importance of macrophages for tissue homeostasis.

1.4 Development and Homeostatic Function of Splenic Macrophages

The spleen is a highly vascularized secondary lymphoid organ. The primary functions of the spleen include the elimination of senescent or damaged red blood cells, storage of platelets, T-cell priming, and B cell maturation. Furthermore, the spleen represents an important filtering system for blood components and the detection of circulating pathogens. Macrophages play crucial roles in all these functions.

Splenic macrophage heterogeneity was first described using conventional microscopy. Histological stainings of spleen sections with F4/80 and Ser-4 antibodies (CD169) indicated that different macrophage subsets occupy discrete spatial areas of the spleen (Rabinowitz and Gordon 1991). Marginal metallophilic macrophages (MMMs) and marginal zone macrophages (MZMs) are located in the marginal zone, which separates the white and red pulp of the spleen. Macrophages in the white pulp (WPMs) are located in a lymphocyte-rich environment, whereas red pulp macrophages (RPMs) are in contact with the cords of Billroth and splenic sinusoids in the red pulp (Fig. 1.3).

Nowadays, single-cell RNA-sequencing (scRNA-Seq) experiments have deeply expanded our knowledge regarding macrophage heterogeneity in various tissues. However scRNA-Seq datasets with a comprehensive representation of splenic macrophage subsets are still missing. This may be because the disaggregation of organs necessary for single-cell isolation often leads to macrophage fragmentation, thereby resulting in the underrepresentation of this particular immune cell subset in scRNA-Seq datasets (Millard et al. 2021). It is also possible that certain enrichment strategies such as negative magnetic cell separation are less efficient in the spleen because of the intrinsic magnetism of splenic macrophages (Franken et al. 2015). More tissue-sensitive approaches, such as single-nucleus RNA-Seq experiments or spatial transcriptomics at the single-cell level, are needed to unravel the heterogeneity of splenic macrophages on a large scale.

Nevertheless, owing to the specific localization of macrophage subsets and the resulting unique cellular microenvironment, each splenic macrophage population is phenotypically distinct and equipped with specialized functions.

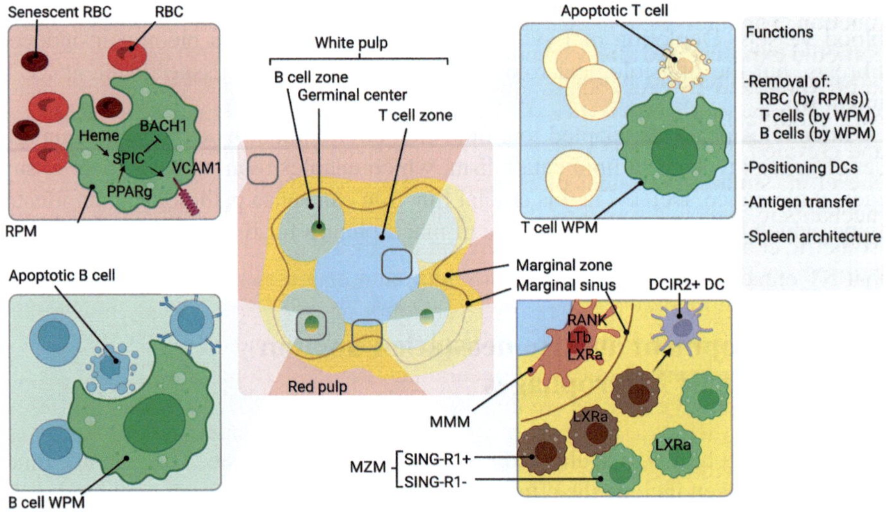

Fig. 1.3 Schematic representation of macrophage localization in the spleen. Five main tissue-resident macrophage subsets have been identified in different areas of the spleen. Red pulp macrophages (RPM) are involved in the clearance of senescent and damaged red blood cells (RBC) via heme, *SpiC*, and the degradation of BACH1. White pulp macrophages (WPM) participate in the phagocytosis of apoptotic T and B cells. Marginal zone macrophages (MZM) and marginal metallophilic macrophages (MMM) are involved in antigen sampling and marginal zone microarchitecture. MZMs influence the positioning of dendritic cells (DC) at the bridging channels. Created with BioRender.com

1.4.1 Marginal Zone Macrophages

MMMs and MZMs share a close functional and developmental relationship, which explains why some gene deletions or cell depletion strategies similarly affect both macrophage populations. Both subsets are involved in apoptotic cell clearance, antigen capture from the blood, and B-cell communication. Therefore, assigning a specific function to one subset, but not to the other, is problematic without mice that allow the specific targeting of individual subsets. However, some specific functions for each subset have emerged over the past years, even though the influence of perturbations on other splenic macrophage subsets has not been investigated in all studies.

Most of the blood that flows into the spleen initially reaches the marginal zone surrounding the B-cell follicles. Accordingly, MZMs in this anatomical niche are suitable for filtering disseminated particles and apoptotic cells from the blood (McGaha et al. 2011), controlling *Listeria monocytogenes* infections (Aichele et al. 2003), and providing antigens to B cells (Prokopec et al. 2016). The role of MZMs was investigated in mice that received low-dose clodronate liposome injections in conjunction with chronic injections of apoptotic cells. Under these circumstances, apoptotic bodies are no longer entrapped in the marginal zone and instead accumulate in the white pulp, which alters adaptive immune responses during lupus erythematosus (McGaha et al. 2011). However, MZMs do not phenotypically form homogenous populations. The middle section of the marginal zone is composed of macrophages that express both SIGN-R1 (encoded by *Cd209a*) and MARCO, whereas the outer area is inhabited by MZMs that only express MARCO (Ito et al. 1999; Pirgova et al. 2020) (Fig. 1.3). Interestingly, compared to MARCO^{+} MZMs, SIGN-R1^{+}MARCO^{+} MZMs seem to be independent of M-CSF signaling (Takahashi et al. 1994; Ito et al. 1999) and only slowly reappear after clodronate liposome depletion (van Rooijen et al. 1989). SIGN-R1^{+} MARCO^{+} MZMs depend on the presence of B cells for differentiation (Nolte et al. 2004; You et al. 2009). This emphasizes that the maintenance of the marginal zone cellular architecture necessitates constant cell-to-cell communication. MZM localization and maintenance depend on *Mertk* (Soni et al. 2018), *Marco*, and *Msr1* (Chen et al. 2005). Furthermore, SIGN-R1^{+} MZMs are important for the correct localization of DCIR2^{+} cDC2 dendritic cells. Upon depletion of SIGN-R1^{+} MZMs in *Cd209a*$^{\text{Dtr-Cre}}$ mice, DCIR2^{+} dendritic cells lost their clustered positioning in the bridging channels, thereby impairing germinal center B cell responses (Pirgova et al. 2020).

1.4.2 Marginal Metallophilic Macrophages

MMMs reside at the border between the marginal zone and the white pulp, and their development is similarly dependent on B cells (Nolte et al. 2004) (Fig. 1.3). These cells express CD169 and MOMA-1, showing similarities to the subcapsular sinus macrophages of the lymph node. MMMs rely on receptor activator of NF-κB (RANK) and lymphotoxin beta signaling. Deletion of either factor in *Cd169*^{Cre} mice leads to a strong reduction in MMMs, whereas MZMs remain unaffected (Camara et al. 2022). Marginal zone reticular cells were identified as the main producers of RANK ligand, thereby influencing MMM differentiation (Camara et al. 2022). In addition, the nuclear receptor LXRα (encoded by *Nr1h3*) and M-CSF are involved in MMM differentiation, even though they also affect MZM development (Takahashi et al. 1994; A-Gonzalez et al. 2013). MMMs can transfer antigens to splenic cDC1 dendritic cells for the cross-presentation and activation of cytotoxic T lymphocytes (Backer et al. 2010), which might represent an efficient vaccination strategy to induce cytotoxic T-cell immunity. It has also been discussed that MMMs participate in oxysterol biosynthesis and thereby affect the migration of activated B cells to the outer follicular niche (Liu et al. 2011), but cell-specific gene deletion approaches are necessary to formally prove this point.

1.4.3 Red Pulp Macrophages

Among all splenic tissue-resident macrophage populations, RPMs form the largest fraction in terms of cell numbers. In mice, RPMs develop from early *Myb*-independent YS precursors (Schulz et al. 2012), but fetal monocyte-derived macrophages may contribute to the RPM pool (Epelman et al. 2014). However, detailed knowledge regarding RPM fate and replenishment during adulthood is still lacking.

RPMs are intrinsically magnetic (Franken et al. 2015), emphasizing their role in iron recycling and red blood cell clearance. The causal link between red blood cell clearance and iron recycling was demonstrated earlier in different knockout mice. CD47 is normally expressed by viable red blood cells and interacts with the inhibitory receptor signal regulatory protein alpha (SIRPα) in RPMs, which prevents phagocytosis. The injection of CD47-deficient red blood cells led to erythrophagocytosis in RPMs, which was independent of red blood cell viability (Oldenborg et al. 2000). However, because *Cd47*- and *Sirpa*-deficient mice do not show an overt phenotype with respect to anemia and splenomegaly under physiological conditions (Bian et al. 2016), the exact *in vivo* contribution of the CD47/SIRPα interaction in RPMs awaits further research.

In contrast, *SpiC*-deficient mice show strongly reduced numbers of RPMs, whereas MZMs and MMMs are present in normal cell numbers (Kohyama et al. 2008). The SPIC-dependent absence of RPMs resulted in splenomegaly with increased iron accumulation in the spleen. Furthermore, proerythroblast numbers decreased in the steady state, supporting the idea that iron recycling is essential to sustain iron-sufficient erythropoiesis. It was postulated that SPIC induced the expression of *Vcam1* (Kohyama et al. 2008), and a direct cell-intrinsic role for VCAM1 in RPM maintenance was shown later (Ulyanova et al. 2016). However, the exact molecular role of *Vcam1* in RPM differentiation requires further investigation. Subsequent work revealed a critical role for heme in the induction of the SPIC-dependent pathway, which involves the proteasomal degradation of the transcription factor BACH1 (Haldar et al. 2014) (Fig. 1.3). The fact that heme plays an important role in RPM biology can also be observed in mice with heme oxygenase-1 (HO-1, encoded by *Hmox1*) deficiency. HO-1 is involved in heme metabolism, thereby releasing free iron, which can then return to circulation. Patients with HO-1 mutations suffer from tissue iron overload and anemia. Similarly, *Hmox1*-deficient mice accumulate cytotoxic heme, which leads to elevated oxidative stress and subsequent cell death in RPMs (Kovtunovych et al. 2010).

However, heme-*SpiC* induction seems to be important for the function of RPMs, but not directly for their development, since young *SpiC*-deficient mice show normal or only slightly reduced cell numbers in RPMs (Ulyanova et al. 2016; Okreglicka et al. 2021). Instead, *Pparg* is a critical transcription factor for the establishment of RPMs (Okreglicka et al. 2021). Lack of PPARγ caused the upregulation of genes involved in the immune response, while pathways related to erythrocyte development and heme biosynthesis were decreased in the remaining spleen macrophages (Okreglicka et al. 2021). Functionally, PPARγ regulates the adhesion and migration of early RPM precursors, which is in line with the PPARγ independence of adult RPMs (Okreglicka et al. 2021). Since RPMs do not express *Pparg2* (Dörr et al. 2022), this phenotype is likely based on *Pparg1* expression, which explains the independence of RPMs from GM-CSF and C/EBPβ (Cain et al. 2013). In addition, the combined deletion of the transcription factors *Irf4* and *Irf8* prevents RPM development, whereas deficiency of each factor alone does not affect RPM cell numbers (Yamamoto et al. 2011). However, the molecular basis of this phenotype remains unknown.

Vcam1 expression by RPMs is also involved in the regulation of extramedullary hematopoiesis. This was investigated in *Cd169*^{Dtr} mice, in which diphtheria toxin injection not only depleted MMMs, but also strongly reduced RPMs. This led to a *Vcam1*-dependent reduction in splenic hematopoietic precursor cells, while simultaneously evaluating their frequency in the circulation (Dutta et al. 2015). These data suggested that VCAM1$^+$ macrophages interact with hematopoietic precursors via the VCAM1 ligand VLA-4, which is important for the formation of an extramedullary hematopoietic niche (Williams et al. 1991). However, as the mouse *Cd169*^{Dtr} depletion system is not specific to RPMs, the contribution of MMMs to this phenotype should be investigated in the future.

1.4.4 White Pulp Macrophages

WPMs are located in the B- and T-cell areas of the spleen and do not express F4/80 (Rabinowitz and Gordon 1991) but are immunoreactive to CD68, MFG-E8, and TIMD4 (Rabinowitz and Gordon 1991; Hanayama et al. 2004; Wong et al. 2010). The B cell area can also contain germinal centers, in which WPMs can accumulate 'tingible bodies' that consist of condensed nuclei (Rabinowitz and Gordon 1991). The intracellular accumulation of nuclei suggests the involvement of WPMs in the clearance of apoptotic cells. In germinal centers, B cells undergo clonal selection, hyperproliferation, and somatic hypermutation. The randomness of receptor editing increases the risk of errors, which are associated with high rates of apoptosis. The efficient removal of cellular corpses is crucial because intracellular components can induce otherwise unwanted immune cell activation and the formation of self-reactive antibodies. This is evident in *Mfge8*-deficient mice, which suffer from autoimmune symptoms, such as glomerulonephritis, as a result of autoantibody production, splenomegaly, and hyperactive germinal centers (Hanayama et al. 2004). A similar phenotype was detected in mice lacking the proto-oncogene tyrosine-protein kinase MER, which exhibited increased non-phagocytized apoptotic cells on the surface of WPMs (Rahman et al. 2010). However, whether these phenotypes are the consequence of *Mfge8*- or *Mertk*-deficiency in WPMs specifically and not related to other splenic macrophage populations requires subset-specific gene deletion approaches. Macrophages in the germinal center are stationary cells that do not migrate in search of apoptotic cells, at least in lymph nodes. Instead, they use long protrusions to scan the environment dynamically (Grootveld et al. 2023; Gurwicz et al. 2023). Interestingly, germinal center macrophages in the lymph node develop via a CD169-lineage independent pathway and do not require M-CSF for differentiation, which emphasizes the existence of a follicle-resident precursor that can be locally activated by the presence of dead cells. This stimulus may be sufficient to induce differentiation into tingible body macrophages (Grootveld et al. 2023; Gurwicz et al. 2023). However, the exact precursor and molecular pathways require further research. It also needs to be shown whether splenic germinal center WPMs follow a similar developmental route.

However, macrophages in the white pulp are not confined to B-cell follicles but are also present in the periarteriolar sheath of the T-cell zone. Currently, detailed knowledge of the biology of T-cell zone macrophages in the spleen is lacking. Notably, a $CX3CR1^{hi}CD64^{+}MERTK^{+}$ macrophage subset was recently identified in the T-cell area of the lymph node (Baratin et al. 2017). These macrophages are derived from BM monocytes after birth and differentiate into long-lived T-cell zone macrophages with a slow replacement during adulthood (Baratin et al. 2017). In the lymph node T-cell zone macrophages are involved in CX3CR1-dependent apoptotic cell clearance, but do not contribute to naïve $CD4^{+}$ T-cell priming or tolerance. If their splenic counterpart fulfills a similar function and follows the same differentiation route, remains to be shown.

The primary functions of splenic macrophages are closely tied to the role of the spleen as an immunological organ. Tissue-resident macrophages in the spleen are uniquely equipped to degrade and recycle senescent red blood cells and apoptotic lymphocytes. Additionally, they are involved in the anatomical microarchitecture of the spleen, sample antigens from circulation, and transfer them to dendritic cells and B cells. However, it has not been shown yet whether these activities promote peripheral tolerance under physiological conditions.

1.5 Development and Homeostatic Function of Kupffer Cells

The liver, the largest solid internal organ in the body, is of crucial importance when it comes to maintaining metabolism, digestion, detoxification, and immune regulation. Hepatocytes, the primary cell type in the liver, perform essential functions, such as detoxifying harmful substances like drugs and alcohol, synthesizing proteins like albumin and clotting factors necessary for blood coagulation, storing glycogen as an energy source, and metabolizing fats, carbohydrates, and proteins. Furthermore, the liver produces bile, which aids in the digestion and absorption of fat in the small intestine. The liver also plays a critical role in metabolizing hormones, cholesterol, and waste products such as bilirubin.

Kupffer cells (KCs), the tissue-resident macrophages of the liver, are among the first described macrophages. Karl Wilhelm von Kupffer observed in 1876 "Sternzellen" that could be stained with a gold chloride solution (Kupffer 1876), years before Ilya Metchnikoff identified macrophages (Metchnikoff 1892). Pathologist Tadeusz Browicz further described KCs (Browicz 1899), and because of his seminal contribution, some authors refer to liver macrophages as Kupffer-Browicz cells.

Evidence that macrophages first appear in the liver during embryogenesis may be attributed to Sorokin et al. (Sorokin et al. 1992). The authors used peroxidase-coupled isolectin B4 to identify macrophages in the rat brain (E12), liver (E12–13), and lungs (E13–14) but not in the spleen, BM, and thymus. Although KCs do not express CX3CR1 in adulthood, they are derived from a CX3CR1^{+} precursor, as shown in *Cx3cr1*^{Cre} reporter mice (Yona et al. 2013). These precursors are YS-derived EMPs and fetal monocytes (Schulz et al. 2012; Hoeffel et al. 2015). Because the liver undergoes significant expansion and reorganization in the first postnatal weeks in mice, BM-derived monocytes can integrate into the KC niche and acquire a KC-specific signature when transferred during this developmental window (Scott et al. 2016). The established KC pool is subsequently able to maintain itself throughout life. Together with microglia in the CNS and Langerhans cells in the skin, KCs are one of the investigated macrophage populations that do not seem to be complemented by monocytes during adulthood, as observed in *Ms4a3*^{Cre} reporter mice (Liu et al. 2019b). However, monocytes can adopt a KC signature when the original embryonic-derived macrophage pool is compromised, particularly after diphtheria toxin-mediated cell depletion in *Clec4f*^{Dtr} mice (Scott et al. 2016; Bonnardel et al. 2019; Sakai et al. 2019). This method was also used to unravel niche-specific signals that are important for instructing KC precursors and contributing to the tissue-specific identity of KCs (Bonnardel et al. 2019; Sakai et al. 2019). KCs line the sinusoidal endothelium of the liver and are in contact with hepatocytes, liver sinusoidal endothelial cells (LSECs), and hepatic stellate cells (Fig. 1.4). Hepatic stellate cells provide M-CSF and bone morphogenic proteins (BMPs) to induce monocyte differentiation into KCs. LSECs further produce BMPs, as well as the Notch pathway ligands DLL1 and DLL4, which induce through RBPJ the transcription factors LXRα and SPIC in KCs (Bonnardel et al. 2019; Sakai et al. 2019).

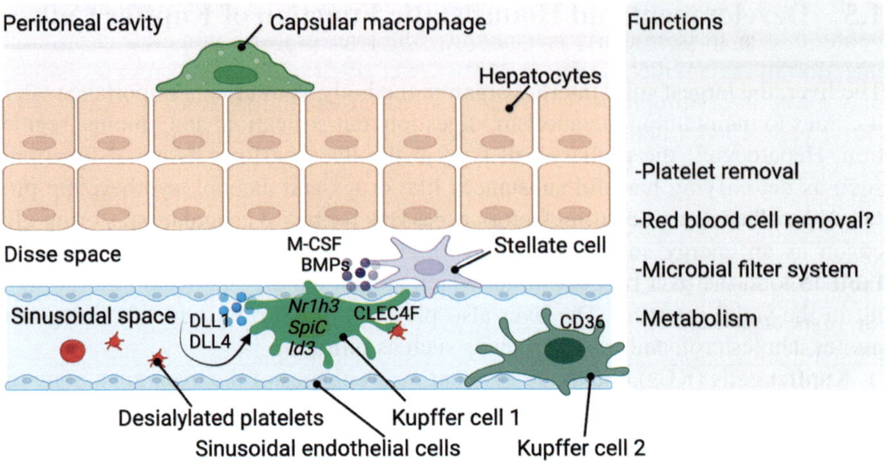

Fig. 1.4 Macrophage localization in the liver. Kupffer cell subsets 1 and 2 are located in the sinusoidal space, where they receive instructive signals from stellate cells (M-CSF, BMPs) and sinusoidal endothelial cells (Notch ligands DLL1 and 4). Kupffer cell subset 1 can remove damaged red blood cells and desialylated platelets via CLEC4F, whereas subset 2 participates in metabolic diseases via CD36. Created with BioRender.com

Deletion of *Nr1h3* in KCs is accompanied by significant transcriptional and phenotypic changes, including the loss of KC signature genes (Sakai et al. 2019) and reduced TIMD4 expression (Scott et al. 2018). Loss of TIMD4^{+} cells in *Nr1h3*-deficient mice indicates that the survival or differentiation of embryonic-derived KCs is LXRα dependent. This is supported by the fact that *Nr1h3*, together with *Id3*, was identified as a possible pre-deterministic candidate for pre-macrophage-to-KC differentiation (Mass et al. 2016). However, the empty hepatic niche in *Nr1h3*-deficient animals is subsequently filled by BM-derived monocytes, which still show significant gene changes compared to wild-type controls (Scott et al. 2018).

SpiC expression suggests that KCs are involved in iron recycling, similar to RPMs in the spleen and BM macrophages (Haldar et al. 2014). Indeed, the intravenous injection of ^{51}Cr-labeled oxidatively damaged red blood cells into mice showed that the main uptake of these cells occurred in the liver (Terpstra and van Berkel 2000). However, whether KCs are also involved in red blood cell clearance under homeostatic conditions remains to be shown. Compared to RPMs, KCs only show partial labeling in *SpiC*^{Gfp} reporter animals, and KC numbers seem to be unchanged in *SpiC*-deficient mice (Haldar et al. 2014). These data suggest that KCs might not use a SPIC-dependent pathway for the removal of senescent red blood cells and may participate in iron homeostasis differently.

KCs also respond to erythropoietin (EPO), which is the main hormone that drives mammalian erythropoiesis. Upon repeated injections of EPO, KC numbers increased in mice, indicating a direct or indirect role of EPO in mediating KC proliferation (Gilboa et al. 2017). Furthermore, KCs have also been shown to scavenge hemoglobin released by erythrocytes in the form of red blood cell-derived vesicles (Willekens et al. 2005). Using radiolabeled vesicles, researchers found that 80% of all radiolabeling is cleared from the bloodstream by the liver in just 5 min after intravenous injection in rats (Willekens et al. 2005). Older observations also suggest that KCs play a role in erythrophagocytosis, particularly in conditions leading to hemolysis (Kaye et al. 1967; Loebl et al. 1973). However, even if these data indicate that KCs can clear damaged red blood cells after injection or hemolysis, the role of KCs in red blood cell clearance under steady-state conditions remains unclear. Instead, KCs may be the main phagocytic cells for the removal of desialylated platelets. Megakaryocyte-derived platelets are the second most common cell type in the blood, and because of their short life span of only a few days, they require constant removal. Platelets express high levels of sialylated N- and O-linked glycans. The level of sialylation can change under pathological conditions or during aging (Goswami and Koner 2002; Li et al. 2015), and desialylated glycans show high binding affinity to receptors such as CLEC4F (Jiang et al. 2021), a marker gene of KCs (Fig. 1.4). Deletion of receptors such as *Clec4f* in KCs (Jiang et al. 2021), macrophage galactose lectin (Deppermann et al. 2020), or genes involved in the generation of O-linked glycans, which reduces the levels of sialic acids on platelets (Li et al. 2017), leads to the accumulation of platelets in the liver.

The functions of KCs may change when the liver becomes the primary organ of extramedullary erythropoiesis, for instance, in splenectomized mice. Under these conditions, KCs upregulate *Vcam1* expression and therefore create a niche for hematopoietic precursor cells in emergency situations (Otsuka et al. 2011), similar to the function of MMMs during extramedullary hematopoiesis (Dutta et al. 2015). It is also possible that KCs can adopt an RPM gene signature in splenectomized mice, thereby securing the alternative removal of red blood cells by KCs.

A recent study explored in deeper detail the functions of KCs in homeostasis, by comparing KC dynamics in steady state with fibrotic livers. Extensive remodeling of the hepatic sinusoids was observed using a model of hepatic fibrosis, leading to the redistribution of blood flow from the sinusoids to collateral vessels (Peiseler et al. 2023). These changes cause deficits in liver sinusoid surveillance by KCs, a decrease in their ability to capture phagocyte-injected bacteria, and partial loss of KC identity (Peiseler et al. 2023). Unexpectedly, despite these changes, the survival of mice after microbial infection is not severely affected (Peiseler et al. 2023). This protection could be attributed to the formation of multinucleated structures (syncytia) in collateral vessels, resulting from the CD36-dependent fusion of BM-derived monocytes that adopt a KC-like signature (Peiseler et al. 2023). These syncytia showed enhanced bacterial clearance and phagocytic capacity compared to KCs and compensated for the loss of KCs due to rarefied sinusoids. These data emphasize that KCs are involved in blood scavenging and represent an important filter system for preventing the systemic distribution of potentially pathogenic materials.

New reporter and fate-mapping mouse models, as well as single-cell analysis, further revealed some heterogeneity in the liver-macrophage pool. More than three decades ago, an MHCII-expressing cell type was detected in the human hepatic capsule, which was classified as dendritic cells, despite their immunoreactivity towards CD68 and CD163 (Prickett et al. 1988). These cells were later identified as liver capsular macrophages, which are constantly replenished by monocytes in parabiotic mice (Sierro et al. 2017) but only partially in the *Ms4a3*^{Cre} fate-mapping model (Liu et al. 2019b). Capsular macrophages are involved in immunosurveillance and recruitment of neutrophils (Sierro et al. 2017), but whether they fulfill physiological and niche-specific functions is unknown. Another hepatic macrophage subset near the bile ducts in humans (Guilliams et al. 2022) partially shares a gene signature with TREM2^{+} lipid-associated macrophages (LAMs) (Jaitin et al. 2019). The enrichment of genes involved in lipid handling suggests that these cells may play a role in metabolic diseases. Indeed, during metabolic-associated fatty liver diseases, monocytes can differentiate into hepatic LAMs and show different abilities to metabolize lipids compared to KCs (Remmerie et al. 2020). Moreover, during steatosis, the hepatic LAM population increases and relocates to steatotic zones (Guilliams et al. 2022); however, their physiological functions are unclear. The liver further harbors central vein macrophages under steady-state conditions, and a CD206^{hi} ESAM^{+} subset termed KC2 (Fig. 1.4). While KC2 are ontogenetically related to KCs, arise from embryonic precursors, occupy the same cellular niche as KCs, and are characterized by the expression of KC marker genes (such as *Clec4f*, *Lyz2*, and *Csf1r*), they also share gene expression patterns with LSECs (Bleriot et al. 2021; De Simone et al. 2021). KC2s are enriched in genes involved in carbohydrate and lipid metabolism, and depletion of KC2 cells or deficiency of the fatty acid transporter CD36 in a mouse model of obesity or liver steatosis revealed that KC2s contribute to the regulation of glucose homeostasis and to the liver oxidative stress response (Bleriot et al. 2021). KC2s further contribute to the cross-presentation of hepatocellular antigens to prime and induce the antiviral functions of CD8^{+} T-cells during hepatitis B virus infection (De Simone et al. 2021). Notably, data suggesting the existence of a KC subset that shows LSEC-related gene expression could also be attributed to the formation of doublets or tissue fragmentation (Millard et al. 2021; Guilliams et al. 2022; Hume et al. 2022).

In summary, KCs act as an indispensable central microbial filter system and are involved in platelet clearance. They may contribute to senescent red blood cell removal under steady-state conditions and participate in metabolism.

1.6 Development and Homeostatic Function of Cardiac Macrophages

The heart is a muscular organ responsible for pumping oxygenated blood throughout the body and deoxygenated blood into the lungs for gas exchange. The rhythmic contractions are based on the electrical conduction system of the heart, which ensures efficient blood flow throughout the body. Central to this are specialized muscle cells called cardiomyocytes, which can rhythmically contract and relax. Cardiomyocytes are long-lived cells and have limited regenerative capacity, with less than 50% being replaced during the normal life span in humans (Bergmann et al. 2009). This longevity combined with the extreme physical requirements (one cardiomyocyte contracts approx. $>10^7$ times per year) exposes cardiomyocytes to constant cellular stress, suggesting that cardiomyocytes require specific protection mechanisms that support their functionality.

Cardiac macrophages are integral components of the immune cell repertoire of the heart but also fulfill tissue homeostatic functions and are involved in developmental aspects of the heart. The origin of cardiac macrophages has been an area of extensive research, leading to the understanding that they are derived from different cellular sources during the development of an organism. In mice, the heart is colonized by fetal YS-derived myeloid precursors around E10 (characterized by an F4/80^{hi}CD11b^{low} phenotype), which are supplemented by F4/80^{low}CD11b^{hi} fetal monocyte-derived cells at E12 (Epelman et al. 2014). Self-maintaining embryo-derived macrophages can further diversify into MHCII^{low} and MHCII^{hi} subsets postnatally (Molawi et al. 2014). While the MHCII^{low} subset is additionally characterized by TIMD4, LYVE1, and FOLR2 expression, the MHCII^{hi} subset only shows a low expression of these markers (Rizzo et al. 2023). During late embryogenesis and after birth, a small fraction of BM monocyte-derived macrophages further contribute to the tissue-resident cardiac pool and coexist with embryonic-derived macrophages (Epelman et al. 2014; Heidt et al. 2014; Leid et al. 2016). These cells can be distinguished by CCR2 expression and slowly outcompete their embryonic counterparts during aging (Molawi et al. 2014). Similar observations were made in patients who received a sex-mismatched heart transplant (Bajpai et al. 2018).

In terms of cardiac development, fetal cardiac macrophages have been shown to play important roles in coronary remodeling during late embryonic stages. The hearts of *Csf1*$^{op/op}$ and *LysM*^{Cre} *Rosa26*^{Dta} mice, which show a general macrophage deficiency, exhibited higher coronary vessel density when compared to controls, and their vascular plexuses failed to remodel into blood vessels of different diameters (Leid et al. 2016). Furthermore, macrophage depletion resulted in diminished growth of perfused blood vessels, whereas an increase in the proliferation of non-perfused blood vessels was observed (Leid et al. 2016) (Fig. 1.5). This implies that cardiac macrophages directly or indirectly influence the proliferation of coronary endothelial cells. Factors that could play a role in vascularization are insulin-like growth factors 1 and 2, which are preferentially expressed by embryonic cardiac macrophages (Leid et al. 2016). This is in agreement with a study that investigated the role of embryonic-derived macrophages versus monocyte-derived cells in neonatal and adult cardiac injury models. Genetic cardiomyocyte ablation in neonatal mice increased the population of embryonic-derived cardiac macrophages, whereas the same injury in adults led to an influx of monocyte-derived macrophages (Lavine et al. 2014). Monocyte-derived cells show an inflammatory signature and are devoid of reparative functions, while embryonic-derived macrophages show a minimal inflammatory response and promote cardiac recovery through cardiomyocyte proliferation and angiogenesis (Lavine et al. 2014). These results suggest that embryonic cardiac macrophages are superior in supporting angiogenesis, whereas monocyte-derived cells are involved in inflammatory reactions that can contribute to tissue damage.

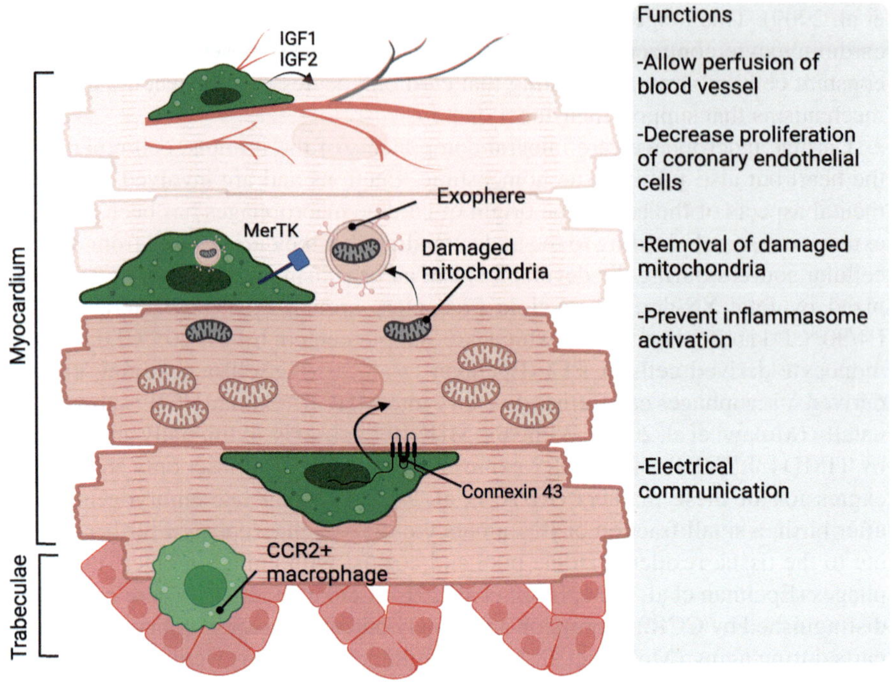

Fig. 1.5 Functions of cardiac macrophages. Embryonic-derived cardiac macrophages (dark green) are involved in perfused cardiac blood vessel formation during development. They are also involved in MerTK-dependent removal of damaged mitochondria, which are released from cardiomyocytes in the form of exophers. Cardiac macrophages also form connexin 43-dependent gap junctions, which positively depolarize the resting membrane potential in cardiomyocytes. $CCR2^+$ monocyte-derived macrophages were specifically identified in the trabecular projections. The function of this subset under homeostatic conditions is currently unknown. Created with BioRender.com

Interestingly, the spatial analysis revealed that $CCR2^-$ embryonic-derived macrophages were almost exclusively found within the myocardium, whereas $CCR2^+$ macrophages were located within the trabecular projections of the endocardium (Leid et al. 2016). This indicates that embryonic-derived macrophages may support cardiomyocyte function. Contractile large cardiomyocytes have a high ATP demand to drive cardiac contraction/relaxation and ionic homeostasis, which is reflected by the high mitochondrial density in this cell population (Barth et al. 1992). Furthermore, cardiomyocytes have a low turnover rate (Bergmann et al. 2009), suggesting that mechanisms are needed to protect cardiomyocytes from malfunction. One layer of protection is represented by embryonic-derived $CX3CR1^+MHCII^+CD169^+$ cardiac macrophages. Depletion of macrophages in *Cd169*^{Dtr} mice increases the number of morphologically altered mitochondria in cardiomyocytes (Nicolás-Ávila et al. 2020). Further experiments have revealed that cardiac macrophages can uptake exophers that contain dysfunctional mitochondria from cardiomyocytes in an

MerTK-dependent manner (Nicolás-Ávila et al. 2020). Depletion of macrophages increases mitochondria deposition in the extracellular space and causes inflammasome activation, leading to defects in ATP generation and diastolic and systolic dysfunctions (Nicolás-Ávila et al. 2020). Therefore, embryonic-derived cardiac macrophages likely protect cardiomyocytes by removing the old and damaged mitochondria (Fig. 1.5).

As cardiomyocytes are on average surrounded by five macrophages (Nicolás-Ávila et al. 2020), it is possible that macrophages also modulate electrical conduction. Disruptions in macrophage function can thus have direct implications for cardiac electrophysiology and arrhythmia development. Indeed, cardiac macrophage depletion in *Cd11b*^{Dtr} mice resulted in complete third-degree atrioventricular blockade (Hulsmans et al. 2017). This phenotype could be partially attributed to the physical interaction between atrioventricular node macrophages and cardiomyocytes. Macrophages directly couple with cardiomyocytes through gap junctions containing connexin 43 to facilitate electrical communication and exhibit rhythmic depolarization, resulting in a more positively depolarized resting membrane potential in cardiomyocytes (Hulsmans et al. 2017) (Fig. 1.5). Therefore, they possibly allow for accelerated repolarization and a shorter action potential. Importantly, the cardiomyocyte support function of macrophages might change under pathological conditions when monocyte-derived cells rather than embryonic-derived macrophages dominate affected cardiac tissue areas. Both subsets execute different functions within the heart (Lavine et al. 2014; Hulsmans et al. 2023). This raises the possibility that an infarcted heart is functionally impaired due to the increased presence of monocyte-derived cells that do not form connexin 43-dependent contacts with cardiomyocytes and can cause atrioventricular blockade.

Collectively, these recent data demonstrate that macrophages are the guardians of cardiomyocytes and influence their function through mitochondrial removal and electrical conduction. Therefore, a detailed understanding of the origin and function of cardiac macrophages is pivotal to identifying the causes of arrhythmias, a common feature of many cardiac diseases.

1.7 Development and Homeostatic Function of Microglia

The CNS is responsible for processing sensory information obtained from the peripheral nervous system, regulating vital processes such as respiration and heartbeat, and also controlling cognitive functions, including memory, learning, and behavior. The ability to perform these functions is a consequence of the interplay between four specialized cell subsets in the CNS: neurons, astrocytes, oligodendrocytes, and microglia.

Research on microglia, the primary immune cells and tissue-resident macrophages of the CNS and once considered the "almost forgotten third glial element" (Streit et al. 1988), attracted considerable attention in the early 2000s, and the number of microglia-related research articles is currently increasing nearly exponentially. There are now approximately 5,000 publications per year on microglia, possibly making them the most studied tissue-resident macrophage population.

Microglia originate from "early" EMPs in the YS during embryogenesis, a process that depends on the transcription factors PU.1 (encoded by *Spi1*). Primitive macrophages colonize the developing brain around E9.5, which is before the blood-brain barrier (BBB) is fully established around E14.5 (Ginhoux et al. 2010; Gomez-Perdiguero et al. 2015; Hoeffel et al. 2015). Microglial development and function are finely tuned by a combination of intrinsic genetic programs that are linked to the expression of *Sall1* (Buttgereit et al. 2016), *Irf8* (Masuda et al. 2012; Kierdorf et al. 2013), *P2ry12* (Haynes et al. 2006), *Hexb* (Masuda et al. 2020), and extrinsic signals from the surrounding glial environment, such as M-CSF (Kana et al., 2019), IL-34 (Wang et al. 2012; Greter et al. 2012), and TGFβ (Qin et al. 2018).

Microglia influx into the developing brain occurs before the peak of neurogenesis and neuronal migration. During this time, microglia help regulate the size of neural precursor cell populations in the cerebral cortex through selective engulfment of apoptotic and/or excess neural precursors (Cunningham et al. 2013), in the hippocampus by clearing dying neuroblasts that fail to transition into mature neurons (Sierra et al. 2010), and in the developing cerebellum by promoting the death and efferocytosis of Purkinje cells (Marín-Teva et al. 2004). It has also been shown that during postnatal development, the CX3CL1-CX3CR1 axis and the supply of insulin growth factor 1 (IGF1) by microglia are important trophic factors supporting neuronal survival (Ueno et al. 2013) (Fig. 1.6). During the formation of neuronal circuits, microglial processes further form contacts with developing neurons, reminiscent of purinergic junctions, which are crucial for neuron-microglia communication. Constitutive deletion of P2RY12 leads to an aberrant cortical cytoarchitecture, spanning from development throughout adulthood (Cserép et al. 2022). Additionally, microglia modulate dopaminergic axon outgrowth and interneuron positioning in the neocortex as well as axon tract fasciculation during corpus callosum development through the adaptor protein DAP12 and CX3CR1 (Pont-Lezica et al. 2014; Squarzoni et al. 2014).

Upon formation of nerve tracts during the early postnatal period, a distinct amoeboid microglial population can be detected in the ventricular zone. This area was termed the "fountain of microglia" (Kershman 1939). This subset, also recently called axon-tract-associated microglia (ATM), has been demonstrated to aid in maintaining the structural integrity of the developing brain at vulnerable boundaries, such as the fetal cortico-striato-amygdalar and cortico-septal boundaries, where they prevent the formation of large cavitary lesions. This protective mechanism involves the pleiotropic ATM factor osteopontin (encoded by *Spp1*) (Lawrence et al. 2024). ATMs further interact with oligodendrocyte progenitor cells (OPCs) before myelination and contribute to oligodendrogenesis by controlling OPC numbers through phagocytosis (Hagemeyer et al. 2017; Nemes-Baran et al. 2020). This process of maintaining normal OPC-to-axon ratio as a homeostatic mechanism for proper myelination relies on the CX3CR1 pathway and IGF1 (Wlodarczyk et al. 2017; Nemes-Baran et al. 2020).

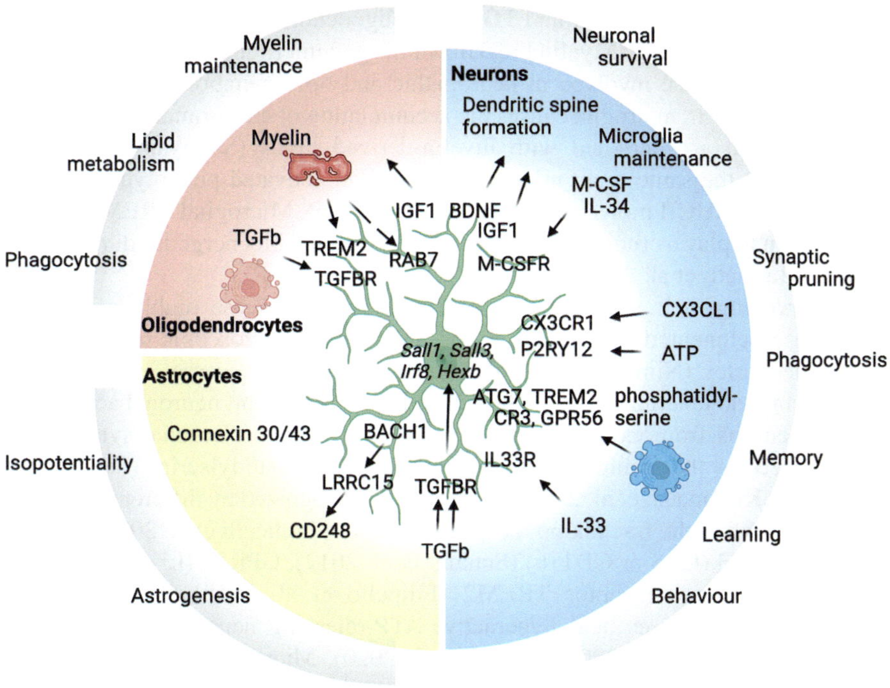

Fig. 1.6 Interaction of microglia with glial cells. Microglia are involved in intercellular communication with neurons, astrocytes, and oligodendrocytes, by contributing crucial factors and activities that are necessary for proper brain function. These functions include apoptotic processes (as synaptic pruning), provision of survival factors, and myelin maintenance. Microglial dysfunction can lead to deficits in learning, memory, and behavior. Created with BioRender.com

However, little is known about how microglia influence astrocyte development and function under steady-state conditions *in vivo*. Microglia regulate astrocyte-mediated control of neuronal structure and function, as evidenced by their role in modulating the astrocyte syncytium and synaptic transmission in the mouse hippocampus. Depletion of microglia disrupts astrocyte syncytial isopotential, decreases the expression of gap junction proteins such as connexins 30 and 43, and weakens synaptic transmission, whereas priming microglia enhances synaptic transmission without affecting astrocyte network function, suggesting context-dependent effects of microglia on astrocytes and neurons (Du et al. 2022) (Fig. 1.6). Recent evidence shows that this microglia-astrocyte crosstalk is potentially mediated by the transcription factor BACH1, known for its involvement in microglial metabolic adaptations during early brain development (Wang et al. 2024).

As the CNS matures, microglia play a crucial role in maintaining myelin integrity throughout adulthood. The absence of microglia in *Csf1r*$^{\Delta FIRE/\Delta FIRE}$ mice leads to hyper- and/or demyelination, confirming their fundamental role in preserving myelin integrity (McNamara et al. 2022). The absence of microglia causes oligodendrocytes to accumulate cholesterol and impairs lipid export. This phenotype is driven by TGFβ1 (microglia)-and TGFβR1 (oligodendrocyte)-dependent pathways (McNamara et al. 2022). Myelin breakdown in aged microglia is partially mediated by RAB7, a GTPase involved in neuropathic and lipid metabolism diseases. The absence of *Rab7* in microglia causes the accumulation of endosomal and lysosomal inclusions that are enriched with myelin-derived lipids (Safaiyan et al. 2016) (Fig. 1.6). At the same time, microglia showed an activated phenotype with high expression of MHCII molecules (Safaiyan et al. 2016). Microglial TREM2 has also been shown to play a crucial role in controlling neuronal bioenergetics during development (Tagliatti et al. 2024).

In surveying their microenvironment, adult microglia use highly motile processes, as documented by seminal in vivo two-photon microscopic imaging of the mouse neocortex (Nimmerjahn et al. 2005). With these processes, microglia can form purinergic junctions at the neuronal soma, which inhibit neuronal activity and protect neurons from excitotoxicity (Cserép et al. 2020). Neuronal synapses are characterized by the accumulation of externalized phosphatidylserine (Scott-Hewitt et al. 2020; Kurematsu et al. 2022), which can be recognized by different microglial receptors, such as the fractalkine receptor CX3CR1 (Paolicelli et al. 2011), complement receptor 3 (CR3 or CD11b) (Schafer et al. 2012), GPR56 (Li et al. 2020), or the innate immune receptor TREM2 (Filipello et al. 2018). The positioning of microglial processes near hyperactive ATP-releasing neurons is mediated by P2RY12 (Haynes et al. 2006; Cserép et al. 2020). Microglia further express the surface ectoenzymes CD39 and CD73, which mediate the rapid breakdown of ATP into adenosine and thereby reduce neuronal firing through adenosine receptors (Badimon et al. 2020). The interaction between neurons and microglia involves brain-derived neurotrophic factor (BDNF) and tropomyosin receptor kinase B (TrkB) signaling. Deletion of microglial BDNF results in a significant decrease in learning-induced formation of postsynaptic dendritic spines in young and mature adult mice (Parkhurst et al. 2013), even though there is some controversy regarding the role of microglia-derived BDNF (Honey et al. 2022). Synapse reorganization in the adult stage also relies on neuronal cytokine IL-33. Disruption of the IL-33R axis in microglia negatively affects the engulfment of the extracellular matrix, affecting the synaptic environment and memory formation (Nguyen et al. 2020). Notably, IL-33 levels decline in aged mice, which contributes to reduced plasticity and memory (Nguyen et al. 2020).

Due to the prominent role of microglia in shaping neuronal connectivity, defects in microglial function can also cause behavioral phenotypes. Deletion of *Csf1* from neuroectodermal-derived cells causes Purkinje cell alterations and deficits in motor learning and social memory in adult mice, suggesting cerebellar dysfunction in these animals (Kana et al. 2019). Deficits in social behavior have also been observed in *Trem2*- (Filipello et al. 2018) and *Atg7*-deficient mice (Kim et al. 2017). Additionally, increased elimination of inhibitory synapses in the ventral thalamus can cause hyperexcitability in the thalamocortical circuits and obsessive grooming behaviors as observed in progranulin-deficient mice (Lui et al. 2016). Other approaches, such as microglial ablation and conditional gene knockout, further emphasize the role of microglia in regulating learning and memory (Parkhurst et al. 2013), compulsive behaviors (Chen et al. 2010), and sleep (Liu et al. 2021; Hristovska et al. 2022). Whether dysregulation of these mechanisms is directly linked to autism and other neurodevelopmental and neuropsychiatric conditions is still a matter of debate (Mordelt and de Witte 2023).

In summary, microglia orchestrate various aspects of early brain development. In addition, microglia are involved in synaptic remodeling, perform intercellular communication with other glial elements, and thereby represent an indispensable integral element of the CNS.

1.8 Development and Homeostatic Function of Kidney Macrophages

The kidneys serve to maintain fluid and electrolyte balance, regulate blood pressure, and eliminate metabolic waste products from the body by producing urine. Nephrons, which are the functional units of the kidney (Fig. 1.7), filter blood to eliminate waste products such as urea and creatinine, as well as excess substances such as electrolytes and water, while simultaneously reabsorbing important nutrients and maintaining the acid-base balance. Additionally, the kidneys play a critical role in hormone production, including erythropoietin and renin production.

Kidney macrophages (KMs) develop from YS-derived precursors during early embryogenesis (Schulz et al. 2012; Epelman et al. 2014; Hoeffel et al. 2015). However, the proportion of YS-derived macrophages decreases with the arrival of fetal monocyte-derived cells around E14–16 (Epelman et al. 2014). Cell proliferation analysis of KMs indicates that a large proportion of the cells undergo cell division during the first three postnatal weeks, which subsequently declines (Liu et al. 2020). Using the *Ms4a3*^{Cre} lineage tracing mouse model, it was shown that 40–50% of KMs at the age of 3 months showed reporter labeling, indicating that a substantial part of the KM pool is derived from monocytes during adulthood (Liu et al. 2019a). Similar results were obtained using the *Cx3cr1*$^{\mathrm{CreERT2}}$ tamoxifen-inducible fate-mapping system. Labeled embryonic-derived KMs declined during the first 2 months of age, which was accompanied by an increase in unlabeled adult monocyte-derived KMs (Liu et al. 2020). Accordingly, two populations of tissue-resident macrophages of different origins might coexist in the kidney: embryo-derived kidney macrophages (E-KMs) and BM-derived kidney macrophages (BM-KMs). E-KMs are characterized by higher expression of F4/80 and CX3CR1, while BM-KMs show increased expression of CCR2 and CD11c (Liu et al. 2020). Unbiased single-cell sequencing also suggested the presence of two distinct KM subsets, of which one expressed high levels of *C1qa*, *Cd81*, *Cd63*, and *Pf4* (Cd63^{+} KMs), while the other subset expressed, in addition to the KM markers *C1qa* and *Cd81*, the genes *Ccr2*, *Mmp12*, and *Clec12a* (Ccr2^{+} KMs) (Yashchenko et al. 2023). However, both subsets were found to be equally labeled in *Ms4a3*-Cre reporter animals,

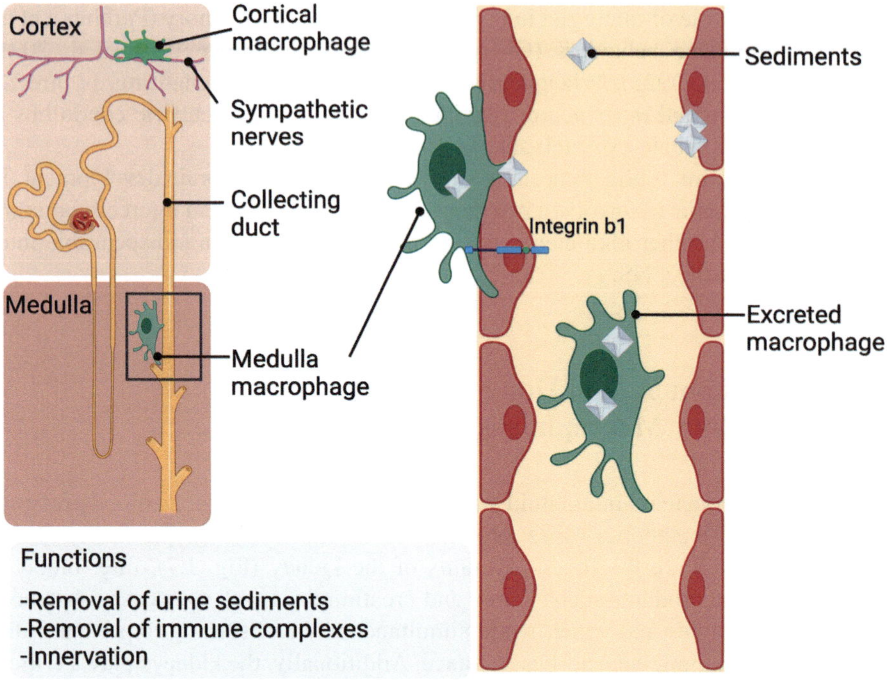

Fig. 1.7 Localization and function of renal macrophages. The nephron spans the cortical and medullary areas of the kidney (left). The culminating collecting ducts harbor macrophages that form transepithelial protrusions to scavenge larger particles in the urine, which can otherwise potentially induce kidney stone formation and obstruction (right). These macrophages can also be excreted in urine. Cortical renal macrophages are associated with sympathetic nerves, innervation control, and nerve activity. Created with BioRender.com

suggesting that both subsets were at least partially derived from monocytes (Yashchenko et al. 2023).

E-KMs were found in higher densities in the medulla compared to the cortex and papilla of the kidney (Liu et al. 2020), where they are located in spatial proximity to renal tubules. These juxtatubular macrophages extend long integrin β1-dependent transepithelial protrusions through the tubular epithelial monolayer, which is less frequently observed in the cortex (He et al. 2024) (Fig. 1.7). These protrusions have been described to penetrate the laminin^{+} basement membrane and epithelial cells, suggesting that they are exposed in the tubular lumen (He et al. 2024). Indeed, transepithelial protrusions were found to show extension and retraction movements, with signs of late endosome/lysosome accumulation. The sampling ability of KMs was demonstrated using fluorescent latex bead injection experiments into the kidney pelvis. Beads associated with macrophage transepithelial protrusions were readily visible, and depletion of macrophages led to longer retention of beads in the lumen of the collecting ducts, suggesting that urine flushing alone was insufficient for the removal of larger particles (He et al. 2024). However, KMs are not only involved in the transepithelial removal of large particles. They are also able to transmigrate into the lumen, associate with particles that otherwise potentially lead to tubule blockade, and be ultimately excreted in the urine (He et al. 2024) (Fig. 1.7). Compared to embryonic-derived macrophages in the brain, heart, lung, or liver, E-KMs in the kidney medulla have higher proliferative rates (He et al. 2024). One possible explanation for this observation might be that the proportion of transmigrated macrophages lost in the urine must be continuously replaced.

KMs localized on the abluminal side of the endothelium of peritubular capillaries are also involved in immune surveillance. By sampling the collection ducts, renal macrophages can instantly detect opsonized particles. These particles can be small immune complexes that can otherwise cause type III hypersensitivity reactions. Renal macrophages are located close to the endothelial cells around the glomeruli and tubules, where they can pick up particles through trans-endothelial transport. The uptake of these foreign particles is dependent on the scavenging receptor FcγRIV and triggers a local immune response characterized by the release of cytokines, which leads to the recruitment of neutrophils and monocytes to the target tissue (Stamatiades et al. 2016).

KMs express CX3CR1 and genetic deletion of this chemokine receptor significantly reduces medullary KM numbers by almost 50% in adult mice (Mass et al. 2016; He et al. 2024), leading to intratubular sediments containing calcium and lipids (He et al. 2024). Since the kidney niche is not replenished by monocytes in *Cx3cr1*-deficient mice, it can be speculated that CX3CR1 is indispensable for KM maintenance, independent of cellular origin. Indeed, *Cx3cr1*-deficiency strongly affected the gene expression of $Cd63^+$ KMs with a shift toward immune activation (Yashchenko et al. 2023). Of note, not medullary $Cd63^+$ KMs but cortical $Ccr2^+$ KMs (identified in $Ccr2^{Rfp/wt}$ $Cx3cr1^{Gfp/wt}$ animals) were numerically reduced in *Cx3cr1*-deficient mice (Yashchenko et al. 2023). However, if $Ccr2^+$ KMs differ in function from medullary E-KM, needs to be shown in the future. Cortical E-KMs are also associated with sympathetic nerves (Fig. 1.6). Deletion of KMs in $Cx3cr1^{CreER}$ $Rosa26^{Dtr}$ mice causes a pronounced reduction in renal innervation and reduced renal sympathetic nerve activity (Zhu et al. 2023). Consequently, a decline in plasma renin concentration, decreased kidney noradrenaline levels, and an increase in glomerular filtration rate could be detected, culminating in excessive Na^+, K^+, and Cl^- excretion and water loss in KM-deficient mice (Zhu et al. 2023).

Collectively, these results indicate that KMs play an important role in immune surveillance and urine monitoring by facilitating the removal of immune complexes and urine particles, suggesting that KMs may critically participate in kidney stone formation. KMs further contribute to sympathetic innervation and ion homeostasis in the kidneys.

1.9 Development and Homeostatic Function of Adipose Tissue Macrophages

Adipose tissue is a specialized connective tissue primarily composed of adipocytes, which are fat cells that store energy in the form of triglycerides. However, it has also other important functions such as insulation to maintain body temperature, thermoregulation, and providing energy during periods of fasting or increased energy demands. Additionally, adipose tissue secretes adipokines, hormones, and cytokines, which regulate metabolism, inflammation, and appetite. Imbalances in these functions can contribute to metabolic disorders such as type 2 diabetes, cardiovascular diseases, and inflammatory conditions.

The earliest observation of adipose tissue macrophages (ATMs) in fat tissue can possibly be attributed to Gilchrist and Ketron in 1916 (1916). They observed macrophages that phagocytosed adipocytes in subcutaneous fat samples. Decades later, studies reported chronic inflammation in multiple tissues and organs of obese and/or diabetic patients (Hartroft 1960; Braillon et al. 1985). It was also observed that macrophage-secreted cytokines could induce insulin resistance in preadipocytes *in vitro* (Pekala et al. 1983) and that *Tnf* expression in the adipocyte fraction correlated with obesity (Hotamisligil et al. 1993; Hotamisligil et al. 1994). Accordingly, neutralization of TNF-signaling partially restored the uptake of glucose in response to insulin, which emphasizes that chronic inflammation of the white adipose tissue (WAT) might contribute to obesity (Uysal et al. 1997). The subsequent discovery that BM monocyte-derived macrophages in irradiated chimeras accumulate in the expanding adipose tissue during obesity further supports the idea that inflammation contributes to the pathology (Weisberg et al. 2003; Xu et al. 2003). These studies also demonstrated that macrophages, but not adipocytes, are the major source of TNFα in WAT. Lumeng and colleagues described phenotypic changes in ATMs as a consequence of obesity, and suggested that BM monocyte-derived macrophages are different cells compared to ATMs in homeostasis (Lumeng et al. 2007a; Lumeng et al. 2007b). During the same year, LYVE1^{+} macrophages were described in early postnatal epididymal WAT (Cho et al. 2007; Han et al. 2011), suggesting the presence of a tissue-resident macrophage population under homeostatic conditions. Nevertheless, it was only in 2017 that the embryonic, BM-independent origin of tissue-resident ATMs was demonstrated in *Cx3cr1*^{CreERT2} fate-mapping mice (Hassnain Waqas et al. 2017). However, WAT contains more than just one homogenous subset of resident macrophages, as shown by scRNA-Seq of cell suspensions derived from multiple tissues (Chakarov et al. 2019). Similar to the situation in the lung, two general macrophage populations could be identified across different tissues: a population characterized by low expression of LYVE1, but high expression of MHCII and CX3CR1 (named Lyve1^{low}MHCII^{hi}), and a population with high LYVE1 expression, but low MHCII and CX3CR1 expression (Lyve1^{hi}MHCII^{low}) (Chakarov et al. 2019). Both subsets were derived from embryonic precursors but were progressively replenished by BM monocyte-derived macrophages in WAT.

1.9.1 White Adipose Tissue Macrophages

LYVE1⁺ macrophages can be identified at a high density in the hypoxic tip region of epididymal WAT, which harbors a dense vascular network (DVN) (Cho et al. 2007) (Fig. 1.8). Systemic depletion of macrophages by clodronate liposome injection reduced vascular sprouting and DVN size in adult mice (Cho et al. 2007). These macrophages expressed CXCR4 and various matrix metalloproteinases (MMPs). Proteolytic inactivation of MMPs not only reduced DVN formation and vascular sprouting but also prevented macrophage infiltration. A similar phenotype was observed when VEGF-VEGFR2 and SDF1-CXCR4 were blocked (Cho et al. 2007). ATMs are also important for angiogenesis during the early postnatal epididymal adipose tissue development. Systemic depletion of macrophages causes avascular regions with reduced adipocyte coverage (Han et al. 2011). These data support the notion that $Lyve1^{hi}MHCII^{low}$ macrophages participate in angiogenesis and WAT formation. Systemic depletion of myeloid cells in $LysM^{Cre}$ $Rosa26^{Dtr}$ mice has also suggested that macrophages play a role in energy homeostasis, contributing to the hypothalamic regulation of appetite (Lee et al. 2014). After depletion, mice lost weight owing to decreased food intake, which was linked to lower STAT3 signaling

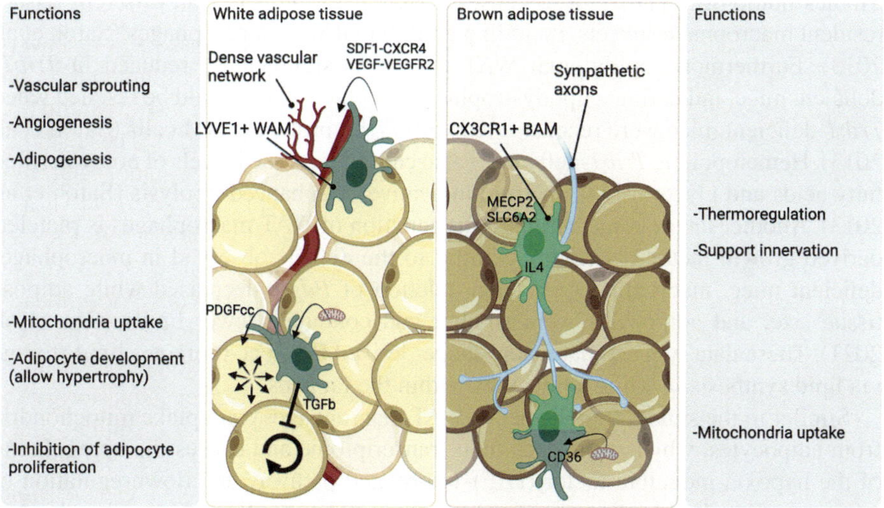

Fig. 1.8 Adipose tissue macrophages. Two main subsets of adipose tissue macrophages can be identified in white (left) and brown adipose tissue (right): $Lyve1^{hi}MHCII^{low}$ (dark green) and $CX3CR1^{+}Lyve1^{low}MHCII^{hi}$ (light green) macrophages. $Lyve1^{hi}MHCII^{low}$ macrophages are involved in angiogenesis and blood vessel sprouting in white adipose tissue. They further contribute to adipocyte maintenance by inhibiting proliferation and mediating hypertrophy. CX3CR1⁺ macrophages in brown adipose tissue are involved in thermoregulation, which is mediated by IL-4, MECP2, and SLC4A1. Adipose macrophages can take up damaged mitochondria from adipocytes. WAM - White Adipose tissue-resident Macrophage; BAM - Brown Adipose tissue-resident Macrophage. Created with BioRender.com

in the hypothalamus. However, the *LysM*^{Cre} *Rosa26*^{Dtr} approach is not specific to WAT macrophages and results in cell death of multiple macrophage subsets. Therefore, it is difficult to assign complex behavioral observations to a single macrophage population.

Other depletion methods, such as *Csf1r*^{Cre} *Csf1r*$^{fl/fl}$ or *Tnfrsf11a*^{Cre} *Spi1*$^{fl/fl}$ mice, have also revealed that embryonic-derived macrophages are important modulators of adipocyte hypertrophy during the first postnatal week. The absence of macrophages in ATMs decreases fat pad size, which harbors smaller adipocytes with low lipid content (Cox et al. 2021). Importantly, the introduction of *Csf1r* deletion to the hematopoietic lineage using *Flt3*^{Cre} mice did not affect adipocyte development and hypertrophy, thereby excluding the role of monocyte-derived cells in neonatal adipose tissue expansion (Cox et al. 2021).

Further evidence came from experiments in which CD206$^+$ WAT macrophages, a subset that likely corresponds to *Cd206*-expressing Lyve1^{hi}MHCII^{low} macrophages (Chakarov et al. 2019), were depleted in adult *Cd206*^{Dtr} mice. Under these circumstances, increased proliferation of adipocyte precursor cells was evident, which was TGFβ-dependent (Nawaz et al. 2017). This resulted in improvements in glucose tolerance and insulin sensitivity under physiological conditions, suggesting that WAT macrophages may regulate the balance between quiescence and proliferation of adipocyte precursor cells in response to nutritional changes. Genetic deletion of tribbles homolog 1 (*Trib1*) also causes a significant reduction in different tissue-resident macrophage subsets, including epididymal WAT macrophages (Satoh et al. 2013). Furthermore, epididymal WAT mass was significantly reduced in *Trib1*-deficient mice, indicating a lipodystrophic phenotype, which could be rescued when *Trib1*-deficient mice were reconstituted with *Trib1*-proficient BM cells (Satoh et al. 2013). Hematopoietic *Trib1*-deficiency also caused increased levels of non-esterified fatty acids and glycerol in the serum, indicative of enhanced lipolysis (Satoh et al. 2013). Another factor contributing to the function of WAT macrophages is platelet-derived growth factor C (*Pdgfc*). Similar to the defects observed in macrophage-deficient mice, macrophage-restricted deletion of *Pdgfc* decreased white adipose tissue size, and adipocytes were smaller and contained fewer lipids (Cox et al. 2021). These data suggest that macrophage-derived PDGFcc promotes lipid storage via lipid synthesis in white adipocytes within the fat pads.

Similar to the situation in the heart, WAT macrophages can uptake mitochondria from adipocytes, which influences their transcriptome and causes the upregulation of the hypoxia-inducible factor (HIF)-1 α-related pathway and downregulation of genes associated with electron transport (Brestoff et al. 2021). Genes involved in heparan sulfate biosynthesis, including *Ext1*, are crucial for this mitochondrial uptake. The deletion of *Ext1* in macrophages increases body weight and absolute epididymal WAT mass, resulting in impaired glucose and insulin tolerance under steady-state conditions (Brestoff et al. 2021). Accordingly, macrophages remain in immunometabolic crosstalk with adipocytes and can adapt to microenvironmental changes through mitochondrial transfer.

1.9.2 Brown Adipose Tissue Macrophages

Brown adipose tissue (BAT) in homeotherms is crucial for maintaining a constant body temperature even in colder environments. Previously, it was thought that this mechanism relies exclusively on the hypothalamus, which triggers sympathetic discharge upon cold temperature exposure, resulting in the release of catecholamines into WAT and BAT. Norepinephrine then binds to the adipocyte β3 adrenergic receptor and activates thermogenesis via the cyclic AMP/protein kinase A (cAMP/PKA) pathway (Chouchani et al. 2019). However, a seminal study showed that ATMs can also participate in the regulation of norepinephrine levels and thereby contribute to thermoregulation (Nguyen et al. 2011). It was first observed that cold exposure increased CD206, CD301, and arginase 1 protein levels in BAT and WAT macrophages, indicative of an alternatively activated macrophage phenotype (Nguyen et al. 2011). Mice with a defective IL-4 signaling pathway, the main cytokine involved in alternative activation, were then shown to be susceptible to cold-induced hypothermia, which was accompanied by decreased expression of cold-inducible thermogenic genes, reduced noradrenaline levels, and low levels of circulating free fatty acids (Nguyen et al. 2011). Injection of IL-4 reversed these effects in wild-type mice. This indicates that alternative "M2-like" activated macrophages contribute to the adaptive and facultative aspects of thermogenesis. Notably, IL-4 is not only required for alternative activation but also regulates the proliferation and cellular aging of some tissue-resident macrophages, such as peritoneal macrophages and splenic RPMs (Jenkins et al. 2011; Jenkins et al. 2013; Okreglicka et al. 2021; Zhou et al. 2024). Therefore, IL-4 may be similarly involved in ATM maintenance.

The role of BAT macrophages in thermogenesis was further studied in mice with a *Cx3cr1*$^{\mathrm{CreERT2}}$-restricted deletion of *Mecp2*, which is related to neurodegenerative Rett syndrome (Wolf et al. 2017). Interestingly, the *Mecp2*-deficiency in CX3CR1^{+} cells (which include microglia) did not cause neurodevelopmental symptoms in mice but resulted in spontaneous obesity. Mice with *Mecp2*-deficient macrophages produced less heat than controls, suggesting impairment of BAT thermogenesis. Indeed, microscopy revealed fewer sympathetic axons innervating BAT upon deletion of *Mecp2* and reduced local titers of norepinephrine, which resulted in lower expression of thermogenic factors by adipocytes (Wolf et al. 2017). The molecular mechanism involves the clearance of norepinephrine by BAT macrophages via the solute carrier family 6 member 2 (SLC6A2) transporter, and monoamine oxidase A, a degradation enzyme (Pirzgalska et al. 2017). Deletion of *Slc6a2* increases BAT content, causes browning of white fat, increases thermogenesis, and leads to weight loss in obese mice. These results indicate that CX3CR1^{+} macrophages play a crucial role in BAT innervation and thermogenesis. Since both studies used a CX3CR1-dependent system for the identification and targeting of BAT macrophages, respectively, the role of Lyve1^{hi}MHCII^{low} macrophages in BAT remains unclear.

BAT macrophages are also able to take up oxidatively damaged mitochondria from adipocytes in a CD36-dependent manner (Rosina et al. 2022). Accordingly, the depletion of BAT macrophages *in vivo* causes the accumulation of extracellular vesicles containing mitochondrial proteins, accumulation of lipids, and failure to initiate the thermogenic response to cold exposure (Rosina et al. 2022).

In summary, adipose macrophages are important for adipose tissue development, angiogenesis, innervation, thermogenesis, and the regulation of lipid metabolism. However, despite significant advances in adipose tissue macrophage research, the exact subset-specific functions of $Lyve1^{hi}MHCII^{low}$ and $CX3CR1^{+}$ $Lyve1^{low}MHCII^{hi}$ macrophages remain to be elucidated.

1.10 Macrophages in Other Tissues and Their Homeostatic Functions

The previous paragraphs represent a selection of some of the most studied tissue-resident macrophage populations to date. However, macrophages inhabit all vertebrate organs and given the crucial role of macrophages for organ and tissue integrity, research on these cells is rapidly increasing and extending to less studied tissue populations. A detailed discussion of all these populations is beyond the scope of this chapter, but to further illustrate the diversity and tissue specificity of macrophages, a few additional populations are (briefly) mentioned below.

1.10.1 Peritoneal Cavity Macrophages

Large peritoneal macrophages (PMs) express the lineage-determining transcription factor *Gata6*, which is induced by retinoic acid, a metabolite of vitamin A secreted by stromal cells in the omentum and visceral adipose tissues and essential for PM identity (Okabe and Medzhitov 2014; Buechler et al. 2019). Embryonic-derived PMs are long-lived cells and are gradually outcompeted by monocyte-derived cells with age (Bain et al. 2016). Primed by the peritoneal microenvironment and regulated by the transcription factors KLF2 and KLF4, PMs monitor visceral organs and related mesothelium and efficiently clear apoptotic cells. This mitigates inflammation caused by self-derived nucleic acids, while retaining responsiveness to infections (Roberts et al. 2017). PMs can also be recruited to inflammatory reactions that occur in peritoneal cavity-associated organs, such as the intestine (Honda et al. 2021) or the heart (Deniset et al. 2019). As a result, they may act as a local source of immune cells in these locations.

1.10.2 Lymph Node Macrophages

Lymph nodes are home to two main types of macrophages: subcapsular sinus macrophages (SSMs) and medullary sinus macrophages (MSMs). The subcapsular sinus is a significant entry point for pathogens where the afferent lymph flows into the lymph nodes. At this location, $CD169^{+}F4/80^{-}SIGNR1^{-}$ SSMs act as a physical barrier that prevents the entry of large molecules into the conduit system and the migration of cells into the parenchyma. Consequently, SSMs are well positioned to filter any material that reaches the lymph nodes via the lymph from the periphery, including viruses, bacteria, pathogen-derived antigens, and immune complexes. They can translocate antigens across the subcapsular sinus and activate B cells (Phan et al. 2007, 2009). Similar to MMMs in the spleen, differentiation of SSMs relies on lymphotoxin signaling (Phan et al. 2009; Moseman et al. 2012). $CD169^{+}F4/80^{+}SIGNR1^{+}$ MSMs, on the other hand, are lymphotoxin-independent and can be found in the cortical interfollicular region between the B cell follicles in the cortex and medullary sinuses. MSMs appear to be highly phagocytic cells with increased endocytic and degradative capabilities compared to SSMs (Phan et al. 2009), but it remains unclear to what extent the two macrophage subsets participate in antigen presentation and contribute to lymph node architecture under steady-state conditions.

1.10.3 Thymic Macrophages

Like the spleen and lymph nodes, the thymus also contains macrophages, which play an essential role in efferocytosis during the negative and positive selection of thymocytes. This metabolic process is driven by the pentose-phosphate pathway (Esashi et al. 2003; Tsai et al. 2022). Two distinct subsets of thymic macrophages have been identified: embryonic-derived $TIMD4^{+}$ cells in the cortex, and monocyte-derived $CX3CR1^{+}$ macrophages at the cortico-medullary junction. Embryonic thymic macrophages bear similarities to splenic red pulp macrophages and Kupffer cells because of their *SpiC* expression (Zhou et al. 2022). However, the precise function of each thymic subset during T-cell development and tolerance formation remains unclear.

1.10.4 Bone Macrophages

The bone contains at least three types of macrophages: erythroblastic island macrophages (EIMs), osteoclasts, and osteal macrophages (sometimes also referred to as osteomacs) (Chang et al. 2008). EIMs show similarities to red pulp macrophages in the spleen and are likewise involved in red blood cell clearance. Accordingly, *Pparg* and *SpiC* crucially orchestrate their development and function (Okreglicka et al. 2021). Osteoclasts are unique multinucleated giant cells in the BM that are involved in bone resorption. This specialized cell type develops from EMP-derived myeloid cells (Jacome-Galarza et al. 2019) that fuse in a M-CSF and RANK ligand-dependent manner (Lacey et al. 1998). Downstream of the RANK ligand pathway, the activation of NF-kB and MAPK signaling results in the metabolic reprogramming of osteoclasts, which involves NFATc1, a crucial regulator of osteoclastogenesis (Takayanagi et al. 2002). In addition, monocytes can contribute to multinucleated giant cell formation during a defined postnatal time window (Jacome-Galarza et al. 2019). The functional ability of osteoclasts to remodel the bone can have consequences for the hematopoietic niche, thereby affecting HSC mobilization, proliferation, and retention (Adams et al. 2006; Kollet et al. 2006; Lymperi et al. 2011). Defective osteoclast activity further contributes to osteopetrosis and BM failure, and significantly impairs erythropoiesis (Chow et al. 2013). Non-multinucleate osteal macrophages can, in conjunction with megakaryocytes, also influence the hematopoietic niche (Mohamad et al. 2017), but are mainly involved in bone homeostasis by promoting bone deposition instead of bone resorption as osteoclasts. Accordingly, osteal macrophages regulate osteoblast function, as observed in *Cd169*^{Dtr} mice. Deletion of osteal macrophages in this model system causes a significant reduction in osteoblasts (Batoon et al. 2019), thereby reducing bone mass (Cho et al. 2014) and demonstrating the role of osteal macrophages in bone anabolism.

1.10.5 Epithelial Barrier Macrophages

Langerhans cells are located in all epithelial barriers. While skin Langerhans cells are of embryonic origin and only minimally replenished by monocyte-derived cells (Hoeffel et al. 2012), other Langerhans cell subsets such as in the oral cavity can be of mixed origin, including monocyte and pre-dendritic cell precursors (Capucha et al. 2015). The development of epidermal Langerhans cells is regulated by the three cytokines TGF-β (Zahner et al. 2011), IL-34 (Greter et al. 2012), RANKL (Barbaroux et al. 2008), the mTOR signaling pathway (Kellersch and Brocker 2013; Sparber et al. 2014), and the transcription factors *Rara* (Hashimoto-Hill et al. 2018), *Runx3* (Chopin et al. 2013), and *Id2* (Hacker et al. 2003), of which the latter two are direct targets of TGF-β. It was shown that Langerhans cells selectively and specifically induce the activation and proliferation of skin-resident regulatory T-cells and thereby maintain tolerance in normal skin (Seneschal et al. 2012). In addition to their role as immunological gatekeepers of tissue barriers, Langerhans cells are involved in apoptotic cell clearance and intercellular communication. Langerhans cells directly communicate with keratinocytes, thereby maintaining a constant epithelial cell-to-immune cell ratio across the basal layer of the epidermis (Park et al. 2021). Accordingly, immune cell density increases when the number of keratinocytes increases (Park et al. 2021). However, the depletion of skin-resident immune cells does not affect epidermal thickness or epithelial basal density, and the role of Langerhans cells in homeostatic skin requires further investigation. Langerhans cells are also involved in skin histiocytosis, a condition characterized by the accumulation of mononuclear phagocytes in granulomatous lesions (Allen et al. 2018). It was shown that activating somatic mutations in genes involved in the mitogen-activated protein kinase pathway is sufficient to drive the development of histiocytosis in a Langerhans cell-dependent manner (Bigenwald et al. 2021).

1.10.6 Intestine Macrophages

The intestine contains short-lived monocyte-derived *Cx3cr1*-expressing macrophages in the mucosal lamina propria below the intestinal epithelium, which serve as a frontline defense against invading pathogens. These cells require constant IL-10-dependent immune silencing, which otherwise results in epithelial hyperplasia and erosion and causes severe spontaneous colitis (Zigmond et al. 2014). Moreover, intestinal macrophages are involved in the efferocytosis of apoptotic material, which contributes to the repression of pro-inflammatory pathways, thereby securing epithelial integrity and mitigating colitis (Cummings et al. 2016). Additional embryonic-derived gut macrophage populations include blood vessel-associated macrophages, which are crucial for vascular integrity (Honda et al. 2020), and neuron-associated macrophages within the muscularis region, which are essential for enteric neuron survival (Gabanyi et al. 2016; De Schepper et al. 2018).

1.10.7 Mammary Gland Macrophages

Two macrophage subsets with distinct phenotypes (F4/80^{hi} and F4/80^{lo} subsets) are evident within mammary glands, where they are involved in various developmental processes, including ductal development and epithelial remodeling (Jäppinen et al. 2019; Dawson et al. 2020). Additionally, a distinct CX3CR1^{+} macrophage subset arises during lactation and may play a role in immune surveillance, and possibly milk production (Cansever et al. 2023).

1.10.8 Other Interstitial Macrophage Subsets

Similar to the situation observed in adipose tissue and the lungs, Lyve1^{hi}MHCII^{low} and Lyve1^{low}MHCII^{hi} macrophages can be found in various organs, including the bladder, testis, ganglia, trachea, tongue, and others (Chakarov et al. 2019; Dick et al. 2022). It seems that both subsets are derived from fetal precursors during embryogenesis, but that the *Cx3cr1*-expressing Lyve1^{low}MHCII^{hi} population is replenished faster by BM-derived monocytes during the postnatal phase (Lyras et al. 2022). Lyve1^{low}MHCII^{hi} macrophages have been suggested to be more frequently associated with nerve bundles, whereas Lyve1^{hi}MHCII^{low} macrophages are located in proximity to blood vessels and are transcriptionally enriched in pathways related to angiogenesis (Chakarov et al. 2019). It is possible that the anatomy of an organ with distinct areas of high vessel density and nerve bundles, respectively, creates distinct macrophage niches. *Cx3cr1*-expressing Lyve1^{low}MHCII^{hi} macrophages in the tongue, for instance, are preferentially located in the highly innervated lamina propria beneath the tongue epidermis, whereas FOLR2^{+} (Lyve1^{hi}MHCII^{low}) macrophages reside in deeper muscular tissues (Lyras et al. 2022). As described above, an association of *Cx3cr1*-expressing Lyve1^{low}MHCII^{hi} macrophages with nerves and of Lyve1^{hi}MHCII^{low} cells with vessels could also be detected in adipose tissue (Fig. 1.8) and skin (Kolter et al. 2019). These data suggest that a general functional division of labor may exist between the two macrophage subsets, which can be found in interstitial or connective tissues. Therefore, it is possible that Lyve1^{low}MHCII^{hi} macrophages are functionally connected to neurons in many tissues and that Lyve1^{low}MHCII^{hi} participates more preferentially in angiogenesis. However, if this holds true for all interstitial tissues must be individually proven.

1.11 Conclusion

For a long time, macrophages have been solely recognized as immune cells that primarily perform functions in immune defense and surveillance. This prevailing view is currently changing, and macrophages are slowly starting to be appreciated for their role in tissue homeostasis and maintenance. Therefore, the fact that functional or developmental defects of macrophages can lead to organ dysfunctions and pathological manifestations is a relevant finding that also needs to be considered in the clinic. Furthermore, it is possible that self-renewing macrophages that arise during embryogenesis cannot be functionally replaced, at least not entirely, by monocytes during adulthood. However, these observations are currently based on experimental inbred mice and await confirmation in humans. Nevertheless, macrophages occupy discrete niches, which provide crucial support to essential cell populations such as cardiomyocytes, neurons, and blood vessels. Breaking intercellular communication during inflammation, drug treatment such as chemotherapy, or organ transplantation can cause, at least temporarily, fluctuations in these delicate functions, which may delay healing and contribute to disease complications.

Acknowledgments We would like to thank Simon Yona for discussion. We are grateful for the support from the Research Council of Finland's Flagship InFLAMES (337530 and 357910), the Jane and Aatos Erkko Foundation, the Sigrid Juselius Foundation, the German Research Foundation (SPP2395; MI1328), and the Research Council of Finland (355727).

References

Adams GB, Chabner KT, Alley IR et al (2006) Stem cell engraftment at the endosteal niche is specified by the calcium-sensing receptor. Nature 439:599–603. https://doi.org/10.1038/nature04247

A-Gonzalez N, Guillen JA, Gallardo G et al (2013) The nuclear receptor LXRα controls the functional specialization of splenic macrophages. Nat Immunol 14:831–839. https://doi.org/10.1038/ni.2622

Aichele P, Zinke J, Grode L et al (2003) Macrophages of the splenic marginal zone are essential for trapping of blood-borne particulate antigen but dispensable for induction of specific T cell responses. J Immunol 171:1148–1155. https://doi.org/10.4049/jimmunol.171.3.1148

Ajami B, Bennett JL, Krieger C et al (2007) Local self-renewal can sustain CNS microglia maintenance and function throughout adult life. Nat Neurosci 10:1538–1543. https://doi.org/10.1038/nn2014

Allen CE, Merad M, McClain KL (2018) Langerhans-cell histiocytosis. N Engl J Med 379:856–868. https://doi.org/10.1056/NEJMra1607548

Alliot F, Godin I, Pessac B (1999) Microglia derive from progenitors, originating from the yolk sac, and which proliferate in the brain. Brain Res Dev Brain Res 117:145–152. https://doi.org/10.1016/s0165-3806(99)00113-3

Backer R, Schwandt T, Greuter M et al (2010) Effective collaboration between marginal metallophilic macrophages and CD8+ dendritic cells in the generation of cytotoxic T cells. Proc Natl Acad Sci USA 107:216–221. https://doi.org/10.1073/pnas.0909541107

Badimon A, Strasburger HJ, Ayata P et al (2020) Negative feedback control of neuronal activity by microglia. Nature 586:417–423. https://doi.org/10.1038/s41586-020-2777-8

Bain CC, Hawley CA, Garner H et al (2016) Long-lived self-renewing bone marrow-derived macrophages displace embryo-derived cells to inhabit adult serous cavities. Nat Commun 7:ncomms11852–14. https://doi.org/10.1038/ncomms11852

Bajpai G, Schneider C, Wong N et al (2018) The human heart contains distinct macrophage subsets with divergent origins and functions. Nat Med 24:1234–1245. https://doi.org/10.1038/s41591-018-0059-x

Baratin M, Simon L, Jorquera A et al (2017) T cell zone resident macrophages silently dispose of apoptotic cells in the lymph node. Immunity 47:349–362.e5. https://doi.org/10.1016/j.immuni.2017.07.019

Barbaroux J-BO, Beleut M, Brisken C et al (2008) Epidermal receptor activator of NF-kappaB ligand controls Langerhans cells numbers and proliferation. J Immunol 181:1103–1108. https://doi.org/10.4049/jimmunol.181.2.1103

Barth E, Stämmler G, Speiser B, Schaper J (1992) Ultrastructural quantitation of mitochondria and myofilaments in cardiac muscle from 10 different animal species including man. J Mol Cell Cardiol 24:669–681. https://doi.org/10.1016/0022-2828(92)93381-s

Bates SR, Tao J-Q, Collins HL et al (2005) Pulmonary abnormalities due to ABCA1 deficiency in mice. Am J Physiol Lung Cell Mol Physiol 289:L980–L989. https://doi.org/10.1152/ajplung.00234.2005

Batoon L, Millard SM, Wullschleger ME et al (2019) CD169+ macrophages are critical for osteoblast maintenance and promote intramembranous and endochondral ossification during bone repair. Biomaterials 196:51–66. https://doi.org/10.1016/j.biomaterials.2017.10.033

Bergmann O, Bhardwaj RD, Bernard S et al (2009) Evidence for cardiomyocyte renewal in humans. Science 324:98–102. https://doi.org/10.1126/science.1164680

Bertrand JY, Jalil A, Klaine M et al (2005) Three pathways to mature macrophages in the early mouse yolk sac. Blood 106:3004–3011. https://doi.org/10.1182/blood-2005-02-0461

Bian Z, Shi L, Guo Y-L et al (2016) Cd47-Sirpα interaction and IL-10 constrain inflammation-induced macrophage phagocytosis of healthy self-cells. Proc Natl Acad Sci USA 113:E5434–E5443. https://doi.org/10.1073/pnas.1521069113

Bigenwald C, Le Berichel J, Wilk CM et al (2021) BRAFV600E-induced senescence drives Langerhans cell histiocytosis pathophysiology. Nat Med 27:851–861. https://doi.org/10.1038/s41591-021-01304-x

Bleriot C, Barreby E, Dunsmore G et al (2021) A subset of Kupffer cells regulates metabolism through the expression of CD36. Immunity 54:2101–2116.e6. https://doi.org/10.1016/j.immuni.2021.08.006

Bonnardel J, T'Jonck W, Gaublomme D et al (2019) Stellate cells, hepatocytes, and endothelial cells imprint the Kupffer cell identity on monocytes colonizing the liver macrophage niche. Immunity 51:638–654.e9. https://doi.org/10.1016/j.immuni.2019.08.017

Braillon A, Capron JP, Hervé MA et al (1985) Liver in obesity. Gut 26:133–139. https://doi.org/10.1136/gut.26.2.133

Brestoff JR, Wilen CB, Moley JR et al (2021) Intercellular mitochondria transfer to macrophages regulates White adipose tissue homeostasis and is impaired in obesity. Cell Metab 33:270–282.e8. https://doi.org/10.1016/j.cmet.2020.11.008

Browicz (1899) Ueber intravasculäre Zellen in den Blutcapillaren der Leberacini. Arch Mikrosk Anat 55:420–426

Buechler MB, Kim K-W, Onufer EJ et al (2019) A stromal niche defined by expression of the transcription factor WT1 mediates programming and homeostasis of cavity-resident macrophages. Immunity 51:119–130.e5. https://doi.org/10.1016/j.immuni.2019.05.010

Buttgereit A, Lelios I, Yu X et al (2016) Sall1 is a transcriptional regulator defining microglia identity and function. Nat Immunol 17:1397–1406. https://doi.org/10.1038/ni.3585

Cain DW, O'Koren EG, Kan MJ et al (2013) Identification of a tissue-specific, C/EBPβ-dependent pathway of differentiation for murine peritoneal macrophages. J Immunol 191:4665–4675. https://doi.org/10.4049/jimmunol.1300581

Camara A, Lavanant AC, Abe J et al (2022) CD169+ macrophages in lymph node and spleen critically depend on dual RANK and LTbetaR signaling. Proc Natl Acad Sci USA 119:e2108540119. https://doi.org/10.1073/pnas.2108540119

Cansever D, Petrova E, Krishnarajah S et al (2023) Lactation-associated macrophages exist in murine mammary tissue and human milk. Nat Immunol 24:1098–1109. https://doi.org/10.1038/s41590-023-01530-0

Capucha T, Mizraji G, Segev H et al (2015) Distinct murine mucosal Langerhans cell subsets develop from pre-dendritic cells and monocytes. Immunity 43:369–381. https://doi.org/10.1016/j.immuni.2015.06.017

Chakarov S, Lim HY, Tan L et al (2019) Two distinct interstitial macrophage populations coexist across tissues in specific subtissular niches. Science 363:eaau0964. https://doi.org/10.1126/science.aau0964

Chang MK, Raggatt L-J, Alexander KA et al (2008) Osteal tissue macrophages are intercalated throughout human and mouse bone lining tissues and regulate osteoblast function in vitro and in vivo. J Immunol 181:1232–1244. https://doi.org/10.4049/jimmunol.181.2.1232

Chen G, Zhuchenko O, Kuspa A (2007) Immune-like phagocyte activity in the social amoeba. Science 317:678–681. https://doi.org/10.1126/science.1143991

Chen S-K, Tvrdik P, Peden E et al (2010) Hematopoietic origin of pathological grooming in Hoxb8 mutant mice. Cell 141:775–785. https://doi.org/10.1016/j.cell.2010.03.055

Chen Y, Pikkarainen T, Elomaa O et al (2005) Defective microarchitecture of the spleen marginal zone and impaired response to a thymus-independent type 2 antigen in mice lacking scavenger receptors MARCO and SR-A. J Immunol 175:8173–8180. https://doi.org/10.4049/jimmunol.175.12.8173

Cho C-H, Koh YJ, Han J et al (2007) Angiogenic role of LYVE-1-positive macrophages in adipose tissue. Circ Res 100:e47–e57. https://doi.org/10.1161/01.RES.0000259564.92792.93

Cho SW, Soki FN, Koh AJ et al (2014) Osteal macrophages support physiologic skeletal remodeling and anabolic actions of parathyroid hormone in bone. Proc Natl Acad Sci USA 111:1545–1550. https://doi.org/10.1073/pnas.1315153111

Chopin M, Seillet C, Chevrier S et al (2013) Langerhans cells are generated by two distinct PU.1-dependent transcriptional networks. J Exp Med 210:2967–2980. https://doi.org/10.1084/jem.20130930

Chouchani ET, Kazak L, Spiegelman BM (2019) New advances in adaptive thermogenesis: UCP1 and beyond. Cell Metab 29:27–37. https://doi.org/10.1016/j.cmet.2018.11.002

Chow A, Huggins M, Ahmed J et al (2013) CD169+ macrophages provide a niche promoting erythropoiesis under homeostasis and stress. Nat Med 19:429–436. https://doi.org/10.1038/nm.3057

Cox N, Crozet L, Holtman IR et al (2021) Diet-regulated production of PDGFcc by macrophages controls energy storage. Science. https://doi.org/10.1126/science.abe9383

Cserép C, Pósfai B, Lénárt N et al (2020) Microglia monitor and protect neuronal function through specialized somatic purinergic junctions. Science 367:528–537. https://doi.org/10.1126/science.aax6752

Cserép C, Schwarcz AD, Pósfai B et al (2022) Microglial control of neuronal development via somatic purinergic junctions. Cell Rep 40:111369. https://doi.org/10.1016/j.celrep.2022.111369

Cummings RJ, Barbet G, Bongers G et al (2016) Different tissue phagocytes sample apoptotic cells to direct distinct homeostasis programs. Nature 539:565–569. https://doi.org/10.1038/nature20138

Cunningham CL, Martínez-Cerdeño V, Noctor SC (2013) Microglia regulate the number of neural precursor cells in the developing cerebral cortex. J Neurosci 33:4216–4233. https://doi.org/10.1523/JNEUROSCI.3441-12.2013

Dawson CA, Pal B, Vaillant F et al (2020) Tissue-resident ductal macrophages survey the mammary epithelium and facilitate tissue remodelling. Nat Cell Biol 22:546–558. https://doi.org/10.1038/s41556-020-0505-0

de Aguiar Vallim TQ, Lee E, Merriott DJ et al (2017) ABCG1 regulates pulmonary surfactant metabolism in mice and men. J Lipid Res 58:941–954. https://doi.org/10.1194/jlr.M075101

De Schepper S, Verheijden S, Aguilera-Lizarraga J et al (2018) Self-maintaining gut macrophages are essential for intestinal homeostasis. Cell 175:400–415.e13. https://doi.org/10.1016/j.cell.2018.07.048

De Simone G, Andreata F, Bleriot C et al (2021) Identification of a Kupffer cell subset capable of reverting the T cell dysfunction induced by hepatocellular priming. Immunity 54:2089–2100.e8. https://doi.org/10.1016/j.immuni.2021.05.005

Defaye A, Evans I, Crozatier M et al (2009) Genetic ablation of drosophila phagocytes reveals their contribution to both development and resistance to bacterial infection. J Innate Immun 1:322–334. https://doi.org/10.1159/000210264

Deniset JF, Belke D, Lee WY et al (2019) Gata6+ pericardial cavity macrophages relocate to the injured heart and prevent cardiac fibrosis. Immunity 51:131–140.e5. https://doi.org/10.1016/j.immuni.2019.06.010

Deppermann C, Kratofil RM, Peiseler M et al (2020) Macrophage galactose lectin is critical for Kupffer cells to clear aged platelets. J Exp Med. https://doi.org/10.1084/jem.20190723

Dick SA, Wong A, Hamidzada H et al (2022) Three tissue resident macrophage subsets coexist across organs with conserved origins and life cycles. Sci Immunol 7:eabf7777. https://doi.org/10.1126/sciimmunol.abf7777

Dörr D, Obermayer B, Weiner JM et al (2022) C/EBPβ regulates lipid metabolism and Pparg isoform 2 expression in alveolar macrophages. Sci Immunol 7:eabj0140. https://doi.org/10.1126/sciimmunol.abj0140

Du Y, Brennan FH, Popovich PG, Zhou M (2022) Microglia maintain the normal structure and function of the hippocampal astrocyte network. Glia 70:1359–1379. https://doi.org/10.1002/glia.24179

Dutta P, Hoyer FF, Grigoryeva LS et al (2015) Macrophages retain hematopoietic stem cells in the spleen via VCAM-1. J Exp Med 212:497–512. https://doi.org/10.1084/jem.20141642

Ebina-Shibuya R, Matsumoto M, Kuwahara M et al (2017) Inflammatory responses induce an identity crisis of alveolar macrophages, leading to pulmonary alveolar proteinosis. J Biol Chem 292:18098–18112. https://doi.org/10.1074/jbc.M117.808535

Ecker E (1847) Ueber die veränderungen, welche die blutkörperchen in der milz erleiden Z. Rationelle Med 6:206
Eddy WE, Gong K-Q, Bell B et al (2017) Stat5 is required for CD103+ dendritic cell and alveolar macrophage development and protection from lung injury. J Immunol 198:4813–4822. https://doi.org/10.4049/jimmunol.1601777
Epelman S, Lavine KJ, Beaudin AE et al (2014) Embryonic and adult-derived resident cardiac macrophages are maintained through distinct mechanisms at steady state and during inflammation. Immunity 40:91–104. https://doi.org/10.1016/j.immuni.2013.11.019
Esashi E, Sekiguchi T, Ito H et al (2003) Cutting edge: a possible role for CD4+ thymic macrophages as professional scavengers of apoptotic thymocytes. J Immunol 171:2773–2777. https://doi.org/10.4049/jimmunol.171.6.2773
Filipello F, Morini R, Corradini I et al (2018) The microglial innate immune receptor TREM2 is required for synapse elimination and Normal brain connectivity. Immunity 48:979–991.e8. https://doi.org/10.1016/j.immuni.2018.04.016
Franken L, Klein M, Spasova M et al (2015) Splenic red pulp macrophages are intrinsically superparamagnetic and contaminate magnetic cell isolates. Sci Rep 5:12940–12910. https://doi.org/10.1038/srep12940
Gabanyi I, Muller PA, Feighery L et al (2016) Neuro-immune interactions drive tissue programming in intestinal macrophages. Cell 164:378–391. https://doi.org/10.1016/j.cell.2015.12.023
Gilboa D, Haim-Ohana Y, Deshet-Unger N et al (2017) Erythropoietin enhances Kupffer cell number and activity in the challenged liver. Sci Rep 7:10379–10313. https://doi.org/10.1038/s41598-017-11082-7
Gilchrist TC, Ketron LW (1916) A unique case of atrophy of the fatty layer of the skin, preceded by the ingestion of the fat by large phagocytic cells—macrophages. Bull Johns Hopkins Hosp 27:291–294
Ginhoux F, Greter M, Leboeuf M et al (2010) Fate mapping analysis reveals that adult microglia derive from primitive macrophages. Science 330:841–845. https://doi.org/10.1126/science.1194637
Godwin JW, Pinto AR, Rosenthal NA (2013) Macrophages are required for adult salamander limb regeneration. Proc Natl Acad Sci USA 110:9415–9420. https://doi.org/10.1073/pnas.1300290110
Gomez-Perdiguero E, Geissmann F (2013) Myb-independent macrophages: a family of cells that develops with their tissue of residence and is involved in its homeostasis. Cold Spring Harb Symp Quant Biol 78:91–100. https://doi.org/10.1101/sqb.2013.78.020032
Gomez-Perdiguero E, Klapproth K, Schulz C et al (2015) Tissue-resident macrophages originate from yolk-sac-derived erythro-myeloid progenitors. Nature 518:547–551. https://doi.org/10.1038/nature13989
Goswami K, Koner BC (2002) Level of sialic acid residues in platelet proteins in diabetes, aging, and Hodgkin's lymphoma: a potential role of free radicals in desialylation. Biochem Biophys Res Commun 297:502–505. https://doi.org/10.1016/s0006-291x(02)02241-6
Greter M, Lelios I, Pelczar P et al (2012) Stroma-derived interleukin-34 controls the development and maintenance of langerhans cells and the maintenance of microglia. Immunity 37:1050–1060. https://doi.org/10.1016/j.immuni.2012.11.001
Grootveld AK, Kyaw W, Panova V et al (2023) Apoptotic cell fragments locally activate tingible body macrophages in the germinal center. Cell 186:1144–1161.e18. https://doi.org/10.1016/j.cell.2023.02.004
Gschwend J, Sherman SPM, Ridder F et al (2021) Alveolar macrophages rely on GM-CSF from alveolar epithelial type 2 cells before and after birth. J Exp Med. https://doi.org/10.1084/jem.20210745
Guilliams M, De Kleer I, Henri S et al (2013) Alveolar macrophages develop from fetal monocytes that differentiate into long-lived cells in the first week of life via GM-CSF. J Exp Med 210:1977–1992. https://doi.org/10.1084/jem.20131199

Guilliams M, Thierry GR, Bonnardel J, Bajenoff M (2020) Establishment and maintenance of the macrophage niche. Immunity 52:434–451. https://doi.org/10.1016/j.immuni.2020.02.015
Guilliams M, Bonnardel J, Haest B et al (2022) Spatial proteogenomics reveals distinct and evolutionarily conserved hepatic macrophage niches. Cell 185:379–396.e38. https://doi.org/10.1016/j.cell.2021.12.018
Gurwicz N, Stoler-Barak L, Schwan N et al (2023) Tingible body macrophages arise from lymph node-resident precursors and uptake B cells by dendrites. J Exp Med. https://doi.org/10.1084/jem.20222173
Hacker C, Kirsch RD, Ju X-S et al (2003) Transcriptional profiling identifies Id2 function in dendritic cell development. Nat Immunol 4:380–386. https://doi.org/10.1038/ni903
Hagemeyer N, Hanft K-M, Akriditou M-A et al (2017) Microglia contribute to normal myelinogenesis and to oligodendrocyte progenitor maintenance during adulthood. Acta Neuropathol 134:441–458. https://doi.org/10.1007/s00401-017-1747-1
Haldar M, Kohyama M, So AY-L et al (2014) Heme-mediated SPI-C induction promotes monocyte differentiation into iron-recycling macrophages. Cell 156:1223–1234. https://doi.org/10.1016/j.cell.2014.01.069
Han J, Lee J-E, Jin J et al (2011) The spatiotemporal development of adipose tissue. Development 138:5027–5037. https://doi.org/10.1242/dev.067686
Hanayama R, Tanaka M, Miyasaka K et al (2004) Autoimmune disease and impaired uptake of apoptotic cells in MFG-E8-deficient mice. Science 304:1147–1150. https://doi.org/10.1126/science.1094359
Hartroft WS (1960) The pathology of obesity. Bull N Y Acad Med 36:313–322
Hashimoto D, Chow A, Noizat C et al (2013) Tissue-resident macrophages self-maintain locally throughout adult life with minimal contribution from circulating monocytes. Immunity 38:792–804. https://doi.org/10.1016/j.immuni.2013.04.004
Hashimoto-Hill S, Friesen L, Park S et al (2018) RARα supports the development of Langerhans cells and langerin-expressing conventional dendritic cells. Nat Commun 9:3896–3813. https://doi.org/10.1038/s41467-018-06341-8
Hassnain Waqas SF, Noble A, Hoang AC et al (2017) Adipose tissue macrophages develop from bone marrow-independent progenitors in Xenopus laevis and mouse. J Leukoc Biol 102:845–855. https://doi.org/10.1189/jlb.1A0317-082RR
Haynes SE, Hollopeter G, Yang G et al (2006) The P2Y12 receptor regulates microglial activation by extracellular nucleotides. Nat Neurosci 9:1512–1519. https://doi.org/10.1038/nn1805
He J, Cao Y, Zhu Q et al (2024) Renal macrophages monitor and remove particles from urine to prevent tubule obstruction. Immunity 57:106–123.e7. https://doi.org/10.1016/j.immuni.2023.12.003
Heidt T, Courties G, Dutta P et al (2014) Differential contribution of monocytes to heart macrophages in steady-state and after myocardial infarction. Circ Res 115:284–295. https://doi.org/10.1161/CIRCRESAHA.115.303567
Hoeffel G, Ginhoux F (2018) Fetal monocytes and the origins of tissue-resident macrophages. Cell Immunol 330:5–15. https://doi.org/10.1016/j.cellimm.2018.01.001
Hoeffel G, Wang Y, Greter M et al (2012) Adult Langerhans cells derive predominantly from embryonic fetal liver monocytes with a minor contribution of yolk sac-derived macrophages. J Exp Med 209:1167–1181. https://doi.org/10.1084/jem.20120340
Hoeffel G, Chen J, Lavin Y et al (2015) C-Myb(+) erythro-myeloid progenitor-derived fetal monocytes give rise to adult tissue-resident macrophages. Immunity 42:665–678. https://doi.org/10.1016/j.immuni.2015.03.011
Honda M, Surewaard BGJ, Watanabe M et al (2020) Perivascular localization of macrophages in the intestinal mucosa is regulated by Nr4a1 and the microbiome. Nat Commun 11:1329–1317. https://doi.org/10.1038/s41467-020-15068-4
Honda M, Kadohisa M, Yoshii D et al (2021) Directly recruited GATA6 + peritoneal cavity macrophages contribute to the repair of intestinal serosal injury. Nat Commun 12:7294–7215. https://doi.org/10.1038/s41467-021-27614-9

Honey D, Wosnitzka E, Klann E, Weinhard L (2022) Analysis of microglial BDNF function and expression in the motor cortex. Front Cell Neurosci 16:961276. https://doi.org/10.3389/fncel.2022.961276

Hotamisligil GS, Shargill NS, Spiegelman BM (1993) Adipose expression of tumor necrosis factor-alpha: direct role in obesity-linked insulin resistance. Science 259:87–91. https://doi.org/10.1126/science.7678183

Hotamisligil GS, Budavari A, Murray D, Spiegelman BM (1994) Reduced tyrosine kinase activity of the insulin receptor in obesity-diabetes. Central role of tumor necrosis factor-alpha. J Clin Invest 94:1543–1549. https://doi.org/10.1172/JCI117495

Hristovska I, Robert M, Combet K et al (2022) Sleep decreases neuronal activity control of microglial dynamics in mice. Nat Commun 13:6273–6215. https://doi.org/10.1038/s41467-022-34035-9

Hu W, Jiang C, Kim M et al (2022) Isoform-specific functions of PPARγ in gene regulation and metabolism. Genes Dev 36:300–312. https://doi.org/10.1101/gad.349232.121

Hulsmans M, Clauss S, Xiao L et al (2017) Macrophages facilitate electrical conduction in the heart. Cell 169:510–522.e20. https://doi.org/10.1016/j.cell.2017.03.050

Hulsmans M, Schloss MJ, Lee I-H et al (2023) Recruited macrophages elicit atrial fibrillation. Science 381:231–239. https://doi.org/10.1126/science.abq3061

Hume DA (2023) Fate-mapping studies in inbred mice: a model for understanding macrophage development and homeostasis? Eur J Immunol 53:e2250242. https://doi.org/10.1002/eji.202250242

Hume DA, Offermanns S, Bonnavion R (2022) Contamination of isolated mouse Kupffer cells with liver sinusoidal endothelial cells. Immunity 55:1139–1140. https://doi.org/10.1016/j.immuni.2022.06.010

Ito S, Naito M, Kobayashi Y et al (1999) Roles of a macrophage receptor with collagenous structure (MARCO) in host defense and heterogeneity of splenic marginal zone macrophages. Arch Histol Cytol 62:83–95. https://doi.org/10.1679/aohc.62.83

Izquierdo HM, Brandi P, Gómez M-J et al (2018) Von Hippel-Lindau protein is required for optimal alveolar macrophage terminal differentiation, self-renewal, and function. Cell Rep 24:1738–1746. https://doi.org/10.1016/j.celrep.2018.07.034

Jacome-Galarza CE, Percin GI, Muller JT et al (2019) Developmental origin, functional maintenance and genetic rescue of osteoclasts. Nature 568:541–545. https://doi.org/10.1038/s41586-019-1105-7

Jaitin DA, Adlung L, Thaiss CA et al (2019) Lipid-associated macrophages control metabolic homeostasis in a Trem2-dependent manner. Cell 178:686–698.e14. https://doi.org/10.1016/j.cell.2019.05.054

Jäppinen N, Félix I, Lokka E et al (2019) Fetal-derived macrophages dominate in adult mammary glands. Nat Commun 10:281–212. https://doi.org/10.1038/s41467-018-08065-1

Jenkins SJ, Ruckerl D, Cook PC et al (2011) Local macrophage proliferation, rather than recruitment from the blood, is a signature of TH2 inflammation. Science 332:1284–1288. https://doi.org/10.1126/science.1204351

Jenkins SJ, Ruckerl D, Thomas GD et al (2013) IL-4 directly signals tissue-resident macrophages to proliferate beyond homeostatic levels controlled by CSF-1. J Exp Med 210:2477–2491. https://doi.org/10.1084/jem.20121999

Jiang Y, Tang Y, Hoover C et al (2021) Kupffer cell receptor CLEC4F is important for the destruction of desialylated platelets in mice. Cell Death Differ 28:3009–3021. https://doi.org/10.1038/s41418-021-00797-w

Jung HS, Lee YJ, Kim YH et al (2012) Peroxisome proliferator-activated receptor gamma/signal transducers and activators of transcription 5A pathway plays a key factor in adipogenesis of human bone marrow-derived stromal cells and 3T3-L1 preadipocytes. Stem Cells Dev 21:465–475. https://doi.org/10.1089/scd.2010.0591

Kana V, Desland FA, Casanova-Acebes M et al (2019) CSF-1 controls cerebellar microglia and is required for motor function and social interaction. J Exp Med 216:2265–2281. https://doi.org/10.1084/jem.20182037

Kaye D, Gill FA, Hook EW (1967) Factors influencing host resistance to Salmonella infections: the effects of hemolysis and erythrophagocytosis. Am J Med Sci 254:205–215. https://doi.org/10.1097/00000441-196708000-00011

Kellersch B, Brocker T (2013) Langerhans cell homeostasis in mice is dependent on mTORC1 but not mTORC2 function. Blood 121:298–307. https://doi.org/10.1182/blood-2012-06-439786

Kershman J (1939) Genesis of microglia in the human brain. Arch NeurPsych 41:24–50. https://doi.org/10.1001/archneurpsyc.1939.02270130034002

Kierdorf K, Erny D, Goldmann T et al (2013) Microglia emerge from erythromyeloid precursors via Pu.1- and Irf8-dependent pathways. Nat Neurosci 16:273–280. https://doi.org/10.1038/nn.3318

Kim H-J, Cho M-H, Shim WH et al (2017) Deficient autophagy in microglia impairs synaptic pruning and causes social behavioral defects. Mol Psychiatry 22:1576–1584. https://doi.org/10.1038/mp.2016.103

Kohyama M, Ise W, Edelson BT et al (2008) Role for Spi-C in the development of red pulp macrophages and splenic iron homeostasis. Nature 457:318–321. https://doi.org/10.1038/nature07472

Kollet O, Dar A, Shivtiel S et al (2006) Osteoclasts degrade endosteal components and promote mobilization of hematopoietic progenitor cells. Nat Med 12:657–664. https://doi.org/10.1038/nm1417

Kolter J, Feuerstein R, Zeis P et al (2019) A subset of skin macrophages contributes to the surveillance and regeneration of local nerves. Immunity 50:1482–1497.e7. https://doi.org/10.1016/j.immuni.2019.05.009

Kovtunovych G, Eckhaus MA, Ghosh MC et al (2010) Dysfunction of the heme recycling system in heme oxygenase 1-deficient mice: effects on macrophage viability and tissue iron distribution. Blood 116:6054–6062. https://doi.org/10.1182/blood-2010-03-272138

Kupffer C (1876) Ueber Sternzellen der Leber. Arch Mikrosk Anat 12:353–358

Kurematsu C, Sawada M, Ohmuraya M et al (2022) Synaptic pruning of murine adult-born neurons by microglia depends on phosphatidylserine. J Exp Med. https://doi.org/10.1084/jem.20202304

Lacey DL, Timms E, Tan HL et al (1998) Osteoprotegerin ligand is a cytokine that regulates osteoclast differentiation and activation. Cell 93:165–176. https://doi.org/10.1016/s0092-8674(00)81569-x

Lavine KJ, Epelman S, Uchida K et al (2014) Distinct macrophage lineages contribute to disparate patterns of cardiac recovery and remodeling in the neonatal and adult heart. Proc Natl Acad Sci USA 111:16029–16034. https://doi.org/10.1073/pnas.1406508111

Lawrence AR, Canzi A, Bridlance C et al (2024) Microglia maintain structural integrity during fetal brain morphogenesis. Cell 187:962–980.e19. https://doi.org/10.1016/j.cell.2024.01.012

Lazarov T, Juarez-Carreño S, Cox N, Geissmann F (2023) Physiology and diseases of tissue-resident macrophages. Nature 618:698–707. https://doi.org/10.1038/s41586-023-06002-x

Lee B, Qiao L, Kinney B et al (2014) Macrophage depletion disrupts immune balance and energy homeostasis. PLoS One 9:e99575. https://doi.org/10.1371/journal.pone.0099575

Leid J, Carrelha J, Boukarabila H et al (2016) Primitive embryonic macrophages are required for coronary development and maturation. Circ Res 118:1498–1511. https://doi.org/10.1161/CIRCRESAHA.115.308270

Li F, Piattini F, Pohlmeier L et al (2022) Monocyte-derived alveolar macrophages autonomously determine severe outcome of respiratory viral infection. Sci Immunol 7:eabj5761. https://doi.org/10.1126/sciimmunol.abj5761

Li J, van der Wal DE, Zhu G et al (2015) Desialylation is a mechanism of Fc-independent platelet clearance and a therapeutic target in immune thrombocytopenia. Nat Commun 6:7737–7716. https://doi.org/10.1038/ncomms8737

Li T, Chiou B, Gilman CK et al (2020) A splicing isoform of GPR56 mediates microglial synaptic refinement via phosphatidylserine binding. EMBO J 39:e104136. https://doi.org/10.15252/embj.2019104136
Li Y, Fu J, Ling Y et al (2017) Sialylation on O-glycans protects platelets from clearance by liver Kupffer cells. Proc Natl Acad Sci USA 114:8360–8365. https://doi.org/10.1073/pnas.1707662114
Liu C, Yang XV, Wu J et al (2011) Oxysterols direct B-cell migration through EBI2. Nature 475:519–523. https://doi.org/10.1038/nature10226
Liu F, Dai S, Feng D et al (2020) Distinct fate, dynamics and niches of renal macrophages of bone marrow or embryonic origins. Nat Commun 11:2280–2216. https://doi.org/10.1038/s41467-020-16158-z
Liu H, Wang X, Chen L et al (2021) Microglia modulate stable wakefulness via the thalamic reticular nucleus in mice. Nat Commun 12:4646–4616. https://doi.org/10.1038/s41467-021-24915-x
Liu Z, Gu Y, Chakarov S, et al (2019a) Fate mapping via Ms4a3 expression history traces monocyte-derived cells. bioRxiv 652032
Liu Z, Gu Y, Chakarov S et al (2019b) Fate mapping via Ms4a3-expression history traces monocyte-derived cells. Cell 178:1509–1525.e19. https://doi.org/10.1016/j.cell.2019.08.009
Loebl EC, Baxter CR, Curreri PW (1973) The mechanism of erythrocyte destruction in the early post-burn period. Ann Surg 178:681–686. https://doi.org/10.1097/00000658-197312000-00001
Lui H, Zhang J, Makinson SR et al (2016) Progranulin deficiency promotes circuit-specific synaptic pruning by microglia via complement activation. Cell 165:921–935. https://doi.org/10.1016/j.cell.2016.04.001
Lumeng CN, Bodzin JL, Saltiel AR (2007a) Obesity induces a phenotypic switch in adipose tissue macrophage polarization. J Clin Invest 117:175–184. https://doi.org/10.1172/JCI29881
Lumeng CN, Deyoung SM, Bodzin JL, Saltiel AR (2007b) Increased inflammatory properties of adipose tissue macrophages recruited during diet-induced obesity. Diabetes 56:16–23. https://doi.org/10.2337/db06-1076
Lund H, Pieber M, Parsa R et al (2018) Competitive repopulation of an empty microglial niche yields functionally distinct subsets of microglia-like cells. Nat Commun 9:4845. https://doi.org/10.1038/s41467-018-07295-7
Lymperi S, Ersek A, Ferraro F et al (2011) Inhibition of osteoclast function reduces hematopoietic stem cell numbers in vivo. Blood 117:1540–1549. https://doi.org/10.1182/blood-2010-05-282855
Lyras EM, Zimmermann K, Wagner LK et al (2022) Tongue immune compartment analysis reveals spatial macrophage heterogeneity. Elife. https://doi.org/10.7554/eLife.77490
Marín-Teva JL, Dusart I, Colin C et al (2004) Microglia promote the death of developing Purkinje cells. Neuron 41:535–547. https://doi.org/10.1016/s0896-6273(04)00069-8
Mass E, Ballesteros I, Farlik M et al (2016) Specification of tissue-resident macrophages during organogenesis. Science. https://doi.org/10.1126/science.aaf4238
Mass E, Nimmerjahn F, Kierdorf K, Schlitzer A (2023) Tissue-specific macrophages: how they develop and choreograph tissue biology. Nat Rev Immunol 23:563–579. https://doi.org/10.1038/s41577-023-00848-y
Masuda T, Tsuda M, Yoshinaga R et al (2012) IRF8 is a critical transcription factor for transforming microglia into a reactive phenotype. Cell Rep 1:334–340. https://doi.org/10.1016/j.celrep.2012.02.014
Masuda T, Amann L, Sankowski R et al (2020) Novel Hexb-based tools for studying microglia in the CNS. Nat Immunol 21:802–815. https://doi.org/10.1038/s41590-020-0707-4
McGaha TL, Chen Y, Ravishankar B et al (2011) Marginal zone macrophages suppress innate and adaptive immunity to apoptotic cells in the spleen. Blood 117:5403–5412. https://doi.org/10.1182/blood-2010-11-320028
McGrath KE, Koniski AD, Malik J, Palis J (2003) Circulation is established in a stepwise pattern in the mammalian embryo. Blood 101:1669–1676. https://doi.org/10.1182/blood-2002-08-2531

McNamara NB, Munro DAD, Bestard-Cuche N et al (2022) Microglia regulate central nervous system myelin growth and integrity. Nature 1–10. https://doi.org/10.1038/s41586-022-05534-y
Medina-Gomez G, Gray SL, Yetukuri L et al (2007) PPAR gamma 2 prevents lipotoxicity by controlling adipose tissue expandability and peripheral lipid metabolism. PLoS Genet 3:e64. https://doi.org/10.1371/journal.pgen.0030064
Metchnikoff E (1892) Leçons sur la pathologie comparée de l"inflammation: faites à l"Institut Pasteur en avril et mai 1891. G. Masson
Metchnikoff E (1905) Immunity in infective diseases. University Press
Mildner A, Schmidt H, Nitsche M et al (2007) Microglia in the adult brain arise from Ly-6ChiCCR2+ monocytes only under defined host conditions. Nat Neurosci 10:1544–1553. https://doi.org/10.1038/nn2015
Millard SM, Heng O, Opperman KS et al (2021) Fragmentation of tissue-resident macrophages during isolation confounds analysis of single-cell preparations from mouse hematopoietic tissues. Cell Rep 37:110058. https://doi.org/10.1016/j.celrep.2021.110058
Mohamad SF, Xu L, Ghosh J et al (2017) Osteomacs interact with megakaryocytes and osteoblasts to regulate murine hematopoietic stem cell function. Blood Adv 1:2520–2528. https://doi.org/10.1182/bloodadvances.2017011304
Molawi K, Wolf Y, Kandalla PK et al (2014) Progressive replacement of embryo-derived cardiac macrophages with age. J Exp Med 211:2151–2158. https://doi.org/10.1084/jem.20140639
Mordelt A, de Witte LD (2023) Microglia-mediated synaptic pruning as a key deficit in neurodevelopmental disorders: hype or hope? Curr Opin Neurobiol 79:102674. https://doi.org/10.1016/j.conb.2022.102674
Moseman EA, Iannacone M, Bosurgi L et al (2012) B cell maintenance of subcapsular sinus macrophages protects against a fatal viral infection independent of adaptive immunity. Immunity 36:415–426. https://doi.org/10.1016/j.immuni.2012.01.013
Nawaz A, Aminuddin A, Kado T et al (2017) CD206+ M2-like macrophages regulate systemic glucose metabolism by inhibiting proliferation of adipocyte progenitors. Nat Commun 8:286–216. https://doi.org/10.1038/s41467-017-00231-1
Neehus A-L, Carey B, Landekic M et al (2024) Human inherited CCR2 deficiency underlies progressive polycystic lung disease. Cell 187:390–408.e23. https://doi.org/10.1016/j.cell.2023.11.036
Nemes-Baran AD, White DR, DeSilva TM (2020) Fractalkine-dependent microglial pruning of viable oligodendrocyte progenitor cells regulates myelination. Cell Rep 32:108047. https://doi.org/10.1016/j.celrep.2020.108047
Neupane AS, Willson M, Chojnacki AK et al (2020) Patrolling alveolar macrophages conceal bacteria from the immune system to maintain homeostasis. Cell 183:110–125.e11. https://doi.org/10.1016/j.cell.2020.08.020
Nguyen KD, Qiu Y, Cui X et al (2011) Alternatively activated macrophages produce catecholamines to sustain adaptive thermogenesis. Nature 480:104–108. https://doi.org/10.1038/nature10653
Nguyen LT, Zimmermann K, Kowenz-Leutz E et al (2024) C/EBPβ-induced lymphoid-to-myeloid transdifferentiation emulates granulocyte-monocyte progenitor biology. Stem Cell Rep 19:112–125. https://doi.org/10.1016/j.stemcr.2023.11.011
Nguyen PT, Dorman LC, Pan S et al (2020) Microglial remodeling of the extracellular matrix promotes synapse plasticity. Cell 182:388–403.e15. https://doi.org/10.1016/j.cell.2020.05.050
Nicolás-Ávila JA, Lechuga-Vieco AV, Esteban-Martínez L et al (2020) A network of macrophages supports mitochondrial homeostasis in the heart. Cell 183:94–109.e23. https://doi.org/10.1016/j.cell.2020.08.031
Nimmerjahn A, Kirchhoff F, Helmchen F (2005) Resting microglial cells are highly dynamic surveillants of brain parenchyma in vivo. Science 308:1314–1318. https://doi.org/10.1126/science.1110647

Nolte MA, Arens R, Kraus M et al (2004) B cells are crucial for both development and maintenance of the splenic marginal zone. J Immunol 172:3620–3627. https://doi.org/10.4049/jimmunol.172.6.3620

Okabe Y, Medzhitov R (2014) Tissue-specific signals control reversible program of localization and functional polarization of macrophages. Cell 157:832–844. https://doi.org/10.1016/j.cell.2014.04.016

Okreglicka K, Iten I, Pohlmeier L et al (2021) PPARγ is essential for the development of bone marrow erythroblastic Island macrophages and splenic red pulp macrophages. J Exp Med. https://doi.org/10.1084/jem.20191314

Oldenborg PA, Zheleznyak A, Fang YF et al (2000) Role of CD47 as a marker of self on red blood cells. Science 288:2051–2054. https://doi.org/10.1126/science.288.5473.2051

Otsuka H, Yagi H, Endo Y et al (2011) Kupffer cells support extramedullary erythropoiesis induced by nitrogen-containing bisphosphonate in splenectomized mice. Cell Immunol 271:197–204. https://doi.org/10.1016/j.cellimm.2011.06.025

Palis J, Robertson S, Kennedy M et al (1999) Development of erythroid and myeloid progenitors in the yolk sac and embryo proper of the mouse. Development 126:5073–5084. https://doi.org/10.1242/dev.126.22.5073

Squarzoni P, Oller G, Hoeffel G, Pont-Lezica L, Rostaing P, Low D, Bessis A, Ginhoux F, Garel S (2014) Microglia modulate wiring of the embryonic forebrain. Cell Rep 8(5):1271–1279. https://doi.org/10.1016/j.celrep.2014.07.042

Paolicelli RC, Bolasco G, Pagani F et al (2011) Synaptic pruning by microglia is necessary for normal brain development. Science 333:1456–1458. https://doi.org/10.1126/science.1202529

Park MD, Silvin A, Ginhoux F, Merad M (2022) Macrophages in health and disease. Cell 185:4259–4279. https://doi.org/10.1016/j.cell.2022.10.007

Park S, Matte-Martone C, Gonzalez DG et al (2021) Skin-resident immune cells actively coordinate their distribution with epidermal cells during homeostasis. Nat Cell Biol 23:476–484. https://doi.org/10.1038/s41556-021-00670-5

Parkhurst CN, Yang G, Ninan I et al (2013) Microglia promote learning-dependent synapse formation through brain-derived neurotrophic factor. Cell 155:1596–1609. https://doi.org/10.1016/j.cell.2013.11.030

Patel AA, Ginhoux F, Yona S (2021) Monocytes, macrophages, dendritic cells and neutrophils: an update on lifespan kinetics in health and disease. Immunology 163:250–261. https://doi.org/10.1111/imm.13320

Peiseler M, Araujo David B, Zindel J et al (2023) Kupffer cell-like syncytia replenish resident macrophage function in the fibrotic liver. Science 381:eabq5202. https://doi.org/10.1126/science.abq5202

Pekala P, Kawakami M, Vine W et al (1983) Studies of insulin resistance in adipocytes induced by macrophage mediator. J Exp Med 157:1360–1365. https://doi.org/10.1084/jem.157.4.1360

Phan TG, Grigorova I, Okada T, Cyster JG (2007) Subcapsular encounter and complement-dependent transport of immune complexes by lymph node B cells. Nat Immunol 8:992–1000. https://doi.org/10.1038/ni1494

Phan TG, Green JA, Gray EE et al (2009) Immune complex relay by subcapsular sinus macrophages and noncognate B cells drives antibody affinity maturation. Nat Immunol 10:786–793. https://doi.org/10.1038/ni.1745

Pirgova G, Chauveau A, MacLean AJ et al (2020) Marginal zone SIGN-R1+ macrophages are essential for the maturation of germinal center B cells in the spleen. Proc Natl Acad Sci USA 117:12295–12305. https://doi.org/10.1073/pnas.1921673117

Pirzgalska RM, Seixas E, Seidman JS et al (2017) Sympathetic neuron-associated macrophages contribute to obesity by importing and metabolizing norepinephrine. Nat Med 23:1309–1318. https://doi.org/10.1038/nm.4422

Pont-Lezica L, Beumer W, Colasse S, Drexhage H, Versnel M, Bessis A (2014) Microglia shape corpus callosum axon tract fasciculation: functional impact of prenatal inflammation. Abstract Eur J Neurosci 39(10):1551–1557. https://doi.org/10.1111/ejn.12508

Prickett TC, McKenzie JL, Hart DN (1988) Characterization of interstitial dendritic cells in human liver. Transplantation 46:754–761. https://doi.org/10.1097/00007890-198811000-00024

Prokopec KE, Georgoudaki A-M, Sohn S et al (2016) Cutting edge: marginal zone macrophages regulate antigen transport by B cells to the follicle in the spleen via CD21. J Immunol 197:2063–2068. https://doi.org/10.4049/jimmunol.1502282

Qin Y, Garrison BS, Ma W et al (2018) A milieu molecule for TGF-β required for microglia function in the nervous system. Cell 174:156–171.e16. https://doi.org/10.1016/j.cell.2018.05.027

Rabinowitz SS, Gordon S (1991) Macrosialin, a macrophage-restricted membrane sialoprotein differentially glycosylated in response to inflammatory stimuli. J Exp Med 174:827–836. https://doi.org/10.1084/jem.174.4.827

Rahman ZSM, Shao W-H, Khan TN et al (2010) Impaired apoptotic cell clearance in the germinal center by Mer-deficient tingible body macrophages leads to enhanced antibody-forming cell and germinal center responses. J Immunol 185:5859–5868. https://doi.org/10.4049/jimmunol.1001187

Rantakari P, Jäppinen N, Lokka E et al (2016) Fetal liver endothelium regulates the seeding of tissue-resident macrophages. Nature 538:392–396. https://doi.org/10.1038/nature19814

Remmerie A, Martens L, Thoné T et al (2020) Osteopontin expression identifies a subset of recruited macrophages distinct from Kupffer cells in the fatty liver. Immunity 53:641–657.e14. https://doi.org/10.1016/j.immuni.2020.08.004

Rizzo G, Gropper J, Piollet M et al (2023) Dynamics of monocyte-derived macrophage diversity in experimental myocardial infarction. Cardiovasc Res 119:772–785. https://doi.org/10.1093/cvr/cvac113

Roberts AW, Lee BL, Deguine J et al (2017) Tissue-resident macrophages are locally programmed for silent clearance of apoptotic cells. Immunity 47:913–927.e6. https://doi.org/10.1016/j.immuni.2017.10.006

Rosina M, Ceci V, Turchi R et al (2022) Ejection of damaged mitochondria and their removal by macrophages ensure efficient thermogenesis in brown adipose tissue. Cell Metab 34:533–548.e12. https://doi.org/10.1016/j.cmet.2022.02.016

Safaiyan S, Kannaiyan N, Snaidero N et al (2016) Age-related myelin degradation burdens the clearance function of microglia during aging. Nat Neurosci 19:995–998. https://doi.org/10.1038/nn.4325

Sakai M, Troutman TD, Seidman JS et al (2019) Liver-derived signals sequentially reprogram myeloid enhancers to initiate and maintain Kupffer cell identity. Immunity 51:655–670.e8. https://doi.org/10.1016/j.immuni.2019.09.002

Satoh T, Kidoya H, Naito H et al (2013) Critical role of Trib1 in differentiation of tissue-resident M2-like macrophages. Nature 495:524–528. https://doi.org/10.1038/nature11930

Schafer DP, Lehrman EK, Kautzman AG et al (2012) Microglia sculpt postnatal neural circuits in an activity and complement-dependent manner. Neuron 74:691–705. https://doi.org/10.1016/j.neuron.2012.03.026

Schneider C, Kopf M (2015) tEMPting fate MaYBe the solution. Immunity 42:597–599. https://doi.org/10.1016/j.immuni.2015.04.001

Schneider C, Nobs SP, Kurrer M et al (2014) Induction of the nuclear receptor PPAR-γ by the cytokine GM-CSF is critical for the differentiation of fetal monocytes into alveolar macrophages. Nat Immunol 15:1026–1037. https://doi.org/10.1038/ni.3005

Schulz C, Gomez-Perdiguero E, Chorro L et al (2012) A lineage of myeloid cells independent of Myb and hematopoietic stem cells. Science 336:86–90. https://doi.org/10.1126/science.1219179

Scott CL, Zheng F, De Baetselier P et al (2016) Bone marrow-derived monocytes give rise to self-renewing and fully differentiated Kupffer cells. Nat Commun 7:10321–10310. https://doi.org/10.1038/ncomms10321

Scott CL, T'Jonck W, Martens L et al (2018) The transcription factor ZEB2 is required to maintain the tissue-specific identities of macrophages. Immunity 49:312–325.e5. https://doi.org/10.1016/j.immuni.2018.07.004

Scott-Hewitt N, Perrucci F, Morini R et al (2020) Local externalization of phosphatidylserine mediates developmental synaptic pruning by microglia. EMBO J 39:e105380. https://doi.org/10.15252/embj.2020105380

Scur M, Mahmoud AB, Dey S et al (2022) Alveolar macrophage metabolic programming via a C-type lectin receptor protects against lipo-toxicity and cell death. Nat Commun 13:7272–7220. https://doi.org/10.1038/s41467-022-34935-w

Seneschal J, Clark RA, Gehad A et al (2012) Human epidermal Langerhans cells maintain immune homeostasis in skin by activating skin resident regulatory T cells. Immunity 36:873–884. https://doi.org/10.1016/j.immuni.2012.03.018

Seymour JF, Presneill JJ (2002) Pulmonary alveolar proteinosis: progress in the first 44 years. Am J Respir Crit Care Med 166:215–235. https://doi.org/10.1164/rccm.2109105

Shemer A, Grozovski J, Tay TL et al (2018) Engrafted parenchymal brain macrophages differ from microglia in transcriptome, chromatin landscape and response to challenge. Nat Commun 9:5206. https://doi.org/10.1038/s41467-018-07548-5

Sierra A, Encinas JM, Deudero JJP et al (2010) Microglia shape adult hippocampal neurogenesis through apoptosis-coupled phagocytosis. Cell Stem Cell 7:483–495. https://doi.org/10.1016/j.stem.2010.08.014

Sierro F, Evrard M, Rizzetto S et al (2017) A liver capsular network of monocyte-derived macrophages restricts hepatic dissemination of Intraperitoneal bacteria by neutrophil recruitment. Immunity 47:374–388.e6. https://doi.org/10.1016/j.immuni.2017.07.018

Siersbæk R, Nielsen R, John S et al (2011) Extensive chromatin remodelling and establishment of transcription factor "hotspots" during early adipogenesis. EMBO J 30:1459–1472. https://doi.org/10.1038/emboj.2011.65

Soni C, Schell SL, Fasnacht MJ et al (2018) Crucial role of Mer tyrosine kinase in the maintenance of SIGN-R1+ marginal zone macrophages. Immunol Cell Biol 96:298–315. https://doi.org/10.1111/imcb.12003

Sorokin SP, Hoyt RF, Blunt DG, McNelly NA (1992) Macrophage development: II. Early ontogeny of macrophage populations in brain, liver, and lungs of rat embryos as revealed by a lectin marker. Anat Rec 232:527–550. https://doi.org/10.1002/ar.1092320410

Sparber F, Scheffler JM, Amberg N et al (2014) The late endosomal adaptor molecule p14 (LAMTOR2) represents a novel regulator of Langerhans cell homeostasis. Blood 123:217–227. https://doi.org/10.1182/blood-2013-08-518555

Stamatiades EG, Tremblay M-E, Bohm M et al (2016) Immune monitoring of trans-endothelial transport by kidney-resident macrophages. Cell 166:991–1003. https://doi.org/10.1016/j.cell.2016.06.058

Streit WJ, Graeber MB, Kreutzberg GW (1988) Functional plasticity of microglia: a review. Glia 1:301–307. https://doi.org/10.1002/glia.440010502

Tagliatti E, Desiato G, Mancinelli S et al (2024) Trem2 expression in microglia is required to maintain normal neuronal bioenergetics during development. Immunity 57:86–105.e9. https://doi.org/10.1016/j.immuni.2023.12.002

Takahashi K, Umeda S, Shultz LD et al (1994) Effects of macrophage colony-stimulating factor (M-CSF) on the development, differentiation, and maturation of marginal metallophilic macrophages and marginal zone macrophages in the spleen of osteopetrosis (op) mutant mice lacking functional M-CSF activity. J Leukoc Biol 55:581–588. https://doi.org/10.1002/jlb.55.5.581

Takayanagi H, Kim S, Koga T et al (2002) Induction and activation of the transcription factor NFATc1 (NFAT2) integrate RANKL signaling in terminal differentiation of osteoclasts. Dev Cell 3:889–901. https://doi.org/10.1016/s1534-5807(02)00369-6

Tan SYS, Krasnow MA (2016) Developmental origin of lung macrophage diversity. Development 143:1318–1327. https://doi.org/10.1242/dev.129122

Terpstra V, van Berkel TJ (2000) Scavenger receptors on liver Kupffer cells mediate the in vivo uptake of oxidatively damaged red blood cells in mice. Blood 95:2157–2163

Todd EM, Zhou JY, Szasz TP et al (2016) Alveolar macrophage development in mice requires L-plastin for cellular localization in alveoli. Blood 128:2785–2796. https://doi.org/10.1182/blood-2016-03-705962

Tsai T-L, Zhou T-A, Hsieh Y-T et al (2022) Multiomics reveal the central role of pentose phosphate pathway in resident thymic macrophages to cope with efferocytosis-associated stress. Cell Rep 40:111065. https://doi.org/10.1016/j.celrep.2022.111065

Ueno M, Fujita Y, Tanaka T et al (2013) Layer V cortical neurons require microglial support for survival during postnatal development. Nat Neurosci 16:543–551. https://doi.org/10.1038/nn.3358

Ulyanova T, Phelps SR, Papayannopoulou T (2016) The macrophage contribution to stress erythropoiesis: when less is enough. Blood 128:1756–1765. https://doi.org/10.1182/blood-2016-05-714527

Uysal KT, Wiesbrock SM, Marino MW, Hotamisligil GS (1997) Protection from obesity-induced insulin resistance in mice lacking TNF-alpha function. Nature 389:610–614. https://doi.org/10.1038/39335

van de Laar L, Saelens W, De Prijck S et al (2016) Yolk sac macrophages, fetal liver, and adult monocytes can colonize an empty niche and develop into functional tissue-resident macrophages. Immunity 44:755–768. https://doi.org/10.1016/j.immuni.2016.02.017

van Rooijen N, Kors N, Kraal G (1989) Macrophage subset repopulation in the spleen: differential kinetics after liposome-mediated elimination. J Leukoc Biol 45:97–104. https://doi.org/10.1002/jlb.45.2.97

Wang Y, Szretter KJ, Vermi W et al (2012) IL-34 is a tissue-restricted ligand of CSF1R required for the development of Langerhans cells and microglia. Nat Immunol 13:753–760. https://doi.org/10.1038/ni.2360

Wang Y, Wang W, Su L et al (2024) BACH1 changes microglial metabolism and affects astrogenesis during mouse brain development. Dev Cell 59:108–124.e7. https://doi.org/10.1016/j.devcel.2023.11.018

Wculek SK, Dunphy G, Heras-Murillo I et al (2022) Metabolism of tissue macrophages in homeostasis and pathology. Cell Mol Immunol 19:384–408. https://doi.org/10.1038/s41423-021-00791-9

Wculek SK, Heras-Murillo I, Mastrangelo A et al (2023) Oxidative phosphorylation selectively orchestrates tissue macrophage homeostasis. Immunity 56:516–530.e9. https://doi.org/10.1016/j.immuni.2023.01.011

Weisberg SP, McCann D, Desai M et al (2003) Obesity is associated with macrophage accumulation in adipose tissue. J Clin Invest 112:1796–1808. https://doi.org/10.1172/JCI19246

Westphalen K, Gusarova GA, Islam MN et al (2014) Sessile alveolar macrophages communicate with alveolar epithelium to modulate immunity. Nature 506:503–506. https://doi.org/10.1038/nature12902

Willekens FLA, Werre JM, Kruijt JK et al (2005) Liver Kupffer cells rapidly remove red blood cell-derived vesicles from the circulation by scavenger receptors. Blood 105:2141–2145. https://doi.org/10.1182/blood-2004-04-1578

Williams DA, Rios M, Stephens C, Patel VP (1991) Fibronectin and VLA-4 in haematopoietic stem cell-microenvironment interactions. Nature 352:438–441. https://doi.org/10.1038/352438a0

Wlodarczyk A, Holtman IR, Krueger M et al (2017) A novel microglial subset plays a key role in myelinogenesis in developing brain. EMBO J 36:3292–3308. https://doi.org/10.15252/embj.201696056

Wolf Y, Boura-Halfon S, Cortese N et al (2017) Brown-adipose-tissue macrophages control tissue innervation and homeostatic energy expenditure. Nat Immunol 18:665–674. https://doi.org/10.1038/ni.3746

Wong K, Valdez PA, Tan C et al (2010) Phosphatidylserine receptor Tim-4 is essential for the maintenance of the homeostatic state of resident peritoneal macrophages. Proc Natl Acad Sci USA 107:8712–8717. https://doi.org/10.1073/pnas.0910929107

Xu H, Barnes GT, Yang Q et al (2003) Chronic inflammation in fat plays a crucial role in the development of obesity-related insulin resistance. J Clin Invest 112:1821–1830. https://doi.org/10.1172/JCI19451

Yamamoto M, Kato T, Hotta C et al (2011) Shared and distinct functions of the transcription factors IRF4 and IRF8 in myeloid cell development. PLoS One 6:e25812. https://doi.org/10.1371/journal.pone.0025812

Yashchenko A, Bland SJ, Song CJ et al (2023) Cx3cr1 controls kidney resident macrophage heterogeneity. Front Immunol 14:1082078. https://doi.org/10.3389/fimmu.2023.1082078

Yona S, Kim K-W, Wolf Y et al (2013) Fate mapping reveals origins and dynamics of monocytes and tissue macrophages under homeostasis. Immunity 38:79–91. https://doi.org/10.1016/j.immuni.2012.12.001

You Y, Zhao H, Wang Y, Carter RH (2009) Cutting edge: primary and secondary effects of CD19 deficiency on cells of the marginal zone. J Immunol 182:7343–7347. https://doi.org/10.4049/jimmunol.0804295

Yu X, Buttgereit A, Lelios I et al (2017) The cytokine TGF-β promotes the development and homeostasis of alveolar macrophages. Immunity 47:903–912.e4. https://doi.org/10.1016/j.immuni.2017.10.007

Zahner SP, Kel JM, Martina CAE et al (2011) Conditional deletion of TGF-βR1 using Langerin-Cre mice results in Langerhans cell deficiency and reduced contact hypersensitivity. J Immunol 187:5069–5076. https://doi.org/10.4049/jimmunol.1101880

Zhang WX, Neupane AS, David BA et al (2024) A functional assessment of fetal liver and monocyte-derived macrophages in the lung alveolar environment. J Immunol. https://doi.org/10.4049/jimmunol.2300626

Zhang X, Zhuchenko O, Kuspa A, Soldati T (2016) Social amoebae trap and kill bacteria by casting DNA nets. Nat Commun 7:10938–10939. https://doi.org/10.1038/ncomms10938

Zhou T-A, Hsu H-P, Tu Y-H et al (2022) Thymic macrophages consist of two populations with distinct localization and origin. elife. https://doi.org/10.7554/eLife.75148

Zhou Z, Yao J, Wu D et al (2024) Type 2 cytokine signaling in macrophages protects from cellular senescence and organismal aging. Immunity 57:513–527.e6. https://doi.org/10.1016/j.immuni.2024.01.001

Zhu Q, Xiao L, Cheng G et al (2023) Self-maintaining macrophages within the kidney contribute to salt and water balance by modulating kidney sympathetic nerve activity. Kidney Int 104:324–333. https://doi.org/10.1016/j.kint.2023.04.023

Zigmond E, Bernshtein B, Friedlander G et al (2014) Macrophage-restricted interleukin-10 receptor deficiency, but not IL-10 deficiency, causes severe spontaneous colitis. Immunity 40:720–733. https://doi.org/10.1016/j.immuni.2014.03.012

Chapter 2
Macrophages and the Extracellular Matrix

William Meza-Morales, Maria Jimenez-Socha, Donald O. Freytes, and Camilo Mora

Abstract Macrophages are critical to the immune response, serving multiple essential roles in maintaining tissue homeostasis and providing immune protection. These cells also interact with and influence the extracellular matrix (ECM) by sensing and responding to its components. Such interactions between macrophages and the ECM are mediated through the secretion and uptake of various biomacromolecules, such as cytokines and the extracellular vesicles, including exosomes and microvesicles. These vesicles are pivotal in regulating cellular behaviors that affect the organism's overall function. Moreover, macrophages are integral to the repair mechanisms that alter tissue structure and functionality during tissue remodeling. This chapter will delineate how macrophages interact with the ECM and discuss potential therapeutic strategies leveraging these interactions. It will conclude with a discussion of the challenges ahead, highlighting the importance of understanding macrophage-ECM dynamics for advancing basic biology and clinical applications.

2.1 Introduction

The interplay between macrophages and the extracellular matrix (ECM) represents a dynamic and multifaceted aspect of cellular biology with far-reaching implications for health and disease. Macrophages, a key orchestrator of the immune system, exhibit diverse functions ranging from immune surveillance to tissue repair

W. Meza-Morales · C. Mora (✉)
Department of Chemical Engineering, University of Puerto Rico-Mayaguez, Mayaguez, PR, USA
e-mail: camilo.mora@upr.edu

M. Jimenez-Socha · D. O. Freytes
The Joint Department of Biomedical Engineering, North Carolina State University/University of North Carolina at Chapel Hill, Raleigh, NC, USA

M. Kloc et al. (eds.), *Monocytes and Macrophages in Development, Regeneration, and Disease*, Results and Problems in Cell Differentiation 74,
https://doi.org/10.1007/978-3-031-65944-7_2

and remodeling (Jain et al. 2019; Osorio-Valencia and Zhou 2024; Alshangiti et al. 2023; Yu et al. 2015; Joyce 2023). Concurrently, the ECM, a complex network of biomolecules and tissue-specific signaling (e.g., proteins, matrix-bound vesicles (MBVs), and polysaccharides), provides structural support to tissues while actively modulating cellular behavior through biomolecular signaling (see Fig. 2.1).

Understanding the relationship between macrophages and the ECM is crucial in biomedical research and offers insights into fundamental biological processes and disease pathogenesis (Dai et al. 2024; Gareev et al. 2024; Angelov et al. 2023; Ramachandran et al. 2020; Engin et al. 2017). Therefore, this chapter will discuss macrophage-ECM interactions to shed light on the mechanisms underlying their dynamic interplay and significance in various physiological and pathological contexts. Also, it will discuss the implications of macrophage-ECM for tissue remodeling, inflammation, and immune response modulation, highlighting their role in physiological processes and disease pathogenesis (Avalos and Forsthoefel 2022; Sánchez-Alonso et al. 2018; Pieters et al. 2019; Niu et al. 2016; Rilla et al. 2019; Nail et al. 2023; Maacha et al. 2019; Sánchez et al. 2021; Liu et al. 2023; Gregory and Rimmer 2023). Finally, we will discuss the translational implications of macrophage-ECM interactions, offering insights into potential therapeutic strategies targeting this interplay. This chapter aims to contribute to the growing body of knowledge in the field, ultimately paving the way for innovative disease intervention and treatment approaches.

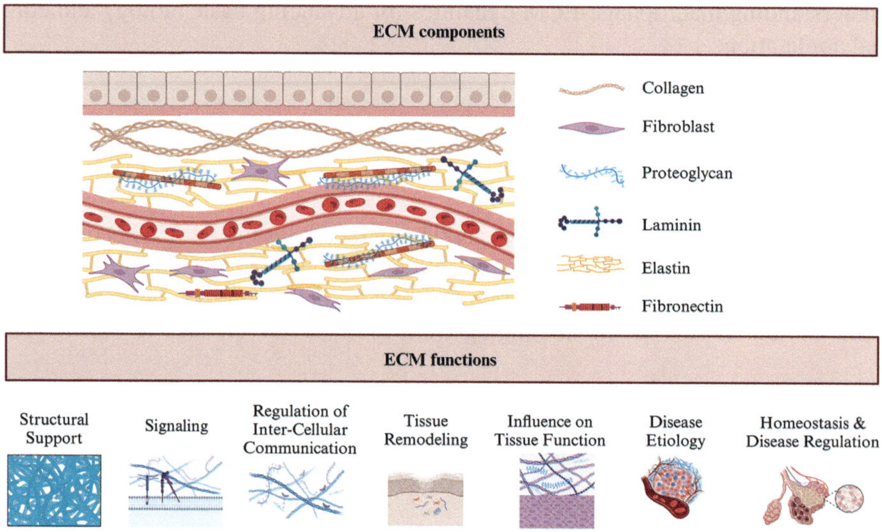

Fig. 2.1 Diagram illustrating the components and functions of the ECM. The ECM, composed of structural proteins such as collagen, laminin, elastin, fibronectin, proteoglycans, and glycoproteins, plays a crucial role in the cellular microenvironment. These functions include structural support, signaling, regulation of inter-cellular communication, tissue remodeling, influence on tissue function, disease etiology and homeostasis, and disease regulation. Created with BioRender.com

2.2 Macrophage-ECM Interactions

Macrophages interact with the ECM by sensing and regulating the tissue environment through key cells such as Fibroblasts. Tissue-resident macrophages or recruited monocyte-derived macrophages play a critical role in driving collagen deposition by fibroblast and resolving tissue repair through ECM degradation, among other functions (Bozso et al. 2021) (see Fig. 2.2). They act as sensors within a homeostatic circuit, monitoring variables of interest (cytokines, inflammatory signaling, apoptosis, etc.) and communicating with other cells, known as effectors, that can "correct" these variables when they deviate from the desired range (Zhou et al. 2022).

During tissue repair, macrophages can indirectly influence ECM production and turnover by secreting cytokines such as interleukin-10 (IL-10) and tumor growth factor beta-1 (TGF-β1). These cytokines recruit and activate fibroblasts, leading to ECM deposition. The paracrine effects of TGF-β1, platelet-derived growth factor (PDGF), and matrix metalloproteinases (MMPs) secretion have also been shown to

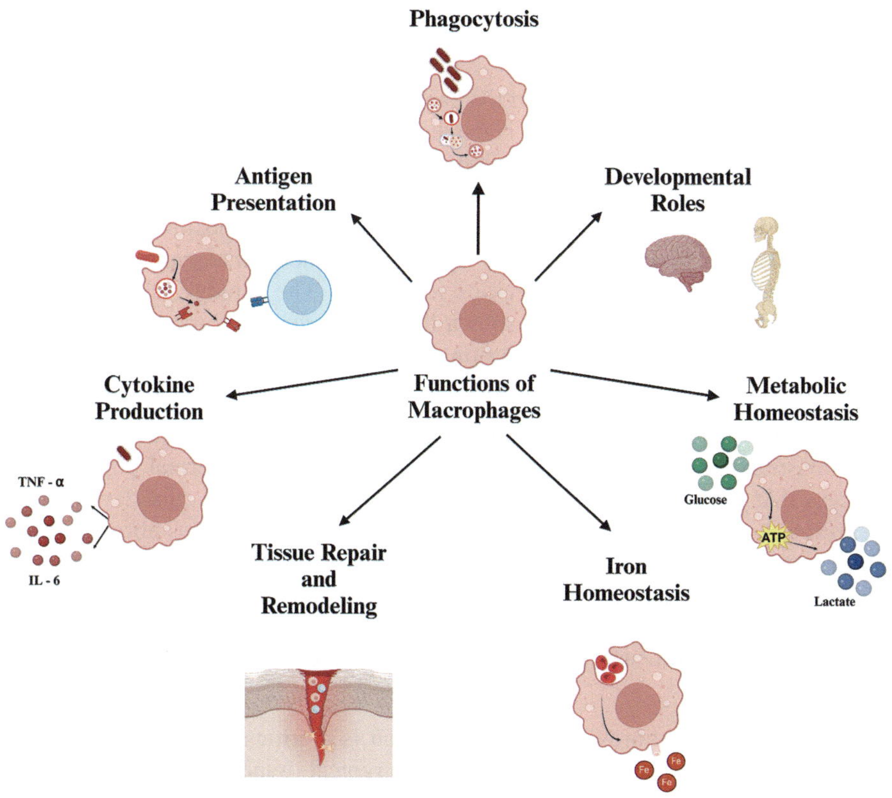

Fig. 2.2 Several macrophage functions are involved in the immune response. These include phagocytosis, developmental roles, metabolic homeostasis, iron homeostasis, tissue repair and remodeling, cytokine production, and antigen presentation. Created with BioRender.com

activate ECM-producing cells, such as stem cells, to favor tissue repair and remodeling (Xiong et al. 2024).

Macrophages can sense the mechanical properties of the ECM, which offers feedback signals to their gene expression program involved in tissue repair. This sensing is thought to occur in an integrin-independent fashion, distinct from established mechanisms of cellular mechanosensing. Alterations in cytoskeletal dynamics control downstream changes in gene expression (Zhou et al. 2022). The local tissue environment, including the ECM composition and mechanical properties of different organs, will also likely provide critical information that may guide macrophages' tissue-specific identities. The study of macrophage mechanosensing of the ECM has established that it can control chromatin availability and gene expression programs, suggesting that these features of the ECM might be critical in specifying the identity of tissue-resident macrophage (Torregrossa et al. 2021).

2.2.1 *How Macrophages Interact with the ECM in Various Tissues*

Macrophages monitor and respond to changes in the ECM, particularly during tissue repair processes. When tissue is damaged, monocytes enter the tissue and differentiate into macrophages, playing a pivotal role in promoting tissue regeneration via a coordinated orchestration of ECM deposition and degradation (Bozso et al. 2021). The local tissue environment, including the ECM composition and mechanical properties, is thought to provide critical information that may guide macrophages in acquiring tissue-specific identities. The study of macrophage mechanosensing of the ECM has shown that it can control chromatin availability and gene expression programs, suggesting that these ECM features might be critical in specifying the identity of tissue-resident macrophage (Torregrossa et al. 2021).

ECM Degradation ECM degradation refers to the process of breaking the ECM into constituent components. Degradation of the ECM is a normal physiological process that occurs during tissue remodeling, wound healing, and development. However, it can also be involved in pathological conditions such as fibrosis, cancer, and chronic wounds. The primary enzymes responsible for ECM degradation are MMPs, a family of zinc-dependent endopeptidases that can degrade various ECM components, including collagen, elastin, and proteoglycans. Macrophages produce MMPs and other proteases that contribute to ECM degradation (Bozso et al. 2021; Zhou et al. 2022).

In the context of fibrosis, ECM degradation is a critical process that can help resolve excessive ECM deposition and restore typical tissue architecture. This process is mediated by enzymes such as MMP-1 and MMP-8, which are primarily secreted by macrophages and neutrophils, respectively. These enzymes target and break down the fibrotic matrix, facilitating the restoration of normal tissue structure and function. Fibrolytic macrophages are a subset of macrophages that express ECM degradation-related genes (MMP-1 and MMP-8) and promote fibrosis resolution by breaking down ECM. The activity of fibrolytic macrophages is essential in counteracting the excessive deposition of ECM components, a hallmark of fibrotic diseases, and promoting tissue repair and remodeling. This subset of macrophages exemplifies the dynamic role of the immune system in tissue homeostasis and disease resolution (Bozso et al. 2021; Zhou et al. 2022).

The interaction between macrophages and the ECM in the tumor microenvironment (TME) is complex in cancer. Matrix stiffness can affect macrophage function and promote tumor progression. Targeting matrix stiffness and the associated macrophage responses is being explored as a strategy for cancer treatment (Xiong et al. 2024; Torregrossa et al. 2021).

In chronic wounds, ECM-inspired materials are crafted to modulate macrophage activity and promote healing. These materials can deliver anti-inflammatory and immunoregulatory factors such as cytokines or extracellular vesicles (EVs) to control macrophage activation and scavenge pro-inflammatory signals from the wound environment (Papachristoforou and Ramachandran 2022; Mushenkova et al. 2021).

ECM Deposition Promoting ECM deposition refers to the process by which cells in the body, including macrophages, stimulate the production and accumulation of ECM components. This process is essential for tissue repair and remodeling.

Macrophages can promote ECM deposition in several ways:

1. **Secretion of Cytokines and Growth Factors**: Macrophages secrete various cytokines and growth factors such as TGF-β1 and PDGF, which can activate fibroblasts and other ECM-producing cells to synthesize and deposit ECM components like collagen, elastin, and proteoglycans.
2. **Activation of Fibroblasts**: Macrophages can mediate the activation of fibroblasts into myofibroblasts phenotype, the primary cells responsible for ECM production. This activation can lead to collagen deposition and other ECM components, contributing to tissue repair. However, ECM's aberrant production and disorganized deposition could lead to fibrosis, negatively impacting tissue functionality (Mora-Navarro et al. 2020; Biehl et al. 2022; Martins et al. 2023).
3. **Collagen Remodeling**: Tumor-associated macrophages (TAMs) can promote fibrosis in pancreatic cancer by mannose receptor-mediated collagen internalization and subsequent lysosomal degradation-induced metabolic reprogramming. TAMs also advance tumor progression by synthesizing and assembling collagenous ECM, specifically collagen types I, VI, and XIV, resulting in remodeling its ECM composition and structure (Bozso et al. 2021).
4. **Immunosuppressive Network**: TAMs can create an immunosuppressive network by secreting immunosuppressive factors, mediating the activation of cancer-associated fibroblasts (CAFs), and ultimately leading to ECM deposition. This network can contribute to the tumor microenvironment's ability to support cancer progression.

In summary, promoting ECM deposition by macrophages is a double-edged sword. While it is necessary for regular tissue repair, excessive or uncontrolled ECM deposition can lead to pathological conditions such as fibrosis and contribute to the stiffness and progression of solid tumors (Bozso et al. 2021; Zhou et al. 2022).

Mechanosensing refers to cells' ability to perceive and respond to mechanical environmental cues. These signals can include forces such as tension, compression, and shear stress, as well as the stiffness and topography of the surrounding ECM. Mechanosensing is a crucial aspect of cellular function, influencing cell behavior, differentiation, migration, and gene expression (Bozso et al. 2021; Zhou et al. 2022). Cells use a variety of mechanoreceptors and mechanotransducers to sense mechanical signals. These can include:

1. **Integrin**: Transmembrane proteins that connect the ECM to the cell's cytoskeleton and can sense changes in ECM stiffness.
2. **Ion Channels**: Proteins such as Piezo1 and transient receptor potential (TRP) channels can be activated by mechanical forces to allow ion fluxes that trigger downstream signaling pathways.
3. **Cytoskeletal Elements**: The actin cytoskeleton, microtubules, and intermediate filaments can respond to mechanical forces and contribute to the cell's ability to sense its mechanical environment.
4. **YAP/TAZ**: Transcriptional coactivators are regulated by mechanical cues and can translocate to the nucleus to influence gene expression in response to ECM stiffness.
5. **Rho/ROCK Signaling**: A pathway that regulates cytoskeletal tension and cell contractility in response to mechanical stimuli.

Macrophages can sense the mechanical properties of the ECM, such as stiffness, through an integrin-independent mechanism that involves changes in the cytoskeleton. This ability allows macrophages to regulate specific gene expression programs involved in tissue repair and inflammation. For example, the gene Fizz1 has been identified as a hallmark of the mechanosensitive gene expression program in macrophages, playing a role in tissue repair and ECM deposition (Bozso et al. 2021; Zhou et al. 2022; Xiong et al. 2024).

2.3 Tissue-Specific Extracellular Vesicles (EVs)

EVs are nanosized lipid membrane particles released by cells into the extracellular environment. They could be found embedded in the ECM as Matrix-bound vesicles (MBVs) or in various body fluids, including blood, urine, saliva, and cerebrospinal fluid (liquid phase EVs). EVs play a key role in intercellular communication by transferring their cargo contents (e.g., RNA, proteins, cryptic peptides, etc) from one cell to another (Gupta et al. 2021; Yokoi and Ochiya 2021; Liangsupree et al. 2021). EVs can be classified into subtypes based on their size, biogenesis, and release pathways: exosomes, microvesicles, and apoptotic bodies (see Fig. 2.3)

(Zhang et al. 2021; Zeng et al. 2022). Furthermore, MBVs have tissu-specific features, immunomodulatory effect, but their biogenesis remains poorly explored.

2.3.1 *Biogenesis and Release of EVs by Macrophages*

As part of the body's immune system, macrophages play a crucial role in the biogenesis and release of EVs. These EVs are formed through several pathways and can carry a variety of bioactive molecules, including proteins, lipids, and nucleic acids, which can influence the function of recipient cells (Gupta et al. 2021; Zhang et al. 2021). The biogenesis of EVs in macrophages involves the endosomal system, where early endosomes mature into multivesicular bodies (MVBs). During this process, intraluminal vesicles (ILVs) are formed within the MVBs by the inward budding of the endosomal membrane. These ILVs are released into the extracellular space as exosomes when the MVBs fuse with the plasma membrane.

Another pathway for macrophages to release EVs is the direct outward budding and fission of the plasma membrane, which leads to the formation of microvesicles or microparticles. This process differs from exosome formation and is typically associated with cell activation or apoptosis (Zhang et al. 2021). Various stimuli can

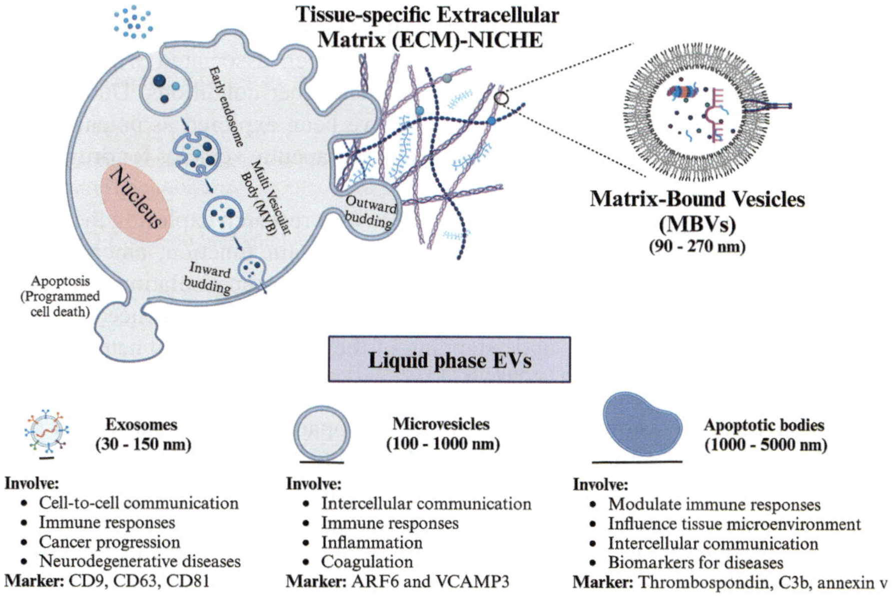

Fig. 2.3 Biogenesis of extracellular vesicles. EVs can be divided into three subtypes: exosomes, microvesicles, and apoptotic bodies. Exosomes are involved in cell-to-cell communication and cancer progression. Microvesicles are implicated in immune responses and inflammation, and apoptotic bodies in modulating immune responses and disease biomarkers (Fyfe et al. 2023; Yu et al. 2024; Aloi et al. 2024; Amin et al. 2024; Liu and Wang 2023). Created with BioRender.com

influence the content and function of EVs released by macrophages. For example, pro-inflammatory stimuli like lipopolysaccharide (LPS) can cause macrophages to release EVs that contain mitochondrial components, which may further stimulate inflammation in already pro-inflammatory settings (Yokoi and Ochiya 2021). This highlights the role of macrophage-derived EVs not only in intercellular communication but also in modulating the immune response.

The biogenesis of EVs involves multiple cellular mechanisms that lead to the formation and release of these vesicles into the extracellular space. EVs can be broadly classified into three main subtypes: exosomes, microvesicles, and apoptotic bodies (see Fig. 2.3), each with distinct biogenesis pathways (Gupta et al. 2021; Zhang et al. 2021).

Exosomes Small membrane vesicles with a size range of approximately 30–150 nm diameter (see Fig. 2.3). They are a subtype of EVs that cells release into the extracellular environment. Exosome biogenesis involves the endosomal pathway, in which early endosomes mature into MVBs containing ILVs. Upon fusion of MVBs with the plasma membrane, these ILVs are released as exosomes.

Exosomes contain a variety of biomolecules, including proteins, lipids, and nucleic acids such as RNA (messenger RNA, microRNA, and other non-coding RNAs) and sometimes DNA. The molecular composition of exosomes reflects the cell of origin and can change in response to various physiological and pathological conditions (Gupta et al. 2021; Zhang et al. 2021). Exosomes play a significant role in cell-to-cell communication by transferring their contents to recipient cells, influencing multiple biological processes. They have been associated with immune responses, cancer, neurodegenerative diseases, and other conditions. Due to their ability to carry and protect cargo, exosomes have been explored as potential biomarkers for disease diagnosis and prognosis and therapeutic vehicles for drug delivery and vaccine development.

The study of exosomes has expanded rapidly, with research exploring their roles in extracellular remodeling, tumor immunity, and brain function, among others (Théry et al. 2018; Nieuwland et al. 2022). However, exosome isolation, characterization, and therapeutic application present challenges, including the need for standardized methods and a better understanding of their biogenesis and heterogeneity (Gupta et al. 2021; Zhang et al. 2021).

Microvesicles Also known as ectosomes or microparticles, are a subtype of EVs whose diameter ranges from approximately 100 nm to 1000 nm (see Fig. 2.3). They are formed through the outward budding and fission of a cell's plasma membrane. Unlike exosomes originating from the endosomal pathway, microvesicles are shed directly from the cell surface. The biogenesis of microvesicles is distinct from that of exosomes and involves the reorganization of the cytoskeleton and changes in the lipid composition of the plasma membrane. Various stimuli can trigger this process, including cell activation, stress, or apoptosis (Zhang et al. 2021; Bazzan et al. 2021).

Microvesicles carry many proteins, lipids, and nucleic acids, including mRNA, microRNA, and sometimes DNA, which can be biomarkers for various diseases.

They are involved in intercellular communication and have been implicated in multiple physiological and pathological processes, such as immune responses, coagulation, inflammation, and cancer progression. Due to the overlap in size, morphology, and molecular contents of microvesicles and exosomes, there are challenges in isolating pure populations of these EVs. The International Society for Extracellular Vesicles (ISEV) recommends using operational terms for EV subtypes based on size, density, composition, isolation conditions, or cell origin rather than strictly using the terms "exosomes" or "microvesicles" (Théry et al. 2018).

Microvesicles have therapeutic potential in regenerative medicine, drug delivery, and vaccine development, and they are also being explored as biomarkers for various diseases. However, further research is needed to overcome challenges in isolation, storage, and drug delivery efficiency (Zhang et al. 2021; Yan et al. 2021).

Apoptotic Bodies Apoptotic bodies are one of the subtypes of EVs released during programmed cell death, known as apoptosis. They are generally larger than exosomes and microvesicles, typically between 1000 and 5000 nm in diameter (see Fig. 2.3). Apoptotic bodies are formed during the later stages of apoptosis and are characterized by their content, including cell organelles, DNA fragments, and cell surface proteins (Zhang et al. 2021; Bazzan et al. 2021; Yan et al. 2021). The formation of apoptotic bodies is a part of the cellular disassembly process in apoptosis. As a cell undergoes apoptosis, it undergoes characteristic morphological changes, like cell shrinkage, chromatin condensation, and membrane blebbing. During blebbing, the plasma membrane forms protrusions that can eventually pinch off and release apoptotic bodies into the extracellular space.

Apoptotic bodies are essential in clearing dying cells by phagocytic cells such as macrophages. Phagocytic cells recognize and engulf apoptotic bodies, allowing the dead cell components to be recycled without initiating an inflammatory response. This is crucial for maintaining tissue homeostasis. Due to their origin in dying cells, apoptotic bodies can carry various signals that modulate the immune response and influence the tissue microenvironment. Apoptotic bodies from Mesenchymal stem cells (MSCs) are reported to stimulate wound healing in macrophages (Liu et al. 2020). They are also being studied for their potential role in intercellular communication and as biomarkers for diseases characterized by increased cell death rates, such as certain types of cancer and neurodegenerative disorders (Zhang et al. 2021; Yan et al. 2021). Research on apoptotic bodies and other EVs continues to expand our understanding of their functions in health and disease and their potential applications in diagnostics and therapeutics.

Matrix-Bound Extracellular Vesicles Matrix-bound extracellular vesicles (MBVs) are a novel class of extracellular vesicles that, unlike secreted vesicles such as exosomes and microvesicles, are attached to the ECM (see Fig. 2.3). These vesicles contain cargo like microRNA, cytokines, and lipids and play a role in modulating cellular communication and function within the ECM environment (Mora-Navarro et al. 2020; Piening and Wachs 2023; Huleihel et al. 2017; Yang et al. 2024; Crum et al. 2023).

MBVs are reported as distinct from other EVs like exosomes and microvesicles. While the precise mechanisms of MBV formation are still being elucidated, it is known that MBVs are secreted and kept in the ECM. This attachment suggests a unique pathway of vesicle formation where the vesicles are directly anchored to the ECM components, possibly during or immediately after their formation. Unlike exosomes, which are formed inside endosomal compartments, or microvesicles, which bud directly from the plasma membrane, MBVs may represent a subgroup of microvesicles given their size, origin, and contents, but with the distinct characteristic of being matrix-bound. Further research is needed to fully understand the biogenesis of MBVs and how they are directed to and integrated within the ECM.

MBVs have shown promise in various applications, particularly in regenerative medicine and therapeutic interventions. MBVs have been reported to steer macrophage responses towards a pro-healing phenotype through interleukin 33 (IL-33), yet more research is needed to fully understand these EVs (Cramer et al. 2022; Bartolacci et al. 2024; Hussey et al. 2019; Badylak et al. 2019). One notable application of MBVs is their therapeutic potential in degenerative nerve diseases, such as retinal degeneration, diabetic neuropathy, or nerve injuries. This potential is due to their unique properties, including their cargo of cell-specific miRNAs, proteins, and lipids, which can be released into the surrounding environment as the ECM degrade, allowing them to be taken up by cells (Piening and Wachs 2023; Boyan et al. 2022).

2.4 Signaling Pathways Involved in Macrophage-ECM Crosstalk

The signaling pathways involved in macrophage-ECM crosstalk are multifaceted and include both direct and indirect mechanisms that allow macrophages to sense and respond to the ECM:

1. **Integrin-Mediated Signaling**: Macrophages interact with the ECM through integrin receptors, such as focal adhesions, which bind to ECM components like collagens, fibronectin, and glycosaminoglycans. This binding triggers signal transduction mechanisms that enable macrophages to detect changes in the ECM microenvironment, including composition, stiffness, and orientation, and to migrate accordingly (Bozso et al. 2021; Vasse et al. 2021).
2. **Direct Signaling**: Macrophages directly sense the physical and biological properties of the ECM through various receptors, including integrins, hyaluronic acid receptors, and toll-like receptors (TLR). This direct signaling influences macrophage activation, differentiation, and functions such as adhesion, migration, and proliferation.
3. **Indirect Signaling via Growth Factors and Cytokines**: The ECM can modulate the availability and activity of growth factors and cytokines, which in turn signal to macrophages. This indirect signaling pathway is crucial for controlling

macrophage functions and can be harnessed to develop immunomodulatory ECM-based wound dressings (Zhou et al. 2022).

4. **Cytoskeletal Integration of ECM Mechanics and Growth Factor Signaling**: Macrophages can sense ECM mechanics through integrin-independent mechanisms mediated by cytoskeletal responses. These responses integrate mechanical signals with biochemical signaling of growth factors such as macrophage colony-stimulating factor (MCSF), affecting chromatin accessibility and gene expression programs involved in tissue repair (Xiong et al. 2024; Torregrossa et al. 2021; Zhang et al. 2022).
5. **Toll-Like Receptor (TLR) Signaling**: TLR signaling, mainly through TLR4, is implicated in ECM remodeling. Macrophages may utilize TLR4 signaling pathways in response to ECM components or ligands, such as LPS from translocated intestinal bacteria, contributing to ECM remodeling in various tissues (Papachristoforou and Ramachandran 2022).
6. **Cytoskeletal Dynamics and Transcriptional Regulation**: In response to ECM mechanics, alterations in the actin cytoskeleton of macrophages are translated into changes in chromatin accessibility and the binding of transcription factors like C/EBP β to regulatory elements of target genes (Bozso et al. 2021).

2.5 Macrophages in Tissue Remodeling

Macrophages play a vital role in tissue repair and remodeling, acting as indispensable agents in the regeneration process across various tissues. Their involvement spans the entire spectrum of the healing process, from the initial inflammatory response to the final tissue remodeling phase. Macrophage phenotypes, such as the pro-inflammatory M1-like and the anti-inflammatory M2-like, are instrumental in different stages of tissue repair and regeneration (Gao et al. 2021; Ruggiero et al. 2021) (see Fig. 2.4).

During the early stages of wound healing, macrophages are primarily involved in the inflammatory phase, where they help to clear the wound of debris and pathogens. This is crucial for preventing infection and creating a clean slate for the subsequent stages of healing. Macrophages secrete various cytokines and growth factors, such as VEGF, IGF-1, TGF-β, TNF-α, IL-6, IL-8, and IL-10, which play pivotal roles in chemotaxis, attracting other immune cells and endothelial cells to the site of injury. These factors are essential for initiating new tissue formation and angiogenesis, promoting the proliferation and differentiation of cells necessary for tissue repair (Thorseth et al. 2022).

In addition to their role in inflammation and tissue formation, macrophages are also key players in the tissue remodeling phase. They secrete enzymes such as matrix MMPs, elastases, and serine proteases that degrade the ECM, facilitating the migration of cells involved in tissue repair and reforming the vessel wall. Furthermore, macrophages regulate angiogenesis during the later stages of healing by secreting factors like thrombospondin-1 (TSP-1), which promotes the regression

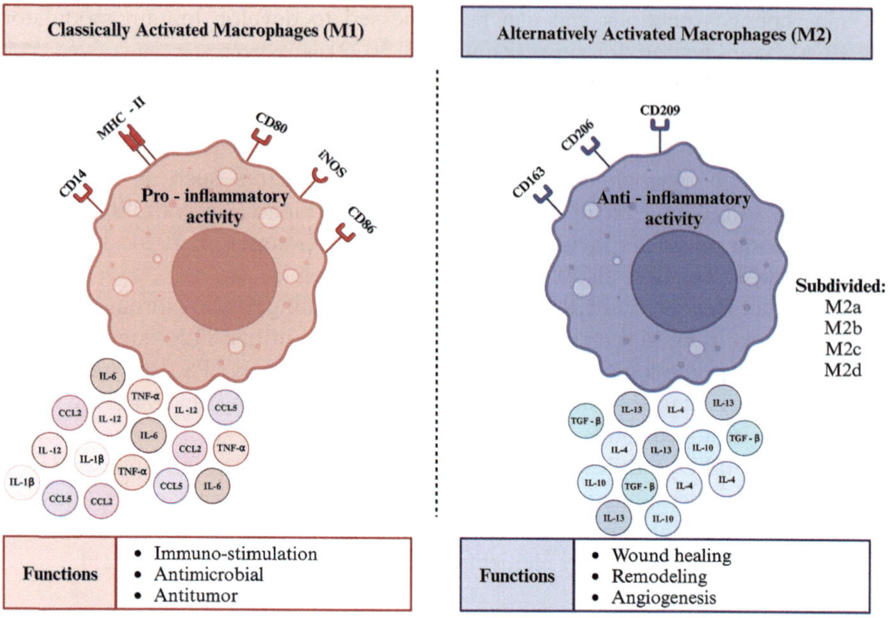

Fig. 2.4 Description of the two main macrophage phenotypes, M1 and M2. M1 macrophages, classically activated macrophages, are identified by the secretion of proinflammatory cytokines such as TNF-α, IL-6, CCL2, IL12, CCL5, and IL-1β, and the expression of receptors like CD14, MHC-II, CD80, iNOS, and CD86. In contrast, M2 macrophages, alternatively activated macrophages, exhibit anti-inflammatory activity and secrete cytokines including IL10, IL4, IL13, and TGF-β1 while expressing receptors such as CD163, CD206, and CD209. Created with BioRender.com

of newly formed capillaries to prevent excessive vascular proliferation (Thorseth et al. 2022).

2.5.1 *Mechanisms of Macrophages in Tissue Repair and Remodeling*

Macrophages are involved in all stages of the wound repair response, from initiating the inflammatory response to transitioning into a pro-resolving phenotype that promotes tissue repair and remodeling. Their ability to adapt and assume different roles makes them potential therapeutic targets for enhancing tissue repair and preventing pathological remodeling. These crucial roles play in critical about tissue repair and remodeling through several mechanisms:

Initiation of the Inflammatory Response The initiation of the inflammatory response is a complex process that involves the rapid mobilization and action of immune cells, particularly macrophages, at the site of tissue injury or infection.

Macrophages play a fundamental role in this early phase by recognizing and responding to pathogenic microbes and damaged cells. Upon encountering such stimuli, macrophages activate and release various pro-inflammatory cytokines and chemokines. These signaling molecules recruit other immune cells, such as neutrophils and additional macrophages, to the site of injury or infection, amplifying the inflammatory response. This recruitment is crucial for containing and eliminating pathogens and initiating the process of tissue repair and healing (Thorseth et al. 2022; Schlundt et al. 2021). In addition to local responses, initiating the inflammatory response involves systemic effects (Apaydin et al. 2023). In general, initiating the inflammatory response by macrophages is a critical first step in the body's defense mechanism against injury and infection, setting the stage for the eventual resolution of inflammation and the restoration of tissue integrity (Thorseth et al. 2022; Schlundt et al. 2021).

Plasticity and Polarization Plasticity and polarization refer to the ability of macrophages to adopt different functional states in response to various environmental signals. This adaptability allows macrophages to perform multiple roles in immune responses, tissue repair, and homeostasis. Macrophages can polarize into at least two major phenotypes: classically activated (M1-like) and alternatively activated (M2-like) macrophages. Be aware that the use of M1 and M2 nomenclature is for simplicity purposes; macrophage polarization comprises a complex and dynamic phenotypic response to cytokine gradients in the microenvironment. The plasticity and polarization of macrophages are crucial for their function in both health and disease. This versatility allows macrophages to respond dynamically to changes in their environment, contributing to their roles in maintaining tissue homeostasis, responding to infection, and facilitating tissue repair and remodeling (Gao et al. 2021; Thorseth et al. 2022; Apaydin et al. 2023).

Indispensability in Regeneration Macrophages' indispensability in regeneration is highlighted by their multifaceted roles in initiating and resolving the inflammatory response and active participation in tissue repair and remodeling. Macrophages are essential for clearing debris and pathogens from the injury site, a critical step in regeneration. They achieve this through phagocytosis, effectively removing dead cells and other debris to prevent infection and allow the regeneration of healthy tissue (Gao et al. 2021; Yu et al. 2022).

The transition from M1 to M2 macrophages is indispensable for successfully resolving inflammation and commencing tissue regeneration. In specific contexts, such as heart regeneration in an animal model and muscle regeneration, macrophages have been shown to control fibroblast activation and modulate the ECM, further underscoring their critical role in regeneration (Ruggiero et al. 2021). Also, in peripheral nerve regeneration, macrophages facilitate the process by removing myelin debris, secreting growth factors, and inducing angiogenesis, which are essential steps for successful nerve repair (Yu et al. 2022).

Secretion of Cytokines The secretion of cytokines is a fundamental process in the immune system, involving the production and release of small proteins by cells that have specific effects on cell-cell interactions and communications. Cytokines can be secreted by various cell types, including immune cells like macrophages, B lymphocytes, T lymphocytes, and mast cells, as well as endothelial cells, fibroblasts, and diverse stromal cells of the spleen and bone marrow (Gao et al. 2021; Thorseth et al. 2022; Mu et al. 2020).

For example, IL-6 is a cytokine that induces inflammation and regulates immunity. It is involved in developing inflammation and is a potent cytokine for inducing acute phase proteins in the liver. In response to infection symptoms and tissue damage, IL-6 is produced by macrophages, endothelial cells, and smooth muscle cells, promoting host defense by stimulating acute phase response, immune response, and hematopoiesis (Thorseth et al. 2022; Gallerand et al. 2021). Another example is IL-10, a key anti-inflammatory cytokine with a wide range of anti-inflammatory effects in vivo. It is involved in all aspects of the inflammatory response, reducing the expression of MHC class II molecules on the surface of dendritic cells and macrophages, weakening antigen-presenting cells' function, and inhibiting T cell proliferation. This effectively inhibits cell-mediated immune responses (Ruggiero et al. 2021; Li et al. 2021a).

2.5.2 How Macrophages Contribute to ECM Turnover and Degradation during Tissue Healing

Macrophages play an essential role in ECM turnover and degradation during tissue healing, contributing to the removal of damaged ECM components and tissue remodeling. During the healing process, macrophages regulate ECM production and organization, crucial for restoring tissue integrity and function (Thorseth et al. 2022; Cox et al. 2021). Macrophages certainly contribute to ECM turnover and degradation during tissue healing in several ways:

Secretion of Matrix Metalloproteinases (MMPs), Elastases, and Serine Proteases The secretion of MMPs, elastases, and serine proteases is a critical aspect of ECM remodeling, impacting various physiological and pathological processes, such as wound healing, inflammation, and tumor progression. MMPs are a family of zinc-dependent endopeptidases capable of degrading ECM proteins. They are secreted as inactive proenzymes and require activation, which can be achieved by proteolytic removal of their propeptide domains, a process facilitated by serine proteases (Gao et al. 2021; Ruggiero et al. 2021; Apaydin et al. 2023).

Elastases and serine proteases also contribute significantly to ECM remodeling. Elastases specifically target and degrade elastin, a key component of the ECM, while serine proteases are involved in the activation of MMPs and the direct degradation of ECM components. This coordinated action between MMPs, elastases, and serine proteases ensures the dynamic turnover of the ECM, allowing for cell migration, tissue repair, and the maintenance of tissue homeostasis (Ruggiero et al. 2021; Apaydin et al. 2023).

Moreover, the activity of these proteases is tightly regulated to prevent excessive ECM degradation that could lead to pathological conditions. The regulation involves complex interactions between various inhibitors and activators within the tissue microenvironment, ensuring that ECM remodeling occurs in a controlled manner conducive to normal physiological processes (Ruggiero et al. 2021; Apaydin et al. 2023; Wong et al. 2021).

Regulation of ECM Production and Organization The regulation of ECM production and organization is a complex process that involves diverse cellular and molecular mechanisms. The ECM is a dynamic structure that surrounds and anchors cellular components in tissues, providing structural support and regulating diverse biological functions, including cell adhesion, proliferation, differentiation, migration, cell-cell interactions, and intracellular signaling events (Ruggiero et al. 2021; Mosser et al. 2021).

1. **Cellular Contributions**: Dermal fibroblasts (dFBs), for example, are one of the primary cellular source of skin ECM. They develop from a common embryonic precursor into highly heterogeneous subpopulations during development and adulthood. Upon injury, dFBs migrate into wound granulation tissue and transdifferentiate into myofibroblasts, critical in wound contraction and dermal ECM regeneration and deposition (Ruggiero et al. 2021).
2. **Molecular Mechanisms**: ECM molecules, including collagens, proteoglycans, fibrillins, and matricellular proteins, are essential for normal skin development, homeostasis, and wound healing. Through anchorage dependence, these molecules support most mammalian cells' growth, survival, and function. They also serve as reservoirs and presenters of growth factors and cytokines, transmitting mechanical and biological signaling intracellularly (Thorseth et al. 2022).
3. **Regulation by ECM Molecules**: The expression of ECM molecules is differentially regulated during skin development and wound regeneration. This regulation ensures that the appropriate ECM components are produced at the right time and place to support tissue repair and regeneration processes (Apaydin et al. 2023).
4. **Pathological Conditions**: Dysregulation of ECM production and organization is associated with the pathogenesis of wound healing-related skin diseases, including chronic wounds and keloid. This underscores the importance of tightly regulated ECM production and organization for proper tissue function and healing (Ruggiero et al. 2021).

Promotion of Angiogenesis The promotion of angiogenesis refers to the process by which new blood vessels are formed from pre-existing vessels, a critical mechanism in tissue repair, wound healing, and various physiological and pathological conditions. Macrophages promote angiogenesis by secretion of multiple growth factors and cytokines. Macrophages, upon activation, can secrete vascular endothelial growth factor (VEGF), which directly stimulates endothelial cells to proliferate, migrate, and form new capillary structures. Additionally, macrophages can release other angiogenic factors, such as platelet-derived growth factor (PDGF), fibroblast growth factor (FGF), and angiopoietins, further contributing to the angiogenic process (Gao et al. 2021; Huang et al. 2022).

The angiogenic process is essential not only for wound healing, where the formation of new blood vessels is necessary to deliver oxygen and nutrients to the healing tissue but also in chronic inflammation and tumor growth, where angiogenesis can have detrimental effects. In tissue repair, promoting angiogenesis by macrophages is a tightly regulated process that ensures adequate vascularization of the healing tissue, thereby facilitating efficient repair and regeneration (Gao et al. 2021).

Prevention of Excessive Vascular Proliferation The prevention of excessive vascular proliferation involves mechanisms that regulate and limit the growth of new blood vessels (angiogenesis) to ensure that vascularization is appropriate for tissue needs without leading to pathological conditions such as tumor growth, diabetic retinopathy, or rheumatoid arthritis. A key aspect of this regulation is the balance between pro-angiogenic factors, such as Vascular Endothelial Growth Factor (VEGF), and anti-angiogenic factors, including thrombospondin-1 (TSP-1) and angiostatin (Gao et al. 2021; Thorseth et al. 2022).

TSP-1 is a well-studied matricellular protein that prevents excessive vascular proliferation. TSP-1 exerts its anti-angiogenic effects by directly binding to and inhibiting the activity of VEGF and interacting with CD36 receptors on endothelial cells, leading to the inhibition of endothelial cell migration and proliferation (Thorseth et al. 2022). This action is essential in the later stages of wound healing and in preventing aberrant blood vessel networks in diseases.

Moreover, the balance between pro-angiogenic and anti-angiogenic factors is tightly regulated by various signaling pathways and cellular interactions within the tissue microenvironment. Macrophages, for instance, can switch from a pro-angiogenic phenotype to an anti-angiogenic phenotype, secreting factors like TSP-1 to ensure that angiogenesis is curtailed once the necessary tissue repair and regeneration have been achieved (Thorseth et al. 2022; Cox et al. 2021).

2.6 Immune Response and ECM

The immune system maintains the homeostasis of healthy tissues, coordinating tissue development and healing and responding to invasive threats. Unresolved inflammation is increasingly recognized as a fundamental cause of a wide range of human diseases, from obesity to cardiovascular disease. Routine tissue healing necessitates an initial inflammatory response and a subsequent anti-inflammatory or pro-healing immune response. The classical activation of the immune system to a heightened state of readiness is essential to thwart infection, treat cancer, and develop a foreign body response to implanted materials. Conversely, immunomodulation or alternative activation may be employed to promote tolerance in autoimmune diseases, prevent graft rejection, and facilitate the integration of biomedical implants (Dixit et al. 2023).

2.6.1 How Macrophages Modulate the Immune Response in the Context of ECM Interactions

Macrophages modulate the immune response, mainly through their interactions with the biomolecular signaling in the ECM. These complex and multifaceted interactions influence macrophage behavior and the broader immune response in various physiological and pathological contexts.

Macrophage Polarization and ECM Macrophage polarization and its interaction with ECM are critical aspects of the immune response and tissue remodeling. The ECM influences macrophage polarization through its composition, mechanical properties, and the presence of specific molecules (Dixit et al. 2023; Soni and Rodell 2021; Wolf et al. 2021).

The ECM can drive macrophages toward an M2-like phenotype associated with tissue repair and remodeling by presenting specific biochemical and mechanical signals. For instance, decellularized ECM scaffolds have been shown to promote the polarization of M2-like macrophages, which promote constructive tissue remodeling (Soni and Rodell 2021; Sridharan et al. 2020). Conversely, specific ECM components and their degradation products can also stimulate macrophages to adopt an M1-like phenotype associated with pro-inflammatory responses and may contribute to tissue damage or chronic inflammation (Dixit et al. 2023).

In addition, macrophages are involved in ECM remodeling by secreting MMPs and other enzymes that degrade ECM components. This activity facilitates tissue repair and modulates the immune response by altering the ECM landscape and influencing the infiltration and behavior of other immune cells (Wolf et al. 2021; Guan et al. 2023).

ECM as a Regulator of Macrophage Function The ECM serves as a critical regulator of macrophage function, influencing their behavior, polarization, and, ultimately, their role in tissue homeostasis, repair, and disease progression. The following provides an overview of how the ECM regulates macrophage function based on the provided citations:

1. **Modulation of Immune Responses**: The degradation of the ECM by macrophages, through the secretion of MMPs and other enzymes, releases matrix-bound growth factors and bioactive fragments. These components can modulate immune responses by affecting macrophage behavior, including their recruitment, migration, and phenotypic polarization. Such activities are crucial for tis-

sue remodeling and repair but can contribute to disease progression in contexts like cancer (Soni and Rodell 2021; Guan et al. 2023).

2. **Interactions with Tumor-Associated Macrophages (TAMs)**: In the tumor microenvironment, the ECM influences the migration, positioning, and phenotype of the TAMs. In turn, TAMs can remodel the ECM to promote tumor growth and metastasis. Targeting ECM components or their interactions with TAMs has been suggested as a potential therapeutic strategy to modulate the tumor microenvironment and inhibit cancer progression (Wolf et al. 2021; Martin and García 2021; Kim et al. 2021).
3. **Role in Regenerative Medicine**: The ECM's interaction with macrophages is also exploited in regenerative medicine. Decellularized ECM materials of human or porcine sources have been shown to favor the M2 phenotype, promoting tissue regeneration. This highlights the potential of using ECM-derived scaffolds to modulate macrophage responses in tissue engineering applications (Li et al. 2021b; Ye et al. 2021).
4. **Dynamic Interplay in Disease Progression**: The dynamic interplay in disease progression, particularly in inflammatory and fibrotic diseases, involves complex interactions among various cell types, signaling molecules, and ECM. This multifaceted communication network drives disease progression and offers potential therapeutic targets.

Feedback Mechanisms Feedback mechanisms are biological processes that regulate various physiological functions within organisms. These mechanisms can be positive or negative, depending on their effect on the stimulus that triggers them.

1. **Negative Feedback Mechanisms** work to reduce or inhibit the effects of the initial stimulus. This type of feedback is crucial for maintaining homeostasis within the body. For example, in the context of immune responses, when a pathogen is detected, the immune system is activated to eliminate the threat. Once the pathogen is eliminated, negative feedback mechanisms, such as the secretion of anti-inflammatory cytokines, dampen the immune response and prevent excessive tissue damage (Sridharan et al. 2020).
2. **Positive Feedback Mechanisms** amplify the effects of the initial stimulus. This type of feedback can be observed in the inflammation process, where the initial presence of a pathogen or injury leads to an increase in the immune response. The activated macrophages and other immune cells release pro-inflammatory cytokines, which recruit more immune cells to the site of infection or injury, intensifying the response until the pathogen is eliminated or the wound is healed (Sridharan et al. 2020).

Both feedback mechanisms are essential for the proper functioning of biological systems. Negative feedback mechanisms are particularly important for maintaining stability and preventing overreaction to stimuli, which could lead to autoimmune diseases or chronic inflammation. Positive feedback mechanisms, while potentially leading to a rapid escalation of responses, are crucial for processes that require a decisive and strong response, such as fighting an infection or healing a wound.

In the context of ECM-macrophage interaction, the ECM can modulate macrophage function through mechanotransduction, influencing macrophage polarization and activity. This interaction can be considered part of a feedback mechanism in which the ECM not only influences macrophage behavior but is also remodeled by macrophages through the secretion of enzymes like MMPs. This remodeling may influence the behavior of macrophages and other cells, contributing to tissue repair or, in some cases, to the progression of fibrosis or cancer (Dixit et al. 2023; Soni and Rodell 2021; Guan et al. 2023; Li et al. 2021b).

2.7 Role of Macrophages in Mediating Inflammation and Tissue Repair

Promoting tissue repair and regeneration is a complex biological process that involves the orchestrated interaction of various cell types, signaling molecules, and the ECM to restore tissue integrity and function after injury. This process can be significantly influenced by the immune system, mainly through the actions of macrophages, which play a pivotal role in the inflammatory response and the subsequent healing phases. The following are key aspects of how tissue repair and regeneration are promoted, drawing on insights from recent research and applications in biomaterials and regenerative medicine:

1. **Biomaterials and Scaffolds**: Using biomaterials and scaffolds can provide structural support and a conducive environment for tissue regeneration. These materials can be engineered to mimic the natural ECM, offering mechanical support and topographical cues that guide cell attachment, proliferation, and differentiation. Additionally, biomaterials can be functionalized to deliver bioactive molecules, such as growth factors and cytokines, in a controlled manner to promote tissue repair (Soni and Rodell 2021; Wolf et al. 2021; Sridharan et al. 2020).
2. **Growth Factor Delivery**: Growth factors play a key role in tissue regeneration by stimulating cell proliferation, migration, and differentiation. The controlled delivery of growth factors using biomaterials or cell-based therapies can enhance the regenerative process. For example, vascular endothelial growth factor (VEGF) promotes angiogenesis, while bone morphogenetic proteins (BMPs) stimulate osteogenesis in bone repair (Li et al. 2021b; Antmen et al. 2021).
3. **Cell Therapy**: The transplantation of stem cells or progenitor cells is a promising strategy for tissue regeneration. For instance, mesenchymal stem cells (MSCs) can differentiate into various cell types and secrete paracrine factors that modulate the immune response, reduce inflammation, and promote tissue repair. MSCs can also "educate" macrophages towards a pro-regenerative phenotype, further enhancing the healing process (Kim et al. 2021; Niu et al. 2021).
4. **Immunomodulation**: Immunomodulation refers to the alteration or regulation of the immune system response through various means, including using drugs, biologics, biomaterials, or other therapies (e.g., EVs are a novel type of effectors

used as immunomodulators). This process can enhance or suppress the activity of the immune system to achieve a desired therapeutic outcome. Immunomodulation is critical to treating a wide range of diseases, from autoimmune disorders and inflammatory conditions to cancer and infectious diseases. The goal is to correct the immune system dysfunction or modulate its response to fight disease or tolerate foreign substances more effectively, such as transplanted tissues or biomaterial implants (Mansour et al. 2022; Salthouse et al. 2023).

Regulation of Fibrosis The regulation of fibrosis involves a complex interplay of cellular and molecular mechanisms that control ECM deposition and remodeling in response to tissue injury. Fibrosis can be seen as an excessive wound healing response where the balance between ECM production and degradation is disrupted, accumulating fibrotic tissue and impairing organ function. Key aspects of fibrosis regulation include:

1. **Inflammatory Response**: The initial inflammatory response to tissue injury plays a crucial role in the development of fibrosis. Pro-inflammatory cytokines such as tumor necrosis factor-alpha (TNF-α) and interleukins (IL-1β, IL-6) can promote fibroblast activation and ECM production. Conversely, anti-inflammatory cytokines like IL-10 can mitigate these effects and reduce fibrosis (Dixit et al. 2023; Guan et al. 2023).
2. **Fibroblast Activation**: Fibroblasts are central to the fibrotic process, producing ECM components such as collagens. Factors like TGF-β1, a potent inducer of fibrosis, can stimulate their activation. Activated fibroblasts, also known as myofibroblasts, exhibit increased ECM production and greater resistance to apoptosis, contributing to the persistence of fibrosis (Dixit et al. 2023; Martin and García 2021).
3. **ECM Remodeling**: MMPs and their inhibitors (TIMPs) are involved in ECM remodeling. An imbalance favoring TIMPs can lead to reduced ECM degradation and fibrotic tissue accumulation. Modulating the activity of MMPs and TIMPs represents a potential therapeutic approach to managing fibrosis (Wolf et al. 2021; Guan et al. 2023).

2.8 Therapeutic Approaches

As discussed, the interactions between macrophages and ECM play a pivotal role in various physiological and pathological processes, including tissue repair, inflammation, and cancer progression. Targeting these interactions presents a promising avenue for therapeutic interventions.

2.8.1 Potential Therapeutic Strategies Targeting Macrophage-ECM Interactions

There are various therapeutic strategies that target the interactions between macrophages and ECM. These strategies aim to modulate the complex interplay between macrophages and ECM to treat diseases that are characterized by abnormal tissue remodeling, inflammation, and cancer. To achieve this, several potential therapeutic strategies have been identified (see Fig. 2.5).

2.8.1.1 Modulation of Macrophage Polarization through Photoresponsive Hydrogels

The modulation of macrophage polarization through photoresponsive hydrogels is achieved by engineering biomimetic and photoresponsive hyaluronan (HA) hydrogel nanocomposite that allows for dynamic immunomodulation of macrophages. This innovative approach utilizes photodegradable alkoxyl phenacyl-based polycarbonate (APP) nanocomposites within the HA hydrogel to permit user-controlled release and conjugation of Arg-Gly-Asp (RGD) adhesive peptides to the HA-based ECM. This process facilitates real-time integrin activation of macrophages encapsulated within the 3D HA-APP nanocomposite hydrogels (Clark et al. 2022).

Photocontrolled conjugation of the 3D peptide ECM-RGD activates αvβ3 integrin on macrophages, and periodic activation of αvβ3 integrin has been shown to enhance the anti-inflammatory polarization of M2-like macrophages. This modulation of macrophage polarization aims to control inflammatory responses and promote tissue repair (Clark et al. 2022; Michael et al. 2022). Thus, the photoresponsive HA hydrogel nanocomposite represents a significant advancement in immunomodulatory therapies, offering a dynamic and controllable platform to study and regulate cell-ECM interactions in real-time (Kubota et al. 2009; Moradinasab et al. 2022).

Certainly, the process of macrophage polarization through photoresponsive hydrogels can be enumerated as follows:

1. **Engineering of Biomimetic and Photoresponsive Hydrogels**: This is achieved by incorporating photodegradable alkoxyl phenacyl-based polycarbonate (APP) nanocomposites within the HA hydrogel, allowing for user-controlled release and conjugation of Arg-Gly-Asp (RGD) adhesive peptides to the HA-based ECM (Clark et al. 2022; Michael et al. 2022).
2. **Activation of Integrins**: The photocontrolled 3D ECM-RGD peptide conjugation activates αvβ3 integrin on macrophages. This integrin activation is crucial for modulating macrophage behavior, particularly in enhancing anti-inflammatory M2 macrophage polarization. The ability to control integrin activation in real-time within a 3D hydrogel environment provides a powerful tool to study and influence macrophage function (Clark et al. 2022; Michael et al. 2022; Kubota et al. 2009).

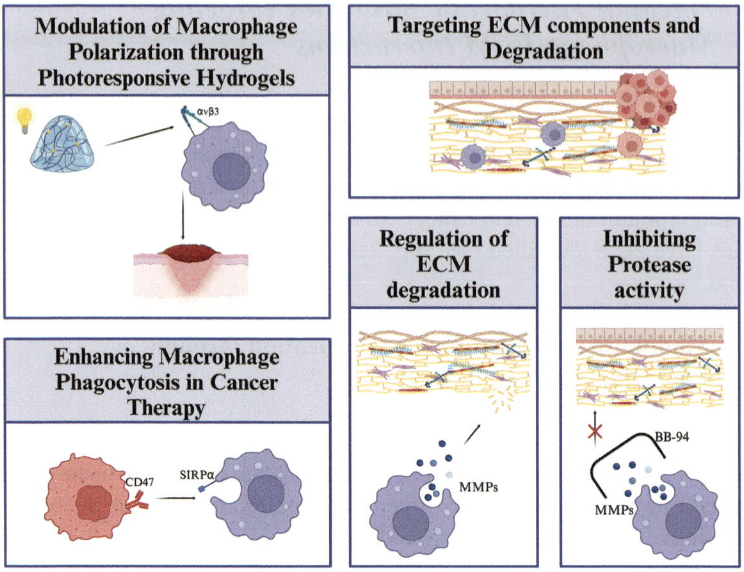

Fig. 2.5 Description of different potential therapeutic strategies in macrophage-ECM interactions. Strategies such as modulating macrophage polarization through photoresponsive hydrogels, enhancing macrophage phagocytosis in cancer therapy, targeting ECM components and degradation, regulating ECM breakdown, and inhibiting protease activity seek to address physiological and pathological processes. Created with BioRender.com

3. **Temporal Control over Cell-ECM Interactions**: The photoresponsive HA hydrogel nanocomposite allows temporal control of cell-ECM interactions. This dynamic control is achieved by on-demand photostimulation, which can be used to model different tissue microenvironments and study macrophage behavior in a biomimetic ECM. Compared to other stimuli systems, the system's stable controllability over RGD peptide release and conjugation ensures reproducibility and therapeutic efficacy (Michael et al. 2022; Moradinasab et al. 2022).
4. **Phototriggered Immunomodulation**: HA-APP hydrogels control macrophage-ECM adhesive interactions and αvβ3 integrin expression upon UV exposure. This photo-triggered modulation upregulates αvβ3 integrin expression and controls macrophage-ECM interactions, enhancing macrophage polarization toward an anti-inflammatory phenotype. The ability to upregulate integrin expression and control macrophage-ECM interactions highlights the potential of photoresponsive hydrogels in immunomodulatory therapies (Kubota et al. 2009; Wang and Chi 2022).
5. **Dynamic Immunomodulatory Platform**: refers to an engineered system that allows real-time control and modulation of immune cell behavior, specifically macrophage polarization, through photoresponsive hydrogel nanocomposites. This platform utilizes the unique properties of these hydrogels to dynamically modulate ECM interactions and macrophage integrin expression, thereby influ-

encing their polarization towards an anti-inflammatory or pro-inflammatory phenotype depending on the therapeutic need (Zhuang et al. 2023).

2.8.1.2 Enhancing Macrophage Phagocytosis in Cancer Therapy

Enhancing macrophage phagocytosis in cancer therapy involves strategies aimed at boosting the ability of macrophages to recognize, engulf, and destroy cancer cells. This approach is part of a broader effort to harness the immune system's power to fight cancer by taking advantage of the natural role of macrophages in the body's defense mechanisms. The main strategies and insights are outlined below:

1. **Targeting Phagocytosis Checkpoints**: Blocking phagocytosis checkpoints refers to the therapeutic strategy to inhibit the signals that prevent macrophages from engulfing and destroying cancer cells. This approach has emerged as a promising avenue in cancer treatment, leveraging the body's immune system to fight malignancies more effectively. The most well-known example of such a checkpoint is the interaction between CD47 on tumor cells and SIRPα on macrophages. CD47 acts as a "don't eat me" signal, inhibiting phagocytosis of cancer cells by macrophages. The phagocytic barrier is lifted by blocking this interaction through antibodies targeting CD47 or SIRPα, recombinant proteins, or suppressing CD47 expression, enabling macrophages to engulf and eliminate tumor cells (Clark et al. 2022).

   Recent breakthroughs have highlighted the potential of anti-immune checkpoint monoclonal antibodies in cancer therapy, affecting various immune cell functions, including phagocytosis. Blockade of anti-phagocytosis checkpoints, particularly in macrophages, has been identified as an attractive target for cancer therapy. Strategies to antagonize this interaction include using anti-CD47 or anti-SIRPα antibodies, recombinant proteins competing with endogenous proteins, and targeting the CD47 transcriptional pathway to suppress CD47 cell surface expression. This blockade allows pro-phagocytic receptors to produce activating signals, restoring the ability of macrophages to phagocytose and eliminate tumor cells (Clark et al. 2022; Huang et al. 2019).

   Moreover, exploration of other phagocytic checkpoints beyond CD47-SIRPα, such as PD-L1-PD-1, MHC I-LILRB1, and CD24-Siglec-10 axes, has introduced additional targets to enhance macrophage phagocytosis and subsequent anti-cancer responses. The diversity of anti-phagocytic checkpoints among different tumor cells and malignancies suggests that combined therapies or bispecific antibodies targeting multiple checkpoints simultaneously might offer more effective treatment options. However, choosing the most suitable immunotherapeutic approach should consider the specific cancer types and patient conditions (Kubota et al. 2009).
2. **Nanomedicine Interventions**: Nanomedicine interventions utilize the unique properties of nanoparticles to improve the delivery and efficacy of cancer therapies. These interventions are particularly well-suited for immunotherapy due to

nanoparticles' ability to penetrate and accumulate in the tumor microenvironment (TME), offering targeted delivery of therapeutic agents with controlled release kinetics. Nanoparticles can be engineered with targeting ligands to induce cell-specific uptake, addressing off-target issues and potentially reducing the doses required for effective treatment. This approach allows for the co-delivery of drugs with diverse features. It enables control over the kinetics of drug release, which can be designed to respond to specific stimuli such as pH changes or external triggers (Clark et al. 2022).

However, compared to conventional treatments, nanomedicines' safety and long-term biosafety remain a concern. Despite promising in vitro biocompatibility demonstrated by several nanomaterials, their potential toxicity to cells, tissues, and overall health necessitates thorough toxicity studies and assessments. In addition, nanomaterials may aggregate in organisms, altering their stability and properties, which underscores the need to develop new materials, better production technologies, and rigorous safety evaluations (Michael et al. 2022).

An innovative approach to enhance the efficiency of drug permeation in the TME involves a "nanolymphatic" strategy. This strategy aims to overcome the challenge posed by the lack of a lymphatic system in tumor tissue, which increases tumor interstitial pressure (TIP) and limits drug delivery efficiency. This approach directly kills tumor cells by targeting the tumor with hyaluronic acid-mediated nanoparticles and applying near-infrared laser irradiation for photothermal therapy. It reduces TIP, enhancing convective diffusion in the tumor stroma. This method demonstrates a novel perspective for improving drug permeation efficiency in the TME (Moradinasab et al. 2022).

3. **CAR-Macrophage Therapy**: CAR-macrophage therapy represents a novel and promising approach to cancer immunotherapy by harnessing the body's innate immune system to fight cancer more effectively. This therapy involves genetically engineering macrophages, a white blood cell responsible for engulfing and digesting cellular debris and pathogens, to express Chimeric Antigen Receptors (CARs). These CARs are designed to recognize specific antigens on the surface of cancer cells, thereby directing the macrophages to target and destroy these malignant cells (Bahr et al. 2022).

   The emergence of CAR-macrophage (CAR-M) therapy has provided new insights into oncology, particularly for treating solid tumors. Preclinical results have been promising, leading to clinical trials to evaluate CAR-Ms' efficacy, including those targeting HER2 for solid tumors and mesothelin for ovarian cancer and peritoneal mesothelioma. Unlike CAR T-cell therapies, CAR-Ms present a lower risk of graft-versus-host disease (GVHD), making them potential candidates for off-the-shelf immunotherapy products. However, challenges such as the limited number of macrophages and their tendency to accumulate in the liver highlight the need for further research and development in this area (Clark et al. 2022; Michael et al. 2022).

   Recent studies have demonstrated the potential of CAR-engineered macrophages to induce phagocytosis of targeted tumor cells, offering a new direction for cancer treatment. For instance, mouse macrophages engineered with CD19-

CAR could engulf antigen-coated decoy particles and human cancer cells under ex vivo conditions. This approach directly targets tumor cells and modulates the tumor microenvironment to enhance T-cell infiltration and efficacy, providing a multifaceted attack against cancer (Kubota et al. 2009; Moradinasab et al. 2022).

4. **Combining Therapies**: The use of combination therapies, in particular, the utilization of anti-CD47 antibodies in conjunction with other therapeutic agents, has emerged as a promising strategy to enhance the efficacy of cancer immunotherapy. Anti-CD47 antibodies target the CD47-SIRPα axis, a critical "don't eat me" signal cancer cells use to evade phagocytosis by macrophages. By blocking this signal, anti-CD47 antibodies promote the immune system's engulfment and destruction of cancer cells. However, the complexity of cancer and the tumor microenvironment often necessitates a multifaceted treatment approach to achieve optimal therapeutic outcomes.

   One notable combination involves anti-CD47 antibodies and rituximab, an anti-CD20 antibody. This combination has been shown to stimulate antitumor activity in patients with rituximab-resistant non-Hodgkin's lymphoma, highlighting the potential of anti-CD47 therapy to overcome resistance to existing treatments. Furthermore, in a phase I trial, the combination of magrolimab (an anti-CD47 antibody) with azacitidine demonstrated treatment efficacy in patients with untreated myelodysplastic syndrome (MDS) and acute myeloid leukemia (AML). Azacitidine exposure leads to the presentation of pro-phagocytic calreticulin on tumor cells, which synergizes with anti-CD47 antibodies to promote tumor cell phagocytosis (Clark et al. 2022).

   Beyond these combinations, bispecific antibodies represent another innovative approach. These antibodies are engineered to have two different antigen-binding sites: one targeting CD47 to block the "don't eat me" signal and the other targeting a specific tumor antigen to enhance the specificity and efficacy of the immune response against cancer cells. For example, HMBD004, a bispecific antibody targeting CD47 and CD33, has shown potent antitumor effects in preclinical AML models, demonstrating this strategy's potential to provide a more targeted and effective treatment option (Moradinasab et al. 2022).

2.8.1.3 Targeting ECM Components and Degradation

Targeting ECM components and their degradation pathways offers a promising avenue for therapeutic strategies, particularly in the context of macrophage-mediated inflammation and tissue remodeling diseases. The ECM plays a crucial role in regulating various cellular functions, including proliferation, migration, and differentiation, through its interactions with cells, such as macrophages. These interactions are mediated by specific ECM receptors on the cell surface and are critical in both physiological and pathological processes, including lung disease and cancer (Clark et al. 2022; Michael et al. 2022).

Moreover, the interaction between macrophages and ECM components, such as versican and hyaluronan, plays a critical role in regulating inflammatory events within the lung. These components increase in various lung diseases and influence leukocyte retention, accumulation, and activation, highlighting their importance in the pathogenesis of lung inflammation. Therapeutic interventions for these ECM components or their interactions with macrophages could help control inflammation and promote tissue repair (Michael et al. 2022).

Additionally, leveraging advanced materials like photoresponsive nanocomposites to control the adhesion sites for macrophages dynamically offers a novel strategy for immunomodulation. By temporally controlling the activation of integrins, such as αvβ3, it is possible to maintain an anti-inflammatory macrophage phenotype, which could accelerate tissue repair processes in inflammatory diseases (Kubota et al. 2009).

2.8.1.4 Regulation of ECM Degradation

Regulation of ECM degradation is a critical aspect of tissue remodeling, inflammation, and disease progression, including lung diseases and cancer. Macrophages play a key role in this process through their ability to secrete MMPs, such as MMP10. This activity affects the structural integrity of tissues and influences cellular behaviors such as proliferation, migration, and differentiation, which are crucial during inflammation and tissue repair (Clark et al. 2022; Kubota et al. 2009; Wang and Chi 2022).

In lung diseases, for example, the interaction between macrophages and ECM components like versican and hyaluronan is known to control leukocyte retention, accumulation, and activation, which are essential aspects of the inflammatory phase of lung disease. The ECM undergoes significant macrophage-mediated remodeling, which impacts normal tissue architecture and lung function. Macrophages facilitate this remodeling through their interaction with specific ECM receptors and the subsequent degradation of the ECM via proteases, including MMP10 (Clark et al. 2022; Kubota et al. 2009).

Given the central role of macrophages and MMPs in ECM degradation and regulation of inflammatory responses, targeting these components presents a promising therapeutic strategy. For instance, modulating the activity of MMPs or their interaction with macrophages could control the extent of ECM degradation, thereby influencing tissue remodeling and inflammation. This approach could be particularly beneficial in diseases characterized by excessive ECM remodeling and fibrosis, such as certain lung diseases and cancer (Clark et al. 2022; Kubota et al. 2009; Wang and Chi 2022).

2.8.1.5 Inhibiting Protease Activity

Inhibiting protease activity, particularly MMPs, represents a potential therapeutic strategy targeting macrophage-ECM interactions in various diseases. Macrophages, key players in these processes, secrete MMPs, including MMP10, which degrade

ECM proteins and regulate specific immune processes such as leukocyte influx and activation (Moradinasab et al. 2022).

The use of broad-spectrum inhibitors against protease families, including cysteine, serine, or metalloproteases, has been explored. However, it has been found that high concentrations of cysteine or serine protease inhibitors fail to inhibit basement membrane remodeling significantly. In contrast, the pan-specific MMP inhibitor, BB-94, blocks basement membrane degradation considerably, highlighting the critical role of MMPs in ECM remodeling (Clark et al. 2022). This suggests that targeting MMP activity, mainly through specific inhibitors like BB-94, could modulate macrophage-mediated ECM degradation, potentially offering therapeutic benefits in diseases characterized by excessive ECM remodeling (Yuan et al. 2024).

Moreover, macrophages deploy the membrane-anchored MMP, MT1-MMP/Mt1-mmp, as the dominant effector of basement membrane proteolysis. This finding underscores the importance of MMPs, especially MT1-MMP, in basement membrane remodeling, a process crucial for embryonic development and potentially for disease progression (Michael et al. 2022). Therefore, targeting MT1-MMP or other MMPs could provide a means to control the extent of ECM degradation and influence tissue remodeling and inflammation.

2.9 Conclusion and Future Directions

In this chapter, a comprehensive exploration of the intricate interplay between macrophages and the ECM is recognized as a crucial component in regulating the response to various pathological conditions, including lung diseases, cancer, and chronic inflammatory states, has been provided. In addition, the multifaceted roles of macrophages in tissue homeostasis, immune responses, and tissue repair have been unraveled, highlighting their pivotal contributions to physiological processes. Moreover, the ECM has underscored its structural significance and regulatory influence on cellular behavior, emphasizing its role in providing structural support to tissues and modulating cellular signaling.

Furthermore, EVs secreted by macrophages have highlighted their functional importance in intercellular communication and disease pathogenesis within the ECM microenvironment. The ability of macrophages to interact with the ECM offers promising avenues for disease intervention and treatment that may improve patient outcomes and quality of life.

One area of focus is the role of MMPs, such as MMP10, which are secreted by macrophages and are key mediators of ECM degradation. MMPs facilitate tissue remodeling and influence macrophage behavior, including proliferation, migration, and differentiation (Clark et al. 2022; Moradinasab et al. 2022). Targeting MMP activity with specific inhibitors or modulating their expression could provide a means to control excessive ECM degradation observed in diseases like lung fibrosis and certain cancers (Moradinasab et al. 2022). Another promising direction is the engineering of immunomodulatory biomaterials, such as EVs or photoresponsive nanocomposites, which can dynamically control macrophage adhesion to the ECM and modulate macrophage phenotype. By leveraging materials that permit on-demand modulation of ECM adhesion sites, it is possible to sustain an anti-inflammatory macrophage phenotype, potentially accelerating tissue repair processes in inflammatory diseases (Michael et al. 2022; Kubota et al. 2009).

Versican and hyaluronan influence leukocyte retention, accumulation, and activation, significantly affecting tissue architecture and function in lung diseases (Moradinasab et al. 2022; Zhuang et al. 2023). Therapeutic strategies that target these ECM components or their interactions with macrophages could help control inflammation and promote tissue repair.

Future research directions include the development of targeted therapies that can specifically modulate macrophage-ECM interactions without affecting the beneficial aspects of ECM remodeling necessary for tissue repair. Additionally, exploring advanced biomaterials and nanotechnology to control the macrophage phenotype and its spatial and temporally precise interaction with the ECM offers a novel approach to treating diseases associated with aberrant macrophage-ECM dynamics (Courvan and Parker 2023).

Acknowledgments We extend acknowledgment to the following projects: CAWT: **1849243**; CMat: **1648035**; COVID PR-DEC **2024-000358**; and NIH Supplemental **3SC1GM131967-04S1**.

References

Aloi N, Drago G, Ruggieri S, Cibella F, Colombo P, Longo V (2024) Extracellular vesicles and immunity: at the crossroads of cell communication. Int J Mol Sci 25:1205. https://doi.org/10.3390/ijms25021205

Alshangiti DM, El-damhougy TK, Zaher A, Madani M, ghobashy MM. (2023) Revolutionizing biomedicine: advancements, applications, and prospects of nanocomposite macromolecular carbohydrate-based hydrogel biomaterials: a review. RSC Adv 13:35251–35291. https://doi.org/10.1039/d3ra07391b

Amin S, Massoumi H, Tewari D, Roy A, Chaudhuri M, Jazayerli C et al (2024) Cell type-specific extracellular vesicles and their impact on health and disease. Int J Mol Sci 25:2730. https://doi.org/10.3390/ijms25052730

Angelov AK, Markov M, Ivanova M, Georgiev T (2023) The genesis of cardiovascular risk in inflammatory arthritis: insights into glycocalyx shedding, endothelial dysfunction, and atherosclerosis initiation. Clin Rheumatol 42:2541–2555. https://doi.org/10.1007/s10067-023-06738-x

Antmen E, Vrana NE, Hasirci V (2021) The role of biomaterials and scaffolds in immune responses in regenerative medicine: macrophage phenotype modulation by biomaterial properties and scaffold architectures. Biomater Sci 9:8090–8110. https://doi.org/10.1039/d1bm00840d

Apaydin O, Altaikyzy A, Filosa A, Sawamiphak S (2023) Alpha-1 adrenergic signaling drives cardiac regeneration via extracellular matrix remodeling transcriptional program in zebrafish macrophages. Dev Cell 58:2460–2476.e7. https://doi.org/10.1016/j.devcel.2023.09.011

Avalos PN, Forsthoefel DJ (2022) An emerging frontier in intercellular communication: extracellular vesicles in regeneration. Front Cell Dev Biol 10:849905. https://doi.org/10.3389/fcell.2022.849905

Badylak SF, Hussey GS, Turnquist H, Dziki JL, Liu Q, Zhang Z (2019) Matrix bound vesicles (MBVS) containing IL-33 and their use. Published online 2019. https://patentscope.wipo.int/search/en/detail.jsf?docId=US321828595

Bahr JC, Li X-Y, Feinberg TY, Jiang L, Weiss SJ (2022) Divergent regulation of basement membrane trafficking by human macrophages and cancer cells. Nat Commun 13:6409. https://doi.org/10.1038/s41467-022-34087-x

Bartolacci JG, Behun MN, Warunek JP, Li T, Sahu A, Dwyer GK et al (2024) Matrix-bound nanovesicle-associated IL-33 supports functional recovery after skeletal muscle injury by initiating a pro-regenerative macrophage phenotypic transition. NPJ Regen Med 9:7. https://doi.org/10.1038/s41536-024-00346-2

Bazzan E, Tinè M, Casara A, Biondini D, Semenzato U, Cocconcelli E et al (2021) Critical review of the evolution of extracellular vesicles' knowledge: from 1946 to today. Int J Mol Sci 22:6417. https://doi.org/10.3390/ijms22126417

Biehl A, Martins AMG, Davis ZG, Sze D, Collins L, Mora-Navarro C et al (2022) Towards a standardized multi-tissue decellularization protocol for the derivation of extracellular matrix materials. Biomater Sci-UK 11:641–654. https://doi.org/10.1039/d2bm01012g

Boyan BD, Asmussen NC, Lin Z, Schwartz Z (2022) The role of matrix-bound extracellular vesicles in the regulation of endochondral bone formation. Cells 11:1619. https://doi.org/10.3390/cells11101619

Bozso SJ, EL-Andari R, Al-Adra D, Moon MC, Freed DH, Nagendran J et al (2021) A review of the immune response stimulated by xenogenic tissue heart valves. Scand J Immunol 93:e13018. https://doi.org/10.1111/sji.13018

Clark GT, Yu Y, Urban CA, Fu G, Wang C, Zhang F et al (2022) Circadian control of heparan sulfate levels times phagocytosis of amyloid beta aggregates. PLoS Genet 18:e1009994. https://doi.org/10.1371/journal.pgen.1009994

Courvan EMC, Parker RR (2023) Post-transcriptional control contributes to hypoxia-induced tumorigenic phenotypes in macrophages. BioRxiv 2023.10.09.561558. https://doi.org/10.1101/2023.10.09.561558

Cox N, Pokrovskii M, Vicario R, Geissmann F (2021) Origins, biology, and diseases of tissue macrophages. Annu Rev Immunol 39:313–344. https://doi.org/10.1146/annurev-immunol-093019-111748

Cramer MC, Molina CP, Hussey GS, Turnquist HR, Badylak SF (2022) Transcriptomic regulation of macrophages by matrix bound nanovesicle associated Interleukin-33. Tissue Eng 0. https://doi.org/10.1089/ten.tea.2022.0006

Crum RJ, Capella-Monsonís H, Chang J, Dewey MJ, Kolich BD, Hall KT et al (2023) Biocompatibility and biodistribution of matrix-bound nanovesicles in vitro and in vivo. Acta Biomater 155:113–122. https://doi.org/10.1016/j.actbio.2022.11.026

Dai Y, Wei X, Jiang T, Wang Q, Li Y, Ruan N et al (2024) Ferroptosis in age-related vascular diseases: molecular mechanisms and innovative therapeutic strategies. Biomed Pharmacother 173:116356. https://doi.org/10.1016/j.biopha.2024.116356

de Visser KE, Joyce JA (2023) The evolving tumor microenvironment: from cancer initiation to metastatic outgrowth. Cancer Cell 41:374–403. https://doi.org/10.1016/j.ccell.2023.02.016

Dixit K, Bora H, Parimi JL, Mukherjee G, Dhara S (2023) Biomaterial mediated immunomodulation: an interplay of material environment interaction for ameliorating wound regeneration. J Biomater Appl 37:1509–1528. https://doi.org/10.1177/08853282231156484

Engin AB, Nikitovic D, Neagu M, Henrich-Noack P, Docea AO, Shtilman MI et al (2017) Mechanistic understanding of nanoparticles' interactions with extracellular matrix: the cell and immune system. Part Fibre Toxicol 14:22. https://doi.org/10.1186/s12989-017-0199-z

Fyfe J, Casari I, Manfredi M, Falasca M (2023) Role of lipid signalling in extracellular vesicles-mediated cell-to-cell communication. Cytokine Growth Factor Rev 73:20–26. https://doi.org/10.1016/j.cytogfr.2023.08.006

Gallerand A, Stunault MI, Merlin J, Luehmann HP, Sultan DH, Firulyova MM et al (2021) Brown adipose tissue monocytes support tissue expansion. Nat Commun 12:5255. https://doi.org/10.1038/s41467-021-25616-1

Gao W-J, Liu J-X, Liu M-N, Yao Y-D, Liu Z-Q, Liu L et al (2021) Macrophage 3D migration: a potential therapeutic target for inflammation and deleterious progression in diseases. Pharmacol Res 167:105563. https://doi.org/10.1016/j.phrs.2021.105563

Gareev I, Shumadalova A, Ilyasova T, Beilerli A, Shi H (2024) Circular RNAs in intracranial aneurysms: emerging roles in pathogenesis, diagnosis and therapeutic intervention. Non-Coding RNA Res 9:211–220. https://doi.org/10.1016/j.ncrna.2023.11.012

Gregory CD, Rimmer MP (2023) Extracellular vesicles arising from apoptosis: forms, functions, and applications. J Pathol 260:592–608. https://doi.org/10.1002/path.6138

Guan Y, Racioppi L, Gerecht S (2023) Engineering biomaterials to tailor the microenvironment for macrophage–endothelium interactions. Nat Rev Mater 8:688–699. https://doi.org/10.1038/s41578-023-00591-9

Gupta D, Zickler AM, Andaloussi SE (2021) Dosing extracellular vesicles. Adv Drug Deliv Rev 178:113961. https://doi.org/10.1016/j.addr.2021.113961

Huang J, Heng S, Zhang W, Liu Y, Xia T, Ji C et al (2022) Dermal extracellular matrix molecules in skin development, homeostasis, wound regeneration and diseases. Semin Cell Dev Biol 128:137–144. https://doi.org/10.1016/j.semcdb.2022.02.027

Huang Y-K, Wang M, Sun Y, Costanzo ND, Mitchell C, Achuthan A et al (2019) Macrophage spatial heterogeneity in gastric cancer defined by multiplex immunohistochemistry. Nat Commun 10:3928. https://doi.org/10.1038/s41467-019-11788-4

Huleihel L, Bartolacci JG, Dziki JL, Vorobyov T, Arnold B, Scarritt ME et al (2017) Matrix-bound nanovesicles recapitulate extracellular matrix effects on macrophage phenotype. Tissue Eng Pt A 23:1283–1294. https://doi.org/10.1089/ten.tea.2017.0102

Hussey GS, Dziki JL, Lee YC, Bartolacci JG, Behun M, Turnquist HR et al (2019) Matrix bound nanovesicle-associated IL-33 activates a pro-remodeling macrophage phenotype via a non-canonical, ST2-independent pathway. J Immunol Regen Med 3:26–35. https://doi.org/10.1016/j.regen.2019.01.001

Jain N, Moeller J, Vogel V (2019) Mechanobiology of macrophages: how physical factors coregulate macrophage plasticity and phagocytosis. Annu Rev Biomed Eng 21:267–297. https://doi.org/10.1146/annurev-bioeng-062117-121224

Kim Y, Nurakhayev S, Nurkesh A, Zharkinbekov Z, Saparov A (2021) Macrophage polarization in cardiac tissue repair following myocardial infarction. Int J Mol Sci 22:2715. https://doi.org/10.3390/ijms22052715

Kubota Y, Takubo K, Shimizu T, Ohno H, Kishi K, Shibuya M et al (2009) M-CSF inhibition selectively targets pathological angiogenesis and lymphangiogenesis. J Exp Med 206:1089–1102. https://doi.org/10.1084/jem.20081605

Li J, Jiang X, Li H, Gelinsky M, Gu Z (2021b) Tailoring materials for modulation of macrophage fate. Adv Mater 33:e2004172. https://doi.org/10.1002/adma.202004172

Li S, Gao H, Hasegawa Y, Lu X (2021a) Fight against fibrosis in adipose tissue remodeling. Am J Physiol-Endocrinol Metab 321:E169–E175. https://doi.org/10.1152/ajpendo.00558.2020

Liangsupree T, Multia E, Riekkola M-L (2021) Modern isolation and separation techniques for extracellular vesicles. J Chromatogr A 1636:461773. https://doi.org/10.1016/j.chroma.2020.461773

Liu J, Qiu X, Lv Y, Zheng C, Dong Y, Dou G et al (2020) Apoptotic bodies derived from mesenchymal stem cells promote cutaneous wound healing via regulating the functions of macrophages. Stem Cell Res Ther 11:507. https://doi.org/10.1186/s13287-020-02014-w

Liu W, Jin M, Chen Q, Li Q, Xing X, Luo Y et al (2023) Insight into extracellular vesicles in vascular diseases: intercellular communication role and clinical application potential. Cell Commun Signal 21:310. https://doi.org/10.1186/s12964-023-01304-z

Liu Y-J, Wang C (2023) A review of the regulatory mechanisms of extracellular vesicles-mediated intercellular communication. Cell Commun Signal 21:77. https://doi.org/10.1186/s12964-023-01103-6

Maacha S, Bhat AA, Jimenez L, Raza A, Haris M, Uddin S et al (2019) Extracellular vesicles-mediated intercellular communication: roles in the tumor microenvironment and anti-cancer drug resistance. Mol Cancer 18:55. https://doi.org/10.1186/s12943-019-0965-7

Mansour AA, Krautter F, Zhi Z, Iqbal AJ, Recio C (2022) The interplay of galectins-1, −3, and −9 in the immune-inflammatory response underlying cardiovascular and metabolic disease. Cardiovasc Diabetol 21:253. https://doi.org/10.1186/s12933-022-01690-7

Martin KE, García AJ (2021) Macrophage phenotypes in tissue repair and the foreign body response: implications for biomaterial-based regenerative medicine strategies. Acta Biomater 133:4–16. https://doi.org/10.1016/j.actbio.2021.03.038

Martins AMG, Snider DB, Popowski KD, Schuchard KG, Tenorio M, Akunuri S et al (2023) Low-dose intrapulmonary drug delivery device for studies on next-generation therapeutics in mice. J Control Release 359:287–301. https://doi.org/10.1016/j.jconrel.2023.05.039

Michael H, Amimo JO, Rajashekara G, Saif LJ, Vlasova AN (2022) Mechanisms of kwashiorkor-associated immune suppression: insights from human, mouse, and pig studies. Front Immunol 13:826268. https://doi.org/10.3389/fimmu.2022.826268

Moradinasab S, Pourbagheri-Sigaroodi A, Ghaffari SH, Bashash D (2022) Targeting macrophage-mediated tumor cell phagocytosis: an overview of phagocytosis checkpoints blockade, nano-medicine intervention, and engineered CAR-macrophage therapy. Int Immunopharmacol 103:108499. https://doi.org/10.1016/j.intimp.2021.108499

Mora-Navarro C, Badileanu A, Martins AMG, Ozpinar EW, Gaffney L, Huntress I et al (2020) Porcine vocal fold lamina Propria-derived biomaterials modulate TGF-β1-mediated fibroblast activation in vitro. ACS Biomater Sci Eng 6:1690–1703. https://doi.org/10.1021/acsbiomaterials.9b01837

Mosser DM, Hamidzadeh K, Goncalves R (2021) Macrophages and the maintenance of homeostasis. Cell Mol Immunol 18:579–587. https://doi.org/10.1038/s41423-020-00541-3

Mu X, Li Y, Fan G-C (2020) Tissue-resident macrophages in the control of infection and resolution of inflammation. Shock 55:14–23. https://doi.org/10.1097/shk.0000000000001601

Mushenkova NV, Bezsonov EE, Orekhova VA, Popkova TV, Starodubova AV, Orekhov AN (2021) Recognition of oxidized lipids by macrophages and its role in atherosclerosis development. Biomedicines 9:915. https://doi.org/10.3390/biomedicines9080915

Nail HM, Chiu C-C, Leung C-H, Ahmed MMM, Wang H-MD (2023) Exosomal miRNA-mediated intercellular communications and immunomodulatory effects in tumor microenvironments. J Biomed Sci 30:69. https://doi.org/10.1186/s12929-023-00964-w

Nieuwland R, Siljander PR-M, Falcón-Pérez JM, Witwer KW (2022) Reproducibility of extracellular vesicle research. Eur J Cell Biol 101:151226. https://doi.org/10.1016/j.ejcb.2022.151226

Niu C, Wang X, Zhao M, Cai T, Liu P, Li J et al (2016) Macrophage foam cell–derived extracellular vesicles promote vascular smooth muscle cell migration and adhesion. J Am Hear Assoc 5:e004099. https://doi.org/10.1161/jaha.116.004099

Niu Y, Wang Z, Shi Y, Dong L, Wang C (2021) Modulating macrophage activities to promote endogenous bone regeneration: biological mechanisms and engineering approaches. Bioact Mater 6:244–261. https://doi.org/10.1016/j.bioactmat.2020.08.012

Osorio-Valencia S, Zhou B (2024) Roles of macrophages and endothelial cells and their crosstalk in acute lung injury. Biomedicines 12:632. https://doi.org/10.3390/biomedicines12030632

Papachristoforou E, Ramachandran P (2022) Macrophages as key regulators of liver health and disease. Int Rev Cell Mol Biol 368:143–212. https://doi.org/10.1016/bs.ircmb.2022.04.006

Piening LM, Wachs RA (2023) Matrix-bound nanovesicles: what are they and what do they do? Cells Tissues Organs 212:111–123. https://doi.org/10.1159/000522575

Pieters BCH, Cappariello A, van den Bosch MHJ, van Lent PLEM, Teti A, van de Loo FAJ (2019) Macrophage-derived extracellular vesicles as carriers of Alarmins and their potential involvement in bone homeostasis. Front Immunol 10:1901. https://doi.org/10.3389/fimmu.2019.01901

Poulis N, Martin M, Hoerstrup SP, Emmert MY, Fioretta ES (2022) Macrophage-extracellular matrix interactions: perspectives for tissue engineered heart valve remodeling. Front Cardiovasc Med 9:952178. https://doi.org/10.3389/fcvm.2022.952178

Puttock EH, Tyler EJ, Manni M, Maniati E, Butterworth C, Ramos MB et al (2023) Extracellular matrix educates an immunoregulatory tumor macrophage phenotype found in ovarian cancer metastasis. Nat Commun 14:2514. https://doi.org/10.1038/s41467-023-38093-5

Ramachandran P, Matchett KP, Dobie R, Wilson-Kanamori JR, Henderson NC (2020) Single-cell technologies in hepatology: new insights into liver biology and disease pathogenesis. Nat Rev Gastroenterol Hepatol 17:457–472. https://doi.org/10.1038/s41575-020-0304-x

Rilla K, Mustonen A-M, Arasu UT, Härkönen K, Matilainen J, Nieminen P (2019) Extracellular vesicles are integral and functional components of the extracellular matrix. Matrix Biol 75:201–219. https://doi.org/10.1016/j.matbio.2017.10.003

Ruggiero AD, Key C-CC, Kavanagh K (2021) Adipose tissue macrophage polarization in healthy and unhealthy obesity. Front Nutr 8:625331. https://doi.org/10.3389/fnut.2021.625331

Salthouse D, Novakovic K, Hilkens CMU, Ferreira AM (2023) Interplay between biomaterials and the immune system: challenges and opportunities in regenerative medicine. Acta Biomater 155:1–18. https://doi.org/10.1016/j.actbio.2022.11.003

Sánchez GB, Bunn KE, Pua HH, Rafat M (2021) Extracellular vesicles: mediators of intercellular communication in tissue injury and disease. Cell Commun Signal 19:104. https://doi.org/10.1186/s12964-021-00787-y

Sánchez-Alonso S, Alcaraz-Serna A, Sánchez-Madrid F, Alfranca A (2018) Extracellular vesicle-mediated immune regulation of tissue Remodeling and angiogenesis after myocardial infarction. Front Immunol 9:2799. https://doi.org/10.3389/fimmu.2018.02799

Schlundt C, Fischer H, Bucher CH, Rendenbach C, Duda GN, Schmidt-Bleek K (2021) The multi-faceted roles of macrophages in bone regeneration: a story of polarization, activation and time. Acta Biomater 133:46–57. https://doi.org/10.1016/j.actbio.2021.04.052

Soni SS, Rodell CB (2021) Polymeric materials for immune engineering: molecular interaction to biomaterial design. Acta Biomater 133:139–152. https://doi.org/10.1016/j.actbio.2021.01.016

Sridharan R, Kelly DJ, O'Brien FJ (2020) Substrate stiffness modulates the crosstalk between mesenchymal stem cells and macrophages. J Biomech Eng 143. https://doi.org/10.1115/1.4048809

Théry C, Witwer KW, Aikawa E, Alcaraz MJ, Anderson JD, Andriantsitohaina R et al (2018) Minimal information for studies of extracellular vesicles 2018 (MISEV2018): a position statement of the International society for extracellular vesicles and update of the MISEV2014 guidelines. J Extracell Vesicles 8:1535750. https://doi.org/10.1080/20013078.2018.1535750

Thorseth M-L, Carretta M, Jensen C, Mølgaard K, Jürgensen HJ, Engelholm LH et al (2022) Uncovering mediators of collagen degradation in the tumor microenvironment. Matrix Biol Plus 13:100101. https://doi.org/10.1016/j.mbplus.2022.100101

Torregrossa M, Kakpenova A, Simon JC, Franz S (2021) Modulation of macrophage functions by ECM-inspired wound dressings—a promising therapeutic approach for chronic wounds. Biol Chem 402:1289–1307. https://doi.org/10.1515/hsz-2021-0145

Vasse GF, Nizamoglu M, Heijink IH, Schlepütz M, Rijn P, Thomas MJ et al (2021) Macrophage–stroma interactions in fibrosis: biochemical, biophysical, and cellular perspectives. J Pathol 254:344–357. https://doi.org/10.1002/path.5632

Wang Q, Chi L (2022) The alterations and roles of glycosaminoglycans in human diseases. Polymers 14:5014. https://doi.org/10.3390/polym14225014

Wilson HM (2023) Modulation of macrophages by biophysical cues in health and beyond. Discov Immunol 2:kyad013. https://doi.org/10.1093/discim/kyad013

Wolf SJ, Melvin WJ, Gallagher K (2021) Macrophage-mediated inflammation in diabetic wound repair. Semin Cell Dev Biol 119:111–118. https://doi.org/10.1016/j.semcdb.2021.06.013

Wong NR, Mohan J, Kopecky BJ, Guo S, Du L, Leid J et al (2021) Resident cardiac macrophages mediate adaptive myocardial remodeling. Immunity 54:2072–2088.e7. https://doi.org/10.1016/j.immuni.2021.07.003

Xiong J, Xiao R, Zhao J, Zhao Q, Luo M, Li F et al (2024) Matrix stiffness affects tumor-associated macrophage functional polarization and its potential in tumor therapy. J Transl Med 22:85. https://doi.org/10.1186/s12967-023-04810-3

Yan H, Li Y, Cheng S, Zeng Y (2021) Advances in analytical technologies for extracellular vesicles. Anal Chem 93:4739–4774. https://doi.org/10.1021/acs.analchem.1c00693

Yang J, Bahcecioglu G, Ronan G, Zorlutuna P (2024) Aged breast matrix bound vesicles promote breast cancer invasiveness. Biomaterials 306:122493. https://doi.org/10.1016/j.biomaterials.2024.122493

Ye J, Xie C, Wang C, Huang J, Yin Z, Heng BC et al (2021) Promoting musculoskeletal system soft tissue regeneration by biomaterial-mediated modulation of macrophage polarization. Bioact Mater 6:4096–4109. https://doi.org/10.1016/j.bioactmat.2021.04.017

Yokoi A, Ochiya T (2021) Exosomes and extracellular vesicles: rethinking the essential values in cancer biology. Semin Cancer Biol 74:79–91. https://doi.org/10.1016/j.semcancer.2021.03.032

Yu J, Sane S, Kim J-E, Yun S, Kim H-J, Jo KB et al (2024) Biogenesis and delivery of extracellular vesicles: harnessing the power of EVs for diagnostics and therapeutics. Front Mol Biosci 10:1330400. https://doi.org/10.3389/fmolb.2023.1330400

Yu Y, Wu R-X, Yin Y, Chen F-M (2015) Directing immunomodulation using biomaterials for endogenous regeneration. J Mater Chem B 4:569–584. https://doi.org/10.1039/c5tb02199e

Yu Y, Yue Z, Xu M, Zhang M, Shen X, Ma Z et al (2022) Macrophages play a key role in tissue repair and regeneration. PeerJ 10:e14053. https://doi.org/10.7717/peerj.14053

Yuan X, Shi J, Kang Y, Dong J, Pei Z, Ji X (2024) Piezoelectricity, pyroelectricity, and ferroelectricity in biomaterials and biomedical applications. Adv Mater 36:e2308726. https://doi.org/10.1002/adma.202308726

Zeng Y, Qiu Y, Jiang W, Shen J, Yao X, He X et al (2022) Biological features of extracellular vesicles and challenges. Front Cell Dev Biol 10:816698. https://doi.org/10.3389/fcell.2022.816698

Zhang X, Zhang H, Gu J, Zhang J, Shi H, Qian H et al (2021) Engineered extracellular vesicles for cancer therapy. Adv Mater 33:e2005709. https://doi.org/10.1002/adma.202005709

Zhang X, Rotllan N, Canfrán-Duque A, Sun J, Toczek J, Moshnikova A et al (2022) Targeted suppression of miRNA-33 using pHLIP improves atherosclerosis regression. Circ Res 131:77–90. https://doi.org/10.1161/circresaha.121.320296

Zhou X, Franklin RA, Adler M, Carter TS, Condiff E, Adams TS et al (2022) Microenvironmental sensing by fibroblasts controls macrophage population size. Proc Natl Acad Sci 119:e2205360119. https://doi.org/10.1073/pnas.2205360119

Zhuang Q, Yang H, Mao Y (2023) The oncogenesis of glial cells in diffuse gliomas and clinical opportunities. Neurosci Bull 39:393–408. https://doi.org/10.1007/s12264-022-00953-3

Chapter 3
Immunometabolic Rewiring: A Tale of Macronutrients and Macrophages

Gerasimos Anagnostopoulos, Camille Blériot, Nicolas Venteclef, and Florent Ginhoux

Abstract Myeloid cells, including monocytes, macrophages, dendritic cells, and polymorphonuclear cells are key components of homeostasis maintenance and immune response. Among the myeloid lineage, macrophages stand out as highly versatile cells that safeguard tissue functions but also sense and respond to potentially harmful microenvironmental cues. Numerous studies have demonstrated that the nutritional status and macronutrient availability affect macrophage identity and function. However, the exact mechanistic links between macronutrient intake and cellular metabolic shifts are only beginning to be understood. In this chapter, we explore how dietary "macros"—carbohydrates, fats, and amino acids—impact the immunomodulatory activity of macrophages in healthy and inflamed tissues.

G. Anagnostopoulos
Institut Gustave Roussy, INSERM U1015, Villejuif, France

C. Blériot
Institut Gustave Roussy, INSERM U1015, Villejuif, France

Institut Necker Enfants Malades (INEM), INSERM, CNRS, Université Paris Cité, Paris, France

N. Venteclef
Institut Necker Enfants Malades (INEM), INSERM, CNRS, Université Paris Cité, Paris, France

F. Ginhoux (✉)
Institut Gustave Roussy, INSERM U1015, Villejuif, France

Singapore Immunology Network (SIgN), Agency for Science, Technology and Research (A∗STAR), Singapore, Singapore

Shanghai Institute of Immunology, Shanghai Jiao Tong University School of Medicine, Shanghai, China

Translational Immunology Institute, SingHealth Duke-NUS, Singapore, Singapore
e-mail: florent.ginhoux@gustaveroussy.fr

M. Kloc et al. (eds.), *Monocytes and Macrophages in Development, Regeneration, and Disease*, Results and Problems in Cell Differentiation 74,
https://doi.org/10.1007/978-3-031-65944-7_3

3.1 Introduction

The immune system plays a central role in maintaining tissue homeostasis. The cells comprising this highly versatile system have evolved to respond to a wide spectrum of environmental stimuli including infection, tissue injury, cellular stress, and tumor invasion. This requires a dynamic *metabolic reprogramming*, a term used to describe the ability of cells to actively favor specific metabolic pathways in order to generate energy, and also particular macromolecules to achieve the desired effector functions. Metabolic reprogramming in response to microenvironmental cues has rapidly become a fundamental concept in immune cell biology introducing the field of *immunometabolism* (Yoshida 2015; Faubert et al. 2020). This relatively new field investigates how metabolic alterations of immune cells may influence host immunity on a broader scale (Buck et al. 2017). In addition, it explores how immune cells can modulate the host metabolism (Pearce and Pearce 2013).

After their maturation in bone marrow (BM), thymus, and lymphoid tissues (with the noticeable exception of most macrophages which mainly emerge during embryonic development), immune cells colonize their tissue of residence to exert their homeostatic or pro-inflammatory functions. Once infiltrating the tissue, the cells (1) get exposed to spatially and temporally dynamic tissue-specific profiles of nutrient, oxygen, and growth factor availability, which dictate their metabolic fate, and (2) acquire a transcriptional signature and specific metabolic profiles reflecting that particular niche (Blériot et al. 2020).

Myeloid cells form a major functional branch of the immune system. They act as effector cells in homeostasis and inflammation, in the context of which they control the initiation of innate and adaptive immunity, resolution of inflammation, and tissue repair. Myeloid cells include mononuclear cells—monocytes, macrophages, and dendritic cells (DCs)—and polymorphonuclear cells—neutrophils, eosinophils, and basophils—along with immature myeloid progenitors. They are located both in the circulation and tissues, where they constantly survey their surroundings for foreign microbial threats, tissue perturbations, and injuries (Soehnlein and Lindbom 2010).

In this chapter, we focus on macrophage function from an immunometabolic perspective. As key tissue sentinel cells, macrophages are present across various organs throughout the body and are involved in complex tissue-specific processes, including neurogenesis, angiogenesis, osteogenesis, and erythropoiesis, which is indicative of their unique ability to adapt and contribute to their place of residence (Ginhoux and Guilliams 2016).

Over the last two decades, the scientific community has been attempting to specify the origin of resident tissue macrophages (RTMs) as a means to decipher their function in healthy and inflamed tissues. In this direction, the traditional notion that RTMs originate from circulating monocytes and rely on input from stem and progenitor cells of the BM has evolved (Ginhoux and Guilliams 2016). Mouse studies demonstrated that certain RTM populations stem exclusively from embryonic progenitors that include yolk-sac macrophages or fetal-liver monocytes (e.g. skin, lung, kidney, and brain macrophages) (Merad et al. 2002; Ajami et al. 2007; Ginhoux et al. 2010; Hashimoto et al. 2013; Guilliams et al. 2013; Jakubzick et al. 2013). When the host is not challenged with a pathological condition, these long-lived populations maintain themselves locally throughout adulthood and participate in normal tissue physiology (Ginhoux et al. 2010; Schulz et al. 2012; Hoeffel et al. 2012; Hashimoto et al. 2013). In the opposite case, when inflammation induces RTM death and disruption of tissue integrity, circulating monocytes contribute to resident pools in order to replace the compromised niche. That said, other macrophage populations depend mostly on monocyte replacement (e.g. intestinal lamina and dermis macrophages) (Tamoutounour et al. 2013; Epelman et al. 2014; Molawi et al. 2014; Bain et al. 2014).

When exposed to varying concentrations of macronutrients and growth factors, RTMs activate tissue-specific transcriptional programs that allow them to maintain homeostasis. In the next sections, we describe how certain dietary macronutrients and their deficiencies can affect cellular metabolism and subsequent immunomodulatory functions (Fig. 3.1).

Box 3.1 Key Catabolic Pathways and Macrophage States

Macrophages need to maintain a fine balance between two states: the anti-inflammatory steady state (during homeostasis) relying on fatty acid oxidation (FAO) and oxidative phosphorylation (OXPHOS) for energy generation, and the pro-inflammatory polarized state mobilizing the glycolytic pathway to rapidly produce energy (Fig. 3.2) (Ryan and O'Neill 2017; Michaudel and Sokol 2020).

- *Glycolysis* is a central metabolic pathway that occurs in the cytoplasm of cells and includes the enzymatic breakdown of glucose into pyruvate. Even though glycolysis is a fundamental process generating ATP for several cellular functions, it is regarded as an inefficient pathway in terms of energy production generating only two ATP molecules per unit of glucose. That said, glycolysis has the advantage of generating fast energy when "time is of the essence". Apart from pyruvate, this pathway provides intermediate products for the synthesis of other macromolecules, such as nucleotides and amino acids. When exposed to pro-inflammatory stimuli, macrophages undergo rapid metabolic reprogramming to generate fast energy and biosynthetic precursors for combating fast-growing microbes. This metabolic shift resembles the Warburg effect observed in tumor cells (Newsholme et al. 1986), favoring aerobic glycolysis over oxidative phosphorylation (see below). In this state, macrophages secrete the excess carbon in the form of lactate instead of using it to produce acetyl-CoA from pyruvate.

- The *tricarboxylic acid cycle* (TCA cycle), also referred to as the *Krebs cycle*, is an essential metabolic pathway of eukaryotes that takes place in the mitochondrial matrix and generates energy in the form of ATP through oxidative phosphorylation (OXPHOS). It includes a series of enzymatic reactions that oxidize acetyl-CoA which derives from glycolysis or β-oxidation to carbon dioxide. It begins with the reaction of acetyl-CoA with oxaloacetate which forms citrate. Citrate then undergoes several enzymatic reactions forming isocitrate, alpha-ketoglutarate, succinyl-CoA, succinate, fumarate, malate, and oxaloacetate. Through this process, NAD^+ and FAD are reduced to NADH and $FADH_2$, respectively, and carbon dioxide is released as a byproduct. The TCA cycle serves as a source of highly efficient ATP production as well as a source of intermediates for amino acid synthesis, gluconeogenesis and lipid metabolism.
- *β-oxidation* is a catabolic pathway that takes place in mitochondria with the primary role of breaking down fatty acids to acetyl-CoA which can then be oxidized in the TCA cycle for energy production (a pathway referred to as fatty acid oxidation, FAO). In this pathway, each fatty acid molecule is cleaved into two-carbon fragments which are released in the form of acetyl-CoA. This pathway is critical particularly for periods of glucose scarcity (e.g. prolonged fasting, exercise, but also in ketogenic regimes where carbohydrates are in very low content) when fatty acids serve as a major fuel source for cells.

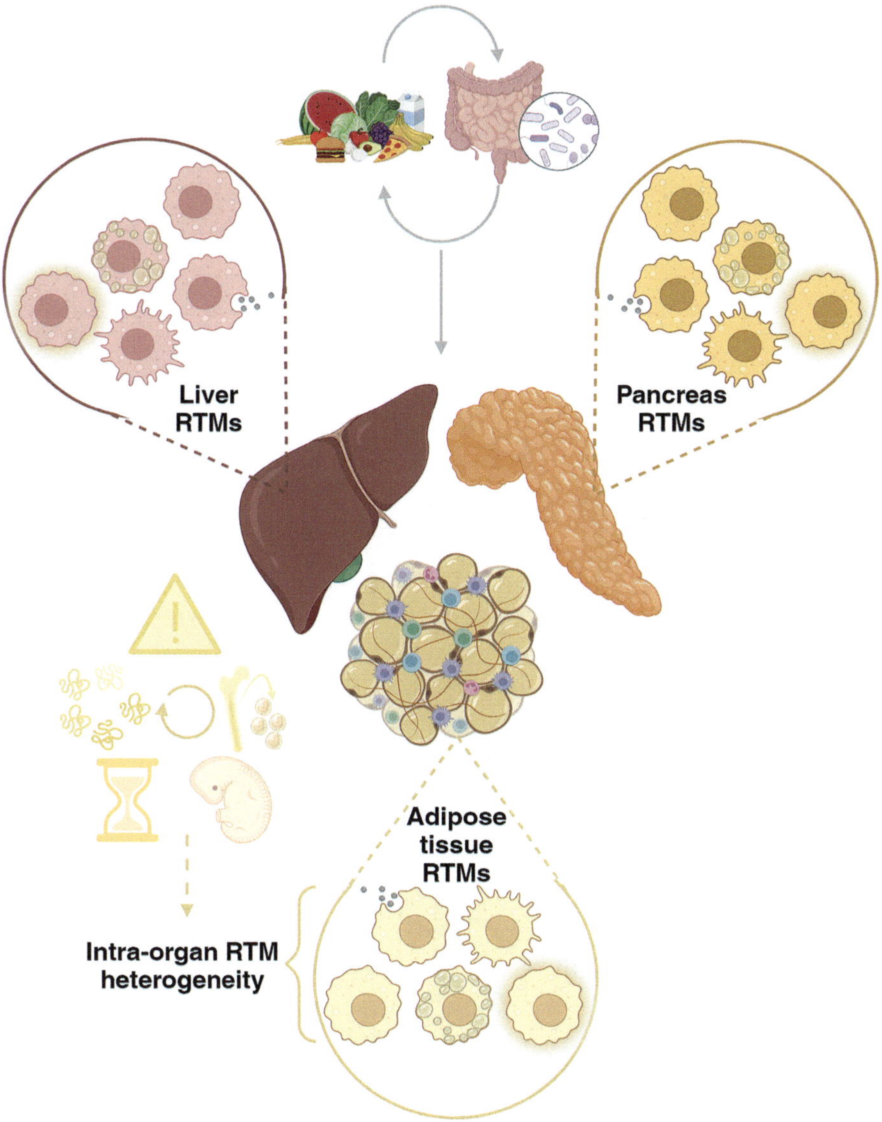

Fig. 3.1 Nutrition affects the functional heterogeneity of resident tissue macrophages (RTMs) with respect to their potential to maintain homeostasis or promote inflammation. This overview highlights the role of *nutrition* (encompassing meal consumption frequency, caloric content within a meal, macronutrient supply, and dietary influence on the commensal microbiome) as a key determinant in shaping the phenotype of RTMs. Nonetheless, beyond the "extrinsic" signals of *nutrition*, there are multiple "intrinsic" factors that dynamically shape RTM identity within a tissue (extensively reviewed elsewhere: Blériot et al. 2020). These include cellular ontogeny, host inflammation status, endocrine/trophic/danger signals to which the cell is exposed, the temporal span of cellular colonization within the tissue of residence, as well as biological aging

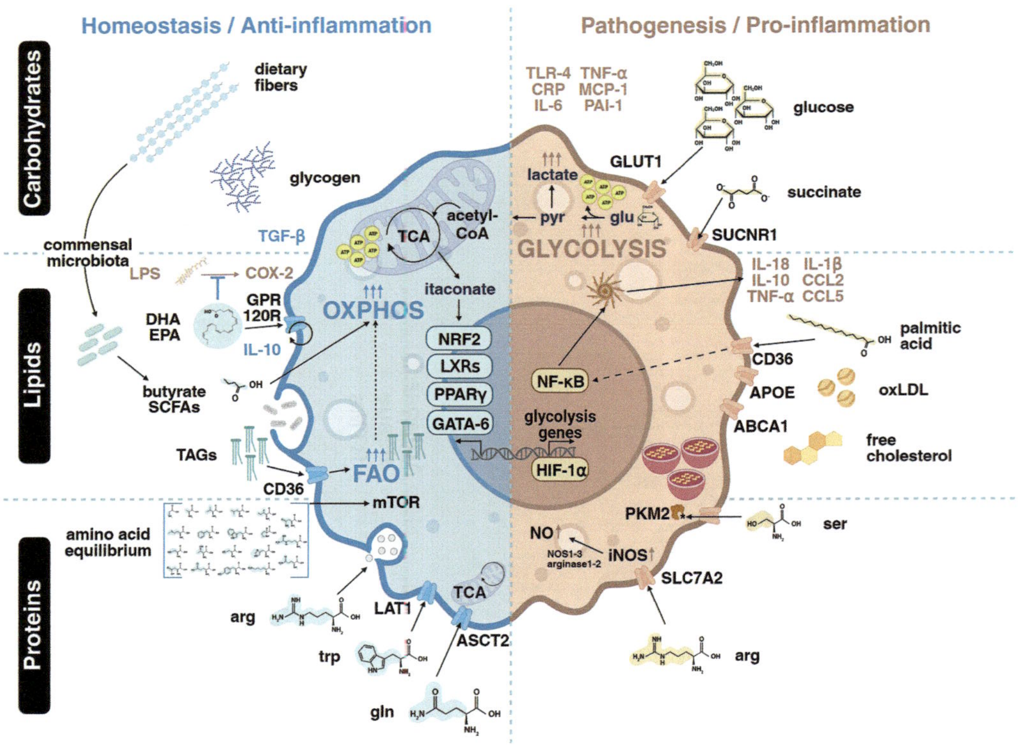

Fig. 3.2 The scheme summarizes the current knowledge on how distinct dietary macronutrients influence the immunometabolic status of resident tissue macrophages (RTMs). Carbohydrate (upper), lipid (middle), and amino acid (lower panel) dietary derivatives dynamically alter the RTM metabolism transitioning between glycolytic and OXPHOS pathways, thereby favoring an inflammatory (right panel) or an anti-inflammatory (left panel) function respectively. This dynamic nature of metabolism serves the immunocompetence dichotomy and is considered essential in safeguarding homeostasis. ABCA1 ATP-binding cas-

sette transporter 1, APOE apolipoprotein E, arg arginine, ASCT2 alanine serine cysteine transporter 2 (also referred to as SLC1A5), CCL2 C–C motif chemokine ligand 2, CCL5 C–C motif chemokine ligand 5, CD36 cluster of differentiation 36, COX-2 cyclooxygenase-2, CRP plasma C-reactive protein, DHA docosahexaenoic acid, EPA eicosapentaenoic acid, FAO fatty acid oxidation, GATA-6 GATA binding protein 6, gln glutamine, glu glucose, GLUT1 glucose transporter 1, GPR120 G protein-coupled receptor 120, HIF-1α hypoxia-Inducible factor-1α, IL-1β interleukin-1β, IL-6 interleukin-6, IL-10 interleukin-10, IL-18 interleukin-18, iNOS inducible nitric oxide synthase, LAT1 L-type amino acid transporter 1 (also referred to as SLC7A5), LPS lipopolysaccharides, LXR liver X-family of receptor, MCP-1 monocyte chemoattractant protein-1, mTOR mammalian target of rapamycin, NF-κB nuclear factor-κB, NO nitric oxide, NOS nitric oxide synthase, NRF2 nuclear factor erythroid 2—related factor 2, oxLDL oxidized low-density lipoprotein, OXPHOS oxidative phosphorylation, PAI1 plasminogen activator inhibitor 1, PKM2 pyruvate kinase M2, PPARγ peroxisome proliferator activated receptor gamma, pyr pyruvate, SCFA short-chain fatty acid, ser serine, SLC7A2 solute carrier family 7 member 2, SUCNR1 succinate receptor 1, TAG triacylglycerol, TCA tricarboxylic acid, TGF-β transforming growth factor beta, TLR4 toll-like receptor 4, TNF-α tumor necrosis factor alpha, trp tryptophan

3.2 Resident Tissue Macrophages in Metabolic Organs

In this section, we present a general framework for understanding the multifaceted functions of macrophages in various tissues, including white adipose tissue (WAT), liver, pancreas, lung, and peritoneal cavity. We describe the interactions between resident macrophages and surrounding parenchymal cells, crucial for responding to a constantly fluctuating nutrient microenvironment.

3.2.1 White Adipose Tissue

The duration and macronutrient content of a diet are etiologically linked to the metabolic health of adipocytes in the WAT. Adipocytes are subjected to changes in number and metabolic status depending on the diet and energetic balance. To respond to the dynamic adipocyte metabolism, resident WAT macrophages subsequently undergo significant shifts in numbers and activation profiles (Sun et al. 2011).

In lean healthy individuals, WAT macrophages constitute approximately 10–15% of stromal cells and express canonical markers such as Arg1, CD301, and CD206 (Lumeng et al. 2007). These macrophages maintain euglycemia and adipocyte insulin sensitivity (Sun et al. 2011) by secreting interleukin 10 (IL-10) which potentiates insulin function in adipocytes. The maintenance of such macrophages requires the function of peroxisome proliferator—activated family of receptor gamma (PPARγ), PPARδ, and Krüppel-like factor 4 (KLF4) (Odegaard et al. 2007, 2008; Kang et al. 2008; Liao et al. 2011).

In contrast, the obese state is characterized by elevated chronic energetic surplus, imposing significant metabolic stress on adipocytes. They increase in size (hypertrophy), and number (hyperplasia), while secreting chemokines such as C–C motif chemokine ligand 2 (CCL2), CCL5, and CCL8 (Rosen and Spiegelman 2006). These chemokines trigger a significant blood-derived Ly6C^{hi} monocyte infiltration which eventually differentiate into CD11c^{+} pro-inflammatory macrophages. Recruited macrophages have been shown to form "crown-like structures" surrounding dead adipocytes (Weisberg et al. 2003; Lumeng et al. 2007). These macrophages are also referred to as lipid-associated macrophages (LAMs) due to their post-phagocytosis increase in intracellular lipid cargo. LAMs have been the subject of intense study and have been reported to secrete numerous pro-inflammatory cytokines such as tumor necrosis alpha (TNF-α) and IL-6. LAMs also facilitate WAT hypertrophy and promote nutrient excess storage and insulin resistance of adipocytes (Sun et al. 2011).

3.2.2 Liver

The liver plays a pivotal role in integrating nutrient, hormonal, and environmental signals to maintain carbohydrate and lipid homeostasis in the body. Kupffer cells (KCs) reside in the liver sinusoidal lumen comprising the most abundant pool of macrophages (Wardle 1987). Even though KCs have been formerly regarded as exclusively innate-immunity entities responding to diverse threats (such as bacterial and viral infections), they have been recently reported to crosstalk with hepatocytes through secreted growth factors, hormones, trophic signals, and cytokines to safeguard hepatic homeostasis (Blériot et al. 2020). In steady-state conditions, KCs are actively involved in lipid metabolism, expressing liver X-family of receptors (LXRs) and PPARs, which regulate a transcriptional network related to cholesterol transport and metabolism (Remmerie and Scott 2018). Studies in mice indicate that KCs act as gatekeepers of low-density lipoprotein (LDL) cholesterol transport to hepatocytes because of their abundant expression of LDL receptors on the cell membrane (Demetz et al. 2020). In addition, Blériot and colleagues recently reported the existence of a specific subpopulation of KCs exhibiting a more elevated expression of genes associated with metabolism regulation (Blériot et al. 2021). Noteworthy characteristics of this subset include increased levels of the KLF6 transcription factor and the CD36 lipid transporter. Furthermore, the proportion of this subset of metabolic KCs increases in mice subjected to a high-fat diet (HFD) while their depletion rescues HFD-induced weight gain, liver steatosis, and insulin sensitivity. This work underlines the fact that even superficially resembling RTM populations (such as KCs) may harbor a functional diversity. Independent genetic studies suggest that decreased KC content is associated with compromised lipid metabolism and age-progressive steatosis in the liver (Keshvari et al. 2021). In conditions of energetic deficit, KCs are key drivers of FAO on the neighboring hepatocytes. In detail, they release TNF in the extracellular space, and once secreted, TNF relays information to hepatocytes, prompting them to switch to a PPARα-dependent transcriptional and metabolic program that favors FAO (Loft et al. 2022).

Finally, in conditions of lipid-dense hypercaloric nutrition, KCs serve as sensors of circulating free fatty acids (FFAs) transmitting this information to hepatocytes via TNF secretion, which in turn alters their lipid metabolism triggering accumulation of triacylglycerols in lipid droplets (Diehl et al. 2020). In such obesogenic conditions, KCs adopt a pro-inflammatory status—characterized by the secretion of inflammatory mediators such as TNF, IL-10, CCL2, and CCL5—which profoundly compromises hepatic lipid metabolism and results in fatty liver disease. Eventually, this state has as a consequence the gradual decline of KCs. To compensate for this loss, but also because of inflammatory signals, the body mobilizes newly recruited macrophages, of circulating monocyte origin, to colonize the affected niche (Remmerie et al. 2020).

3.2.3 Pancreas

The pancreas harbors Langerhans islets which serve as a crucial endocrine hub, producing both insulin and glucagon to maintain euglycemia, thereby impacting the metabolic health of all tissues. The pancreatic tissue harbors a diverse pool of resident macrophages which, under homeostatic conditions, are long-lived and self-maintained (Calderon et al. 2015). The stroma of the exocrine pancreas is inhabited by a diverse mixture of immune cells including macrophages, DCs, and B and T cells. Mouse studies illustrate that within pancreatic Langerhans islets, the only myeloid cells are $CD11c^{+}MHC\text{-}II^{+}F4/80^{lo}$ macrophages (referred to as *intra-islet macrophages*) while the area surrounding the islet harbors a separate group of $F4/80^{hi}$ $CD11c^{-}$ macrophages (referred to as *peri-islet macrophages*) (Ying et al. 2019). The pool of islet macrophages relies on colony-stimulating factor 1 (CSF-1)—a key determinant of macrophage differentiation and function—with CSF-1 deficiency being associated with fewer islet macrophages and reduced islet size (Calderon et al. 2015).

In obesogenic conditions, the energetic surplus profoundly affects macrophage number and function. This effect compromises the glycemic control and drives the onset of numerous islet-derived local stress signals (e.g., ATP released from stressed β-cells, and islet amyloid polypeptide accumulation) compromising the health of pancreatic β-cells (Westwell-Roper et al. 2014; Weitz et al. 2018; Ying et al. 2020). In this state, we observe a significant infiltration of highly proliferating macrophages into pancreatic islets (Ehses et al. 2007; Ying et al. 2019) secreting proinflammatory cytokines including IL-1β (Ying et al. 2020) which are likely to lead to further β-cell apoptosis and suppression of glucose-stimulated insulin secretion (GSIS) (Donath et al. 2010). IL-1β suppresses the expression of genes important for β-cell differentiation, therefore leading to further β-cell apoptosis and dedifferentiation (Talchai et al. 2012; Nordmann et al. 2017). Finally, elevated levels of saturated fatty acids have been postulated to contribute to islet inflammation and β-cell dysfunction and proliferation through a toll-like receptor 4 (TLR4)-mediated mechanism (Eguchi et al. 2012; Ye et al. 2018). Saturated fatty acids bind to TLR4 receptors on islet macrophages and elicit secretion of IL-1β. In addition, they activate TLR4 receptors on β-cells, prompting the secretion of chemokines that attract macrophages to the islets (Eguchi et al. 2012).

3.2.4 Lung

In the lungs, alveolar RTMs are characterized by elevated expression of PPARγ (Gautier et al. 2012) which is considered the master regulator of lipid metabolism across tissues, regulating genes involved in lipid storage, and cholesterol efflux. Such pathways are particularly significant for alveolar macrophages, which continuously need to generate energetic blocks from lipid-rich substrates within the alveolar surfactant. The latter contains an increased concentration of phospholipoproteins, cholesterol, triacylglycerol, and FFAs. Genetic studies suggest that PPARγ-deficient alveolar macrophages result in pulmonary alveolar proteinosis as alveoli become filled with surfactant. The loss of function of *Pparg* in alveolar RTMs favors an intracellular accumulation of lipid droplets due to impaired cholesterol efflux, attributed to diminished expression of the cholesterol transporter ATP-binding cassette (ABC) sub-family G member 1 (ABCG1) (Baker et al. 2010). Furthermore, genetic ablation of *Abca1* and *Abcg1* in mice results in the accumulation of pro-inflammatory macrophage foam cells in the lung (but also in other tissues) (Yvan-Charvet et al. 2010). That said, mice lacking ABCA1 were characterized by lower circulating high-density lipoprotein (HDL) than mice lacking ABCG1, accompanied by pulmonary lipidosis, and chronic inflammation. The scarcity of glucose in alveoli hampers alveolar macrophage activation, likely serving as a control mechanism to mitigate potentially harmful effector functions within this delicate niche (Svedberg et al. 2019).

3.2.5 Peritoneal Cavity

Another example of a tissue-specific transcriptional program is one driven by the transcription factor GATA-6 in resident peritoneal cavity macrophages (Gautier et al. 2014). GATA-6-deficient peritoneal RTMs show compromised expression of the aspartoacylase enzyme which, in turn, hampers cellular proliferation. Aspartoacylase catalyzes the conversion of N-acetyl aspartate, an abundant nutrient in the peritoneal cavity, into aspartate and acetate, which in turn increases acetyl-CoA and fuels OXPHOS. The enrichment of N-acetyl aspartate in the peritoneal cavity relative to its serum concentration raises the possibility that resident peritoneal macrophages utilize this substrate to generate aspartate and acetate. This is particularly intriguing as acetate, a precursor of acetyl-CoA, can fuel the TCA cycle, acting as a precursor for fatty acid synthesis and as an acetate donor for protein acetylation (Davies et al. 2017). Resident cavity macrophages exhibit high mitochondrial oxygen consumption rates, potentially fueled by N-acetyl aspartate. At rest, they utilize fatty acids and glutamine to fuel OXPHOS, like resting BM-derived macrophages. Notably, the higher concentration of glutamate in the peritoneal cavity allows it to supplement glutamine and support enhanced respiration in resident peritoneal macrophages stimulated with zymosan particles. This process involves the production of reactive oxygen species (ROS) through increased respiration, which are then utilized to damage phagocytosed pathogens. Such findings show that RTMs are metabolically primed to offer protection against pathogens within the peritoneal cavity, utilizing locally available fuels to implement specific metabolic programs. Notably, the diverse array of metabolites in various tissues may underlie distinct metabolic requirements in different situations (Orliaguet et al. 2022). That said, not all peritoneal cavity macrophages exist due to GATA-6 signaling. For instance, Randolph and colleagues recently identified two populations of F4/80^{hi} ICAM-2^{-} CD206^{+} peritoneal macrophages: Lyve-1^{lo} MHC-II^{hi} and Lyve-1^{hi} MHC-II^{lo} cells, that were not dependent on GATA-6 (Zhang et al. 2021).

To conclude, RTMs exhibit a remarkable inter-tissular plasticity undertaking various roles within their niche of residence. In the following paragraphs, we provide insight into how specific dietary macronutrients—carbohydrates, fats, and amino acids—can influence macrophage function, consequently shaping RTM heterogeneity in healthy and inflamed tissues.

3.3 Dietary Carbohydrate Intake and Macrophage Function

Carbohydrates represent a significant category of macronutrients and can be classified into simple carbohydrates, which are derived from oligo- and polysaccharides commonly referred to as "sugars," and complex carbohydrates, which originate from starch and fibers. They are the primary energy source for the body and are deemed indispensable for the proper functioning of various cell types including RTMs (Nieman and Mitmesser 2017). Nevertheless, their excessive intake has been associated with pro-inflammatory cytokine secretion—e.g. TLR-4, plasma C-reactive protein (CRP), IL-6, TNF-α, monocyte chemotactic protein 1 (MCP-1), E-selectin, and plasminogen activator inhibitor 1 (PAI-1)—which in turn favors a low-grade chronic inflammation across tissues (Cox et al. 2011; Vasiljević et al. 2014; Charrez et al. 2015; Pavlou et al. 2018).

In a steady state, macrophages exert numerous housekeeping functions favoring FAO and OXPHOS for ATP production while in a pro-inflammatory state, macrophages need fast-released energy from aerobic glycolysis. In the former case, macrophages catabolize triglycerides (Huang et al. 2014), and generate fatty acids for β-oxidation, while producing acetyl-CoA, NADH, and $FADH_2$. Acetyl-CoA enters the TCA cycle, while NADH and $FADH_2$ contribute to ATP production through OXPHOS. In the latter case, macrophages rely on the glucose transporter (GLUT1) for glucose import within the cells (Freemerman et al. 2014). The combination of highly active glycolysis and increased glucose uptake enhances the availability of glycolytic intermediates, fulfilling the energy demand associated with an inflammatory response (Caslin et al. 2020).

When in excess, glucose can be stored in the form of glycogen, which has been shown to exhibit immunomodulatory activity (Ma et al. 2020). It does so by interacting with TLR2 to upregulate inducible nitric oxide synthase (iNOS) expression, leading to the production of nitric oxide (NO) and inflammatory cytokines including IL-6 and TNF-α.

TCA cycle metabolic intermediates act as signal transducers and have been shown to modulate macrophage function (Choi et al. 2021). An important example is the extracellular succinate, a molecule linked to the inflammatory activity of macrophages. Succinate activates the succinate receptor SUCNR1 (or GPR91), inducing IL-1β production (Tannahill et al. 2013). Conversely, macrophage-secreted succinate can interact with neural stem cell SUCNR1 to promote prostaglandin E2 (PGE2) secretion, contributing to the amelioration of chronic neuroinflammation (Minhas et al. 2021). During carcinogenesis, tumor cells secrete succinate, which in turn activates the succinate receptor and hypoxia-inducible factor 1-alpha (HIF-1α) signaling axis in tumor-associated macrophages (TAMs) (Wu et al. 2020). This activation favors immunosuppression and, as a result, facilitates tumorigenesis. Elevated succinate levels also lead to the stabilization of the transcription factor hypoxia-inducible factor 1-alpha (HIF-1α). Succinate inhibits prolyl hydroxylase domain proteins, which hydroxylate HIF-1α, and lead to its ubiquitination and proteasomal degradation (Tannahill et al. 2013). The stabilization of HIF-1α results in its binding to hypoxia response elements on target genes, thereby inducing their expression. HIF-1α regulates genes associated with angiogenesis, proliferation, inflammation, and cellular metabolism. Specifically, HIF-1α promotes the shift to glycolysis by inducing the expression of glycolytic enzymes such as hexokinase 2, glucose-6-phosphate isomerase, and pyruvate kinase M2 (molecules involved in the first, second, and tenth reactions of glycolysis, respectively). This ensures that cells can produce ATP even in oxygen-limited conditions.

Another important intermediate of the TCA cycle is the *itaconate*, which has been shown to exhibit anti-inflammatory effects in RTMs. Mechanistically, itaconate (1) halts the function of pro-inflammatory cytokines (e.g., IL-1β, IL-6 and IL-12) (Lampropoulou et al. 2016). (2) Itaconate also structurally resembles phosphoenolpyruvate and has been shown to bind to the glycolytic enzyme fructose-6-phosphate 2-kinase, thereby acting as an allosteric inhibitor of glycolysis. This results in reduced levels of fructose 2,6-bisphosphate, consequently inhibiting glycolysis in macrophages and likely contributing to itaconate's anti-inflammatory properties (Sakai et al. 2004). In addition, (3) itaconate triggers the activation of nuclear factor erythroid 2-related factor 2 (NRF2) transcription factor by alkylation of Kelch ECH associating protein 1 (KEAP1), thus limiting inflammation and type I interferon production, leading to an anti-inflammatory response. (4) Endogenous itaconate modifies L-lactate dehydrogenase A (LDHA) and annexin A1, which inhibits glycolysis and prostaglandin production, thereby resulting in inflammation amelioration. Finally, (5) itaconate disrupts succinate dehydrogenase (SDH) activity, resulting in the accumulation of succinate and the remodeling of the inflammatory response (O'Neill and Artyomov 2019; Li et al. 2023).

3.4 Dietary Lipid Intake and Macrophage Function

Fatty acids (FAs) play pivotal roles in cellular homeostasis and the overall function of the human body. They constitute key components of biological membranes (contributing to the formation of sphingolipid, phospholipid, glycolipid, and lipoprotein pools) and they serve as an energy source stored in the form of triacylglycerols. Finally, FAs give rise to essential intra- and extracellular lipid mediators and hormones. They possess extensive capabilities to modulate immune cell functions by influencing (1) cellular structure, (2) metabolism, and (3) activity, exerting their effects through surface proteins (such as G-protein-coupled receptors; GPRs), intranuclear receptors, or membrane transporters (Brown and Marnett 2011; Vannice and Rasmussen 2014; Imamura et al. 2016).

In terms of chemical structure, FAs are carboxylic acids with varying carbon and double bond configurations and can be classified into two principal groups: saturated FAs (SFAs) and unsaturated FAs (UFAs). SFAs include palmitic acid (PA; C16:0), lauric acid (C12:0), myristic acid (MA; C14:0), and stearic acid (STA; C18:0), all of which lack double bonds in their carbon backbone (Fahy et al. 2009). UFAs can be subdivided into monounsaturated FAs (MUFAs) with one double bond and polyunsaturated FAs (PUFAs) with multiple double bonds. MUFAs encompass FAs such as n-9 oleic acid (OA; C18:1) while the PUFA family includes α-Linolenic acid (ALA; C18:3), linoleic acid (LA; C18:2), as well as their long-chain derivatives (Vannice and Rasmussen 2014).

While synthesized within the cells, certain FAs have to be supplemented through dietary sources. In particular, meat, dairy, palm, and coconut oil products are primary sources of SFAs, while nuts, fish, and vegetable oils constitute essential sources of UFAs. That is due to the fact that mammals lack the enzymatic machinery required to introduce double bonds beyond C-9 in the FA chain. For this reason, mammals cannot synthesize linoleic acid (LA), an n-6 PUFA derived from meat, or α-Linolenic acid (ALA), an n-3 PUFA derived from vegetable oils. LA and ALA are in fact deemed essential and must be obtained through diet, serving as precursors for the synthesis of various other UFAs (Whelan and Rust 2006). Long-chain derivatives of essential PUFAs, such as n-6 PUFAs γ-linolenic acid (GLA; C18:3) and arachidonic acid (AA; C20:4), as well as n-3 PUFAs docosahexaenoic acid (DHA; C22:6) and eicosapentaenoic acid (EPA; C20:5), are also crucial for diverse physiological functions.

The dysregulated balance between SFAs and UFAs as well as the compromised ratio of n-6/n-3 PUFAs carries substantial implications for immune system homeostasis. It can contribute to the onset of numerous allergic, autoimmune, and metabolic diseases. Examining this dysregulated balance from a dietary perspective, it can happen through elevated consumption of SFAs (derived from processed foods) which has been linked to elevated risks of atherosclerosis, coronary heart disease, obesity, and metabolic syndrome (extensively reviewed in Micha and Mozaffarian 2010). The impact of UFAs on the development of immune-related and metabolic diseases varies, with outcomes depending on factors such as the balance between n-6 and n-3 polyunsaturated FAs, dietary diversity, underlying genetics, and environmental factors such as the microbiome.

FAs are potent modulators of macrophage functions. Notably, PUFAs such as DHA, favor the anti-inflammatory potency of the cells. DHA acts on the GPR120 receptor and drives the secretion of anti-inflammatory IL-10, while macrophages maintain their phagocytic nature (Oh et al. 2010; Oliver et al. 2012). In contrast to DHA, saturated FAs, such as palmitic acid induce the secretion of pro-inflammatory IL-18 and TNF-α cytokines (Samblas et al. 2019).

In conditions of energetic excess, FAs predominantly accumulated in the adipose tissue (Chait and den Hartigh 2020). For this reason, adipose RTMs have been the subject of intense study. Nutritional regimes high in PUFAs (DHA and EPA, fish oil) reduce the population of adipose RTMs while favoring an anti-inflammatory shift (i.e., increased expression of IL-10, arginase, YM-1, CLEC7a, and MMR). *In vitro* studies using RAW264.7 cells suggest that n-3 (DHA) and n-6 AA PUFAs enhance the phagocytic and microbicidal properties of the cells (Adolph et al. 2012), implying a translational significance of PUFA supplementation in the case of immunocompromised patients. *In vivo* studies of n-3 PUFA-rich diets confirm enhanced phagocytosis and decreased apoptosis of alveolar macrophages (Saini et al. 2013). Notably,

DHA and EPA exhibit a striking capacity to modulate TLR4-mediated immune signaling, inhibiting lipopolysaccharide (LPS)-mediated cyclooxygenase-2 (COX-2) activation and subsequent prostaglandin release (Lee et al. 2003; Honda et al. 2015). On the other hand, SFAs (such as PA) bind on macrophage TLR4 and trigger inflammation through the nuclear factor-κB (NF-κB) and inflammasome pathway (Suganami et al. 2007). PA-induced TLR2 activation drives NLRP3 inflammasome activation and the secretion of IL-1β in the THP-1 cell line (Snodgrass et al. 2013).

Apart from the FAs originating from dietary catabolism, *de novo* lipid synthesis in macrophages is critical and has been reported to induce a pro-inflammatory phenotype. Fatty Acid Synthetase (FASN), a key enzyme of lipid anabolism (Gonzalez-Bohorquez et al. 2022), utilizes acetyl-CoA and malonyl-CoA to produce palmitate acid (C16:0), stearic acid (C18:0), myristic acid (C14:0) and other shorter FAs (Fhu and Ali 2020). Reports suggest that FASN plays a critical role in promoting a pro-inflammatory status in macrophages (Xiong et al. 2021). Elevated expression of *Fasn* significantly enhances inflammation in BM-derived macrophages, while its deficiency suffices to inhibit the inflammatory phenotype (Carroll et al. 2018). Mechanistic insight suggests that *Fasn* loss-of-function disrupts JNK signaling—a major mediator of the pro-inflammatory switch in RTMs (Wei et al. 2016). FASN is also involved in regulating NLRP-3-mediated caspase-1 activation, and ROS production in macrophages (Moon et al. 2015). Finally, the FASN enzyme has been reported to facilitate the formation of oxidized LDL-induced macrophage foam cells, thereby triggering the pro-inflammatory nature of macrophages and contributing to the pathogenesis of atherosclerosis (Kumar et al. 2020).

Cholesterol, an important lipid that functions as a precursor for steroid hormones and bile acids rather than a substrate for energy production, serves as a structural element in cellular membranes and significantly influences RTM biology (Luo et al. 2020). In its esterified form, cholesterol constitutes a primary component of cellular membranes and plays a crucial role as a structural constituent in lipid droplets, thereby supporting cytoprotection and cell proliferation (Ikonen 2008). However, it is important to note that free cholesterol has been observed to induce cytotoxic effects on hepatocytes (Gan et al. 2014) and KCs (Repa et al. 2000). Mechanistically, LXRs stimulate the expression of ABCA1 and apolipoprotein E, which subsequently promote the absorption of cholesterol in macrophages. In a modified lipoprotein form, cholesterol has been demonstrated to bind to scavenger receptors (SR) and CD36 transporter on KCs, thereby enhancing the inflammatory response associated with non-alcoholic steatohepatitis (NASH) (Bieghs et al. 2012). Additionally, cholesterol can be stored in lysosomes of KCs and hepatocytes, generating crystal precipitates characteristic of NASH histology, and contributing to inflammasome activation (Ioannou et al. 2013). The presence of free cholesterol activates the pro-inflammatory characteristics of KCs and hepatic stellate cells, exacerbating steatohepatitis and worsening fibrosis (Teratani et al. 2012). Consequently, the pharmacological reduction of free cholesterol using atorvastatin in mice has been associated with decreased inflammation and liver damage, without altering levels of FFAs (Van Rooyen et al. 2013).

3.5 Protein Intake and Macrophage Function

Dietary amino acid intake holds an important role in immune cell ATP generation, nucleotide synthesis, and maintenance of redox balance, thereby regulating their metabolic status in healthy and pathogenic conditions (Kelly and Pearce 2020). In mechanistic terms, amino acid availability controls the mammalian Target of Rapamycin (mTOR) signaling and nitric oxide (NO) production in RTMs (Rath et al. 2014). Protein malnutrition, on the other hand, has potentially detrimental effects on host immunity with many amino acid deficiencies being etiologically associated with compromised immunological status.

Glutamine—the most abundant amino acid in the human plasma—has been extensively examined for its immunomodulatory effects in various pathologies, especially in inflammatory bowel disease (IBD) (Severo et al. 2021). It participates in the immunoregulation of intestinal RTMs inhibiting their pro-inflammatory signature (Ryan and O'Neill 2017). Additional evidence suggests that the exogenous *in vivo* administration of glutamine exhibits protective anti-fibrotic effects in multiple organs, including the liver (Shrestha et al. 2016). Of note, Glutamine can undergo glutaminolysis, the enzymatic conversion of glutamine to α-ketoglutarate or fumarate, both of which exert anti-inflammatory functions in macrophages (Arts et al. 2016; Liu et al. 2017). In anti-inflammatory macrophages, α-ketoglutarate generated from glutaminolysis is essential for OXPHOS and FAO through JMJD3, a key enzyme for H3K27 demethylation. At the same time, the effect of glutamine has not been fully assessed in human IBD patients who are characterized by a significant intestinal fibrosis index. Recent observations propose a potential dual role for glutamine, as it may also contribute to lung fibrosis by promoting the production of proline and glycine, the two principal amino acids constituting collagen protein (Hamanaka et al. 2019). When examining fibrosis, a more comprehensive understanding of the interplay among glutamine, macrophages, and fibroblasts is imperative for a thorough evaluation of glutamine supplementation from a therapeutic standpoint.

Arginine metabolism has emerged as a critical hub in innate and adaptive immunity. This amino acid is regarded as indispensable for embryonic survival, fetal development, and hemodynamics (Yeramian et al. 2006; Yurdagul et al. 2020). In homeostatic steady-state conditions, the cells employ the SLC7A6 and SLC7A7 transporters for arginine uptake (Yeramian et al. 2006) and arginine uptake has been shown to protect against intestinal oxidative stress and inflammation, thus protecting mucosal equilibrium. When responding to apoptotic cells within the tissue, RTMs rely on arginine uptake, which they catabolize with the help of arginase 1. Consequently, RTMs activate the actin regulatory protein RAC1 in a putrescine-dependent manner, which then further assists apoptotic cell internalization, ensuring a supply of metabolites for the macrophage (Yurdagul et al. 2020). In pathological conditions, when macrophages need to favor a pro-inflammatory phenotype, they uptake arginine through the SLC7A2 transporter and activate the NF-κB pathway which renders them capable of converting the arginine to nitric oxide (NO) and citrulline via inducible NO synthase (iNOS) activity. NO in turn increases macrophage cytotoxicity, which proves to be essential when combating pathogenic antigens. During inflammation, NOS1–3 and arginase 1–2 constitute the central arginine-catabolizing enzymes. Strikingly, the regulated oscillations in the transcription levels of the aforementioned enzymes can significantly shape the immune response. Of note, NOS2 and arginase 1 are also crucial in shaping the function of specific DC subpopulations. Arginine depletion also dampens IFN-γ production and proliferation of human NK cells (Kwak et al. 2015). TAMs expressing arginase are associated with pro-tumor growth, whereas TAMs with low arginase expression drive tumor regression by promoting macrophage and NK cell-mediated anti-tumor immunity (Hagemann et al. 2008).

Dietary *serine* has been shown to facilitate glycolytic flux, leading to a pro-inflammatory state. It allosterically activates pyruvate kinase M2 (PKM2) within RTMs (Chaneton et al. 2012). This activation facilitates the conversion of phosphoenolpyruvate to pyruvate in the final stage of glycolysis, similar to the effect of LPS. As a result, this glycolytic shift leads to elevated expression of IL-1β (Palsson-McDermott et al. 2015).

Finally, *tryptophan* (Trp) is considered essential because humans cannot synthesize it, and has to be received from dietary sources. Trp is essential for the immune system function mostly due to its contribution to protein synthesis. It has been involved in the kynurenine (Kyn) metabolic pathway and has been highlighted for its association with the biosynthesis of nicotinamide adenine dinucleotide (NAD) and kynurenic acid. Enzymes like indoleamine 2,3-dioxygenase (IDO and IDO2) and tryptophan 2,3-dioxygenase (TDO) play crucial roles in these pathways. IDO is particularly induced by inflammatory stimuli (such as interferons) and forms a key initiator during inflammation. In terms of gene expression, IFN-γ is recruited to the promoter region of *Ido* and regulates its expression in various cell types, most notably DCs and macrophages. *IDO* expression is of crucial clinical relevance in cases of tumors, infections, auto-immune diseases, and cardiovascular complications. Especially in the case of cancer, Mulder and colleagues highlighted the presence of $IL4I1^{+}PD\text{-}L1^{+}IDO1^{+}$ TAMs in the majority of human tumors exerting anti-inflammatory features (Mulder et al. 2021).

3.6 Dietary Patterns Affect Gastrointestinal Microbiome and RTM Fate

Numerous studies highlight the key role of the gastrointestinal (GI) tract microbiota in shaping local and systemic immunity (Rakoff-Nahoum and Medzhitov 2006; Li et al. 2008; Belkaid and Hand 2014). GI-residing microorganisms secrete various metabolites able to rewire RTM metabolism, thereby safeguarding host immunity but also contributing to the onset of inflammatory disorders (Michaudel and Sokol 2020). The specific nature of such metabolites is influenced by two key factors: the dynamic "microbiome signature" (different species produce distinct arrays of metabolites as products of their catabolic pathways), and the type of "host lifestyle" (certain dietary patterns favor the growth of particular microorganisms, leading to the secretion of distinct metabolite spectra) (Penders et al. 2005; Turnbaugh et al. 2009).

Residing in barrier microenvironments, RTMs are not only exposed to signals from the tissue of residence but also to signals originating from commensal microorganisms. They are capable of sensing microbial-derived products, and, by nature, serve as key players in the dialogue between the host and commensals.

Studies suggest that the microbiome can influence both pre-natal and adult microglia (embryonically "seeded" macrophages with key roles in brain development and homeostasis) by secreting short-chain fatty acids (SCFAs) (Thion et al. 2018; Erny et al. 2021). In steady-state conditions, butyrate—a byproduct of microbiota-mediated SCFA fermentation—regulates the control of GI balance. Similar to other SCFAs, butyrate freely diffuses into cells (including intestinal epithelial cells, but also RTMs) and potentially acts as a substrate for the TCA cycle. As part of this pathway, butyrate directs macrophage metabolism towards FAO (instead of the pro-inflammatory glycolytic pathways), thereby promoting the anti-inflammatory state (Ji et al. 2016). On that note, butyrate enhances the production of antimicrobial peptides by macrophages further reducing inflammatory potential (Meijer et al. 2010). Antibiotic treatment, on the other hand, diminishes the microbiota-mediated butyrate production, thereby polarizing intestinal macrophages toward the pro-inflammatory fate (Parada Venegas et al. 2019). This dynamic activation state of RTMs represents a butyrate-dependent adaptive mechanism that enables the control of pathogenic enteric bacteria (limiting the potentially harmful pro-inflammatory cytokines associated with classical macrophage activation).

In the case of "diet-induced dysbiosis"—a term that is used to describe GI microbiome imbalance resulting from nutrition—the altered microbiota signature affects tissue function and contributes to systemic inflammation. This effect may disrupt the homeostasis of various metabolic organs, including the adipose tissue and the liver. The disrupted microbiome can affect glycemic control, perturb the levels of gut-secreted hormones such as glucagon-like peptide 1 (GLP-1) and peptide YY (PYY), and promote the circulation of LPS and microbial metabolites, all of which contribute to systemic inflammation (Brial et al. 2018). In nutritional regimes that resemble the eating habits of "Western countries" (elevated saturated fat, cholesterol, and carbohydrate concentration), there is an observed impairment in the GI barrier integrity, hyperlipidemia, and hypercholesterolemia. Due to the direct connection between the liver and the GI tract through the portal vein system, liver RTMs are the first responders to the gut-derived signals. Such signals can lead to an elevated flux of microbial-derived innate immune activators and modified dietary lipids through the portal vein to the liver, resulting in the activation of KCs and the release of pro-inflammatory factors. Consequently, an inflammatory response is initiated, recruiting BM-derived $CCR2^+$ inflammatory monocytes, neutrophils, and cells of the adaptive immune system. The activation of cells further propagates inflammation, potentially disturbing liver function through the activation of stellate cells, increased collagen release, and the formation of hepatic fibrosis (Talukdar et al. 2012; Koyama and Brenner 2017).

3.7 "Quantifying" Immunometabolism

From the previous sections, one appreciates the need to address how immune cell metabolism is mechanistically affected by certain dietary patterns (Munteanu and Schwartz 2022). For this reason, the evaluation of how particular nutritional regimes impact macrophage biology is actively explored through various nutrition models in murine and human cohorts. These dietary models vary from restricted (hypocaloric diets/functional amino-acid restriction) to excessive (hypercaloric carbohydrate-rich, fat-rich, cholesterol-rich regime western diet) macronutrient concentration in order to highlight the respective effects on RTMs (Orillion et al. 2018; Jordan et al. 2019; Christ et al. 2019).

Despite the significant progress in the field of nutritional models for preclinical research, developing the appropriate methodological *readout* to analyze the metabolic signature of immune cell populations has historically presented challenges. For this reason, the significant advancements in mass-spectrometry-based metabolomic profiling (extensively analyzed in Buescher et al. 2015; Jang et al. 2018; Yuan et al. 2019; Rinschen et al. 2019) as well as the progress in flow cytometry-based and high-throughput methodologies have revolutionized the field of immunometabolism (Fig. 3.3). This chapter provides examples of "quantifying immunometabolism" from the perspective of flow cytometry and transcriptomics (while in-depth mass-spectrometry metabolomics approaches are available elsewhere, Fig. 3.3c, O'Neill et al. 2016; Singer and Chandel 2019; Hartmann and Bendall 2020; Pålsson-McDermott and O'Neill 2020; Voss et al. 2021).

3.7.1 *Flow-Cytometry*

Flow cytometry is a widely utilized technique for quantifying metabolic parameters in immune cell populations, offering precision in the detection and quantification of intracellular and cell surface proteins (Fig. 3.3a). It has been shown to provide reliable insight into glucose (Palmer et al. 2016), lipid (Pacella et al. 2018), as well as cysteine uptake (Siska et al. 2016) within immune cells, both in steady state and inflammatory contexts. Cytometry by time of flight (CyTOF) mass spectrometry is a revolutionary tool that quantifies not only cell surface markers but also up to 40 intracellular metabolically relevant proteins and cytokine levels. It can also quantify fluctuations in the epigenetic level by providing insight into chromatin acetylation

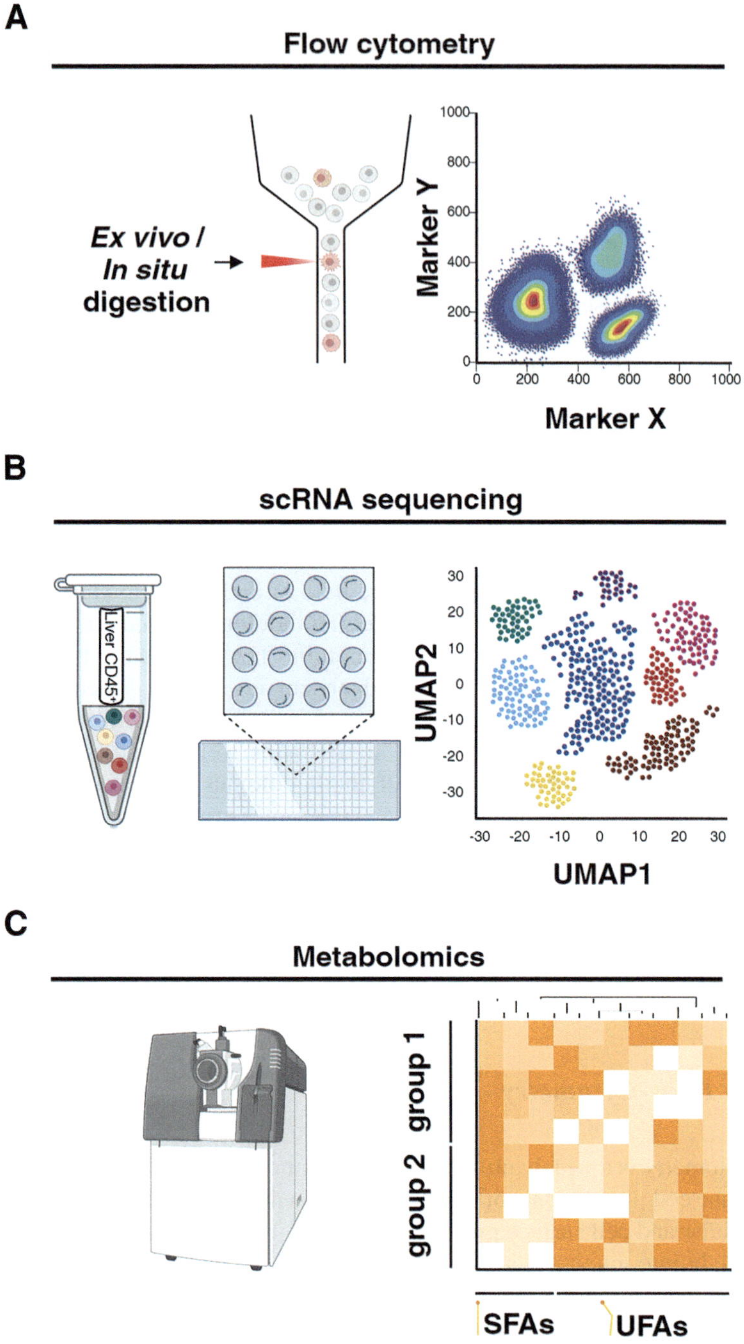

Fig. 3.3 An overview of currently utilized methodologies to address the immunometabolic profile in populations of interest. The function of RTMs can be addressed in terms of number fluctuations (**a**), transcriptomic profiling (**b**), and metabolic profiling (**c**). Notably, the synergistic use of these applications offers unique insight into the phenotype and function of RTMs

modifiers (Yucel et al. 2019). CyTOF workflow provides insight into the metabolic (Beckermann et al. 2020; Levine et al. 2021; Hartmann et al. 2021) and effector states (Subrahmanyam and Maecker 2017; Hartmann and Bendall 2020) of immune cells. However, the limited number of identified markers and metabolic heterogeneity in immune cells pose challenges. These limitations are addressed by innovative approaches like single-cell metabolic regulome profiling (scMEP) and Met-Flow (Ahl et al. 2020; Hartmann et al. 2021). In detail, Hartmann et al. developed a mass-cytometry-based workflow called single-cell metabolic regulome profiling (scMEP) which classifies cell phenotypes based on the levels of central metabolic regulators. In addition, Met-Flow determines the metabolic state of a cell based on the expression levels of rate-limiting enzymes of principal metabolic pathways. Met-Flow accurately quantifies numerous proteins such as GLUT1 and hexokinase 1, the elevated levels of which suggest increased glycolytic activity in the respective immune cells of interest. Notably, Met-Flow also quantifies glucose-6-phosphate dehydrogenase (G6PD) levels, whose increase suggests an activation in the oxidative pentose phosphate pathway. In parallel with these attempts, given that the cell translation rate is indicative of the energetic status, "protein synthesis" kinetics have been successfully utilized as a means to classify immune populations. Argüello et al. developed a novel approach termed "single-cell energetic metabolism by profiling translation inhibition" (SCENITH), which quantifies protein translation with puromycin staining as a surrogate to assess metabolic changes in heterogeneous cell populations at single-cell resolution (Argüello et al. 2020). Finally, the flow cytometry readouts can be made more comprehensive by the inclusion of markers for DNA and RNA alongside proteins (Kimmey et al. 2019).

3.7.2 Single-Cell Transcriptomic Profiling

Quantifying the levels of metabolites and metabolism-related enzymes is essential for understanding the metabolic status of immune cells. However, this can be challenging due to the unstable nature of certain metabolites and enzymes. Overcoming these limitations, recent technological and biocomputational advancements have propelled single-cell RNA sequencing (scRNA-seq) into a revolutionary tool for immunometabolic profiling. This approach provides a precise view of cell population transcriptomic heterogeneity by profiling thousands of cells per sample, accurately detecting two to four thousand transcripts per cell, and enabling a direct examination of transcriptomic activity (Fig. 3.3b).

Numerous examples underscore the effectiveness of scRNA-seq in identifying new metabolic targets that modulate immune cell responses in conditions of altered metabolism (e.g., cancer, non-alcoholic fatty liver disease, and others). For instance, in murine sarcomas, macrophages boost *Glut1* expression, thereby increasing their glycolytic activity when the host is subjected to immune checkpoint blockade antibodies (Ma et al. 2020). In the same model, macrophage oxidative stress response is enriched (as suggested by activated *Nrf2* expression). Parallel studies indicate that leptin administration (an adipokine with metabolic effects) (Pérez-Pérez et al.

2017), enhances basal oxygen consumption and mitochondrial reserve, thereby increasing the inflammatory capacity and antitumor effector functions of T cells. Moreover, single-cell analysis of tumor-infiltrating $CD45^+$ lymphocytes from mice fed a control diet and an HFD demonstrated that a high-fat diet reprograms tumor metabolism, particularly altering fat utilization between tumor and $CD8^+$ T cells (Ringel et al. 2020).

3.8 Concluding Remarks

Over the years, seminal studies have provided evidence demonstrating that RTMs possess a unique potential for metabolic adaptability. In parallel to their reaction to pathogens, macrophages are required to respond to nutrient and metabolite oscillations of their unique microenvironments to safeguard homeostasis. For this reason, an exploration of their metabolic features becomes crucial in order to address their broader contribution to tissue physiology.

The field of immunometabolism provides distinctive insights, coupled with significant clinical relevance to these cells in the context of human health and disease. The definition and exploration of the distinctive metabolic features of immune cells hold the potential to uncover novel therapeutic targets (Lercher et al. 2020). This ongoing investigation into immunometabolism not only enhances our understanding of immune system functionality but also opens avenues for innovative therapeutic interventions.

Acknowledgements The authors acknowledge the support of the French Agence Nationale de la Recherche (ANR), under grant ANR-23-CE14-0072-02 (project MIAM). All figures were created with Biorender.com.

Author Contributions G.A. wrote the manuscript with the input of all co-authors. All authors reviewed and approved the final version of the manuscript.

References

Adolph S, Fuhrmann H, Schumann J (2012) Unsaturated fatty acids promote the phagocytosis of P. aeruginosa and R. equi by RAW264.7 Bieghs macrophages. Curr Microbiol 65(6):649–655

Ahl PJ, Hopkins RA, Xiang WW, Au B, Kaliaperumal N, Fairhurst AM et al (2020) Met-flow, a strategy for single-cell metabolic analysis highlights dynamic changes in immune subpopulations. Commun Biol 3(1):305

Ajami B, Bennett JL, Krieger C, Tetzlaff W, Rossi FM (2007) Local self-renewal can sustain CNS microglia maintenance and function throughout adult life. Nat Neurosci 10(12):1538–1543

Argüello RJ, Combes AJ, Char R, Gigan JP, Baaziz AI, Bousiquot E et al (2020) SCENITH: a flow cytometry-based method to functionally profile energy metabolism with single-cell resolution. Cell Metab 32(6):1063–1075.e7

Arts RJ, Novakovic B, Ter Horst R, Carvalho A, Bekkering S, Lachmandas E et al (2016) Glutaminolysis and fumarate accumulation integrate immunometabolic and epigenetic programs in trained immunity. Cell Metab 24(6):807–819

Bain CC, Bravo-Blas A, Scott CL, Perdiguero EG, Geissmann F, Henri S et al (2014) Constant replenishment from circulating monocytes maintains the macrophage pool in the intestine of adult mice. Nat Immunol 15(10):929–937

Baker AD, Malur A, Barna BP, Ghosh S, Kavuru MS, Malur AG et al (2010) Targeted PPAR{gamma} deficiency in alveolar macrophages disrupts surfactant catabolism. J Lipid Res 51(6):1325–1331

Beckermann KE, Hongo R, Ye X, Young K, Carbonell K, Healey DCC et al (2020) CD28 costimulation drives tumor-infiltrating T cell glycolysis to promote inflammation. JCI Insight 5(16):e138729

Belkaid Y, Hand TW (2014) Role of the microbiota in immunity and inflammation. Cell 157(1):121–141

Bieghs V, Verheyen F, van Gorp PJ, Hendrikx T, Wouters K, Lütjohann D et al (2012) Internalization of modified lipids by CD36 and SR-A leads to hepatic inflammation and lysosomal cholesterol storage in Kupffer cells. PLoS One 7(3):e34378

Blériot C, Chakarov S, Ginhoux F (2020) Determinants of resident tissue macrophage identity and function. Immunity 52(6):957–970

Blériot C, Barreby E, Dunsmore G, Ballaire R, Chakarov S, Ficht X et al (2021) A subset of Kupffer cells regulates metabolism through the expression of CD36. Immunity 54(9):2101–2116.e6

Brial F, Le Lay A, Dumas ME, Gauguier D (2018) Implication of gut microbiota metabolites in cardiovascular and metabolic diseases. Cell Mol Life Sci 75(21):3977–3990

Brown HA, Marnett LJ (2011) Introduction to lipid biochemistry, metabolism, and signaling. Chem Rev 111(10):5817–5820

Buck MD, Sowell RT, Kaech SM, Pearce EL (2017) Metabolic instruction of immunity. Cell 169(4):570–586

Buescher JM, Antoniewicz MR, Boros LG, Burgess SC, Brunengraber H, Clish CB et al (2015) A roadmap for interpreting (13)C metabolite labeling patterns from cells. Curr Opin Biotechnol 34:189–201

Calderon B, Carrero JA, Ferris ST, Sojka DK, Moore L, Epelman S et al (2015) The pancreas anatomy conditions the origin and properties of resident macrophages. J Exp Med 212(10):1497–1512

Carroll RG, Zasłona Z, Galván-Peña S, Koppe EL, Sévin DC, Angiari S et al (2018) An unexpected link between fatty acid synthase and cholesterol synthesis in proinflammatory macrophage activation. J Biol Chem 293(15):5509–5521

Caslin HL, Bhanot M, Bolus WR, Hasty AH (2020) Adipose tissue macrophages: unique polarization and bioenergetics in obesity. Immunol Rev 295(1):101–113

Chait A, den Hartigh LJ (2020) Adipose tissue distribution, inflammation and its metabolic consequences, including diabetes and cardiovascular disease. Front Cardiovasc Med 7:22

Chaneton B, Hillmann P, Zheng L, Martin ACL, Maddocks ODK, Chokkathukalam A et al (2012) Serine is a natural ligand and allosteric activator of pyruvate kinase M2. Nature 491(7424):458–462

Charrez B, Qiao L, Hebbard L (2015) The role of fructose in metabolism and cancer. Horm Mol Biol Clin Investig 22(2):79–89

Choi I, Son H, Baek JH (2021) Tricarboxylic acid (TCA) cycle intermediates: regulators of immune responses. Life (Basel) 11(1):69

Christ A, Lauterbach M, Latz E (2019) Western diet and the immune system: an inflammatory connection. Immunity 51(5):794–811

Cox CL, Stanhope KL, Schwarz JM, Graham JL, Hatcher B, Griffen SC et al (2011) Circulating concentrations of monocyte chemoattractant protein-1, plasminogen activator inhibitor-1, and soluble leukocyte adhesion molecule-1 in overweight/obese men and women consuming fructose- or glucose-sweetened beverages for 10 weeks. J Clin Endocrinol Metab 96(12):E2034–E2038

Davies LC, Rice CM, Palmieri EM, Taylor PR, Kuhns DB, McVicar DW (2017) Peritoneal tissue-resident macrophages are metabolically poised to engage microbes using tissue-niche fuels. Nat Commun 8(1):2074

Demetz E, Tymoszuk P, Hilbe R, Volani C, Haschka D, Heim C et al (2020) The haemochromatosis gene Hfe and Kupffer cells control LDL cholesterol homeostasis and impact on atherosclerosis development. Eur Heart J 41(40):3949–3959

Diehl KL, Vorac J, Hofmann K, Meiser P, Unterweger I, Kuerschner L et al (2020) Kupffer cells sense free fatty acids and regulate hepatic lipid metabolism in high-fat diet and inflammation. Cells 9(10):2258

Donath MY, Böni-Schnetzler M, Ellingsgaard H, Halban PA, Ehses JA (2010) Cytokine production by islets in health and diabetes: cellular origin, regulation and function. Trends Endocrinol Metab 21(5):261–267

Eguchi K, Manabe I, Oishi-Tanaka Y, Ohsugi M, Kono N, Ogata F et al (2012) Saturated fatty acid and TLR signaling link β cell dysfunction and islet inflammation. Cell Metab 15(4):518–533

Ehses JA, Perren A, Eppler E, Ribaux P, Pospisilik JA, Maor-Cahn R et al (2007) Increased number of islet-associated macrophages in type 2 diabetes. Diabetes 56(9):2356–2370

Epelman S, Lavine KJ, Beaudin AE, Sojka DK, Carrero JA, Calderon B et al (2014) Embryonic and adult-derived resident cardiac macrophages are maintained through distinct mechanisms at steady state and during inflammation. Immunity 40(1):91–104

Erny D, Dokalis N, Mezö C, Castoldi A, Mossad O, Staszewski O et al (2021) Microbiota-derived acetate enables the metabolic fitness of the brain innate immune system during health and disease. Cell Metab 33(11):2260–2276.e7

Fahy E, Subramaniam S, Murphy RC, Nishijima M, Raetz CR, Shimizu T et al (2009) Update of the LIPID MAPS comprehensive classification system for lipids. J Lipid Res 50(Suppl):S9–S14

Faubert B, Solmonson A, DeBerardinis RJ (2020) Metabolic reprogramming and cancer progression. Science 368(6487):eaaw5473

Fhu CW, Ali A (2020) Fatty acid synthase: an emerging target in cancer. Molecules 25(17):3935

Freemerman AJ, Johnson AR, Sacks GN, Milner JJ, Kirk EL, Troester MA et al (2014) Metabolic reprogramming of macrophages: glucose transporter 1 (GLUT1)-mediated glucose metabolism drives a proinflammatory phenotype. J Biol Chem 289(11):7884–7896

Gan LT, Van Rooyen DM, Koina ME, McCuskey RS, Teoh NC, Farrell GC (2014) Hepatocyte free cholesterol lipotoxicity results from JNK1-mediated mitochondrial injury and is HMGB1 and TLR4-dependent. J Hepatol 61(6):1376–1384

Gautier EL, Shay T, Miller J, Greter M, Jakubzick C, Ivanov S et al (2012) Gene-expression profiles and transcriptional regulatory pathways that underlie the identity and diversity of mouse tissue macrophages. Nat Immunol 13(11):1118–1128

Gautier EL, Ivanov S, Williams JW, Huang SC, Marcelin G, Fairfax K et al (2014) Gata6 regulates aspartoacylase expression in resident peritoneal macrophages and controls their survival. J Exp Med 211(8):1525–1531

Ginhoux F, Guilliams M (2016) Tissue-resident macrophage ontogeny and homeostasis. Immunity 44(3):439–449

Ginhoux F, Greter M, Leboeuf M, Nandi S, See P, Gokhan S et al (2010) Fate mapping analysis reveals that adult microglia derive from primitive macrophages. Science 330(6005):841–845

Gonzalez-Bohorquez D, Gallego López IM, Jaeger BN, Pfammatter S, Bowers M, Semenkovich CF et al (2022) FASN-dependent de novo lipogenesis is required for brain development. Proc Natl Acad Sci USA 119(2):e2112040119

Guilliams M, De Kleer I, Henri S, Post S, Vanhoutte L, De Prijck S et al (2013) Alveolar macrophages develop from fetal monocytes that differentiate into long-lived cells in the first week of life via GM-CSF. J Exp Med 210(10):1977–1992

Hagemann T, Lawrence T, McNeish I, Charles KA, Kulbe H, Thompson RG et al (2008) "Re-educating" tumor-associated macrophages by targeting NF-kappaB. J Exp Med 205(6):1261–1268

Hamanaka RB, O'Leary EM, Witt LJ, Tian Y, Gökalp GA, Meliton AY et al (2019) Glutamine metabolism is required for collagen protein synthesis in lung fibroblasts. Am J Respir Cell Mol Biol 61(5):597–606

Hartmann FJ, Bendall SC (2020) Immune monitoring using mass cytometry and related high-dimensional imaging approaches. Nat Rev Rheumatol 16(2):87–99

Hartmann FJ, Mrdjen D, McCaffrey E, Glass DR, Greenwald NF, Bharadwaj A et al (2021) Single-cell metabolic profiling of human cytotoxic T cells. Nat Biotechnol 39(2):186–197

Hashimoto D, Chow A, Noizat C, Teo P, Beasley MB, Leboeuf M et al (2013) Tissue-resident macrophages self-maintain locally throughout adult life with minimal contribution from circulating monocytes. Immunity 38(4):792–804

Hoeffel G, Wang Y, Greter M, See P, Teo P, Malleret B et al (2012) Adult Langerhans cells derive predominantly from embryonic fetal liver monocytes with a minor contribution of yolk sac-derived macrophages. J Exp Med 209(6):1167–1181

Honda KL, Lamon-Fava S, Matthan NR, Wu D, Lichtenstein AH (2015) EPA and DHA exposure alters the inflammatory response but not the surface expression of toll-like receptor 4 in macrophages. Lipids 50(2):121–129

Huang SC, Everts B, Ivanova Y, O'Sullivan D, Nascimento M, Smith AM et al (2014) Cell-intrinsic lysosomal lipolysis is essential for alternative activation of macrophages. Nat Immunol 15(9):846–855

Ikonen E (2008) Cellular cholesterol trafficking and compartmentalization. Nat Rev Mol Cell Biol 9(2):125–138

Imamura F, Micha R, Wu JH, de Oliveira Otto MC, Otite FO, Abioye AI et al (2016) Effects of saturated fat, polyunsaturated fat, monounsaturated fat, and carbohydrate on glucose-insulin homeostasis: a systematic review and meta-analysis of randomised controlled feeding trials. PLoS Med 13(7):e1002087

Ioannou GN, Haigh WG, Thorning D, Savard C (2013) Hepatic cholesterol crystals and crown-like structures distinguish NASH from simple steatosis. J Lipid Res 54(5):1326–1334

Jakubzick C, Gautier EL, Gibbings SL, Sojka DK, Schlitzer A, Johnson TE et al (2013) Minimal differentiation of classical monocytes as they survey steady-state tissues and transport antigen to lymph nodes. Immunity 39(3):599–610

Jang C, Chen L, Rabinowitz JD (2018) Metabolomics and isotope tracing. Cell 173(4):822–837

Ji J, Shu D, Zheng M, Wang J, Luo C, Wang Y et al (2016) Microbial metabolite butyrate facilitates M2 macrophage polarization and function. Sci Rep 6:24838

Jordan S, Tung N, Casanova-Acebes M, Chang C, Cantoni C, Zhang D et al (2019) Dietary intake regulates the circulating inflammatory monocyte pool. Cell 178(5):1102–1114.e17

Kang K, Reilly SM, Karabacak V, Gangl MR, Fitzgerald K, Hatano B et al (2008) Adipocyte-derived Th2 cytokines and myeloid PPARdelta regulate macrophage polarization and insulin sensitivity. Cell Metab 7(6):485–495

Kelly B, Pearce EL (2020) Amino assets: how amino acids support immunity. Cell Metab 32(2):154–175

Keshvari S, Caruso M, Teakle N, Batoon L, Sehgal A, Patkar OL et al (2021) CSF1R-dependent macrophages control postnatal somatic growth and organ maturation. PLoS Genet 17(6):e1009605

Kimmey SC, Borges L, Baskar R, Bendall SC (2019) Parallel analysis of tri-molecular biosynthesis with cell identity and function in single cells. Nat Commun 10(1):1185

Koyama Y, Brenner DA (2017) Liver inflammation and fibrosis. J Clin Invest 127(1):55–64

Kumar A, Gupta P, Rana M, Chandra T, Dikshit M, Barthwal MK (2020) Role of pyruvate kinase M2 in oxidized LDL-induced macrophage foam cell formation and inflammation. J Lipid Res 61(3):351–364

Kwak Y, Kim HE, Park SG (2015) Insights into myeloid-derived suppressor cells in inflammatory diseases. Arch Immunol Ther Exp 63(4):269–285

Lampropoulou V, Sergushichev A, Bambouskova M, Nair S, Vincent EE, Loginicheva E et al (2016) Itaconate links inhibition of succinate dehydrogenase with macrophage metabolic remodeling and regulation of inflammation. Cell Metab 24(1):158–166

Lee JY, Plakidas A, Lee WH, Heikkinen A, Chanmugam P, Bray G et al (2003) Differential modulation of toll-like receptors by fatty acids: preferential inhibition by n-3 polyunsaturated fatty acids. J Lipid Res 44(3):479–486

Lercher A, Baazim H, Bergthaler A (2020) Systemic immunometabolism: challenges and opportunities. Immunity 53(3):496–509

Levine LS, Hiam-Galvez KJ, Marquez DM, Tenvooren I, Madden MZ, Contreras DC et al (2021) Single-cell analysis by mass cytometry reveals metabolic states of early-activated CD8+ T cells during the primary immune response. Immunity 54(4):829–844.e5

Li M, Wang B, Zhang M, Rantalainen M, Wang S, Zhou H et al (2008) Symbiotic gut microbes modulate human metabolic phenotypes. Proc Natl Acad Sci USA 105(6):2117–2122

Li Z, Zheng W, Kong W, Zeng T (2023) Itaconate: a potent macrophage immunomodulator. Inflammation 46(4):1177–1191

Liao X, Sharma N, Kapadia F, Zhou G, Lu Y, Hong H et al (2011) Krüppel-like factor 4 regulates macrophage polarization. J Clin Invest 121(7):2736–2749

Liu PS, Wang H, Li X, Chao T, Teav T, Christen S et al (2017) α-Ketoglutarate orchestrates macrophage activation through metabolic and epigenetic reprogramming. Nat Immunol 18(9):985–994

Loft A, Schmidt SF, Caratti G, Stifel U, Havelund J, Sekar R et al (2022) A macrophage-hepatocyte glucocorticoid receptor axis coordinates fasting ketogenesis. Cell Metab 34(3):473–486.e9

Lumeng CN, Bodzin JL, Saltiel AR (2007) Obesity induces a phenotypic switch in adipose tissue macrophage polarization. J Clin Invest 117(1):175–184

Luo J, Yang H, Song BL (2020) Mechanisms and regulation of cholesterol homeostasis. Nat Rev Mol Cell Biol 21(4):225–245

Ma J, Wei K, Liu J, Tang K, Zhang H, Zhu L et al (2020) Glycogen metabolism regulates macrophage-mediated acute inflammatory responses. Nat Commun 11(1):1769

Meijer K, de Vos P, Priebe MG (2010) Butyrate and other short-chain fatty acids as modulators of immunity: what relevance for health. Curr Opin Clin Nutr Metab Care 13(6):715–721

Merad M, Manz MG, Karsunky H, Wagers A, Peters W, Charo I et al (2002) Langerhans cells renew in the skin throughout life under steady-state conditions. Nat Immunol 3(12):1135–1141

Micha R, Mozaffarian D (2010) Saturated fat and cardiometabolic risk factors, coronary heart disease, stroke, and diabetes: a fresh look at the evidence. Lipids 45(10):893–905

Michaudel C, Sokol H (2020) The gut microbiota at the service of immunometabolism. Cell Metab 32(4):514–523

Minhas PS, Latif-Hernandez A, McReynolds MR, Durairaj AS, Wang Q, Rubin A et al (2021) Restoring metabolism of myeloid cells reverses cognitive decline in ageing. Nature 590(7844):122–128

Molawi K, Wolf Y, Kandalla PK, Favret J, Hagemeyer N, Frenzel K et al (2014) Progressive replacement of embryo-derived cardiac macrophages with age. J Exp Med 211(11):2151–2158

Moon JS, Lee S, Park MA, Siempos II, Haslip M, Lee PJ et al (2015) UCP2-induced fatty acid synthase promotes NLRP3 inflammasome activation during sepsis. J Clin Invest 125(2):665–680

Mulder K, Patel AA, Kong WT, Piot C, Halitzki E, Dunsmore G et al (2021) Cross-tissue single-cell landscape of human monocytes and macrophages in health and disease. Immunity 54(8):1883–1900.e5

Munteanu C, Schwartz B (2022) The relationship between nutrition and the immune system. Front Nutr 9:1082500

Newsholme P, Curi R, Gordon S, Newsholme EA (1986) Metabolism of glucose, glutamine, long-chain fatty acids and ketone bodies by murine macrophages. Biochem J 239(1):121–125

Nieman DC, Mitmesser SH (2017) Potential impact of nutrition on immune system recovery from heavy exertion: a metabolomics perspective. Nutrients 9(5):513

Nordmann TM, Dror E, Schulze F, Traub S, Berishvili E, Barbieux C et al (2017) The role of inflammation in β-cell dedifferentiation. Sci Rep 7(1):6285

Odegaard JI, Ricardo-Gonzalez RR, Goforth MH, Morel CR, Subramanian V, Mukundan L et al (2007) Macrophage-specific PPARgamma controls alternative activation and improves insulin resistance. Nature 447(7148):1116–1120

Odegaard JI, Ricardo-Gonzalez RR, Red Eagle A, Vats D, Morel CR, Goforth MH et al (2008) Alternative M2 activation of Kupffer cells by PPARdelta ameliorates obesity-induced insulin resistance. Cell Metab 7(6):496–507
Oh DY, Talukdar S, Bae EJ, Imamura T, Morinaga H, Fan W et al (2010) GPR120 is an omega-3 fatty acid receptor mediating potent anti-inflammatory and insulin-sensitizing effects. Cell 142(5):687–698
Oliver E, McGillicuddy FC, Harford KA, Reynolds CM, Phillips CM, Ferguson JF et al (2012) Docosahexaenoic acid attenuates macrophage-induced inflammation and improves insulin sensitivity in adipocytes-specific differential effects between LC n-3 PUFA. J Nutr Biochem 23(9):1192–1200
O'Neill LAJ, Artyomov MN (2019) Itaconate: the poster child of metabolic reprogramming in macrophage function. Nat Rev Immunol 19(5):273–281
O'Neill LA, Kishton RJ, Rathmell J (2016) A guide to immunometabolism for immunologists. Nat Rev Immunol 16(9):553–565
Orillion A, Damayanti NP, Shen L, Adelaiye-Ogala R, Affronti H, Elbanna M et al (2018) Dietary protein restriction reprograms tumor-associated macrophages and enhances immunotherapy. Clin Cancer Res 24(24):6383–6395
Orliaguet L, Ejlalmanesh T, Humbert A, Ballaire R, Diedisheim M, Julla JB et al (2022) Early macrophage response to obesity encompasses interferon regulatory factor 5 regulated mitochondrial architecture remodelling. Nat Commun 13(1):5089
Pacella I, Procaccini C, Focaccetti C, Miacci S, Timperi E, Faicchia D et al (2018) Fatty acid metabolism complements glycolysis in the selective regulatory T cell expansion during tumor growth. Proc Natl Acad Sci USA 115(28):E6546–E6555
Palmer CS, Anzinger JJ, Butterfield TR, McCune JM, Crowe SM (2016) A simple flow cytometric method to measure glucose uptake and glucose transporter expression for monocyte subpopulations in whole blood. J Vis Exp (114):54255
Pålsson-McDermott EM, O'Neill LAJ (2020) Targeting immunometabolism as an anti-inflammatory strategy. Cell Res 30(4):300–314
Palsson-McDermott EM, Curtis AM, Goel G, Lauterbach MA, Sheedy FJ, Gleeson LE et al (2015) Pyruvate kinase M2 regulates Hif-1α activity and IL-1β induction and is a critical determinant of the Warburg effect in LPS-activated macrophages. Cell Metab 21(1):65–80
Parada Venegas D, De la Fuente MK, Landskron G, González MJ, Quera R, Dijkstra G et al (2019) Short chain fatty acids (SCFAs)-mediated gut epithelial and immune regulation and its relevance for inflammatory bowel diseases. Front Immunol 10:277
Pavlou S, Lindsay J, Ingram R, Xu H, Chen M (2018) Sustained high glucose exposure sensitizes macrophage responses to cytokine stimuli but reduces their phagocytic activity. BMC Immunol 19(1):24
Pearce EL, Pearce EJ (2013) Metabolic pathways in immune cell activation and quiescence. Immunity 38(4):633–643
Penders J, Vink C, Driessen C, London N, Thijs C, Stobberingh EE (2005) Quantification of Bifidobacterium spp., Escherichia coli and Clostridium difficile in faecal samples of breast-fed and formula-fed infants by real-time PCR. FEMS Microbiol Lett 243(1):141–147
Pérez-Pérez A, Vilariño-García T, Fernández-Riejos P, Martín-González J, Segura-Egea JJ, Sánchez-Margalet V (2017) Role of leptin as a link between metabolism and the immune system. Cytokine Growth Factor Rev 35:71–84
Rakoff-Nahoum S, Medzhitov R (2006) Role of the innate immune system and host-commensal mutualism. Curr Top Microbiol Immunol 308:1–18
Rath M, Müller I, Kropf P, Closs EI, Munder M (2014) Metabolism via arginase or nitric oxide synthase: two competing arginine pathways in macrophages. Front Immunol 5:532
Remmerie A, Scott CL (2018) Macrophages and lipid metabolism. Cell Immunol 330:27–42
Remmerie A, Martens L, Thoné T, Castoldi A, Seurinck R, Pavie B et al (2020) Osteopontin expression identifies a subset of recruited macrophages distinct from Kupffer cells in the fatty liver. Immunity 53(3):641–657.e14

Repa JJ, Turley SD, Lobaccaro JA, Medina J, Li L, Lustig K et al (2000) Regulation of absorption and ABC1-mediated efflux of cholesterol by RXR heterodimers. Science 289(5484):1524–1529

Ringel AE, Drijvers JM, Baker GJ, Catozzi A, García-Cañaveras JC, Gassaway BM et al (2020) Obesity shapes metabolism in the tumor microenvironment to suppress anti-tumor immunity. Cell 183(7):1848–1866.e26

Rinschen MM, Ivanisevic J, Giera M, Siuzdak G (2019) Identification of bioactive metabolites using activity metabolomics. Nat Rev Mol Cell Biol 20(6):353–367

Rosen ED, Spiegelman BM (2006) Adipocytes as regulators of energy balance and glucose homeostasis. Nature 444(7121):847–853

Ryan DG, O'Neill LAJ (2017) Krebs cycle rewired for macrophage and dendritic cell effector functions. FEBS Lett 591(19):2992–3006

Saini A, Harjai K, Chhibber S (2013) Inhibitory effect of polyunsaturated fatty acids on apoptosis induced by Streptococcus pneumoniae in alveolar macrophages. Indian J Med Res 137(6):1193–1198

Sakai A, Kusumoto A, Kiso Y, Furuya E (2004) Itaconate reduces visceral fat by inhibiting fructose 2,6-bisphosphate synthesis in rat liver. Nutrition 20(11–12):997–1002

Samblas M, Carraro JC, Martínez JA, Milagro FI (2019) The regulation of inflammation-related genes after palmitic acid and DHA treatments is not mediated by DNA methylation. J Physiol Biochem 75(3):341–349

Schulz C, Gomez Perdiguero E, Chorro L, Szabo-Rogers H, Cagnard N, Kierdorf K et al (2012) A lineage of myeloid cells independent of Myb and hematopoietic stem cells. Science 336(6077):86–90

Severo JS, da Silva Barros VJ, Alves da Silva AC, Luz Parente JM, Lima MM, Moreira Lima AÂ et al (2021) Effects of glutamine supplementation on inflammatory bowel disease: a systematic review of clinical trials. Clin Nutr ESPEN 42:53–60

Shrestha N, Chand L, Han MK, Lee SO, Kim CY, Jeong YJ (2016) Glutamine inhibits CCl4 induced liver fibrosis in mice and TGF-β1 mediated epithelial-mesenchymal transition in mouse hepatocytes. Food Chem Toxicol 93:129–137

Singer BD, Chandel NS (2019) Immunometabolism of pro-repair cells. J Clin Invest 129(7):2597–2607

Siska PJ, Kim B, Ji X, Hoeksema MD, Massion PP, Beckermann KE et al (2016) Fluorescence-based measurement of cystine uptake through xCT shows requirement for ROS detoxification in activated lymphocytes. J Immunol Methods 438:51–58

Snodgrass RG, Huang S, Choi IW, Rutledge JC, Hwang DH (2013) Inflammasome-mediated secretion of IL-1β in human monocytes through TLR2 activation; modulation by dietary fatty acids. J Immunol 191(8):4337–4347

Soehnlein O, Lindbom L (2010) Phagocyte partnership during the onset and resolution of inflammation. Nat Rev Immunol 10(6):427–439

Subrahmanyam PB, Maecker HT (2017) CyTOF measurement of immunocompetence across major immune cell types. Curr Protoc Cytom 82:9.54.1–9.54.12

Suganami T, Tanimoto-Koyama K, Nishida J, Itoh M, Yuan X, Mizuarai S et al (2007) Role of the toll-like receptor 4/NF-kappaB pathway in saturated fatty acid-induced inflammatory changes in the interaction between adipocytes and macrophages. Arterioscler Thromb Vasc Biol 27(1):84–91

Sun K, Kusminski CM, Scherer PE (2011) Adipose tissue remodeling and obesity. J Clin Invest 121(6):2094–2101

Svedberg FR, Brown SL, Krauss MZ, Campbell L, Sharpe C, Clausen M et al (2019) The lung environment controls alveolar macrophage metabolism and responsiveness in type 2 inflammation. Nat Immunol 20(5):571–580

Talchai C, Xuan S, Lin HV, Sussel L, Accili D (2012) Pancreatic β cell dedifferentiation as a mechanism of diabetic β cell failure. Cell 150(6):1223–1234

Talukdar S, Oh DY, Bandyopadhyay G, Li D, Xu J, McNelis J et al (2012) Neutrophils mediate insulin resistance in mice fed a high-fat diet through secreted elastase. Nat Med 18(9):1407–1412

Tamoutounour S, Guilliams M, Montanana Sanchis F, Liu H, Terhorst D, Malosse C et al (2013) Origins and functional specialization of macrophages and of conventional and monocyte-derived dendritic cells in mouse skin. Immunity 39(5):925–938
Tannahill GM, Curtis AM, Adamik J, Palsson-McDermott EM, McGettrick AF, Goel G et al (2013) Succinate is an inflammatory signal that induces IL-1β through HIF-1α. Nature 496(7444):238–242
Teratani T, Tomita K, Suzuki T, Oshikawa T, Yokoyama H, Shimamura K et al (2012) A high-cholesterol diet exacerbates liver fibrosis in mice via accumulation of free cholesterol in hepatic stellate cells. Gastroenterology 142(1):152–164.e10
Thion MS, Low D, Silvin A, Chen J, Grisel P, Schulte-Schrepping J et al (2018) Microbiome influences prenatal and adult microglia in a sex-specific manner. Cell 172(3):500–516.e16
Turnbaugh PJ, Ridaura VK, Faith JJ, Rey FE, Knight R, Gordon JI (2009) The effect of diet on the human gut microbiome: a metagenomic analysis in humanized gnotobiotic mice. Sci Transl Med 1(6):6ra14
Van Rooyen DM, Gan LT, Yeh MM, Haigh WG, Larter CZ, Ioannou G et al (2013) Pharmacological cholesterol lowering reverses fibrotic NASH in obese, diabetic mice with metabolic syndrome. J Hepatol 59(1):144–152
Vannice G, Rasmussen H (2014) Position of the academy of nutrition and dietetics: dietary fatty acids for healthy adults. J Acad Nutr Diet 114(1):136–153
Vasiljević A, Bursać B, Djordjevic A, Milutinović DV, Nikolić M, Matić G et al (2014) Hepatic inflammation induced by high-fructose diet is associated with altered 11βHSD1 expression in the liver of Wistar rats. Eur J Nutr 53(6):1393–1402
Voss K, Hong HS, Bader JE, Sugiura A, Lyssiotis CA, Rathmell JC (2021) A guide to interrogating immunometabolism. Nat Rev Immunol 21(10):637–652
Wardle EN (1987) Kupffer cells and their function. Liver 7(2):63–75
Wei X, Song H, Yin L, Rizzo MG, Sidhu R, Covey DF et al (2016) Fatty acid synthesis configures the plasma membrane for inflammation in diabetes. Nature 539(7628):294–298
Weisberg SP, McCann D, Desai M, Rosenbaum M, Leibel RL, Ferrante AW (2003) Obesity is associated with macrophage accumulation in adipose tissue. J Clin Invest 112(12):1796–1808
Weitz JR, Makhmutova M, Almaça J, Stertmann J, Aamodt K, Brissova M et al (2018) Mouse pancreatic islet macrophages use locally released ATP to monitor beta cell activity. Diabetologia 61(1):182–192
Westwell-Roper CY, Ehses JA, Verchere CB (2014) Resident macrophages mediate islet amyloid polypeptide-induced islet IL-1β production and β-cell dysfunction. Diabetes 63(5):1698–1711
Whelan J, Rust C (2006) Innovative dietary sources of n-3 fatty acids. Annu Rev Nutr 26:75–103
Wu JY, Huang TW, Hsieh YT, Wang YF, Yen CC, Lee GL et al (2020) Cancer-derived succinate promotes macrophage polarization and cancer metastasis via succinate receptor. Mol Cell 77(2):213–227.e5
Xiong W, Sun KY, Zhu Y, Zhang X, Zhou YH, Zou X (2021) Metformin alleviates inflammation through suppressing FASN-dependent palmitoylation of Akt. Cell Death Dis 12(10):934
Ye R, Gordillo R, Shao M, Onodera T, Chen Z, Chen S et al (2018) Intracellular lipid metabolism impairs β cell compensation during diet-induced obesity. J Clin Invest 128(3):1178–1189
Yeramian A, Martin L, Arpa L, Bertran J, Soler C, McLeod C et al (2006) Macrophages require distinct arginine catabolism and transport systems for proliferation and for activation. Eur J Immunol 36(6):1516–1526
Ying W, Lee YS, Dong Y, Seidman JS, Yang M, Isaac R et al (2019) Expansion of islet-resident macrophages leads to inflammation affecting β cell proliferation and function in obesity. Cell Metab 29(2):457–474.e5
Ying W, Fu W, Lee YS, Olefsky JM (2020) The role of macrophages in obesity-associated islet inflammation and β-cell abnormalities. Nat Rev Endocrinol 16(2):81–90
Yoshida GJ (2015) Metabolic reprogramming: the emerging concept and associated therapeutic strategies. J Exp Clin Cancer Res 34:111

Yuan M, Kremer DM, Huang H, Breitkopf SB, Ben-Sahra I, Manning BD et al (2019) Ex vivo and in vivo stable isotope labelling of central carbon metabolism and related pathways with analysis by LC-MS/MS. Nat Protoc 14(2):313–330

Yucel N, Wang YX, Mai T, Porpiglia E, Lund PJ, Markov G et al (2019) Glucose metabolism drives histone acetylation landscape transitions that dictate muscle stem cell function. Cell Rep 27(13):3939–3955.e6

Yurdagul A, Subramanian M, Wang X, Crown SB, Ilkayeva OR, Darville L et al (2020) Macrophage metabolism of apoptotic cell-derived arginine promotes continual efferocytosis and resolution of injury. Cell Metab 31(3):518–533.e10

Yvan-Charvet L, Wang N, Tall AR (2010) Role of HDL, ABCA1, and ABCG1 transporters in cholesterol efflux and immune responses. Arterioscler Thromb Vasc Biol 30(2):139–143

Zhang N, Kim SH, Gainullina A, Erlich EC, Onufer EJ, Kim J et al (2021) LYVE1+ macrophages of murine peritoneal mesothelium promote omentum-independent ovarian tumor growth. J Exp Med 218(12):e20210924

Chapter 4
Macrophage Cell Cycle

Louis Dillac, Leon El Dika, Rahim Ullah, Jacek Z. Kubiak, and Malgorzata Kloc

Abstract Macrophages are dynamic and plastic immune cells essential for tissue homeostasis and pathogen defense. Their cell cycle regulation is highly influenced by intrinsic and extrinsic signals facilitating rapid responses to infections and tissue damage. Dysregulation of their cell cycle leads to diseases like cancer and HIV. This chapter highlights aspects of the macrophage cell cycle crucial for the development of targeted therapies.

L. Dillac (✉)
Department of Molecular Biosciences, University of Texas at Austin, Austin, TX, USA

L. El Dika
Laboratoire Huckert's International, Wavre, Belgium

R. Ullah
Department of Biochemistry, Larner College of Medicine, UVM Cancer Center, University of Vermont, Burlington, VT, USA

J. Z. Kubiak
Laboratory of Molecular Oncology and Innovative Therapies, Military Institute of Medicine-National Research Institute (WIM-PIB), Warsaw, Poland

Dynamics and Mechanics of Epithelia Group, Faculty of Medicine, Institute of Genetics and Development of Rennes, University of Rennes, CNRS, UMR 6290, Rennes, France

M. Kloc
Houston Methodist Research Institute, Transplant Immunology, Houston, TX, USA

Houston Methodist Hospital, Department of Surgery, Houston, TX, USA

University of Texas, Md Anderson Cancer Center, Department of Genetics, Houston, TX, USA

M. Kloc et al. (eds.), *Monocytes and Macrophages in Development, Regeneration, and Disease*, Results and Problems in Cell Differentiation 74,
https://doi.org/10.1007/978-3-031-65944-7_4

4.1 Introduction

Macrophages are highly dynamic immune cells that play a key role in tissue health, homeostasis, defense against pathogens, chronic rejection of transplants, and wound healing (Lucas et al. 2010; Wynn et al. 2013; Jiang et al. 2014; Gordon and Plüddemann 2017; Gosselin et al. 2014, 2017). They are distributed in almost all tissues of the body and originate from bone marrow-derived monocytes and embryonic precursors (Yona et al. 2013; Hoeffel and Ginhoux 2018). Their specific cell cycle regulation and ability to proliferate allow a rapid response to infections, organ transplantations, or tissue damage (Sieweke and Allen 2013).

Macrophages, like other proliferating cells, progress through the cell cycle phases—G1, S, G2, and M (Matsushime 1995; El Dika 2013; Ferreira et al. 2020; El Dika et al. 2022; Uzbekov and Prigent 2022). In tissues, the cell cycle is tightly regulated by extracellular signals that control cell proliferation, thus influencing tissue growth and expansion. Disruption of the cell cycle and cell proliferation is frequently implicated in various pathologies, including cancer (Malumbres and Barbacid 2009; Metcalfe et al. 2022). Macrophages in G1 and possibly in G2 phase can enter a permanent state of cell cycle arrest called senescence, which is characterized by irreversible growth arrest and changes in cellular morphology and function (Fig. 4.1) (Wang et al. 2024). Key molecular changes associated with macrophage senescence include increased levels of Senescence-Associated

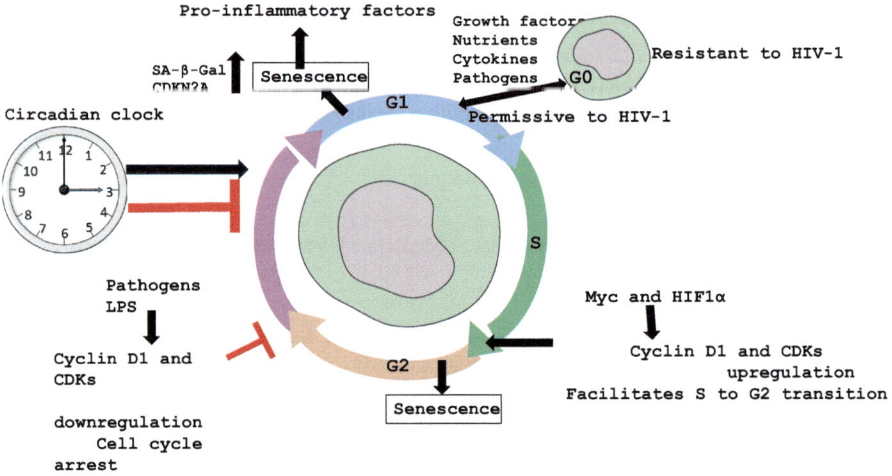

Fig. 4.1 Macrophage functions and processes depend on the phase of the cell cycle. The phases of the cell cycle are: G1, S, G2, M. The cell cycle is also affected by the circadian clock that regulates the expression of cyclins and CDKs. Under the effect of growth factors, nutrients, cytokines, and microbes, G1 can reversibly transit to the resting phase G0. The macrophages in G0 are more resistant, and in G1, more permissive to HIV-1 infection. G1 and probably also G2 macrophages can enter senescence, a durable cell cycle arrest accompanied by the production of inflammatory factors

Beta-Galactosidase (SA-β-Gal) and upregulation of the cyclin-dependent kinase inhibitor 2A (also known as CDKN2A or p16INK4a, Wang et al. (2024)]. Senescent macrophages exhibit a Senescence-Associated Secretory Phenotype (SASP) and secrete pro-inflammatory factors. They are characterized by specific gene expression alterations, imbalanced polarization states, dysregulation of cholesterol homeostasis, and oxidative respiration (Wang et al. 2024).

Macrophages can also enter a resting stage or G0 phase. In the G0 phase macrophages are primarily in a non-dividing state of quiescence until prompted by specific stimuli such as growth factors, nutrients, pathogen invasion, or cytokines (Weiss and Schaible 2015). This state is crucial for maintaining a pool of ready-to-activate macrophages that can respond quickly to infection or tissue damage. When activated, macrophages can exit the G0 phase to proliferate or differentiate into specialized forms, which are more effective at dealing with specific threats or injuries. Their ability to exit the G0 phase and become active is vital for effective immune responses and the resolution of inflammation. Recent research indicates that the shift from G0 to G1 creates an opportunity for HIV-1 infection (Ferreira et al. 2020). Macrophages are key targets for HIV-1, but they produce high levels of SAM domain and HD domain-containing protein 1 (SAMHD1), a deoxynucleotide-triphosphate hydrolase that restricts viral DNA synthesis by reducing dNTP levels (Laguette et al. 2011; Ferreira et al. 2020). While the G0 state is resistant to HIV-1 infection, the G1 state becomes highly permissive (Fig. 4.1). In the G1 state, macrophages deactivate the antiviral restriction factor SAMHD1 through the phosphorylation process, thereby facilitating effective HIV-1 infection (Laguette et al. 2011; Kim et al. 2013; Mlcochova et al. 2017; Antonucci et al. 2017; Schott et al. 2018; Ferreira et al. 2020). The macrophages' internal molecular machinery regulating cell cycle progression is affected by extracellular signals that dictate proliferation, quiescence, or differentiation (Sieweke and Allen 2013; Pearce and Pearce 2013). These signals include interactions with microbial products such as lipopolysaccharides (LPS) or cytokines (e.g. IFN-γ), which can promote macrophages to exit the G0 phase and resume the cell cycle (Kawai and Akira 2010). These stimuli are essential for macrophages to respond effectively to environmental challenges, thus affirming their role as guardians of tissue health and integrity.

Transition in the macrophage cell cycle is also rigorously controlled to ensure proper DNA replication and integrity (Ferreira et al. 2020; Mlcochova et al. 2024). This regulation is vital in normal physiological conditions as well as in pathological contexts, such as mentioned HIV viral infections where the macrophage cell cycle greatly influences vulnerability to the infection of the organism (Dupont and Sattentau 2020; Ferreira et al. 2020; Mlcochova et al. 2024). In the context of cancer, the dynamics of macrophage proliferation and their interactions with tumor cells can influence tumor growth, thus highlighting the complexity of macrophage roles in diseases such as, for example, periodontal and oral cancer (Gosselin et al. 2014, 2017; Metcalfe et al. 2022).

4.2 Regulation of Macrophage Cell Cycle

The regulation of the macrophage cell cycle involves a network of signaling pathways, transcription factors, and external stimuli that govern their proliferation and differentiation. The ability of macrophages to join the cell cycle and undergo self-renewal shows their dynamic nature and plasticity.

4.2.1 Signaling Pathways

Macrophages have the unique ability to mature and specialize under the influence of two important growth factors: Macrophage Colony Stimulating Factor (M-CSF) and Granulocyte-Macrophage Colony Stimulating Factor (GM-CSF) (Mehta et al. 2015; Ushach and Zlotnik 2016; Munsif et al. 2023). M-CSF is found in most tissues and can be easily detected in the blood. In contrast, GM-CSF is primarily produced in the lungs and increases during inflammatory responses. These molecules are essential for the survival and growth of macrophages. When macrophages differentiate in response to M-CSF or GM-CSF, they exhibit different behaviors during the cell cycle, especially during the transition from the G1 to the S phase.

Recent research has shown that GM-CSF and M-CSF activate distinct pathways in macrophages—one responsible for proliferation and the other for survival (Vico et al. 2022). M-CSF is very effective in promoting macrophage proliferation, significantly increasing the growth and differentiation of monocytic/macrophage lineage cells when present. In contrast, although GM-CSF induces proliferation, it does so less efficiently than M-CSF. However, GM-CSF excels in protecting macrophages from DNA damage. After DNA breaks caused by substances such as etoposide or hydrogen peroxide, GM-CSF contributes to faster recovery and reduced proliferation, probably by activating genes that pause the cell cycle. The ability of GM-CSF to protect macrophages against DNA damage exceeds that of M-CSF.

Macrophages differentiated with M-CSF multiply faster than those with GM-CSF, as shown by intracellular Ki67 staining (also known as MKI67, a marker of proliferation) (Badia et al. 2016). The M-CSF-stimulation results in 7% of Ki67+ macrophages compared to GM-CSF-stimulation which results in 0.5% of Ki67+ macrophages (Badia et al. 2016). Also, the M-CSF-stimulated macrophage population has more cells in the active stages of the cell cycle (S/G2M). These differences also affect macrophages' vulnerability to HIV-1 infection. The M-CSF-stimulated macrophages are about ten times more likely to be infected with HIV-1 compared to GM-CSF-stimulated (Badia et al. 2016). There are additionally higher levels of dNTPs in actively cycling M-CSF-treated macrophages, hinting that the dynamics of the cell cycle, especially progression through the S phase, are crucial for HIV-1 to infect and replicate within these cells.

Additionally, SAMHD1, a protein that restricts HIV-1 replication, plays a significant role in this process (Badia et al. 2016). Significant differences in SAMHD1 activity were observed between macrophages treated with M-CSF and those treated with GM-CSF. M-CSF-treated ones show changes in SAMHD1, which could make them more susceptible to HIV-1. When Viral Protein X (HIV-2 Vpx) degrades SAMHD1, HIV-1 replication increases, demonstrating the importance of SAMHD1 in limiting viral infection (Badia et al. 2016). This emphasizes how factors related to the cell cycle and proliferation can affect SAMHD1 and, consequently, the susceptibility of macrophages to HIV-1 infection.

4.2.2 Transcription Factors

Transcription factors are important in coordinating the gene expression programs that govern macrophage proliferation, differentiation, and self-renewal. They interact with specific DNA sequences to control the transcription of genes involved in cell cycle progression and maintenance of macrophage identity.

Several transcription factors are involved in this process:

1. Kruppel-like factor 4 (KLF4), a zinc finger transcription factor critical for macrophage proliferation and differentiation (Liao et al. 2018). KLF4 binds to specific DNA sequences to control the expression of genes involved in the cell cycle progression and self-renewal. Its Inhibition results in aberrant macrophage proliferation and impaired differentiation.
2. Nuclear receptor signaling pathways such as liver X receptors (LXRs) and retinoid X receptors (RXRs) (Chen and Ross 2004; Pascual-García et al. 2011; A-Gonzalez et al. 2013; Menéndez-Gutiérrez et al. 2015). LXRs are lipid- activated transcription factors associated with lipid metabolism and inflammation. They inhibit macrophage cell cycle progression and form heterodimers with LXRs to modulate the macrophage gene expression during the cell cycle. Interactions of LXRs, RXRs, and other nuclear receptors also provide a complex regulatory network that regulates macrophage function and self-renewal.
3. Sirtuin-1 (SIRT1), which is a member of the sirtuin family of NAD + -dependent deacetylases and is implicated in the regulation of macrophage self-renewal and proliferation (Imperatore et al. 2017; Röszer 2018). Its Inhibition reduces alveolar and peritoneal macrophage self-renewal, thus suggesting a role for this transcriptional regulator in macrophage homeostasis. SIRT1 exerts its effects by deacetylating histone and non-histone proteins, thus modulating gene expression and cell cycle progression. However, the precise mechanisms by which SIRT1 regulates macrophage proliferation and differentiation warrant further investigation.
4. Nuclear Factor κB (NF-κB) and Signal Transducer and Activator of Transcription (STAT), which are important transcription factors that regulate the cell cycle and inflammatory responses (Aravindaram and Yang 2010; Pugazhenthi et al. 2013; Dinasarapu et al. 2013; Pinheiro et al. 2019). They are activated by cytokines and other signals. Indeed, NF-κB can both promote and inhibit the cell cycle, which

affects many genes involved in cell proliferation and survival, like cyclin D1, crucial for the G1 phase of the cell cycle (Hinz et al. 1999; Joyce et al. 2001; Witzel et al. 2010; Hwang et al. 2010). Similarly, STAT, particularly STAT3, influences the cell cycle by enhancing the expression of genes like cyclin D1 and Myc (myelocytomatosis oncogene), which drive the transition from the G1 to the S phase (Sai et al. 2012; Zhou et al. 2018; Yu et al. 2022).

5. Myc and Hypoxia-Inducible Factor Alpha (HIF1α), a subunit of HIF-1. Myc facilitates the transition from the G1 phase to the S phase of the macrophage cell cycle by upregulating the expression of cyclins (like cyclin D1) and cyclin-dependent kinases (CDKs), (Liu et al. 2016). Under hypoxia, macrophages undergo HIF1α-mediated transcriptional reprogramming that alters macrophage metabolism and function toward pathogen elimination and inflammatory responses, thus resulting in disruptions in cell cycle and proliferation. Moreover, mitogenic stimulation, such as by CSF-1, activates a Myc-dependent transcriptional program in macrophages to favor cell cycle entry and upregulate glucose and glutamine metabolism supporting growth and proliferation (Liu et al. 2016). Conversely, proinflammatory signals like LPS and IFN-γ shift metabolism towards glycolysis via a HIF1α-dependent program, thus suppressing Myc-driven proliferation while enhancing the proinflammatory response (Liu et al. 2016). M1 macrophages, under these conditions, experience cell cycle arrest, ensuring they maintain their effector functions without excessive proliferation (Fig. 4.1).

4.2.3 Cyclins, CDKs and Circadian Rhythm

Cyclins and cyclin-dependent kinases (CDKs) are key regulators of the cell cycle that control the progression through the different phases of cell division. In macrophages, the expression and activity of cyclins and CDKs are highly controlled to ensure proper cell cycle progression and proliferation (Matsushime 1995). Cyclins and CDKs are also important for the expression of cell cycle checkpoint genes. It has been demonstrated that the reduced gene expression of cyclin D1 in response to LPS blocks the macrophage cell cycle (Matsushime 1995; Röszer 2018). Thus, the dysregulation of the cyclin gene expression (or CDK activity) results in cell cycle arrest.

All cells and organisms function according to the circadian rhythm. The circadian rhythm is endogenous but is adjusted to the local environment by external cues (zeitgebers) such as light, temperature, and redox cycles. Because circadian rhythm oscillates cell processes over a 24-hour cycle, it is not surprising that macrophage functions and responses to pathogens and other stimuli also oscillate according to their circadian rhythm (Fig. 4.2).

It has been shown that, in macrophages, the circadian clock influences the expression of genes involved in cell cycle progression, such as cyclins and cyclin-dependent kinases (CDKs) (Fig. 4.1); (Timmons et al. 2020; Shirato and Sato 2022;

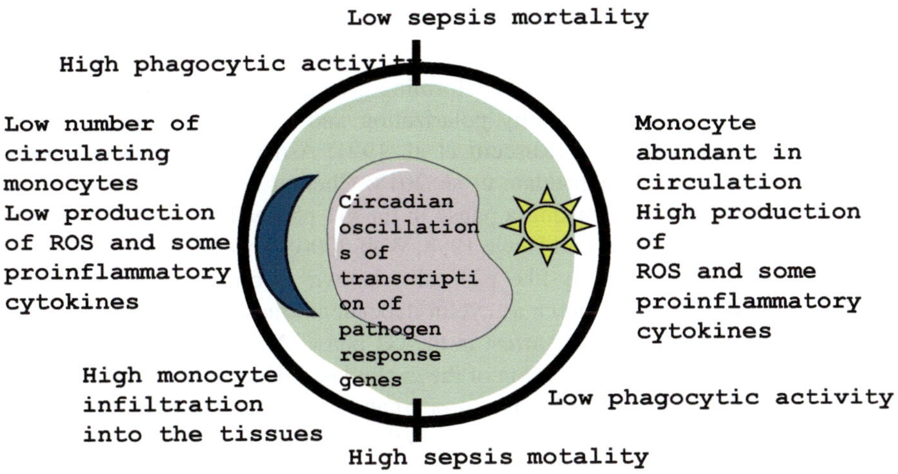

Fig. 4.2 Circadian rhythm regulation of macrophage functions. Macrophage/monocyte migratory behavior, phagocytosis, metabolism, pro- and anti-inflammatory cytokine production, and sepsis mortality (in the mouse model) oscillate within 24 h

Lellupitiyage Don et al. 2022). The rhythmic expression of these regulators is essential for coordinating cell division and proliferation in a timely manner. Disruption of the circadian rhythms can then lead to dysregulation of the cell cycle, thus affecting macrophage proliferation and differentiation.

Furthermore, the circadian clock impacts the response of macrophages to external stimuli, such as pathogens or inflammatory signals, by modulating cell cycle progression and activation of immune responses (Timmons et al. 2020; Shirato and Sato 2022; Lellupitiyage Don et al. 2022). It has been demonstrated that deletion of the circadian gene *BMAL1* in macrophages alters their phagocytic activity and antimicrobial response. Disruptions in circadian rhythms, such as those caused by endotoxin exposure, affect the cell cycle of macrophages through the generation of reactive oxygen species. Additionally, the clearance of apoptotic cells by macrophages, known as efferocytosis, is subject to circadian regulation. This process is crucial for maintaining tissue homeostasis and resolving inflammation. The activity of TAM receptors, a family of receptor tyrosine kinases that includes Tyro3, Axl, and Mer and known as efferocytosis mediators, shows circadian patterns, potentially influenced by oscillations in glucocorticoid levels and core clock genes like *BMAL1*.

4.2.4 TGF-β's Interactions

The complex interaction between transforming growth factor-β (TGF-β) and macrophages impacts their functionality, polarization, and functions in tissue regeneration and immune regulation (Ruscetti et al. 1991; Ashcroft 1999; Vannella and Wynn 2017; Shapouri-Moghaddam et al. 2018; Chen et al. 2023). This cytokine impacts the cell cycle by causing a pause in the G1 phase (Derynck 1994; Ravitz and Wenner 1997; Hocevar and Howe 1998; Wolf 2000). This is primarily achieved by activating CDK inhibitors like p21 (also known as cyclin-dependent kinase inhibitor 1) and p15 (also known as cyclin-dependent kinase 4 inhibitor B), which play a critical role in cell cycle arrest in the G1 phase (Nagahara et al. 1999). The p21 and p15 suppress the expression of the cyclin D-CDK4/6 (D-type cyclins complexed with cyclin-dependent kinases 4 or 6) and cyclin E-CDK2 (E-type cyclins complexed with cyclin-dependent kinases 2) that both are required for progressing towards the S phase (Nagahara et al. 1999; El Dika 2013).

The Smad (Mothers Against Decapentaplegic Homolog), a protein complex induced by TGF-β, plays an essential role in controlling cell growth by suppressing specific cellular activities essential for cell cycle progression (Syed 2016). This suppression arrests the cell cycle in the G1 phase, preventing cells from progressing to the S phase, where DNA synthesis occurs. Both Smad-dependent and -independent pathways are involved in the TGF-β signaling cascade, contributing to its role in cell cycle regulation. As an example, TGF-β can trigger dephosphorylation Ribosomal protein S6 kinase beta-1 (S6K1, also known as p70S6) through protein phosphatase 2A, resulting in cell cycle disruption (Petritsch et al. 2000). This highlights the variety of signaling pathways that TGF-β can stimulate to exert its actions and the importance of TGF-β in cell cycle regulation by balancing proliferation and cell cycle arrest.

4.2.5 Inflammatory Signals

Inflammatory cytokines include tumor necrosis factor alpha (TNFα) and interferon gamma (IFNγ) (Holden et al. 2017; Shapouri-Moghaddam et al. 2018; Mily et al. 2020; Edris et al. 2022). They inhibit macrophage proliferation and induce cell cycle arrest by altering the expression of cell cycle checkpoint genes. These inflammatory mediators shift the transcriptional program of macrophages from a self-renewal state to an inflammatory response state. M1 macrophages (pro-inflammatory) are activated by signaling via Toll-like receptors (TLRs) and cytokines such as TNF-α and IFN-γ, resulting in the expression of pro-inflammatory genes controlled by transcription factors such as NF-κB and STAT1 and release of antimicrobial and tumoricidal cytokines such as TNF-α, IL-1, and IL-6 (Holden et al. 2017; Shapouri-Moghaddam et al. 2018; Mily et al. 2020; Edris et al. 2022). However, the cell cycle regulation of M2 macrophages (anti-inflammatory) involves activating specific signaling pathways such as JAK/STAT, the IL-4, and IL-13 that impact macrophage proliferation, which in turn promotes not only anti-inflammatory responses but also cell growth and cell survival. Besides, STAT6 is activated by IL-4 and IL-13 and plays an important role in stimulating the growth of M2 macrophages. Advanced techniques are essential for investigating the nuanced roles and cell cycle behaviors of macrophages, especially in understanding how M1 and M2 phenotypes influence their proliferation and functionality.

4.3 Cell Cycle-Dependent Plasticity of Macrophages

Macrophage plasticity is critical for their roles in host protection, tissue repair, and immune regulation. Recent research on macrophage plasticity has shown that different phases of the cell cycle influence the macrophage activation states and functional properties (Daniel et al. 2023). For example, macrophages in the G1 phase are more likely to turn on a pro-inflammatory response, while those in the S or G2/M phases might adopt anti-inflammatory characteristics. So, this cell cycle-dependent plasticity permits macrophages to fine-track their responses to changing environmental stimuli, therefore ensuring effective immune function and tissue homeostasis.

Interestingly, single-cell approaches have been recently used to characterize the heterogeneity in macrophage responses based on their cell cycle state (Daniel et al. 2023). It has been shown that macrophages in different phases exhibit distinct gene expression profiles and functional properties. For example, macrophages analyzed at the single-cell level show a wide range of activation states beyond the traditional M1 and M2 classifications. This diversity is influenced by local cytokine milieus, tissue-specific signals, and the cell cycle phase. The authors also showed that IL-4 can induce S-G2/M-biased gene expression, including tissue remodeling genes, in macrophages (Daniel et al. 2023).

These mechanisms underlying cell cycle-dependent plasticity are complex and multifaceted (EL Kasmi and Stenmark 2015; Viola et al. 2019; Chambers et al. 2021; Daniel et al. 2023). As described above, cyclins and CDKs coordinate cell cycle progression and macrophage activation states, while transcription factors such as NF-κB and STATs play crucial roles in macrophage activation and polarization, with their expression and activity varying across different cell cycle phases. These regulators drive dynamic changes in macrophage gene expression profiles, thus influencing their functional properties. Moreover, the cell cycle influences the metabolic and bioenergetic status of macrophages, which in turn affects their activation state (EL Kasmi and Stenmark 2015; Viola et al. 2019; Chambers et al. 2021; Daniel et al. 2023). Metabolic reprogramming is a hallmark of macrophage activation, with distinct metabolic pathways being associated with different activation states. The cell cycle regulates metabolic pathways such as glycolysis, oxidative phosphorylation, and fatty acid metabolism, which all are essential for macrophage function.

Epigenetic modifications affect the macrophage cell cycle and activation (Daniel et al. 2023). They include changes in chromatin structure and histones modifications that alter gene expression patterns in a cell cycle-dependent manner, thus influencing macrophage responses. For instance, during certain cell cycle phases, macrophages may go through chromatin remodeling that enhances or suppresses the expression of genes involved in inflammation or tissue repair. These epigenetic modifications permit macrophages to evolve their activation states to specific environmental cues, thus fine-tuning their responses to inflammatory stimuli and tissue-specific signals.

4.4 Novel Techniques Advancing Macrophage Regulation

Advancing our understanding of the regulatory processes and developing targeted therapies can be greatly aided by innovative techniques in macrophage research (Shimi et al. 2004; Ishikawa-Ankerhold et al. 2012; De Los Santos et al. 2015; Na et al. 2015; Wu et al. 2020; Zhang et al. 2021; Song et al. 2022; Fan et al. 2023; Miljkovic et al. 2023) (1) High-Throughput Screening (HTS) rapidly test thousands of conditions to identify modulators of macrophage function that are relevant to diseases (2) Single-Cell RNA Sequencing (scRNA-seq) generate new data about the individual cellular responses at the transcriptomic level, thus shedding light on eventual population heterogeneity (3) Live cell imaging techniques, including Fluorescence Recovery After Photobleaching (FRAP) and Fluorescence Resonance Energy Transfer (FRET), allow analyzing real-time dynamics of macrophage related proteins and the signaling pathway activation (4) Mass Cytometry (CyTOF) that uses metal-labeled antibodies for detailed cellular analysis and CRISPR-Cas9 gene editing enable precise manipulations of such genes to study their roles in response to cytokines and growth factors (Wu et al. 2020) (5) proteomic screens like Tandem Mass Tag (TMT) and Liquid Chromatography-Tandem Mass Spectrometry (LC-MS/MS) can identify quantifying proteomic changes and discovering new signaling molecules involved in macrophage regulation. All these approaches have recently become pivotal for advancing our understanding of macrophage biology and developing targeted therapies.

4.5 Conclusions

The dynamic regulation of the macrophage cell cycle in response to different stimuli and circadian clock, and the mechanisms underlying cell cycle-dependent plasticity will have significant clinical implications. Dysregulation of macrophage activation states and cell cycle control is associated with various inflammatory diseases, autoimmune disorders, and cancer. Thus, targeting cell cycle regulators, transcription factors, or epigenetic modifiers in macrophages presents promising therapeutic opportunities for modulating immune responses and treating inflammatory disorders. By harnessing the dynamic nature of macrophage plasticity, the cell cycle, and circadian rhythm-regulated responses researchers can develop novel strategies for precision medicine and personalized immunotherapy.

Conflict of Interest The authors declare no conflict of interest.

Funding Dr. JZ Kubiak was funded by the subvention of the Polish Ministry of Science and Higher Education, project # 612/2023 of Wojskowy Instytut Medyczny-Panstwowy Instytut Badawczy, Warsaw, Poland.

References

A-Gonzalez N, Guillen JA, Gallardo G, Diaz M, De La Rosa JV, Hernandez IH, Casanova-Acebes M, Lopez F, Tabraue C, Beceiro S, Hong C, Lara PC, Andujar M, Arai S, Miyazaki T, Li S, Corbi AL, Tontonoz P, Hidalgo A, Castrillo A (2013) The nuclear receptor LXRα controls the functional specialization of splenic macrophages. Nat Immunol 14(8):831–839. https://doi.org/10.1038/NI.2622

Antonucci JM, St Gelais C, Wu L (2017) The dynamic interplay between HIV-1, SAMHD1, and the innate antiviral response. Front Immunol 8:1541. https://doi.org/10.3389/fimmu.2017.01541

Aravindaram K, Yang NS (2010) Anti-inflammatory plant natural products for cancer therapy. Planta Med 76(11):1103–1117. https://doi.org/10.1055/s-0030-1249859

Ashcroft GS (1999) Bidirectional regulation of macrophage function by TGF-beta. Microbes Infect 1(15):1275–1282. https://doi.org/10.1016/S1286-4579(99)00257-9

Badia R, Pujantell M, Riveira-Muñoz E, Puig T, Torres-Torronteras J, Martí R, Clotet B, Ampudia RM, Vives-Pi M, Esté JA, Ballana E (2016) The G1/S specific cyclin D2 is a regulator of HIV-1 restriction in non-proliferating cells. PLoS Pathog 12(8):e1005829. https://doi.org/10.1371/JOURNAL.PPAT.1005829

Chambers M, Rees A, Cronin JG, Nair M, Jones N, Thornton CA (2021) Macrophage plasticity in reproduction and environmental influences on their function. Front Immunol 11:607328. https://doi.org/10.3389/fimmu.2020.607328

Chen Q, Ross AC (2004) Retinoic acid regulates cell cycle progression and cell differentiation in human monocytic THP-1 cells. Exp Cell Res 297(1):68–81. https://doi.org/10.1016/j.yexcr.2004.02.017

Chen S, Saeed AFUH, Liu Q, Jiang Q, Xu H, Xiao GG, Rao L, Duo Y (2023) Macrophages in immunoregulation and therapeutics. Signal Transduct Target Ther 8(1):207. https://doi.org/10.1038/s41392-023-01452-1

Daniel B, Belk JA, Meier SL, Chen AY, Sandor K, Czimmerer Z, Varga Z, Bene K, Buquicchio FA, Qi Y, Kitano H, Wheeler JR, Foster DS, Januszyk M, Longaker MT, Chang HY, Satpathy AT (2023) Macrophage inflammatory and regenerative response periodicity is programmed by cell cycle and chromatin state. Mol Cell 83(1):121–138.e7. https://doi.org/10.1016/J.MOLCEL.2022.11.017

De Los Santos C, Chang CW, Mycek MA, Cardullo RA (2015) FRAP, FLIM, and FRET: detection and analysis of cellular dynamics on a molecular scale using fluorescence microscopy. Mol Reprod Dev 82(7–8):587–604. https://doi.org/10.1002/mrd.22501

Derynck R (1994) TGF-β-receptor-mediated signaling. Trends Biochem Sci 19(12):548–553. https://doi.org/10.1016/0968-0004(94)90059-0

Dinasarapu AR, Gupta S, Maurya MR, Fahy E, Min J, Sud M, Gersten MJ, Glass CK, Subramaniam S (2013) A combined omics study on activated macrophages - enhanced role of STATs in apoptosis, immunity and lipid metabolism. Bioinformatics 29(21):2735–2743. https://doi.org/10.1093/bioinformatics/btt469

Dupont M, Sattentau QJ (2020) Macrophage cell-cell interactions promoting HIV-1 infection. Viruses 12(5):492. https://doi.org/10.3390/v12050492

Edris NA, Aboulhoda BE, El-Din SS, Fouad AM, Albadawi E, Rashed LA, Elessawy KB (2022) Toll-like receptor signaling in the pathogenesis of chronic dacryocystitis: implication of c-FOS transcription factor and its downstream effector chemokine genes CCL2, CCL4, CXCL3, CXCR4 with a shift of the M1/M2 macrophage phenotype. Comb Chem High Throughput Screen 26(3):489–497. https://doi.org/10.2174/1386207325666220509150457

El Dika M (2013) Régulation de la phase M du cycle cellulaire par CDK1, PP2A et CDC6. Ph.D. thesis, University of Rennes 1, Rennes

El Dika M, Fritz AJ, Toor RH, Rodriguez PD, Foley SJ, Ullah R, Nie D, Banerjee B, Lohese D, Tracy KM, Glass KC, Frietze S, Ghule PN, Heath JL, Imbalzano AN, van Wijnen A, Gordon J, Lian JB, Stein JL, Stein GS (2022) Epigenetic-mediated regulation of gene expression for biological control and cancer: fidelity of mechanisms governing the cell cycle. Results Probl Cell Differ 70:375–396. https://doi.org/10.1007/978-3-031-06573-6_13

El Kasmi KC, Stenmark KR (2015) Contribution of metabolic reprogramming to macrophage plasticity and function. Semin Immunol 27(4):267–275. https://doi.org/10.1016/j.smim.2015.09.001

Fan P, Zhang Y, Ding S, Du Z, Zhou C, Du X (2023) Integrating RNA-seq and scRNA-seq to explore the mechanism of macrophage ferroptosis associated with COPD. Front Pharmacol 14:1139137. https://doi.org/10.3389/fphar.2023.1139137

Ferreira IATM, Zachary Porterfield J, Gupta RK, Mlcochova P (2020) Cell cycle regulation in macrophages and susceptibility to HIV-1. Viruses 12(8):839. https://doi.org/10.3390/v12080839

Gordon S, Plüddemann A (2017) Tissue macrophages: heterogeneity and functions. BMC Biol 15(1):53. https://doi.org/10.1186/s12915-017-0392-4

Gosselin D, Link VM, Romanoski CE, Fonseca GJ, Eichenfield DZ, Spann NJ, Stender JD, Chun HB, Garner H, Geissmann F, Glass CK (2014) Environment drives selection and function of enhancers controlling tissue-specific macrophage identities. Cell 159(6):1327–1340. https://doi.org/10.1016/J.CELL.2014.11.023

Gosselin D, Skola D, Coufal NG, Holtman IR, Schlachetzki JCM, Sajti E, Jaeger BN, O'Connor C, Fitzpatrick C, Pasillas MP, Pena M, Adair A, Gonda DD, Levy ML, Ransohoff RM, Gage FH, Glass CK (2017) An environment-dependent transcriptional network specifies human microglia identity. Science 356(6344):1248–1259. https://doi.org/10.1126/SCIENCE.AAL3222

Hinz M, Krappmann D, Eichten A, Heder A, Scheidereit C, Strauss M (1999) NF-κB function in growth control: regulation of Cyclin D1 expression and G 0 /G 1 -to-S-phase transition. Mol Cell Biol 19(4):2690–2698. https://doi.org/10.1128/mcb.19.4.2690

Hocevar BA, Howe PH (1998) Mechanisms of TGF-β-induced cell cycle arrest. Miner Electrolyte Metab 24(2–3):131–135. https://doi.org/10.1159/000057360

Hoeffel G, Ginhoux F (2018) Fetal monocytes and the origins of tissue-resident macrophages. Cell Immunol 330:5–15. https://doi.org/10.1016/j.cellimm.2018.01.001

Holden JA, O'Brien-Simpson NM, Lenzo JC, Orth RKH, Mansell A, Reynolds EC (2017) Porphyromonas gulae activates unprimed and gamma interferon-primed macrophages via the pattern recognition receptors toll-like receptor 2 (TLR2), TLR4, and NOD2. Infect Immun 85(9):e00282-17. https://doi.org/10.1128/IAI.00282-17

Hwang I, Choi YS, Jeon MY, Jeong S (2010) NF-κB p65 represses β-catenin-activated transcription of cyclin D1. Biochem Biophys Res Commun 403(1):79–84. https://doi.org/10.1016/j.bbrc.2010.10.118

Imperatore F, Maurizio J, Vargas Aguilar S, Busch CJ, Favret J, Kowenz-Leutz E, Cathou W, Gentek R, Perrin P, Leutz A, Berruyer C, Sieweke MH (2017) SIRT1 regulates macrophage self-renewal. EMBO J 36(16):2353–2372. https://doi.org/10.15252/EMBJ.201695737

Ishikawa-Ankerhold HC, Ankerhold R, Drummen GPC (2012) Advanced fluorescence microscopy techniques-FRAP, FLIP, FLAP, FRET and FLIM. Molecules 17(4):4047–4132. https://doi.org/10.3390/molecules17044047

Jiang X, Tian W, Sung YK, Qian J, Nicolls MR (2014) Macrophages in solid organ transplantation. Vasc Cell 6(1):5. https://doi.org/10.1186/2045-824X-6-5

Joyce D, Albanese C, Steer J, Fu M, Bouzahzah B, Pestell RG (2001) NF-κB and cell-cycle regulation: the cyclin connection. Cytokine Growth Factor Rev 12(1):73–90. https://doi.org/10.1016/S1359-6101(00)00018-6

Kawai T, Akira S (2010) The role of pattern-recognition receptors in innate immunity: update on toll-like receptors. Nat Immunol 11(5):373–384. https://doi.org/10.1038/ni.1863

Kim ET, White TE, Brandariz-Núñez A, Diaz-Griffero F, Weitzman MD (2013) SAMHD1 restricts herpes simplex virus 1 in macrophages by limiting DNA replication. J Virol 87(23):12949–12956. https://doi.org/10.1128/jvi.02291-13

Laguette N, Sobhian B, Casartelli N, Ringeard M, Chable-Bessia C, Ségéral E, Yatim A, Emiliani S, Schwartz O, Benkirane M (2011) SAMHD1 is the dendritic- and myeloid-cell-specific HIV-1 restriction factor counteracted by Vpx. Nature 474(7353):654–657. https://doi.org/10.1038/NATURE10117

Lellupitiyage Don SS, Mas-Rosario JA, Lin HH, Nguyen EM, Taylor SR, Farkas ME (2022) Macrophage circadian rhythms are differentially affected based on stimuli. Integr Biol (Camb) 14(3):62–75. https://doi.org/10.1093/INTBIO/ZYAC007

Liao X, Shen Y, Zhang R, Sugi K, Vasudevan NT, Amer Alaiti M, Sweet DR, Zhou L, Qing Y, Gerson SL, Fu C, Wynshaw-Boris A, Hu R, Schwartz MA, Fujioka H, Richardson B, Cameron MJ, Hayashi H, Stamler JS, Jain MK (2018) Distinct roles of resident and nonresident macrophages in nonischemic cardiomyopathy. Proc Natl Acad Sci USA 115(20):E4661–E4669. https://doi.org/10.1073/PNAS.1720065115

Liu L, Luc Y, Martinez J, Bi Y, Lian G, Wang T, Milasta S, Wang J, Yang M, Liu G, Green DR, Wang R (2016) Proinflammatory signal suppresses proliferation and shifts macrophage metabolism from Myc-dependent to HIF1α-dependent. Proc Natl Acad Sci USA 113(6):1564–1569. https://doi.org/10.1073/PNAS.1518000113

Lucas T, Waisman A, Ranjan R, Roes J, Krieg T, Müller W, Roers A, Eming SA (2010) Differential roles of macrophages in diverse phases of skin repair. J Immunol 184(7):3964–3977. https://doi.org/10.4049/jimmunol.0903356

Malumbres M, Barbacid M (2009) Cell cycle, CDKs and cancer: a changing paradigm. Nat Rev Cancer 9(3):153–166. https://doi.org/10.1038/NRC2602

Matsushime H (1995) Macrophage cell cycle control by M-CSF/CSF-1. Rinsho Ketsueki 36(5):406–409

Mehta HM, Malandra M, Corey SJ (2015) G-CSF and GM-CSF in neutropenia. J Immunol 195(4):1341–1349. https://doi.org/10.4049/JIMMUNOL.1500861

Menéndez-Gutiérrez MP, Rõszer T, Fuentes L, Núnez V, Escolano A, Redondo JM, De Clerck N, Metzger D, Valledor AF, Ricote M (2015) Retinoid X receptors orchestrate osteoclast differentiation and postnatal bone remodeling. J Clin Invest 125(2):809–823. https://doi.org/10.1172/JCI77186

Metcalfe S, Anselmi N, Escobar A, Visser MB, Kay JG (2022) Innate phagocyte polarization in the oral cavity. Front Immunol 12:768479. https://doi.org/10.3389/fimmu.2021.768479

Miljkovic M, Lozano S, Castellote I, de Cózar C, Villegas-Moreno AI, Gamallo P, Jimenez-Alfaro Martinez D, Fernández-Álvaro E, Ballell L, Garcia GA (2023) Novel inhibitors that target bacterial virulence identified via HTS against intra-macrophage survival of Shigella flexneri. mSphere 8(5):e0015423. https://doi.org/10.1128/msphere.00154-23

Mily A, Kalsum S, Loreti MG, Rekha RS, Muvva JR, Lourda M, Brighenti S (2020) Polarization of m1 and m2 human monocyte-derived cells and analysis with flow cytometry upon mycobacterium tuberculosis infection. J Vis Exp 2020(163):1–20. https://doi.org/10.3791/61807

Mlcochova P, Heilig R, Fischer R, Gupta RK (2024) CD4 T cell contact drives macrophage cell cycle progression and susceptibility to lentiviral transduction. bioRxiv:2024.01.15.575666. https://doi.org/10.1101/2024.01.15.575666

Mlcochova P, Sutherland KA, Watters SA, Bertoli C, de Bruin RA, Rehwinkel J, Neil SJ, Lenzi GM, Kim B, Khwaja A, Gage MC, Georgiou C, Chittka A, Yona S, Noursadeghi M, Towers

GJ, Gupta RK (2017) A G1-like state allows HIV -1 to bypass SAMHD 1 restriction in macrophages. EMBO J 36(5):604–616. https://doi.org/10.15252/embj.201696025

Munsif M, Sweeney D, Leong TL, Stirling RG (2023) Nebulised granulocyte–macrophage colony-stimulating factor (GM-CSF) in autoimmune pulmonary alveolar proteinosis: a systematic review and meta-analysis. Eur Respir Rev 32(170):230080. https://doi.org/10.1183/16000617.0080-2023

Na YR, Hong JH, Lee MY, Jung JH, Jung D, Kim YW, Son D, Choi M, Kim KP, Seok SH (2015) Proteomic analysis reveals distinct metabolic differences between granulocyte-macrophage colony stimulating factor (GM-CSF) and macrophage colony stimulating factor (M-CSF) grown macrophages derived from murine bone marrow cells. Mol Cell Proteomics 14(10):2722–2732. https://doi.org/10.1074/mcp.M115.048744

Nagahara H, Ezhevsky SA, Vocero-Akbani AM, Kaldis P, Solomon MJ, Dowdy SF (1999) Transforming growth factor beta targeted inactivation of cyclin E:cyclin-dependent kinase 2 (Cdk2) complexes by inhibition of Cdk2 activating kinase activity. Proc Natl Acad Sci USA 96(26):14961–14966. https://doi.org/10.1073/PNAS.96.26.14961

Pascual-García M, Carbó JM, León T, Matalonga J, Out R, Van Berkel T, Sarrias M-R, Lozano F, Celada A, Valledor AF (2011) Liver X receptors inhibit macrophage proliferation through downregulation of cyclins D1 and B1 and cyclin-dependent kinases 2 and 4. J Immunol 186(8):4656–4667. https://doi.org/10.4049/JIMMUNOL.1000585

Pearce EL, Pearce EJ (2013) Metabolic pathways in immune cell activation and quiescence. Immunity 38(4):633–643. https://doi.org/10.1016/j.immuni.2013.04.005

Petritsch C, Beug H, Baimain A, Oft M (2000) TGF-beta inhibits p70 S6 kinase via protein phosphatase 2A to induce G(1) arrest. Genes Dev 14(24):3093–3101. https://doi.org/10.1101/GAD.854200

Pinheiro DML, de Oliveira AHS, Coutinho LG, Fontes FL, de Medeiros Oliveira RK, Oliveira TT, Faustino ALF, Lira da Silva V, de Melo Campos JTA, Lajus TBP, de Souza SJ, Agnez-Lima LF (2019) Resveratrol decreases the expression of genes involved in inflammation through transcriptional regulation. Free Radic Biol Med 130:8–22. https://doi.org/10.1016/j.freeradbiomed.2018.10.432

Pugazhenthi S, Zhang Y, Bouchard R, Mahaffey G (2013) Induction of an inflammatory loop by interleukin-1β and tumor necrosis factor-α involves NF-kB and STAT-1 in differentiated human neuroprogenitor cells. PLoS One 8(7):e69585. https://doi.org/10.1371/journal.pone.0069585

Ravitz MJ, Wenner CE (1997) Cyclin-dependent kinase regulation during G 1 phase and cell cycle regulation by TGF-β. Adv Cancer Res 71:165–207. https://doi.org/10.1016/s0065-230x(08)60099-8

Röszer T (2018) Understanding the biology of self-renewing macrophages. Cells 7(8):103. https://doi.org/10.3390/cells7080103

Ruscetti FW, Dubois C, Falk LA, Jacobsen SE, Sing G, Longo DL, Wiltrout RH, Keller JR (1991) In vivo and in vitro effects of TGF-beta 1 on normal and neoplastic haemopoiesis. Ciba Found Symp 157:212–227. https://doi.org/10.1002/9780470514061.ch14

Sai K, Wang S, Balasubramaniyan V, Conrad C, Lang FF, Aldape K, Szymanski S, Fokt I, Dasgupta A, Madden T, Guan S, Chen Z, Yung WKA, Priebe W, Colman H (2012) Induction of cell-cycle arrest and apoptosis in glioblastoma stem-like cells by WP1193, a novel small molecule inhibitor of the JAK2/STAT3 pathway. J Neuro-Oncol 107(3):487–501. https://doi.org/10.1007/s11060-011-0786-z

Schott K, Fuchs NV, Derua R, Mahboubi B, Schnellbächer E, Seifried J, Tondera C, Schmitz H, Shepard C, Brandariz-Nuñez A, Diaz-Griffero F, Reuter A, Kim B, Janssens V, König R (2018) Dephosphorylation of the HIV-1 restriction factor SAMHD1 is mediated by PP2A-B55α holoenzymes during mitotic exit. Nat Commun 9(1):2227. https://doi.org/10.1038/s41467-018-04671-1

Shapouri-Moghaddam A, Mohammadian S, Vazini H, Taghadosi M, Esmaeili SA, Mardani F, Seifi B, Mohammadi A, Afshari JT, Sahebkar A (2018) Macrophage plasticity, polarization, and function in health and disease. J Cell Physiol 233(9):6425–6440. https://doi.org/10.1002/JCP.26429

Shimi T, Koujin T, Segura-Totten M, Wilson KL, Haraguchi T, Hiraoka Y (2004) Dynamic interaction between BAF and emerin revealed by FRAP, FLIP, and FRET analyses in living HeLa cells. J Struct Biol 147(1):31–41. https://doi.org/10.1016/j.jsb.2003.11.013

Shirato K, Sato S (2022) Macrophage meets the circadian clock: implication of the circadian clock in the role of macrophages in acute lower respiratory tract infection. Front Cell Infect Microbiol 12:826738. https://doi.org/10.3389/FCIMB.2022.826738

Sieweke MH, Allen JE (2013) Beyond stem cells: self-renewal of differentiated macrophages. Science 342(6161):1242974. https://doi.org/10.1126/SCIENCE.1242974

Song J, Jiang J, Kuai L, Luo Y, Xing M, Luo Y, Ru Y, Sun X, Zhang H, Liu T, Li X, Li B (2022) TMT-based proteomics analysis reveals the protective effect of Jueyin granules on imiquimod-induced psoriasis mouse model by causing autophagy: inducing autophagy for Jueyin granules treatment of psoriasis-like inflammation. Phytomedicine 96:153846. https://doi.org/10.1016/j.phymed.2021.153846

Syed V (2016) TGF-β signaling in cancer. J Cell Biochem 117(6):1279–1287. https://doi.org/10.1002/jcb.25496

Timmons GA, O'Siorain JR, Kennedy OD, Curtis AM, Early JO (2020) Innate rhythms: clocks at the Center of Monocyte and Macrophage Function. Front Immunol 11:1743. https://doi.org/10.3389/FIMMU.2020.01743

Ushach I, Zlotnik A (2016) Biological role of granulocyte macrophage colony-stimulating factor (GM-CSF) and macrophage colony-stimulating factor (M-CSF) on cells of the myeloid lineage. J Leukoc Biol 100(3):481–489. https://doi.org/10.1189/jlb.3ru0316-144r

Uzbekov R, Prigent C (2022) A journey through time on the discovery of cell cycle regulation. Cells 11(4):704. https://doi.org/10.3390/CELLS11040704

Vannella KM, Wynn TA (2017) Mechanisms of organ injury and repair by macrophages∗. Annu Rev Physiol 79:593–617. https://doi.org/10.1146/annurev-physiol-022516-034356

Vico T, Youssif C, Zare F, Comalada M, Sebastian C, Lloberas J, Celada A (2022) GM-CSF protects macrophages from DNA damage by inducing differentiation. Cells 11(6):935. https://doi.org/10.3390/CELLS11060935

Viola A, Munari F, Sánchez-Rodríguez R, Scolaro T, Castegna A (2019) The metabolic signature of macrophage responses. Front Immunol 10:1462. https://doi.org/10.3389/fimmu.2019.01462

Wang L, Hong W, Zhu H, He Q, Yang B, Wang J, Weng Q (2024) Macrophage senescence in health and diseases. Acta Pharm Sin B 14(4):1508–1524. https://doi.org/10.1016/J.APSB.2024.01.008

Weiss G, Schaible UE (2015) Macrophage defense mechanisms against intracellular bacteria. Immunol Rev 264(1):182–203. https://doi.org/10.1111/IMR.12266

Witzel II, Koh LF, Perkins ND (2010) Regulation of cyclin D1 gene expression. Biochem Soc Trans 38(1):217–222. https://doi.org/10.1042/BST0380217

Wolf G (2000) Cell cycle regulation in diabetic nephropathy. Kidney Int Suppl 58(77):S59–S66. https://doi.org/10.1046/j.1523-1755.2000.07710.x

Wu K, Lin K, Li X, Yuan X, Xu P, Ni P, Xu D (2020) Redefining tumor-associated macrophage subpopulations and functions in the tumor microenvironment. Front Immunol 11:1731. https://doi.org/10.3389/fimmu.2020.01731

Wynn TA, Chawla A, Pollard JW (2013) Macrophage biology in development, homeostasis and disease. Nature 496(7446):445–455. https://doi.org/10.1038/NATURE12034

Yona S, Kim KW, Wolf Y, Mildner A, Varol D, Breker M, Strauss-Ayali D, Viukov S, Guilliams M, Misharin A, Hume DA, Perlman H, Malissen B, Zelzer E, Jung S (2013) Fate mapping reveals origins and dynamics of monocytes and tissue macrophages under homeostasis. Immunity 38(1):79–91. https://doi.org/10.1016/J.IMMUNI.2012.12.001

Yu D, Qi S, Guan X, Yu W, Yu X, Cai M, Li Q, Wang W, Zhang W, Qin J-J (2022) Inhibition of STAT3 signaling pathway by terphenyllin suppresses growth and metastasis of gastric cancer. Front Pharmacol 13:870367. https://doi.org/10.3389/fphar.2022.870367

Zhang L, Jiang X, Pfau D, Ling Y, Nathan CF (2021) Type I interferon signaling mediates mycobacterium tuberculosis-induced macrophage death. J Exp Med 218(2):e20200887. https://doi.org/10.1084/JEM.20200887

Zhou C, Ma J, Su M, Shao D, Zhao J, Zhao T, Song Z, Meng Y, Jiao P (2018) Down-regulation of STAT3 induces the apoptosis and G1 cell cycle arrest in esophageal carcinoma ECA109 cells. Cancer Cell Int 18(1):53. https://doi.org/10.1186/s12935-018-0549-4

Part II
Tissue/Organ Specific Functions of Macrophages

Chapter 5
Origin and Role of Testicular Macrophages in Testis Development, Steroidogenesis, and Spermatogenesis

Rafal P. Piprek, Malgorzata Kloc, Klaudia Porebska, Paulina C. Mizia, Izabela Rams-Pociecha, and Jacek Z. Kubiak

Abstract Testicular macrophages are the principal immune cells in the testis. In addition to their classical immune roles, they regulate male hormone synthesis by Leydig cells, regeneration of Leydig cells, spermatogonia proliferation and differentiation, maintenance of testis-specific environment for sperm formation, and testis development. The juvenile and adult testes contain two distinct macrophage populations with unique tissue localization, genetic markers, morphology, and function. The interstitial macrophages are physically and functionally connected to Leydig cells, while the peritubular macrophages localize around the seminiferous

R. P. Piprek (✉)
Department of Comparative Anatomy, Institute of Zoology and Biomedical Research, Jagiellonian University, Krakow, Poland
e-mail: rafal.piprek@uj.edu.pl

M. Kloc
Transplant Immunology, The Houston Methodist Research Institute, Houston, TX, USA

Department of Surgery, The Houston Methodist Hospital, Houston, TX, USA

Department of Genetics, The University of Texas, MD Anderson Cancer Center, Houston, TX, USA

K. Porebska
Laboratory of Molecular Oncology and Innovative Therapies, Department of Oncology, Military Institute of Medicine, Warsaw, Poland

P. C. Mizia · I. Rams-Pociecha
Department of Comparative Anatomy, Institute of Zoology and Biomedical Research, Jagiellonian University, Krakow, Poland

Doctoral School of Exact and Natural Sciences, Jagiellonian University, Krakow, Poland

J. Z. Kubiak
Laboratory of Molecular Oncology and Innovative Therapies, Department of Oncology, Military Institute of Medicine, Warsaw, Poland

Dynamics and Mechanics of Epithelia Group, Faculty of Medicine, Institute of Genetics and Development of Rennes, University of Rennes, CNRS, UMR 6290, Rennes, France

M. Kloc et al. (eds.), *Monocytes and Macrophages in Development, Regeneration, and Disease*, Results and Problems in Cell Differentiation 74,
https://doi.org/10.1007/978-3-031-65944-7_5

tubules and are crucial for spermatogonia differentiation. Macrophages in the fetal testes regulate testis vasculature formation and clearance of mislocated cells. The origin of testicular macrophages is unclear. Some studies suggest their origin from the yolk sac and others from the bone marrow-derived monocytes. We discuss this issue at the end of this review article.

5.1 Introduction

The testis is composed of two histological compartments with distinct functions: numerous seminiferous tubules that contain germ cells and their precursors (including spermatogonial stem cells (SSCs), spermatogonia, spermatocytes, and spermatids), which, during spermatogenesis give rise to spermatozoa, and somatic Sertoli cells that serve as scaffolding in seminiferous tubules, enclose and nourish germ cells, and regulate spermatogenesis. The seminiferous tubules are surrounded by a basement membrane to which Sertoli cells and spermatogonia adhere (Fig. 5.1). The basement membrane is also a boundary separating the interior of the seminiferous tubule from another compartment of the testis, called the interstitium. The interstitium consists of connective tissue, blood vessels, and a variety of cells, including contractile peritubular myoid cells surrounding the seminiferous tubules and responsible for their peristalsis, steroidogenic Leydig cells producing testosterone, and immune cells, including macrophages.

Macrophages are the most abundant immune cells in the testis and comprise a major portion of cells (up to 25%) in the interstitium (Niemi et al. 1986; Winnall et al. 2011; Bhushan et al. 2015; Bhushan et al. 2020; Gu et al. 2023). Although macrophage roles have been traditionally associated with phagocytosis and antigen-presentation, they also have specialized roles in the development and functioning of many organs including gonads (Stefater 3rd et al. 2011; Wynn et al. 2013). Already in early studies, testicular macrophages have been associated with the regulation of Leydig cell steroidogenesis (Gaytan et al. 1994a, b, c; Hales 2002; Hutson 2006) and maintenance of the immunosuppressive environment in the testis (Kern et al. 1995; Fijak and Meinhardt 2006; Wang et al. 2017). However, more recently, the peritubular macrophages were also shown to promote spermatogonial proliferation and differentiation (DeFalco et al. 2015).

Two major populations of testicular macrophages have been described. They differ in markers, localization within the gonad, roles, and developmental origins: (1) interstitial macrophages (CSF1R$^+$ CD206$^+$ MHCII$^-$) and (2) peritubular macrophages (CSF1R$^-$ CD206$^-$ MHCII$^+$). Even early studies have shown that macrophages, defined as F4/80$^+$ cells, are often associated with developing Leydig cells, and sometimes are present in the myoid layer (Pollard et al. 1997). Interstitial macrophages are spherical and embedded among Leydig cells and blood vessels in the

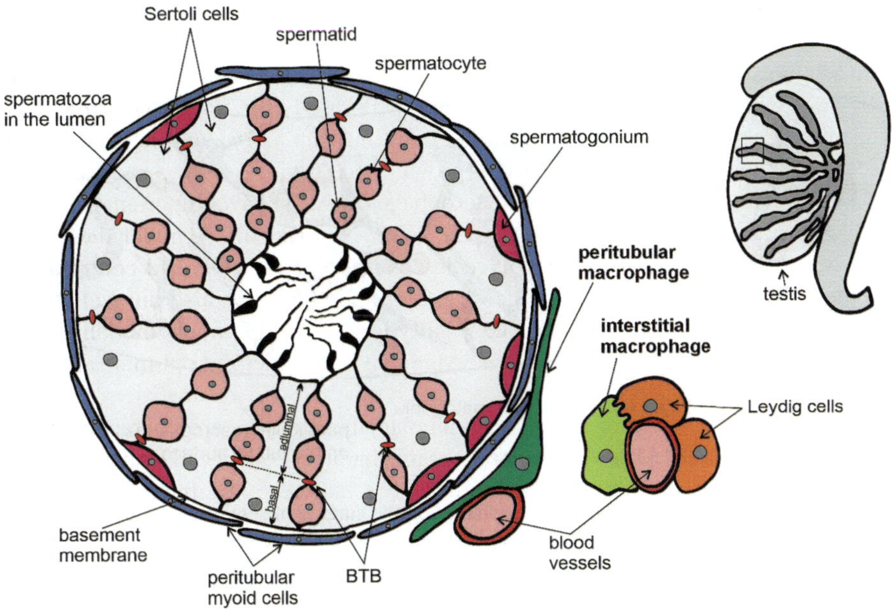

Fig. 5.1 Diagram of the adult testis structure. The seminiferous tubule contains Sertoli cells and germ cells at various stages, including spermatogonia, spermatocytes, spermatids, and spermatozoa, which will be released into the lumen. Sertoli cells and undifferentiated spermatogonia adhere to the basement membrane which separates the interior of the seminiferous tubule from the interstitium. Sertoli cells form tight intercellular connections in some places, separating the basal zone from the adluminal zone and forming the blood-testis barrier (BTB). The seminiferous tubule is surrounded by a layer of peritubular myoid cells, next to which peritubular macrophages, with long processes, are located. In groups of Leydig cells, interstitial macrophages are present. These cells are spherical and adhere to Leydig cells with numerous short projections. Both types of macrophages also contact with blood vessels located in the interstitium

interstitium; they are in contact with blood vessels within Leydig cell clusters (Fig. 5.1) (DeFalco et al. 2015). Peritubular macrophages are distributed exclusively on the external surface of seminiferous tubules and surround the tubules (Figs. 5.1 and 5.2). These cells have long processes (up to 100 μm) and a stellate shape (DeFalco et al. 2015). They stretch over the tubule surface and are closely associated with vasculature. However, some remain independent and intermingle with PMCs (peritubular myoid cells) to form rosette-like cellular structures (DeFalco et al. 2015).

Thus, there are two distinct subpopulations of testicular macrophages with different tissue localization, morphology, and surface marker expression emerging at different times during postnatal development, and they might also have a distinct developmental origin (Mossadegh-Keller et al. 2017).

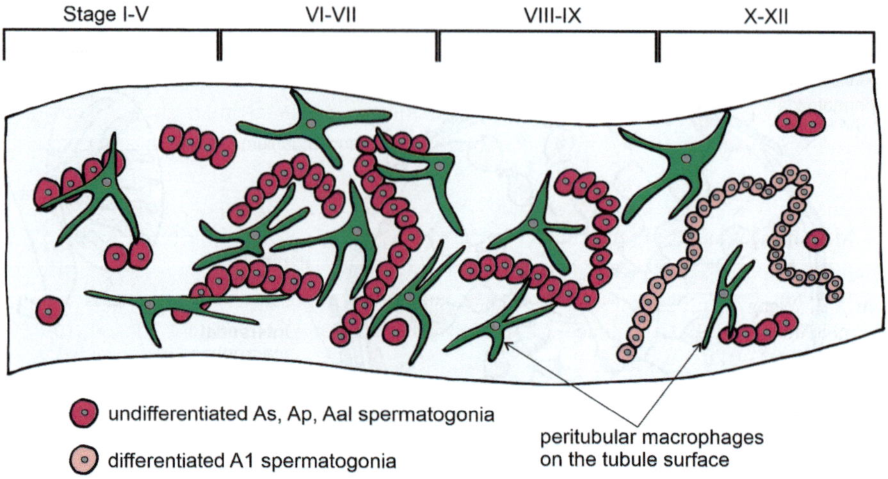

Fig. 5.2 The density of peritubular macrophages on the surface of seminiferous tubules in mice depends on the spermatogenic cycle. The highest density of macrophages occurs in stages VI–VII, when the density of undifferentiated spermatogonia, including A_{single}, A_{paired}, and $A_{aligned}$ spermatogonia, is also the highest. The lowest density of macrophages is in stages X-XII, which correlates with the lowest number of undifferentiated spermatogonia and the highest number of differentiating spermatogonia. The peritubular macrophages are stellate-shaped with long processes (inspired by Fig. 2 in DeFalco et al. 2015)

5.2 Role of Testicular Macrophages in Steroidogenic Activity and Regeneration of Leydig Cells

Macrophages and Leydig cells are in close contact and there is approximately one resident macrophage for every four Leydig cells (Bergh 1985).

Early studies showed intimate associations between testicular macrophages and Leydig cells suggesting the functional interactions (Miller et al. 1983; Hutson 1994; Hales 1996). Direct contact between processes of Leydig cells and invaginations of macrophages were noticed in the rat testis (Christensen and Gillman 1969). It was already known that products of macrophages, especially IL-1 (interleukin-1) and TNFα (tumor necrosis factor α), modulate the steroid biosynthesis by Leydig cells (Hales 1992; Hutson 1993; Xiong and Hales 1993, 1994). Pharmacological depletion of testicular macrophages disrupted Leydig cell function and reduced testosterone biosynthesis (Gaytan et al. 1994a, 1995a; van Rooijen and Sanders 1994; Wang et al. 1994).

Intratesticular injection of liposome-entrapped dichloromethylene diphosphonate (Cl2MDP-Lp) was used to study the functions of macrophages in the testes (Bergh et al. 1993). Cl2MDP also known as DMDP is toxic; its liposome-entrapped form is ingested by macrophages and liposome membrane is digested by macrophage phospholipases, inducing apoptosis *in vivo* and *in vitro* (Naito et al. 1996). Injection of Cl2MDP-Lp led to a 97% reduction in the number of macrophages 10 days after treatment (Gaytan et al. 1994b). In such macrophage-depleted testes, Leydig cell size and the level of testicular testosterone were reduced, and lipid droplets accumulated in the Leydig cells' cytoplasm (Bergh et al. 1993).

The same method of macrophage depletion showed a role of macrophages in the prepubertal development of Leydig cells in the rat (Gaytan et al. 1994b). In macrophage-depleted testes, Leydig cell development was inhibited in the rats injected at postnatal day 5, 10, or 15, however, at day 35, the testis was repopulated with macrophages, followed by Leydig cell development.

The following study by the same group showed that macrophages are essential for the regeneration of the Leydig cell population after their chemical ablation with EDS (ethane dimethane sulphonate). 30 days after EDS treatment, new and regenerating Leydig cells were abundant in the macrophage-containing testes. However, in the macrophage-depleted testes (injected with Cl2MDP-lp), the number of Leydig cells was related to the time elapsed between EDS ablation and the loss of macrophages (Gaytan et al. 1994c). This indicated that testicular macrophages are crucial for the proliferation and differentiation of new Leydig cells after EDS treatment (Gaytan et al. 1994b).

In a more recent study, a depletion of mouse testicular macrophages was achieved using the Cre-inducible diphtheria toxin receptor (DTR) driven by *Cx3cr1*-Cre (DeFalco et al. 2015). This led to a > 90% reduction of macrophages in the testes. Although the expression of HSD3B1 and CYP17A1 was detected, indicating ongoing steroidogenesis, the intratesticular testosterone was reduced by approximately 50% in macrophage-depleted samples compared to controls (DeFalco et al. 2015). No statistically significant reductions in serum testosterone levels were observed. These results suggest that the basic functions and the production of testosterone by Leydig cells were not drastically affected by the absence of (DeFalco et al. 2015).

In the rat testis, after the induction of Leydig cell apoptosis with EDS treatment, there was a twofold increase in the number of macrophages, which is consistent with their role in the removal of apoptotic cells from a tissue (Gaytan et al. 1995b). However, macrophage-depleted testes showed Leydig cell death, greater influx of small lymphocyte-like cells and circulating monocytes, and inflammatory infiltration (Gaytan et al. 1995b). These data suggest that testicular macrophages prevent the inflammatory reaction induced by Leydig cell depletion (Gaytan et al. 1995b).

It has been shown that conditioned medium from cultures of testicular macrophages was capable of stimulating testosterone production when added to *in vitro* culture of Leydig cells in a dose-dependent manner (Yee and Hutson 1985a). Conditioned medium from cultures of peritoneal macrophages had less effect on testosterone production by Leydig cells than testicular macrophage-conditioned medium (TMCM) (Yee and Hutson 1985a). Thus, it was shown that secretory products specifically from testicular macrophages play an important role in testicular function, and the existence of a lipophilic factor, named macrophage-derived factor (MDF), secreted by macrophages and stimulates testosterone production by Leydig cells, has been postulated (Nes et al. 2000). Hypothetical MDF was purified from culture media containing testicular macrophages. Gas chromatography and nuclear magnetic resonance spectrum analysis indicated that the MDF is a 25-hydroxycholesterol (Nes et al. 2000). Intratesticular injection of 25-hydroxycholesterol led to the increase of testosterone production indicating that this is at least one of the factors produced by macrophages and stimulating testosterone production by Leydig cells (Nes et al. 2000).

The intratesticular injection of liposomes containing Cl2MDP led to a significant increase in serum concentrations of LH (luteinizing hormone) (Gaytan et al. 1995a), which indicated the relevance of testicular macrophages in the regulation of the pituitary-testicular axis. Testicular macrophages have follicle-stimulating hormone (FSH) receptors, and in response to FSH, they produce a factor that stimulates Leydig cell testosterone secretion *in vitro* (Yee and Hutson 1983, 1985b, c). It was postulated that immune responses are inhibited in the testis (Hedger et al. 1990; Maddocks and Setchell 1990; Setchell et al. 1990), however, an inflammatory reaction can be provoked in the testis by luteinizing hormone (LH) or hCG treatment (Bergh et al. 1988, 1993).

An important discovery was the presence of a macrophage-associated mutation in mice that disrupted Leydig cell function. Individuals homozygous for a null mutation in the gene of the colony-stimulating factor-1 (CSF1) have a significantly lower density of macrophages in the testis, cauda, caput epididymis, prostate, seminal vesicles, and vas deferens (Cohen et al. 1996; Pollard et al. 1997). Additionally, such mutant mice showed a significant reduction in serum and intratesticular testosterone concentrations, and although the mice were fertile, they had reduced libido and sperm numbers. This finding proved the role of testicular macrophages in the positive regulation of Leydig cell function.

In mice, prolonged treatment with anti-CSF1R antibodies led to the complete removal of macrophages in the testicular interstitium. However, the testicular architecture and spermatogenesis appeared unaffected. There was no effect on the level of circulating testosterone. This finding suggests that macrophage support of the steady-state testis steroidogenesis in adulthood is dispensable or can be compensated through the LH/testosterone feedback (Sauter et al. 2014).

In the supernatant of the testicular macrophage culture medium, the concentration of granulocyte-macrophage colony-stimulating factor (GM-CSF, also known as CSF2) was approximately 740% of that from peritoneal macrophages. IL-1, IL-6 and TNFα concentrations from testicular macrophages were like those from peritoneal macrophages (50%, 60%, and 130% of peritoneal macrophage levels) (Kern et al. 1995). Testicular macrophages are also a rich source of prostaglandins (Kern and Maddocks 1995).

5.3 Role of Macrophages in Spermatogonia Proliferation and Differentiation

The continuity of spermatogenesis from puberty throughout the lifespan is ensured by the presence of spermatogonial stem cells (SSCs) (DeFalco et al. 2015). These cells undergo mitotic divisions, giving rise to a pool of undifferentiated spermatogonia, which proliferate and eventually differentiate into spermatocytes I, and enter meiotic division. SSCs and undifferentiated spermatogonia reside on the basement membrane of seminiferous tubules. Peritubular macrophages have been suggested to participate in maintaining the SSC niche and play a crucial role in spermatogonia proliferation and differentiation (DeFalco et al. 2015).

A transient ablation of macrophages in the adult testis does not affect SSC maintenance, but results in fewer spermatogonia, likely by affecting spermatogonial differentiation (DeFalco et al. 2015). The authors showed that peritubular macrophages are spatially associated with spermatogonial precursors and express spermatogonial proliferation- and differentiation-inducing molecules such as colony-stimulating factor 1 (CSF1) and enzymes involved in the retinoic acid (RA) biosynthetic pathway (DeFalco et al. 2015).

They found a relationship between undifferentiated spermatogonia (ZBTB16-positive cells) and peritubular macrophages in adult mouse testes (DeFalco et al. 2015). Spermatogonia were often adjacent to macrophages surrounding the seminiferous tubules, and large groups of spermatogonia were in seminiferous tubule regions enriched in macrophages (Fig. 5.2). The strongest macrophage signal was detected in the vicinity of undifferentiated spermatogonia (including most undifferentiated spermatogonia A_{single}, syncytial cyst of 2 spermatogonia A_{paired}, and syncytial cysts of 4, 8, and 6 spermatogonia $A_{aligned\text{-}4,8,16}$). A weaker signal was detected near differentiated spermatogonia A1–A3. Macrophages were overlying the basement membrane in the layer of peritubular myoid cells (PMCs) and did not penetrate deeply into the tubules (Fig. 5.1). However, macrophages were often found in close proximity to blood vessels and undifferentiated spermatogonia (DeFalco et al. 2015).

The authors ablated macrophages using the Cre-inducible diphtheria toxin receptor (DTR) method driven by *Cx3cr1*-Cre and obtained a significant reduction (>90%) in macrophage number (DeFalco et al. 2015). Importantly, in macrophage-depleted testes the population of spermatogonial stem cells (SSCs) was maintained, however, the number of undifferentiated and differentiated spermatogonia was significantly reduced. This was accompanied by a drop in the proliferation of spermatogonia (BrdU-incorporating cells). After 14 days macrophages recolonized the gonads, and undifferentiated spermatogonia reappeared, which indicated that SSCs persisted and could produce spermatogonia again once macrophages recolonized the testis (DeFalco et al. 2015). Although the number of spermatogonia was reduced, the number of somatic (Sertoli) cells was normal in macrophage-depleted testes. Sertoli cells had normal morphology and expressed both *Sox9* and *Gata4*, and the vasculature and the Sertoli cell blood-testis barrier were unchanged (DeFalco et al.

2015), indicating that peritubular macrophages are not important for the structure and function of Sertoli cells, integrity of seminiferous epithelium and blood-testis barrier.

Both interstitial and peritubular macrophages express CSF1 (Colony stimulating factor 1; DeFalco et al. 2015). The CSF1 receptor—CSF1R—is expressed in the SSC/undifferentiated spermatogonial population, and CSF1 promotes proliferation and/or self-renewal of *ex vivo* cultured spermatogonia (Kokkinaki et al. 2009; Oatley et al. 2009). This indicates that CSF1 is a factor (or one of the factors) *via* which macrophages influence spermatogonia.

Retinoic acid (RA) is a paracrine factor important for the differentiation of spermatogonia and entry into meiosis (Griswold et al. 1989; Gewiss et al. 2021). In the testes of RA-depleted (by a specific diet) animals, the germ cells are arrested at the spermatogonial stage, like in macrophage-depleted testes. Aldehyde dehydrogenase ALDH1A2 (RALDH2), an enzyme essential for RA synthesis, was expressed in Leydig cells and interstitial macrophages, and only at a low level in peritubular macrophages. However, the expression of retinol dehydrogenase RDH10, another key enzyme for RA synthesis, was detected in peritubular macrophages, and only in a few interstitial macrophages. RDH10 expression in the interstitium was almost completely absent when peritubular macrophages were absent. Thus, peritubular macrophages might contribute to the RA concentration needed for spermatogonia differentiation and entry into meiosis.

Altogether, these findings showed that macrophages locate alongside seminiferous tubules in regions with a high density of spermatogonial precursors and that they impact the ability of undifferentiated spermatogonia to differentiate and proceed through spermatogenesis (DeFalco et al. 2015). They act likely (although not exclusively) through the regulation of CSF1 expression and RA synthesis.

5.4 Role of Macrophages in Testis Development

Early studies of mouse postnatal and adult testis reported the presence of F4/80$^+$ macrophages exclusively in the interstitium, but not within the seminiferous tubules (Hume et al. 1984). Using immune cell markers, such as ED1 (CD68), ED2 (CD163), and OX-42 (CD11b), the presence of macrophages only in the interstitial compartment was also shown in rat fetal (from stage E17 onwards), postnatal, and adult testes (Hutson 1990). In the rat and mouse fetal testes, the analysis with anti-ED1 antibodies showed the presence of macrophages in the interstitium at stage E16.5, followed, in the later stages, by an increase in their size and number (Livera et al. 2006).

DeFalco et al. (2014) reported for the first time the presence of macrophages inside the testis cords of fetal mouse testis. The authors demonstrated that macrophages were present at the site of gonad formation and that they played an integral part in the testicular organogenesis program (DeFalco et al. 2014). Firstly, the large numbers of macrophages accumulated in the vascular plexus at the gonad–mesonephros boundary between E11.5 and E13.5, i.e., in the phase of vascular remodelling and neovascularization of the testis (Coveney et al. 2008; DeFalco et al. 2014). The formation of the coelomic vessel near the coelomic epithelium as well as branches extending from this vessel that separate the developing testis cords is essential for the vasculogenesis of fetal testes (Coveney et al. 2008). Macrophages in the fetal testes were shown to be in blood vessel proximity: in the vascular plexus and along the coelomic vessel, and in its branches separating developing testis cords. During the extensive vascular remodelling in the fetal testis, VEGFR2-positive endothelial cells were surrounded by macrophages, both in the mesonephric vascular plexus and in the coelomic vessel (Larina et al. 2009).

The fetal testis macrophages expressed Neuropilin 1 (NRP1) and endothelial-specific receptor tyrosine kinase (TEK), both well-known as endothelial markers and factors involved in neovascularization (DeFalco et al. 2014). This indicated the role of macrophages in the vascularization of fetal testes.

An experiment depleting gonadal vascularization through inhibition of vascular endothelial growth factor receptor (VEGFR) led to a decrease (55–65%) in the expression of macrophage markers. The authors concluded that macrophage localization and/or migration into the gonad depends on VEGF-mediated vascularization of the fetal testis (DeFalco et al. 2014).

It was also proposed that one of the roles of macrophages in gonadal development is to remove the misplaced cells (DeFalco et al. 2014). Firstly, some of the primordial germ cells (PGCs) migrating to gonadal primordia may get lost along the long way from the yolk sac vicinity to the gonadal ridges. It has been demonstrated that such misplaced PGCs found in the mesonephros cluster into groups, which are surrounded by macrophages. Second, pre-Sertoli cells may become misplaced during the formation of testis cords, and as a result, they may end up in the interstitium. It has been observed that such misplaced pre-Sertoli cells are also surrounded and phagocytosed by macrophages. Thus, macrophages may be responsible for the proper formation of the gonad by phagocyting the mislocated PGCs and somatic cells. This seems of particular significance because mislocated germ cells were shown to give rise to teratomas or other types of tumors (De Felici et al. 2021), and the clearance function of macrophages can prevent such pathologies.

During the extensive vascular remodelling in the developing mouse gonads, the apoptotic endothelial cells were engulfed and phagocytized by macrophages (Larina et al. 2009). It was also postulated that macrophages may physically break down the vascular plexus to free individual endothelial cells, which then can migrate from the mesonephros inward to the developing gonad (DeFalco et al. 2014).

In the mouse fetal testes, besides the interstitium, many macrophages were observed inside the testis cords (DeFalco et al. 2014). The macrophages in the testis cords were much more stellate than the interstitial macrophages and were located between Sertoli cells and germ cells, contacting multiple cells in the testis cord (DeFalco et al. 2014). Postnatally, macrophages were restricted to the interstitium, and, like in adult testes, were completely excluded from testis cords (Hume et al. 1984).

To study the role of macrophages in the developing mouse gonads, a transgenic model with DTA (Diphtheria Toxin A) under the control of the *Cx3cr1* gene allowed specific depletion of macrophages (DeFalco et al. 2014). In experimental animals with macrophage-depleted testes, the vascularization of developing testes was affected; the vascular plexus at the boundary of the mesonephros and gonad and the overall vasculature remaining in the mesonephros were dramatically enlarged. However, inside the gonad, the vasculature was poorly organized with a lower number of endothelial cells, the testis cords were irregularly branched and fused, and the groups of mislocated pre-Sertoli cells were in the interstitium, indicating the loss of macrophage clearing function (DeFalco et al. 2014). Thus, fetal testicular macrophages mediate the remodelling of vasculature and are critical for the normal organogenesis of the testis.

5.5 Proliferation and Turnover of the Testis Macrophages

In adult testes, the proliferation rate of testicular macrophages is usually low. Their proliferation is under the tight control of the core macrophage transcription factors MafB and cMaf. These factors inhibit the proliferative capacity of macrophages (Aziz et al. 2009). However, testicular macrophage proliferation can periodically escalate under specific conditions. For example, the ablation of Leydig cells by EDS in the rat caused a transient increase in interstitial testicular macrophages (Schlatt et al. 1999).

It is possible that under non-inflammatory conditions, testicular macrophages undergo natural turnover through a slow infiltration of circulating monocytes/macrophages followed by a local conversion into an M2 resident phenotype (Sauter et al. 2014).

However, during fetal development, when a massive expansion of embryonic macrophages occurs, for example in the brain and testes, the extensive proliferation of macrophages and increase in macrophage number, has been observed. In mouse fetuses, the proliferation starts at E12.5 and progresses until the early postnatal period (DeFalco et al. 2014; Mossadegh-Keller et al. 2017). In the postnatal period, the decrease in proliferation of interstitial testicular macrophages is compensated by the infiltration of bone marrow-derived macrophages (Mossadegh-Keller et al. 2017; Mossadegh-Keller and Sieweke 2018). Once the populations of peritubular and interstitial macrophages are established, they exhibit a long life span and a low proliferation rate in the steady state (Mossadegh-Keller et al. 2017).

5.6 Immune Aspects and Immune-Suppressive Functions of Testicular Macrophages

5.6.1 Immunosuppressive Role of Testicular Macrophages in the Protection of Spermatogenesis from Autoimmune Attack

Early studies indicated that testicular macrophages may be involved in maintaining the proper environment for spermatogenesis (Hedger et al. 1990; Maddocks and Setchell 1990; Pöllänen et al. 1990; Setchell et al. 1990).

It was shown that antibodies against sperm cells can contribute to male infertility (Wilson 1954). The strong immunogenic properties of the sperm are explained by the fact that these haploid cells, with their repertoire of novel antigens, begin to emerge during the onset of spermatogenesis at puberty, long after the education of the immune system and the development of immune tolerance (Mossadegh-Keller and Sieweke 2018). Thus, to prevent contact of the highly immunogenic sperm with immune cells, which could induce autoimmune infertility, the process of spermatogenesis must be protected from the immune system (Mossadegh-Keller and Sieweke 2018). The blood-testis barrier (BTB), the physical barrier formed by a belt of tight junctions between Sertoli cells, seals off cells in the haploid phase of spermatogenesis inside the seminiferous tubules isolating them from the rest of the testis (Mossadegh-Keller and Sieweke 2018). Testicular macrophages have been indicated as guardians of fertility and crucial for maintaining the immunosuppressive environment in the testis. They create a unique immune-privileged environment by producing immunosuppressive cytokines, such as IL-10 and TGF-β, and low expression of proinflammatory cytokines (Kern et al. 1995; Fijak and Meinhardt 2006; Wang et al. 2017; Mossadegh-Keller and Sieweke 2018). However, two studies have suggested that macrophages are dispensable for spermatogenesis. Lokka et al. (2020) found normal spermatogenesis in the postnatal testis lacking macrophages. The authors concluded that macrophage presence after birth is not needed for normal spermatogenesis. Similarly, DeFalco et al. (2015) did not find any effects on spermatogenesis progression in macrophage-depleted mouse testes. Therefore, testicular macrophages may be dispensable during normal physiological conditions but may play some role during stress or other pathological situations.

5.6.2 Role of Testicular Macrophages in Immune Response

Macrophages are classified into a classical inflammatory (M1) and an alternatively activated immunosuppressive (M2). M1 macrophages participate in the clearance of pathogens, whereas M2 macrophages are involved in tissue homeostasis, immune surveillance, and inflammation resolution (Wang et al. 2017). Testicular macrophages are categorized as M2 macrophages and have immunosuppressive properties. They locally produce small amounts of corticosterone, which suppresses the expression of inflammatory genes and renders them refractory to inflammatory stimuli (Wang et al. 2017). The corticosterone in the testicular environment shapes the immunosuppressive function and phenotype of testicular macrophages. This steroid may play an important role in the establishment and sustenance of the immune privilege of the testis (Wang et al. 2017).

Both peritubular and interstitial macrophages express a series of macrophage-specific genes at similar levels, however, some genes are differentially regulated (Mossadegh-Keller and Sieweke 2018). Both peritubular and interstitial macrophages are positive for CX3CR1, and both are negative for CD11c but express different levels of CD64 and MerTK, and both subpopulations do not express the proinflammatory genes but express many immunosuppressive and M2-type gens (Mossadegh-Keller et al. 2017). The expression of only a few genes differed in these two subpopulations, including the immunosuppressive M2-type genes: IL10 and macrophage receptor with collagenous structure (MARCO), which were highly expressed in interstitial macrophages, and IL1b, which was highly expressed in peritubular macrophages. Peritubular macrophages express high levels of genes regulating antigen processing and presentation such as *H2.Dmb*, *H2.Eb1*, *Nlrp3*, *H2.K1* and *Icam1* (Mossadegh-Keller and Sieweke 2018).

Whole-genome expression profiling using RNA-seq showed that, compared to epididymal and brain macrophages, testicular macrophages expressed significantly higher levels of immunosuppressive genes, which maintains the immune privilege of the testes (Wang et al. 2021).

In response to inflammatory stimuli, testicular macrophages induce the secretion of large amounts of IL10 and a low amount of TNF-α (Bhushan et al. 2015) and activate the anti-inflammatory STAT3 signaling pathway (Bhushan et al. 2015). Testicular macrophages exhibit an attenuated inflammatory response by expressing low levels of Toll-Like Receptor (TLR) signalling pathway genes (compared with peritoneal macrophages) and inhibiting the activation of the NF-κB signaling pathway by impairing the ubiquitination of phosphorylated IκBα (Bhushan et al. 2015; Wang et al. 2017). Although in testicular macrophages the NF-κB signaling pathway is blocked, they paradoxically secrete proinflammatory cytokines through the AP-1 and CREB signaling pathways, albeit at lower levels than peritoneal macrophages (Bhushan et al. 2015; Wang et al. 2017). Testicular macrophages are thought to have a blunted secretory response compared to peritoneal macrophages (Hales 2002). In noninflammatory conditions, testicular macrophages display ideal properties to maintain the normal homeostasis of the testis, while being able to adequately respond to infections by retaining bactericidal and phagocytic activities (Wang et al. 2017).

5.7 Macrophages in Human Fetal and Adult Testes

The global transcriptome analysis using single-cell RNA sequencing identified two populations of fetal testis-specific macrophages (ftMs) in human fetal testes (Garcia-Alonso et al. 2022). The first type, SIGLEC15$^+$ ftMs had an osteoclast-like signature (SIGLEC15, ACP5, ATP6V0D2) gene expression. The second one, TREM2$^+$ ftMs had a microglia-like signature (TREM2, P2RY12, SALL1). They also identified the tissue-repair macrophages (CD68$^+$, F13A1$^+$) in the human fetal testes. The latter comprised 92.2% of all macrophages in fetal testes, the SIGLEC15$^+$ ftMs 2.8%, and TREM2$^+$ ftMs 5%. Thus, testis-specific macrophages constituted only a small fraction of the testis macrophage population. Single-molecule fluorescence *in situ* hybridization (smFISH) showed the presence of all three types of macrophages in the interstitium of human fetal testes, and additionally, TREM2$^+$ ftMs were detected in testis cords (Garcia-Alonso et al. 2022).

SIGLEC15$^+$ ftMs were found in the proximity of endothelial cells in the human fetal testes and showed also an expression of *COL1A2*, which can potentially interact with integrins expressed by endothelial cells (Garcia-Alonso et al. 2022). SIGLEC15$^+$ ftMs also expressed metalloproteinase *MMP9*, which can potentially be a factor involved in promoting mesonephric endothelial cell migration. Moreover, SIGLEC15$^+$ ftMs expressed *LGALS9* and *SPP1*, indicating their potential immunoregulatory role (Garcia-Alonso et al. 2022).

TREM2$^+$ ftMs communicate with Sertoli and germ cells inside the testis cords through the interaction between TREM2 and apolipoproteins (CLU, APOA1, APOE) (Garcia-Alonso et al. 2022). The gene expression analysis also suggested an active phagocytosis machinery in the TREM2$^+$ ftMs (genes *MERTK, AXL, CYBB, BECN1, MTOR*) and their immunomodulatory abilities (genes *HAVCR2, ENTPD1, CD276, IL10, TREM2*) (Garcia-Alonso et al. 2022). It is worth noting that previously it was believed that in mammals, testicular macrophages were only present outside the seminiferous tubules. This study revealed the presence of macrophages inside the tubules.

An early study of adult human testes showed the presence of testicular macrophages in the interstitium and around the seminiferous tubules (Pöllänen and Niemi 1987). These two subpopulations expressed CD14, CD68, and CD163 (Frungieri et al. 2002; Ponte et al. 2018). It remains unknown what proteins regulate the differentiation of peritubular *vs.* interstitial macrophages in the human testis.

5.8 Debate on the Origin of Macrophages in Embryonic Gonads

During mouse fetal development the tissue-resident macrophages originate in three waves. In the first one, termed "primitive haematopoiesis", macrophages originate from the yolk sac. In the second wave, the multipotent erythro-myeloid progenitors

(EMPs) originate from the yolk sac and migrate and colonize the fetal liver, where they generate myeloid progenitors, including fetal liver monocytes, which give rise to most tissue macrophage populations. In the third wave, haematopoiesis proceeds in the aorta-gonad-mesonephros region and contributes to immature hematopoietic stem cells (HSCs), which then migrate and colonize the fetal liver to establish definitive haematopoiesis. Shortly before birth, the site of haematopoiesis shifts from the liver to the bone marrow, and after the first week of life, bone marrow-derived monocytes are produced. They circulate in the blood and are ready to respond to infection (Sieweke and Allen 2013; Bain et al. 2014; Heidt et al. 2014; Ginhoux and Guilliams 2016; Perdiguero and Geissmann 2016; Mossadegh-Keller and Sieweke 2018).

Several studies, presenting contradictory results on the origin of testicular macrophages, have been published to date (DeFalco et al. 2014; Mossadegh-Keller et al. 2017; Lokka et al. 2020; Wang et al. 2021; Gu et al. 2023).

DeFalco et al. (2014) reported that yolk-sac–derived macrophages are the only major leucocytes in the fetal gonad during morphogenesis of the testis. They showed that prior to the beginning of the gonad formation (at embryonic day E9.5) there were no macrophages at the site of the urogenital ridge (UGR; aka gonad–mesonephros primordium) or along the path of primordial germ cell (PGCs) migration from the gut to UGR. At E10.5, macrophages (F4/80$^+$ cells) were detected in the UGR and were specifically concentrated in the mesonephros near the gonad border at E10.5. The analyzed cells had a stellate morphology characteristic of macrophages, were positive for macrophage marker F4/80, and strongly expressed chemokine (C-X3-C motif) receptor 1 (*Cx3cr1*)-GFP at all examined stages of gonadogenesis. The first macrophages within mouse embryonic gonads were observed at stages E11.5 and E12.5, i.e., at the beginning of the sexual differentiation. However, at this stage, they were still concentrated near the gonad-mesonephros boundary.

In the same paper, the authors traced the origin of gonadal–mesonephric macrophages from primitive yolk-sac–derived hematopoietic progenitors (DeFalco et al. 2014). Cells expressing Tomato red signal under the *Csf1r* promoter and originating during primitive haematopoiesis were observed in the embryonic gonads. The Tomato signal overlapped with the signal of macrophage and proliferation markers which suggested that macrophages derived from yolk-sac progenitors proliferate locally in the fetal gonads rather than being recruited from other hematopoietic sources, such as the aorta-gonad-mesonephros complex (AGM). In contrast, macrophages in the adult testes were Ki67-negative suggesting a lack of proliferation.

It was unclear whether the macrophages colonizing fetal gonads persist only during fetal development, or if they also give rise to macrophages in the adult testis. Gu et al. (2023) showed that fetal macrophages have only a minor contribution to adult macrophages in the adult gonads and are gradually replaced during life time. They also showed that in mouse fetuses during late gestation, monocytes originating from hematopoietic stem cells (HSCs) deriving from the aorta-gonad-mesonephros (AGM) invade the fetal testis and give rise to two adult testicular macrophage subsets: interstitial and peritubular macrophages, albeit with distinct developmental timing (Gu et al. 2023). Erythro-Myeloid Progenitors (EMP)-derived macrophages in the fetal testis are replaced in early adulthood by fetal-monocyte-derived macrophages. Moreover, HSCs arising from the bone marrow do not contribute to adult testicular macrophages.

After the depletion of yolk-sac-derived macrophage progenitors by injecting anti-CSF1R, the fetal testes were fully repopulated by stage E18.5 with CD11b + monocytes (Gu et al. 2023). This showed that YS-derived EMPs are dispensable for the formation of late fetal testicular macrophages and that other hematopoietic precursors may replace them functionally (Gu et al. 2023).

The fetal testis colonization by monocytes depends on Sertoli cells. The ablation of Sertoli cells with diphtheria toxin (DTA) does not affect the number of F4/80^{+} macrophages in the testes. However, the number of CD45^{+}IBA1^{-} cells (possibly monocytes) in the fetal testes was significantly reduced compared to the control (Gu et al. 2023). This indicates that Sertoli cells play a major role in the recruitment of fetal testicular monocytes, but not macrophages.

In contrast, in mice with the loss of germ cells in the testes (*Kit*$^{W/W\text{-}v}$ and *Dnd1*$^{Ter/Ter}$), testicular macrophage were not affected. This indicated that germ cells are not critical for macrophage recruitment or differentiation (Gu et al. 2023).

Lokka et al.' (2020) study altered previously established concepts in the macrophage biology of the testis. They showed the presence of many more monocyte and macrophage types in the testis than was thought before. The authors concluded that fetal macrophages persist in the postnatal testis and that it is unlikely that bone marrow-derived monocytes would significantly contribute to the generation of macrophages in the adult testis. They also indicated that fetal-derived CD206^{-}MHCII^{-} macrophages that originate directly from yolk sac macrophages and fetal liver-derived monocytes sequentially give rise to perivascular CD206^{+}MHCII^{-} and peritubular CD206^{-}MHCII^{+} macrophages. These authors found that the macrophages infiltrating the fetal testis were both of yolk sac and non-yolk sac origin. In contrast, DeFalco et al. (2014), using different methods, showed that gonadal macrophages exclusively derived from yolk sac progenitors. In E14.5–E17.5 testis, Lokka et al. (2020) found a definitive yolk-sac-derived macrophage population (CD115^{+}CD206^{+}Ly6C$^{-/Low}$). At this time point, they also found macrophages whose profile CD115^{-}CD206^{-}Ly6C^{Hi} strongly suggested their origin from fetal liver monocytes. Lokka et al. (2020) suggested that the discrepancy between these two studies can be attributed to the fact that only converted cells and only stage E14.5 and earlier time points were studied making impossible the detection of the fetal liver-derived macrophages.

Mossadegh-Keller et al. (2017) showed that the peritubular macrophages in the testis were entirely derived from bone marrow-derived cells, and first appeared during the prepubertal period. They also showed that among the interstitial macrophages, the fetal-derived cells were increasingly superseded by bone marrow-derived cells. In contrast, Lokka et al. (2020) showed no significant contribution from bone marrow-derived cells, and these cells were also unable to fill an empty testicular niche postnatally.

Lokka et al. (2020) also showed for the first time that macrophages were in contact with seminiferous tubules already at birth. In 1-week-old mice, these macrophages displayed the elongated, branching morphology along the tubules, but still did not express MHCII. The authors showed that $CD206^{-}MHCII^{+}$ peritubular macrophages represent fetal cell types, which undergo extensive phenotypic maturation postnatally rather than newly generated descendants of bone marrow-derived monocytes. High-dimensional mass cytometric analyses proved that all testicular monocyte/macrophage cell populations had $MHCII^{-}$ phenotype at birth. The $MHCII^{+}$ population gradually evolved in the testis after birth.

They also showed that fetal liver-derived macrophages rather than yolk sac-derived macrophages or bone marrow-derived ones were the dominant cells among both interstitial and peritubular macrophages (Lokka et al. 2020). Additionally, both these types of macrophages were present already at birth. They indicated fetal-derived $CD206^{-}MHCII^{-}$ macrophages as precursor cells, which sequentially give rise to both interstitial $CD206^{+}MHCII^{-}$ and peritubular $CD206^{-}MHCII^{+}$ macrophages. When repopulating an empty testicular niche postnatally, locally proliferating $CD206^{+}MHCII^{-}$ cells likely differentiate into $CD206^{-}MHCII^{+}$, which then take their peritubular position.

Wang et al. (2021) showed that two testicular macrophage populations $CD64^{hi}MHCII^{lo}$ (interstitial) and $CD64^{lo}MHCII^{hi}$ (peritubular) are self-maintaining in adult mouse testes and monocyte-independent, both arising either embryonically from fetal monocytes or immediately after birth from neonatal nascent blood monocytes. According to their observations, at first week of age, testis contained a single distinct population of $CD64^{hi}MHCII^{lo}$ macrophages, suggesting their embryonic origin (Wang et al. 2021). From the second week onward, a population of $CD64^{lo}MHCII^{hi}$ cells appeared, which progressively increased until it represented most of the cells at 12 weeks of age. These authors also demonstrated that definitive hematopoiesis plays an important role in the development of the testicular macrophage pool. This contribution is stable during early adulthood pointing to the importance of the embryonic and/or neonatal monocyte contribution (Wang et al. 2021). Moreover, testicular macrophages are long-lived and neither circulatory monocytes nor adult definitive HSCs contribute significantly to their pool in adult life.

There is a significant turnover of the testicular macrophage population at first and second week of age, which is reduced to almost zero from fourth week onward (Wang et al. 2021). Thus, the embryonically or neonatally seeded monocytes proliferate rapidly to fill the niche, then stop when the niche is full, and do not proliferate again in the steady state except for a minimal level of self-maintenance, and the replacement by circulatory monocytes is dispensable for the maintenance of both populations (Wang et al. 2021). Irradiation followed by bone marrow transplantation with EGFP-expressing cells showed bone marrow-derived cells can repopulate the testis when the niche is empty, but not in a steady state when the niche is full (Wang et al. 2021).

The conclusions of Wang et al. (2021) and Lokka et al. (2020) on testicular macrophage origin contrast with those of Mossadegh-Keller et al. (2017). Mossadegh-Keller et al. (2017), using different methods, found that the $CD64^{hi}MHCII^{lo}$

(peritubular) macrophages arise from embryonic progenitors, while the $CD64^{lo}MHCII^{hi}$ (interstitial) macrophage population exclusively arose from bone marrow-derived progenitors.

Mossadegh-Keller et al. (2017) used genetic lineage-tracing models and adoptive transfer protocols to study the origin and mode of homeostatic maintenance of testicular macrophages. They suggested that embryonic progenitors give rise to the interstitial macrophage population, whereas peritubular macrophages are exclusively seeded postnatally in the prepuberty period from bone marrow-derived progenitors. Bone marrow progenitors also contribute to the population of interstitial macrophages. Both populations of testicular macrophages have distinct origins during postnatal development. Once established, both the peritubular and interstitial macrophage populations have a long life span and a low turnover in the steady state. At birth and at later time points, there are only $M\text{-}CSFR^{+}$ $MHCII^{-}$ macrophages with interstitial localization. Macrophages at the peritubular location appear at the age of 2 weeks and further increase in number. They initially do not express MHCII. The expression of MHCII becomes visible in the sixth week, suggesting a potential maturation phase after localization to the peritubular niche. Using confocal imaging analysis, the authors showed that interstitial macrophages had a spherical shape while the peritubular macrophages were elongated.

The authors suggested that the lack of embryonic progenitor-derived macrophages in the peritubular population and their emergence only 2 weeks after birth indicates that this population developed from definitive hematopoiesis (from the bone marrow) in the juvenile animal (Mossadegh-Keller et al. 2017). They showed that peritubular macrophages develop exclusively from bone marrow-derived progenitors to populate a dedicated niche that only becomes available after birth (Mossadegh-Keller et al. 2017). However, the interstitial macrophages have mixed embryonic and bone marrow origin, as most macrophage populations in an organism (Mossadegh-Keller et al. 2017).

5.9 Conclusions

Despite decades of studies on testicular macrophages, several issues remain unexplained or yield contradictory results. It is known that in juvenile and adult testes, there are at least two populations of macrophages. In mice, only interstitial testicular macrophages are present at birth, while peritubular testicular macrophages appear 2 weeks after birth. Over the years, a paradigm has been established that testicular macrophages play a crucial role in reproduction, particularly in steroidogenesis and spermatogenesis. They are believed to support the function of Leydig cells, create an immune-privileged environment for sperm production, and influence the proliferation and differentiation of undifferentiated spermatogonia. However, some studies have demonstrated that macrophage depletion only reduces intratesticular testosterone levels by 50% and has no impact on blood testosterone levels. Furthermore, mutant mice with a reduced density of macrophages in the testes are still fertile. Another unclear issue is the origin of testicular macrophages during development. Many studies have provided conflicting results, suggesting that testicular macrophages may originate from yolk sac-derived or bone marrow-derived monocytes. This indicates that the research on the function and origin of testicular macrophages must be continued.

Acknowledgements JZK was supported by subvention of the Polish Ministry of Education and Science, project no 612/2023 to JZK at WIM-PIB.

References

Aziz A, Soucie E, Sarrazin S, Sieweke MH (2009) MafB/c-Maf deficiency enables self-renewal of differentiated functional macrophages. Science 326(5954):867–871

Bain CC, Bravo-Blas A, Scott CL, Perdiguero EG, Geissmann F, Henri S, Malissen B, Osborne LC, Artis D, Mowat AM (2014) Constant replenishment from circulating monocytes maintains the macrophage pool in the intestine of adult mice. Nat Immunol 15(10):929–937

Bergh A (1985) Effect of cryptorchidism on the morphology of testicular macrophages: evidence for a Leydig cell-macrophage interaction in the rat testis. Int J Androl 8(1):86–96

Bergh A, Damber JE, Widmark A (1988) Hormonal control of testicular blood flow, microcirculation and vascular permeability. In: Cooke BA, Sharpe RM (eds) The molecular and cellular endocrinology of the testis, Serono Symposia Publication, vol 50. Raven Press, New York, pp 123–134

Bergh A, Damber JE, van Rooijen N (1993) Liposome-mediated macrophage depletion: an experimental approach to study the role of testicular macrophages in the rat. J Endocrinol 136(3):407–413

Bhushan S, Tchatalbachev S, Lu Y, Fröhlich S, Fijak M, Vijayan V, Chakraborty T, Meinhardt A (2015) Differential activation of inflammatory pathways in testicular macrophages provides a rationale for their subdued inflammatory capacity. J Immunol 194(11):5455–64. https://doi.org/10.4049/jimmunol.1401132

Bhushan S, Theas MS, Guazzone VA, Jacobo P, Wang M, Fijak M, Meinhardt A, Lustig L (2020) Immune cell subtypes and their function in the testis. Front Immunol 11:583304

Christensen AK, Gillman SW (1969) The correlation of fine structure and function in steroid-secreting cells, with emphasis on those of the gonads. In: McKerns KW (ed) The gonads. Appleton Century-Crofts, New York, pp 415–488

Cohen PE, Chisholm O, Arceci RJ, Stanley ER, Pollard JW (1996) Absence of colony-stimulating factor-1 in osteopetrotic (csfmop/csfmop) mice results in male fertility defects. Biol Reprod 55(2):310–317

Coveney D, Cool J, Oliver T, Capel B (2008) Four-dimensional analysis of vascularization during primary development of an organ, the gonad. Proc Natl Acad Sci USA 105(20):7212–7217

De Felici M, Klinger FG, Campolo F, Balistreri CR, Barchi M, Dolci S (2021) To be or not to be a germ cell: the extragonadal germ cell tumor paradigm. Int J Mol Sci 22(11):5982

DeFalco T, Bhattacharya I, Williams AV, Sams DM, Capel B (2014) Yolk-sac-derived macrophages regulate fetal testis vascularization and morphogenesis. Proc Natl Acad Sci USA 111(23):E2384–E2393

DeFalco T, Potter SJ, Williams AV, Waller B, Kan MJ, Capel B (2015) Macrophages contribute to the spermatogonial niche in the adult testis. Cell Rep 12(7):1107–1119

Fijak M, Meinhardt A (2006) The testis in immune privilege. Immunol Rev 213:66–81

Frungieri MB, Calandra RS, Lustig L, Meineke V, Köhn FM, Vogt HJ, Mayerhofer A (2002) Number, distribution pattern, and identification of macrophages in the testes of infertile men. Fertil Steril 78(2):298–306
Garcia-Alonso L, Lorenzi V, Mazzeo CI, Alves-Lopes JP, Roberts K, Sancho-Serra C, Engelbert J, Marečková M, Gruhn WH, Botting RA, Li T, Crespo B, van Dongen S, Kiselev VY, Prigmore E, Herbert M, Moffett A, Chédotal A, Bayraktar OA, Surani A, Haniffa M, Vento-Tormo R (2022) Single-cell roadmap of human gonadal development. Nature 607(7919):540–547
Gaytan F, Bellido C, Aguilar E, van Rooijen N (1994a) Requirement for testicular macrophages in Leydig cells proliferation and differentiation during prepubertal development in rats. J Reprod Fertil 102:393–399
Gaytan F, Bellido C, Morales C, Reymundo C, Aguilar E, van Rooijen N (1994b) Selective depletion of testicular macrophages and prevention of Leydig cell repopulation after treatment with ethylene dimethane sulfonate in rats. J Reprod Fertil 101(1):175–182
Gaytan F, Bellido C, Morales C, Reymundo C, Aguilar E, Van Rooijen N (1994c) Effects of macrophage depletion at different times after treatment with ethylene dimethane sulfonate (EDS) on the regeneration of Leydig cells in the adult rat. J Androl 15(6):558–564
Gaytan F, Bellido C, Aguilar E, van Rooijen N (1995a) Pituitary-testicular axis in rats lacking testicular macrophages. Eur J Endocrinol 132:218–222
Gaytan F, Bellido C, Morales C, Reymundo C, Aguilar E, van Rooijen N (1995b) Response to Leydig cell apoptosis in the absence of testicular macrophages. J Reprod Immunol 29(1):81–94
Gewiss RL, Schleif MC, Griswold MD (2021) The role of retinoic acid in the commitment to meiosis. Asian J Androl 23(6):549–554
Ginhoux F, Guilliams M (2016) Tissue-resident macrophage ontogeny and homeostasis. Immunity 44(3):439–449
Griswold MD, Bishop PD, Kim KH, Ping R, Siiteri JE, Morales C (1989) Function of vitamin A in normal and synchronized seminiferous tubules. Ann N Y Acad Sci 564:154–172
Gu X, Heinrich A, Li SY, DeFalco T (2023) Testicular macrophages are recruited during a narrow fetal time window and promote organ-specific developmental functions. Nat Commun 14(1):1439
Hales DB (1992) Interleukin-l inhibits Leydig cell steroidogenesis primarily by decreasing 17alpha-hydroxylase/C17-20 lyase cytochrome P450 expression. Endocrinology 131:2165–2172
Hales DB (1996) Leydig cell-macrophage interactions: an overview. In: Payne AH, Hardy MP, Russell LD (eds) The Leydig cell. Cache River Press, Vienna, pp 451–466
Hales DB (2002) Testicular macrophage modulation of Leydig cell steroidogenesis. J Reprod Immunol 57(1-2):3–18
Hedger MP, Qin JX, Robertson DM, de Kretser DM (1990) Intragonadal regulation of immune system functions. Reprod Fertil Dev 2:263–280
Heidt T, Courties G, Dutta P, Sager HB, Sebas M, Iwamoto Y, Sun Y, Da Silva N, Panizzi P, van der Laan AM, Swirski FK, Weissleder R, Nahrendorf M (2014) Differential contribution of monocytes to heart macrophages in steady-state and after myocardial infarction. Circ Res 115(2):284–295
Hume DA, Halpin D, Charlton H, Gordon S (1984) The mononuclear phagocyte system of the mouse defined by immunohistochemical localization of antigen F4/80: macrophages of endocrine organs. Proc Natl Acad Sci USA 81(13):4174–4177
Hutson JC (1990) Changes in the concentration and size of testicular macrophages during development. Biol Reprod 43:885–890
Hutson JC (1993) Secretion of tumor necrosis factor alpha by testicular macrophages. J Reprod Immunol 23:63–72
Hutson JC (1994) Testicular macrophages. Int Rev Cytol 149:99–143
Hutson JC (2006) Physiologic interactions between macrophages and Leydig cells. Exp Biol Med (Maywood) 231:1–7
Kern S, Maddocks S (1995) Indomethacin blocks the immunosuppressive activity of rat testicular macrophages cultured in vitro. J Reprod Immunol 28(3):189–201

Kern S, Robertson SA, Mau VJ, Maddocks S (1995) Cytokine secretion by macrophages in the rat testis. Biol Reprod 53:1407–1416

Kokkinaki M, Lee TL, He Z, Jiang J, Golestaneh N, Hofmann MC, Chan WY, Dym M (2009) The molecular signature of spermatogonial stem cells in the 6-day-old mouse testis. Biol Reprod 80:707–717

Larina IV, Shen W, Kelly OG, Hadjantonakis AK, Baron MH, Dickinson ME (2009) A membrane associated mCherry fluorescent reporter line for studying vascular remodeling and cardiac function during murine embryonic development. Anat Rec (Hoboken) 292(3):333–341

Livera G, Delbes G, Pairault C, Rouiller-Fabre V, Habert R (2006) Organotypic culture, a powerful model for studying rat and mouse fetal testis development. Cell Tissue Res 324(3):507–521

Lokka E, Lintukorpi L, Cisneros-Montalvo S, Mäkelä JA, Tyystjärvi S, Ojasalo V, Gerke H, Toppari J, Rantakari P, Salmi M (2020) Generation, localization and functions of macrophages during the development of testis. Nat Commun 11(1):4375

Maddocks S, Setchell BP (1990) Recent evidence for immune privilege in the testis. J Reprod Immunol 18:9–18

Miller SC, Bowman BM, Rowland HG (1983) Structure, cytochemistry, endocytic activity and immunoglobulin(Fc) receptors of rat testicular interstitial-tissue macrophages. Am J Anat 168:1–13

Mossadegh-Keller N, Sieweke MH (2018) Testicular macrophages: guardians of fertility. Cell Immunol 330:120–125

Mossadegh-Keller N, Gentek R, Gimenez G, Bigot S, Mailfert S, Sieweke MH (2017) Developmental origin and maintenance of distinct testicular macrophage populations. J Exp Med 214(10):2829–2841

Naito M, Hirotaka N, Kawano S, Umezu H, Zhu H, Moriyama H, Yamomoto T, Takatsuka H, Takei Y (1996) Liposome-encapsulated dichloro-methylene diphosphonate induces macrophage apoptosis in vivo and in vitro. J Leukoc Biol 60:337–344

Nes WD, Lukyanenko YO, Jia ZH, Quideau S, Howald WN, Pratum TK, West RR, Hutson JC (2000) Identification of the lipophilic factor produced by macrophages that stimulates steroidogenesis. Endocrinology 141(3):953–958

Niemi M, Sharpe RM, Brown WR (1986) Macrophages in the interstitial tissue of the rat testis. Cell Tissue Res 243:337–344

Oatley JM, Oatley MJ, Avarbock MR, Tobias JW, Brinster RL (2009) Colony stimulating factor 1 is an extrinsic stimulator of mouse spermatogonial stem cell self-renewal. Development 136(7):1191–1199

Perdiguero EG, Geissmann F (2016) The development and maintenance of resident macrophages. Nat Immunol 17(1):2–8

Pöllänen P, Niemi M (1987) Immunohistochemical identification of macrophages, lymphoid cells and HLA antigens in the human testis. Int J Androl 10(1):37–42

Pöllänen P, von Euler M, Söder O (1990) Testicular immunoregulatory factors. J Reprod Immunol 18:51–76

Pollard JW, Dominguez MG, Mocci S, Cohen PE, Stanley ER (1997) Effect of the colony-stimulating factor-1 null mutation, osteopetrotic (csfm(op)), on the distribution of macrophages in the male mouse reproductive tract. Biol Reprod 56(5):1290–1300

Ponte R, Dupuy FP, Brimo F, Mehraj V, Brassard P, Belanger M, Yurchenko E, Jenabian MA, Bernard NF, Routy JP, ORCHID Study Group (2018) Characterization of myeloid cell populations in human testes collected after sex reassignment surgery. J Reprod Immunol 125:16–24

Sauter KA, Pridans C, Sehgal A, Tsai YT, Bradford BM, Raza S, Moffat L, Gow DJ, Beard PM, Mabbott NA, Smith LB, Hume DA (2014) Pleiotropic effects of extended blockade of CSF1R signaling in adult mice. J Leukoc Biol 96(2):265–274

Schlatt S, de Kretser DM, Hedger MP (1999) Mitosis of resident macrophages in the adult rat testis. J Reprod Fertil 116:223–228

Setchell BP, Uksila J, Maddocks S, Pöllänen P (1990) Testis physiology relevant to immunoregulation. J Reprod Immunol 18:19–32

Sieweke MH, Allen JE (2013) Beyond stem cells: self-renewal of differentiated macrophages. Science 342(6161):1242974

Stefater JA 3rd, Ren S, Lang RA, Duffield JS (2011) Metchnikoff's policemen: macrophages in development, homeostasis and regeneration. Trends Mol Med 17:743–752

van Rooijen N, Sanders A (1994) Liposome mediated depletion of macrophages: mechanism of action, preparation of liposomes and applications. J Immunol Methods 174:83–93

Wang J, Wreford NGM, Lan HY, Atkins R, Hedger MP (1994) Leukocyte populations of the adult rat testis following removal of the Leydig cells by treatment with ethane dimethane sulfonate and subcutaneous testosterone implants. Biol Reprod 51:551–561

Wang M, Fijak M, Hossain H, Markmann M, Nüsing RM, Lochnit G, Hartmann MF, Wudy SA, Zhang L, Gu H, Konrad L, Chakraborty T, Meinhardt A, Bhushan S (2017) Characterization of the micro-environment of the testis that shapes the phenotype and function of testicular macrophages. J Immunol 198(11):4327–4340

Wang M, Yang Y, Cansever D, Wang Y, Kantores C, Messiaen S, Moison D, Livera G, Chakarov S, Weinberger T, Stremmel C, Fijak M, Klein B, Pleuger C, Lian Z, Ma W, Liu Q, Klee K, Händler K, Ulas T, Schlitzer A, Schultze JL, Becher B, Greter M, Liu Z, Ginhoux F, Epelman S, Schulz C, Meinhardt A, Bhushan S (2021) Two populations of self-maintaining monocyte-independent macrophages exist in adult epididymis and testis. Proc Natl Acad Sci USA 118(1):e2013686117

Wilson L (1954) Sperm agglutinins in human semen and blood. Proc Soc Exp Biol Med 85(4):652–655

Winnall WR, Muir JA, Hedger MP (2011) Rat resident testicular macrophages have an alternatively activated phenotype and constitutively produce interleukin-10 in vitro. J Leukoc Biol 90:133–143

Wynn TA, Chawla A, Pollard JW (2013) Macrophage biology in development, homeostasis and disease. Nature 496:445–455

Xiong Y, Hales DB (1993) The role of tumor necrosis factor-alpha in the regulation of mouse Leydig cell steroidogenesis. Endocrinology 132:2438–2444

Xiong Y, Hales DB (1994) Immune-endocrine interactions in the mouse testis: cytokine-mediated inhibition of Leydig cell steroidogenesis. Endocr J 9:1277–1283

Yee JB, Hutson JC (1983) Testicular macrophages: isolation, characterization and hormonal responsiveness. Biol Reprod 29:1319–1326

Yee JB, Hutson JC (1985a) Effects of testicular macrophage-conditioned medium on Leydig cells in culture. Endocrinology 116(6):2682–2684

Yee JB, Hutson JC (1985b) In vivo effects of FSH on testicular macrophages. Biol Reprod 32:880–883

Yee JB, Hutson JC (1985c) Biochemical consequences of FSH binding to testicular macrophages in culture. Biol Reprod 32:872–879

Chapter 6
Adipose Tissue Macrophages

Tamás Röszer

Abstract In obesity, adipose tissue macrophages (ATMs) are abundant immune cells in the adipose tissue and are known as inducers of metabolic inflammation that may lead to insulin resistance and immune disorders associated with obesity. However, much less is known about the ontogeny and physiological functions of ATMs in lean adipose tissue. ATMs are present at birth and actively participate in the synthesis of mediators that induce lipolysis, mitobiogenesis, and thermogenesis in adipocytes. Later in life ATMs limit the thermogenic competence of the adipocytes and favor lipid storage. ATMs respond to lipid overload of adipocytes in obesity with a sequence of pro-inflammatory events, including inflammasome activation and pyroptosis, as well as stimulation of nuclear factor kappa B and interferon regulatory factors that evoke an uncontrolled inflammation. ATMs are life-long constituents of the adipose tissue and hence signals that control ATM development and ATM-adipocyte interactions determine adipose tissue health.

6.1 Brief History of Adipose Tissue Macrophages

Adipose tissue macrophages (ATMs) are resident innate immune cells of the adipose tissue, that become abundant with the progression of obesity (Boutens and Stienstra 2016; Epelman et al. 2014; Glass and Olefsky 2012; Lee and Lee 2014; Li et al. 2013; Osborn and Olefsky 2012; Qiu et al. 2014; Röszer 2020e, 2022; Xu et al. 2003) (Fig. 6.1a). The presence of ATMs—termed as "fixed histiocytes"—in the adipose tissue is mentioned in a histology book from 1914 (Kopsch 1914), and a comparative histology textbook from the 1950s also mentions the existence of

T. Röszer (✉)
Research Division of Pediatric Obesity, Department of Pediatrics, Faculty of Medicine, University of Debrecen, Debrecen, Hungary

Institute of Neurobiology, Ulm University, Ulm, Germany
e-mail: roszer.tamas@med.unideb.hu

M. Kloc et al. (eds.), *Monocytes and Macrophages in Development, Regeneration, and Disease*, Results and Problems in Cell Differentiation 74,
https://doi.org/10.1007/978-3-031-65944-7_6

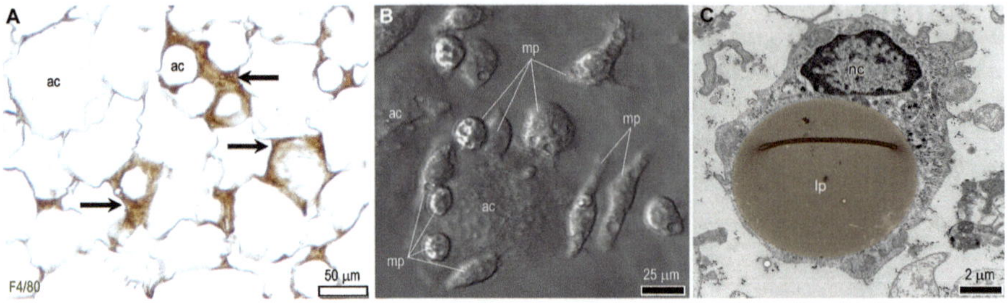

Fig. 6.1 Obese adipose tissue in mouse. (**a**) Adipose tissue macrophages (ATMs) are labeled with an antibody against the mouse macrophage marker F4/80 antigen. ATMs form crown-like structures (indicated with arrows) around adipocytes (ac). (**b**) Apoptotic fat cells (ac) and macrophages (mp) *in vitro*. Such as in the adipose tissue, macrophages form clusters around the apoptotic fat cell and clear their remnants and lipid molecules. (**c**) Transmission electron microscopy of a mouse ATM, engulfing a large lipid droplet. nc nucleus, lp lipid droplet. For better visibility, the lipid droplet is labeled with pseudo-color (Text and images are reprinted from Röszer 2020e)

phagocytic immune cells in the amphibian adipose tissue (Ábrahám 1952). Accumulation of ATMs was first described in obese adipose tissue of mice in the 1960s (Hausberger 1966) and later ATMs were defined as phagocytosing myeloid cells that express CD11c and F4/80 antigen in mice, and CD68 and IBA1 (ionized calcium-binding adaptor molecule 1) in human, and various monocyte-macrophage proteins in other mammals (Röszer 2020e). Since adipose tissue is a specialized connective tissue, and since connective tissues harbor various immune cells, including macrophages, ATMs were considered as patrolling immune cells in the fat depots. Thus, initially, ATMs were not assigned to a dedicated physiological function beyond immune surveillance. However, the central role of adipose tissue inflammation in metabolic syndrome and the clustering of obesity-associated immune pathologies became evident in the 2000s (Lumeng et al. 2007, 2008; Weisberg et al. 2003; Xu et al. 2003), that instigated a growing interest in understanding immune component of obesity.

The health impact of adipose tissue inflammation in obesity led to the rediscovery of ATMs in obese human adipose tissue in the early 2000s, catalyzing the rapid development of an ATM-centered paradigm of obesity-associated pathologies (Lumeng et al. 2007; Weisberg et al. 2003; Xu et al. 2003). ATMs, however, appear in the lean adipose tissue as well and are constituents of the adipose tissue stroma in amphibians, rodents, ruminants, carnivores, primates, and humans (Akter et al. 2012; Ampem et al. 2016; Waqas et al. 2017b; Yu et al. 2019). ATMs secrete cytokines, lipid mediators, and hormones and are sources of reactive oxygen species. The local and systemic effects of these ATM-derived biomolecules determine adipocyte functioning and systemic metabolism.

6.2 Adipose Tissue Macrophages: Friend or Foe?

ATMs are mostly known for their metabolically adverse effects associated with obesity (Lumeng et al. 2007; Rosen and Spiegelman 2014). Obesity is hallmarked with a significant increase in ATM number and pro-inflammatory macrophage activation (Boutens and Stienstra 2016; Feng et al. 2011; Lumeng et al. 2007; Zeyda et al. 2007), which is concomitant with the development of chronic adipose tissue inflammation (Weisberg et al. 2003; Xu et al. 2003). Inflammation impairs insulin sensitivity and deteriorates kidney and liver function, eventually, establishing a bad reputation for ATMs (Boutens and Stienstra 2016). There are several original articles and reviews, detailing the metabolic harm of ATMs residing in the obese adipose tissue (Boutens and Stienstra 2016; Hill et al. 2014; Jacks and Lumeng 2024; Rosen and Spiegelman 2014; Russo and Lumeng 2018). Historically, when the abundance of ATMs in the obese adipose tissue was shown, the so-called M1/M2 polarization model was the central paradigm of macrophage biology (Röszer 2020e). In brief, M1, or classical macrophage activation is elicited by lipopolysaccharides (LPS) and interferon gamma (IFNγ) in macrophages cultured *in vitro*. LPS is an activator of Toll like receptor 4 (TLR4), and IFNγ acts through its membrane receptor. Eventually, LPS and IFNγ stimulation of macrophages leads to the activation of nuclear factor kappa B (NFκB) and interferon regulatory factors (IRFs) controlled pro-inflammatory gene expression (Röszer 2020c). This, in turn, increases pro-inflammatory cytokine secretion and nitric oxide (NO) synthesis by macrophages. In turn, an interleukin-4 (IL-4) or interleukin-10 (IL-10)-induced, so-called M2 or alternative macrophage activation mitigates pro-inflammatory gene expression and favors an immune suppressive, pro-resolving, or pro-fibrotic macrophage phenotype (Röszer 2020c). Initial studies indicated that ATMs of the obese adipose tissue express various M1 macrophage-associated cytokines, such as TNFα and IL-6, that ignite inflammation and impair insulin responsiveness (Röszer 2020e).

For about two decades, the mainstream of ATM studies corroborated the concept of a metabolically harmful M1 activation of ATMs (Boutens and Stienstra 2016; Glass and Olefsky 2012; Jacks and Lumeng 2024; Lackey and Olefsky 2016; Lumeng et al. 2008; Morinaga et al. 2015; Osborn and Olefsky 2012; Shapiro et al. 2011). Albeit obese adipose tissue is rich in ATMs that secrete pro-inflammatory mediators, the transcriptional landscape of ATMs and the *in vitro* activated M1 macrophages are distinct (Aron-Wisnewsky et al. 2009; Coats et al. 2017; Li et al. 2019; Lumeng et al. 2008; Morris et al. 2011; Russo and Lumeng 2018; Shaul et al. 2010). ATM activation can be elicited by various pathogen recognition receptors apart from TLR4 and IFNγ receptors, leading to a pro-inflammatory macrophage phenotype (Röszer 2020b, e). Of note, adipocytes also express pro-inflammatory genes and have inflammasome activation in obesity (Chan et al. 2020; Giordano et al. 2013). Thus, adipose tissue inflammation may be mitigated independently from macrophage activation (Nickl et al. 2021).

Nevertheless, obesity leads to a pro-inflammatory ATM activation and the underlying mechanism may be associated with a macrophage response to lipid-overloaded adipocytes and pro-inflammatory signals of other immune cells of the adipose tissue (Lindhorst et al. 2021). In obesity, adipocyte volume and adipocyte number increase, along with the apoptotic death of lipid-overloaded adipocytes (Fig. 6.1a). Weight loss due to caloric restriction and a diet rich in short-chain fatty acids reduce pro-inflammatory ATM activation (Eslick et al. 2022; Kováčiková et al. 2011), plausibly by eliminating the trigger—i.e., lipid overload of adipocytes—of a pro-inflammatory ATM activation.

ATMs, such as any other tissue-resident macrophages, may engulf apoptotic fat cells and fat cell-derived apoptotic bodies (Fig. 6.1b). Indeed, ATMs form multi-nucleated syncytia, so-called crown-like structures around dying adipocytes in the obese adipose tissue (Amano et al. 2014; Boutens and Stienstra 2016; Epelman et al. 2014; Fuentes et al. 2010; Glass and Olefsky 2012; Lee and Lee 2014; Li et al. 2013; Osborn and Olefsky 2012; Qiu et al. 2014; Waqas et al. 2017a) (Fig. 6.1a). Apoptotic cell uptake promotes anti-inflammatory macrophage activation in most tissues, however, the uptake of apoptotic adipocytes triggers a pro-inflammatory ATM activation (Röszer 2021).

Under physiological conditions, adipocyte differentiation is associated with an increase in the expression of survival factors, rendering the mature adipocytes long-lived cells (Sorisky et al. 2000). In mice, lipid-storing "white" adipocytes are more resistant to apoptosis than the thermogenic "brown" adipocytes (Nisoli et al. 2006). Human abdominal subcutaneous preadipocytes are more resistant to apoptosis than are omental preadipocytes, and the distinct fat depots contain at least two different adipocyte populations based on their resistance to apoptosis (Tchkonia et al. 2005). Adipocyte apoptosis and necrosis become prevalent under pathological conditions, such as cytokine-induced or congenital lipodystrophy (Birk and Rubinstein 2006; Domingo et al. 1999; Vogel et al. 2011), and in obesity.

Hypertrophic adipocytes undergo apoptosis or secondary necrosis in obese adipose tissue (Röszer 2021). Adipocyte death may be a response to lipid overload, mitochondrial damage, and inflammatory cytokines. Lipid overload in hypertrophic adipocytes can lead to the "spillover" of lipids into the cytosol, leading to the so-called lipotoxicity and, ultimately, apoptosis (Prieur et al. 2010). Failure of fatty acid oxidation and oxidative phosphorylation initiates the mitochondrial pathway of apoptosis in adipocytes (Qian et al. 2020), and also triggers inflammatory cell death, termed pyroptosis (Giordano et al. 2013; Qian et al. 2020). Impaired lipolysis and hypertrophy are hence powerful triggers of adipocyte apoptosis (Osuga et al. 2000). Lipotoxicity also triggers apoptosis of adipose tissue stem cells in aged fat (Guo et al. 2007).

Adipocyte cell death is an inflammation-generating process and is a prelude to a sequence of events leading to obesity-associated metabolic diseases (Lindhorst et al. 2021; Vandanmagsar et al. 2011; West 2009). It is plausible that the first wave of apoptosis of adipocytes during the development of obesity is well controlled by the ATMs, and they are able to maintain an M2-like activation state (Shaul et al. 2010), and remove excess lipids (Kosteli et al. 2010). However, the prevalence of damage-associated molecules and pro-inflammatory lipid species in the dying adipocytes—that are engulfed by the ATMs—can switch the function of ATMs towards the release of inflammatory cytokines and reactive oxygen species, probably due to NFκB, interferon regulatory factors (IRFs) and inflammasome activation. Moreover, the capacity of ATMs to clear apoptotic cells may be exhausted in obesity (Luo et al. 2019), and the apoptotic or necrotic cell debris aggravates local inflammation.

ATMs are not the sole immune cells of the adipose tissue; indeed, there is a complex adipose tissue immune cell niche that contains mast cells, innate lymphoid cells, natural killer cells, T cells, and B cells. A pro-inflammatory ATM activation may initiate a cascade of intercellular signaling events within this immune cell niche, leading to uncontrolled inflammation. In turn, immune cells of the adipose tissue release cytokines that further augment the pro-inflammatory ATM activation. This may result in adipose tissue inflammation (Boutens and Stienstra 2016; Coats et al. 2017).

It was initially thought that M2-activated (pro-resolving, anti-inflammatory) ATMs would be beneficial in obesity, by reducing adipose tissue inflammation and increasing insulin sensitivity (Röszer 2020e). It was also assumed, that the lean adipose tissue was rich in M2 ATMs (Rosen and Spiegelman 2014), albeit only a limited number of M2 marker proteins are expressed by ATMs in the lean state (Röszer 2020e). Moreover, the number of M2 ATMs is limited by their removal by type I innate lymphoid cells (Boulenouar et al. 2017), and the abundance of M2 ATMs may cause adipose tissue fibrosis and worsen the metabolic impairment in obesity (Spencer et al. 2010). Fibrosis in the breast adipose tissue may increase the risk for tumor development and tumor aggressiveness (Kuziel et al. 2023). Pro-fibrotic ATMs in the perivascular fat layer also aggravate fibrosis in heart failure and increase the progression of coronary atherosclerosis (Fu et al. 2023). There are various negative feedback mechanisms (Zheng et al. 2015), including IL-4/STAT6 signaling itself (Arpa et al. 2009) that limit the abundance of M2 macrophages in tissues.

Single-cell sequencing data, and the growing body of studies using flow cytometry, and macrophage-specific knockout models show that there are various ATM subsets, which show varying degrees of M1 and M2 traits (Blaszczak et al. 2019; Huang Stanley et al. 2016; Li et al. 2019). Depending on the state of the obesity, the ATM population may be dominated by M2 ATMs, a mixture of M1 and M2 ATMs, and eventually can be dominated by M1, inflammatory, "metabolically activated" ATMs (Aron-Wisnewsky et al. 2009; Basinska et al. 2015; Chung et al. 2014; Coats et al. 2017). There are also depot-specific differences between ATM pools (Bigornia et al. 2012).

Immune-suppressive (often termed as M2) macrophages support tumor growth (Röszer 2020c), and accordingly, ATMs in the breast adipose tissue expressing M2 macrophage genes increase tumor progression (Li et al. 2023). Independent of obesity, an increased number of ATMs of the breast adipose tissue worsens breast cancer prognosis (Lin et al. 2021). It is also postulated, that ATM-derived cytokines, such as IL-6, IL-10, TGFβ, and mediators such as vascular endothelial growth factor, and enzymes, such as metalloproteinases, may support the development of malignant cell proliferation (Singh et al. 2022). ATMs thereby have more complex roles in adipose tissue functioning than inflammation control and interfering with insulin sensitivity.

The simplified concept of antagonistic M1/M2 ATM effects on systemic metabolism is further challenged by the role of pro-inflammatory signal mechanisms that are required for physiological adipose tissue development (Babaei et al. 2018; Sun et al. 2018) (Fig. 6.2a). ATMs stimulate thermogenic and fat catabolizing adipocyte activities after birth by releasing IL-6 and ablation of ATMs in newborn mice leads to the loss of thermogenic fat cells in the subcutaneous fat depot (Yu et al. 2019). Loss of thermogenic fat is associated with adiposity (Gyurina et al. 2023; Honecker et al. 2022; Sacks and Symonds 2013). Increased adiposity at the first year of life and an early adiposity rebound before 5.5 years of age increase the probability of obesity as a young adult (Carolan-Olah et al. 2015; Charney et al. 1976; Dietz 1994; Eriksson et al. 2003; Fall 2006; Geserick et al. 2018; Landgraf et al. 2015; Pietrobelli et al. 2017; Rolland-Cachera et al. 1984; Siervogel et al. 1991). Importantly, ATMs are already present in the fat depots after birth (Waqas et al. 2017b; Yu et al. 2019), concomitantly with the abundance of thermogenic adipocytes in both mouse and human (Gyurina et al. 2023; Hoang et al. 2021). It is plausible that ATMs support thermogenic fat development in early postnatal life (Fig. 6.2a).

6.3 Ontogeny of Adipose Tissue Macrophages

The origin of ATMs is still poorly understood, despite an uncontrolled expansion of the ATM pool worsens obesity-associated diseases. In the visceral and subcutaneous adipose tissues—that are relevant in obesity—the number of ATMs may increase as a result of monocyte infiltration and local proliferation of ATMs (Amano et al. 2014; Haase et al. 2014; Waqas et al. 2017b). Hypertrophic adipocytes release chemotactic and pro-inflammatory signals, that stimulate bone marrow monocyte development and attract monocytes to the obese fat depots, eventually increasing

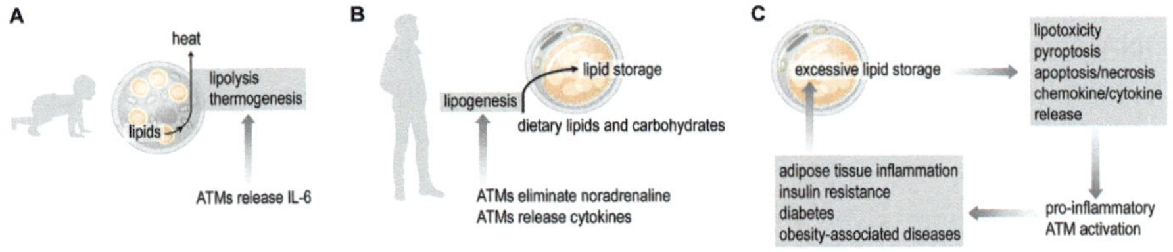

Fig. 6.2 ATM functions under physiological and pathological conditions. (**a**) ATMs promote thermogenesis and lipolysis in the early postnatal development of the adipose tissue. (**b**) In the adult and aging adipose tissue, the ATMs support energy storage in neutral lipids. (**c**) In obesity, the lipid-overloaded adipocytes induce a pro-inflammatory ATM activation that may induce adipose tissue inflammation, insulin resistance and obesity-associated pathologies. As a result, fat accumulation may further increase, initiating a vicious cycle

the monocyte-derived ATM pool (Lumeng et al. 2007; Nagareddy et al. 2014). The first wave of ATMs however develops from hematopoietic cells of the yolk sac—and probably of the fetal liver—and fetal ATMs persist in the newborn (Waqas et al. 2017b). ATMs retain the ability of self-replenishment and proliferate in response to interleukin-6 (IL-6) and neuropeptide FF (NPFF) (Röszer 2018, 2020a; Waqas et al. 2017a, b). NPFF is an appetite-regulating hormone, that is released from the endocrine pancreas and from the hypothalamus (Waqas et al. 2017a). Later in infancy, the fetal ATMs are accompanied—and plausibly gradually replaced—by monocyte-derived ATMs [reviewed in Röszer (2020a)]. Interestingly, the interscapular brown adipose tissue, the largest thermogenic fat depot in rodents, is scarce in macrophages (Alcalá et al. 2017; Waqas et al. 2017b). ATMs of the breast may derive from circulating monocytes, or resident macrophages of the gland stroma (Linde et al. 2018). The origin of ATMs in the perivascular fat depots is still to be elucidated (Röszer 2020a).

6.4 Physiological Functions of ATMs

In contrast to the abundance of studies on the metabolic harms caused by pro-inflammatory ATMs of the obese adipose tissue, only a small fraction of the ATM-related literature deals with ATM functions in lean adipose tissue. In brief, ATMs appear to control the changes in physiological functions of the developing adipose tissue. After birth, when the adipose tissue depots have active lipolysis and thermogenesis, ATMs promote thermogenic "brown" or "beige" adipocyte development (Hoang et al. 2022; Yu et al. 2019) (Fig. 6.2a).

In adulthood and with aging, when fat storage becomes more relevant than thermogenesis and lipolysis, ATMs antagonize lipolysis and inhibit cold-induced thermogenesis (Camell et al. 2017; Pirzgalska et al. 2017) (Fig. 6.2b). It is generally assumed, that the lean fat contains anti-inflammatory ATMs, that express some markers of the M2 macrophages, such as arginase-1 and CD206. However, ATM-derived pro-inflammatory signals are necessary for adipose tissue development, since transient inflammatory signaling is essential for physiological adipogenesis (Chung et al. 2017).

It has been shown that the elimination of ATMs with clodronate liposome uptake shifts subcutaneous adipose tissue development towards a fat-storing state in newborn mice, suggesting that ATMs support the maintenance of the lean state (Yu et al. 2019). Mechanistically, ATMs synthesize platelet-activating factor (PAF) that stimulates IL-6 release, which eventually activates the transcription of genes required for mitochondrial biogenesis and thermogenesis in neighboring adipocytes (Yu et al. 2019). This process is termed adaptive thermogenesis, or adipocyte browning: a process that allows the dissipation of energy stored in fat in the form of heat. Albeit cold exposure increases the number of M2-activated ATMs (Hui et al. 2015), these ATMs do not affect adaptive thermogenesis (Fischer et al. 2017). In the aging adipose tissue that stores fat rather than metabolizing it, ATMs and macrophages in the sympathetic ganglia eliminate noradrenaline, a key signal of lipolysis and thermogenesis (Camell et al. 2017; Pirzgalska et al. 2017).

Safe disposal of apoptotic cells is another M2 function for ATMs, and obesity-inducing substances may also impair this function (Röszer 2017). The long life span of adipocytes however does not keep ATMs busy with this task. ATMs also release neuropeptide Y (Singer et al. 2013), and insulin-like growth factor 1 (Chang et al. 2016), suggesting an endocrine function for ATMs. However, that idea has not been corroborated by recent research. For instance, an initial finding on noradrenaline production by ATMs has been disproved by a more recent study (Fischer et al. 2017).

Albeit emphasis was given to the impact of ATM activation, and M1 macrophages were tagged as "metabolically harmful" and their M2 counterparts as "metabolically beneficial", it is plausible that the metabolic effect of ATMs depends on specific enzymes they possess (e.g., enzymatic synthesis of PAF from dietary lipids and enzymatic degradation of noradrenaline), and signals (e.g., hormones) they release, rather their activation state.

Expression of M1 and M2 markers is regulated by a wide range of signals such as hormones, lipid metabolites, fatty acids, cytokines, and microbial products (Röszer 2020b). Physiologically it appears that the effect of macrophage-activating signals is antagonized by counter-regulatory mechanisms. For instance, LPS released by the gut microbiota to the bloodstream induces the expression of some M1 marker genes in ATMs under physiological conditions (Caesar et al. 2012). In turn, *Lactobacilli* may migrate into the adipose tissue, and mitigate a pro-inflammatory ATM activation (Huang et al. 2017). Moreover, ATMs and regulatory T cells produce IL-10 in the lean adipose tissue, which is a possible endogenous signal for suppressing an immune response in ATMs against gut microbiota (Liu and Nikolajczyk 2019; Russo and Lumeng 2018).

6.5 Blind Spots in ATM Biology

There are several unexplored aspects of ATM biology, especially in the therapeutic or diagnostic use of ATMs. Increased ATM number and inflammatory hyperactivation of ATMs are core traits that make ATMs potentially harmful to the systemic metabolism (Fig. 6.2c). Having pharmacological intervention for a targeted reduction of ATM number is however a yet-to-be-achieved goal. It is known that bariatric surgery, regular exercise, and a diet rich in polyunsaturated fatty acids reduce ATM number and induce profound changes in the immune cell composition of the adipose tissue (Frikke-Schmidt et al. 2017). Weight loss due to caloric restriction induces a transient increase in ATM number in mice, which is followed by a steady reduction in ATM number (Kosteli et al. 2010). It appears that lipolysis and the eventual release of free fatty acids are the signals for the transient accumulation of ATMs (Kosteli et al. 2010). The underlying mechanisms, such as the cell cycle regulation and life span of tissue macrophages are still incompletely understood.

Moreover, we lack ATM-specific markers, that would allow a targeted elimination of the ATMs or a cell-specific drug delivery to ATMs.

Histopathology evaluation of adipose tissue specimens obtained with needle biopsy or removed during elective surgery may be used for the assessment of adipocyte size and number, adipocyte apoptosis, thermogenic and lipogenic potential of the adipocytes; number, and activation state of ATMs, and may provide an accurate prognosis in obesity (Gyurina et al. 2023; Kováčiková et al. 2011). Adipose tissue-based personalized histopathology is however still not in use in clinical settings. As for technical limitations to studying ATMs, it is worth mentioning that we lack immortalized ATM cell lines, and methods for a lineage-specific ablation of ATMs. Moreover, proper discrimination of monocytes, mast cells, and ATMs is challenging due to the overlap of their marker proteins (Röszer 2020d). How ATMs respond to obesogenic signals of the environment (Ampem et al. 2019), or to viruses that can profoundly affect adipose tissue functioning (e.g., thermogenesis and lipolysis) (Atkinson 2007) are still largely unexplored aspects of ATM biology. And lastly, despite the enormous functional and developmental differences between the distinct fat depots of the human body (Cohen and Kajimura 2021), we lack insight into depot-specific ATM functions. It is also plausible that ATMs have functions that are dependent on the developmental stage of the adipose tissue, thus ATMs may have distinct effects after birth, in adulthood, and during aging (Röszer 2022).

Acknowledgements Work in the Author's laboratory is supported by grants from the Hungarian Research Fund (NKFI-OTKA 142939), a Bolyai Research Scholarship of the Hungarian Academy of Sciences, the Dean's Research Fund (Faculty of Medicine, University of Debrecen), a "Momentum" proof-of-concept fund (Faculty of Medicine, University of Debrecen) and by the UNKP-23-5 New National Excellence Program of the Ministry for Culture and Innovation from the source of the National Research, Development and Innovation Fund.

Authorship Conceptualization and writing: TR.

References

Ábrahám A (1952) Comparative animal anatomy. Tankönyvkiadó, Budapest

Akter SH, Häussler S, Germeroth D, von Soosten D, Dänicke S, Südekum KH, Sauerwein H (2012) Immunohistochemical characterization of phagocytic immune cell infiltration into different adipose tissue depots of dairy cows during early lactation. J Dairy Sci 95(6):3032–3044. https://doi.org/10.3168/jds.2011-4856

Alcalá M, Calderon-Dominguez M, Bustos E, Ramos P, Casals N, Serra D, Viana M, Herrero L (2017) Increased inflammation, oxidative stress and mitochondrial respiration in brown adipose tissue from obese mice. Sci Rep 7(1):16082. https://doi.org/10.1038/s41598-017-16463-6

Amano SU, Cohen JL, Vangala P, Tencerova M, Nicoloro SM, Yawe JC, Shen Y, Czech MP, Aouadi M (2014) Local proliferation of macrophages contributes to obesity-associated adipose tissue inflammation. Cell Metab 19(1):162–171. https://doi.org/10.1016/j.cmet.2013.11.017

Ampem G, Azegrouz H, Bacsadi A, Balogh L, Schmidt S, Thuroczy J, Röszer T (2016) Adipose tissue macrophages in non-rodent mammals: a comparative study. Cell Tissue Res 363(2):461–478. https://doi.org/10.1007/s00441-015-2253-1

Ampem G, Junginger A, Yu H, Balogh L, Thuróczy J, Schneider ME, Röszer T (2019) The environmental obesogen bisphenol A increases macrophage self-renewal. Cell Tissue Res 378(1):81–96. https://doi.org/10.1007/s00441-019-03019-5

Aron-Wisnewsky J, Tordjman J, Poitou C, Darakhshan F, Hugol D, Basdevant A, Aissat A, Guerre-Millo M, Clement K (2009) Human adipose tissue macrophages: m1 and m2 cell surface markers in subcutaneous and omental depots and after weight loss. J Clin Endocrinol Metab 94(11):4619–4623. https://doi.org/10.1210/jc.2009-0925

Arpa L, Valledor AF, Lloberas J, Celada A (2009) IL-4 blocks M-CSF-dependent macrophage proliferation by inducing p21Waf1 in a STAT6-dependent way. Eur J Immunol 39(2):514–526. https://doi.org/10.1002/eji.200838283

Atkinson RL (2007) Viruses as an etiology of obesity. Mayo Clin Proc 82(10):1192–1198. https://doi.org/10.4065/82.10.1192

Babaei R, Schuster M, Meln I, Lerch S, Ghandour RA, Pisani DF, Bayindir-Buchhalter I, Marx J, Wu S, Schoiswohl G, Billeter AT, Krunic D, Mauer J, Lee Y-H, Granneman JG, Fischer L, Müller-Stich BP, Amri E-Z, Kershaw EE, Heikenwälder M, Herzig S, Vegiopoulos A (2018) Jak-TGFβ cross-talk links transient adipose tissue inflammation to beige adipogenesis. Sci Signal 11(527):eaai7838. https://doi.org/10.1126/scisignal.aai7838

Basinska K, Marycz K, Śmieszek A, Nicpoń J (2015) The production and distribution of IL-6 and TNF-α in subcutaneous adipose tissue and their correlation with serum concentrations in Welsh ponies with equine metabolic syndrome. J Vet Sci 16(1):113–120. https://doi.org/10.4142/jvs.2015.16.1.113

Bigornia SJ, Farb MG, Mott MM, Hess DT, Carmine B, Fiscale A, Joseph L, Apovian CM, Gokce N (2012) Relation of depot-specific adipose inflammation to insulin resistance in human obesity. Nutr Diabetes 2:e30. https://doi.org/10.1038/nutd.2012.3

Birk RZ, Rubinstein M (2006) IFN-alpha induces apoptosis of adipose tissue cells. Biochem Biophys Res Commun 345(2):669–674. https://doi.org/10.1016/j.bbrc.2006.04.139

Blaszczak AM, Jalilvand A, Liu J, Wright VP, Suzo A, Needleman B, Noria S, Lafuse W, Hsueh WA, Bradley D (2019) Human visceral adipose tissue macrophages are not adequately defined by standard methods of characterization. J Diabetes Res 2019:8124563–8124563. https://doi.org/10.1155/2019/8124563

Boulenouar S, Michelet X, Duquette D, Alvarez D, Hogan AE, Dold C, O'Connor D, Stutte S, Tavakkoli A, Winters D, Exley MA, O'Shea D, Brenner MB, von Andrian U, Lynch L (2017) Adipose type one innate lymphoid cells regulate macrophage homeostasis through targeted cytotoxicity. Immunity 46(2):273–286. https://doi.org/10.1016/j.immuni.2017.01.008

Boutens L, Stienstra R (2016) Adipose tissue macrophages: going off track during obesity. Diabetologia 59(5):879–894. https://doi.org/10.1007/s00125-016-3904-9

Caesar R, Reigstad CS, Backhed HK, Reinhardt C, Ketonen M, Lunden GO, Cani PD, Backhed F (2012) Gut-derived lipopolysaccharide augments adipose macrophage accumulation but is not essential for impaired glucose or insulin tolerance in mice. Gut 61(12):1701–1707. https://doi.org/10.1136/gutjnl-2011-301689

Camell CD, Sander J, Spadaro O, Lee A, Nguyen KY, Wing A, Goldberg EL, Youm YH, Brown CW, Elsworth J, Rodeheffer MS, Schultze JL, Dixit VD (2017) Inflammasome-driven catecholamine catabolism in macrophages blunts lipolysis during ageing. Nature 550(7674):119–123. https://doi.org/10.1038/nature24022

Carolan-Olah M, Duarte-Gardea M, Lechuga J (2015) A critical review: early life nutrition and prenatal programming for adult disease. J Clin Nurs 24(23-24):3716–3729. https://doi.org/10.1111/jocn.12951

Chan CC, Damen MSMA, Moreno-Fernandez ME, Stankiewicz TE, Cappelletti M, Alarcon PC, Oates JR, Doll JR, Mukherjee R, Chen X, Karns R, Weirauch MT, Helmrath MA, Inge TH, Divanovic S (2020) Type I interferon sensing unlocks dormant adipocyte inflammatory potential. Nat Commun 11(1):2745. https://doi.org/10.1038/s41467-020-16571-4

Chang HR, Kim HJ, Xu X, Ferrante AW Jr (2016) Macrophage and adipocyte IGF1 maintain adipose tissue homeostasis during metabolic stresses. Obesity (Silver Spring) 24(1):172–183. https://doi.org/10.1002/oby.21354

Charney E, Goodman HC, McBride M, Lyon B, Pratt R (1976) Childhood antecedents of adult obesity. Do chubby infants become obese adults? N Engl J Med 295(1):6–9. https://doi.org/10.1056/nejm197607012950102

Chung S, Cuffe H, Marshall SM, McDaniel AL, Ha JH, Kavanagh K, Hong C, Tontonoz P, Temel RE, Parks JS (2014) Dietary cholesterol promotes adipocyte hypertrophy and adipose tissue inflammation in visceral, but not in subcutaneous, fat in monkeys. Arterioscler Thromb Vasc Biol 34(9):1880–1887. https://doi.org/10.1161/ATVBAHA.114.303896

Chung K-J, Chatzigeorgiou A, Economopoulou M, Garcia-Martin R, Alexaki VI, Mitroulis I, Nati M, Gebler J, Ziemssen T, Goelz SE, Phieler J, Lim J-H, Karalis KP, Papayannopoulou T, Blüher M, Hajishengallis G, Chavakis T (2017) A self-sustained loop of inflammation-driven inhibition of beige adipogenesis in obesity. Nat Immunol 18(6):654–664. https://doi.org/10.1038/ni.3728

Coats BR, Schoenfelt KQ, Barbosa-Lorenzi VC, Peris E, Cui C, Hoffman A, Zhou G, Fernandez S, Zhai L, Hall BA, Haka AS, Shah AM, Reardon CA, Brady MJ, Rhodes CJ, Maxfield FR, Becker L (2017) Metabolically activated adipose tissue macrophages perform detrimental and beneficial functions during diet-induced obesity. Cell Rep 20(13):3149–3161. https://doi.org/10.1016/j.celrep.2017.08.096

Cohen P, Kajimura S (2021) The cellular and functional complexity of thermogenic fat. Nat Rev Mol Cell Biol 22(6):393–409. https://doi.org/10.1038/s41580-021-00350-0

Dietz WH (1994) Critical periods in childhood for the development of obesity. Am J Clin Nutr 59(5):955–959. https://doi.org/10.1093/ajcn/59.5.955

Domingo P, Matias-Guiu X, Pujol RM, Francia E, Lagarda E, Sambeat MA, Vázquez G (1999) Subcutaneous adipocyte apoptosis in HIV-1 protease inhibitor-associated lipodystrophy. AIDS 13(16):2261–2267

Epelman S, Lavine KJ, Randolph GJ (2014) Origin and functions of tissue macrophages. Immunity 41(1):21–35. https://doi.org/10.1016/j.immuni.2014.06.013

Eriksson JG, Forsén T, Tuomilehto J, Osmond C, Barker DJ (2003) Early adiposity rebound in childhood and risk of type 2 diabetes in adult life. Diabetologia 46(2):190–194. https://doi.org/10.1007/s00125-002-1012-5

Eslick S, Williams EJ, Berthon BS, Wright T, Karihaloo C, Gately M, Wood LG (2022) Weight loss and short-chain fatty acids reduce systemic inflammation in monocytes and adipose tissue macrophages from obese subjects. Nutrients 14(4):765. https://doi.org/10.3390/nu14040765

Fall C (2006) Developmental origins of cardiovascular disease, type 2 diabetes and obesity in humans. Adv Exp Med Biol:8–28

Feng B, Jiao P, Nie Y, Kim T, Jun D, van Rooijen N, Yang Z, Xu H (2011) Clodronate liposomes improve metabolic profile and reduce visceral adipose macrophage content in diet-induced obese mice. PLoS One 6(9):e24358. https://doi.org/10.1371/journal.pone.0024358

Fischer K, Ruiz HH, Jhun K, Finan B, Oberlin DJ, van der Heide V, Kalinovich AV, Petrovic N, Wolf Y, Clemmensen C, Shin AC, Divanovic S, Brombacher F, Glasmacher E, Keipert S, Jastroch M, Nagler J, Schramm KW, Medrikova D, Collden G, Woods SC, Herzig S, Homann D, Jung S, Nedergaard J, Cannon B, Tschop MH, Muller TD, Buettner C (2017) Alternatively activated macrophages do not synthesize catecholamines or contribute to adipose tissue adaptive thermogenesis. Nat Med 23(5):623–630. https://doi.org/10.1038/nm.4316

Frikke-Schmidt H, Zamarron BF, O'Rourke RW, Sandoval DA, Lumeng CN, Seeley RJ (2017) Weight loss independent changes in adipose tissue macrophage and T cell populations after sleeve gastrectomy in mice. Mol Metab 6(4):317–326. https://doi.org/10.1016/j.molmet.2017.02.004

Fu M, Shu S, Peng Z, Liu X, Chen X, Zeng Z, Yang Y, Cui H, Zhao R, Wang X, Du L, Wu M, Feng W, Song J (2023) Single-cell RNA sequencing of coronary perivascular adipose tissue from end-stage heart failure patients identifies SPP1(+) macrophage subpopulation as a target for alleviating fibrosis. Arterioscler Thromb Vasc Biol 43(11):2143–2164. https://doi.org/10.1161/atvbaha.123.319828

Fuentes L, Röszer T, Ricote M (2010) Inflammatory mediators and insulin resistance in obesity: role of nuclear receptor signaling in macrophages. Mediat Inflamm 2010:219583. https://doi.org/10.1155/2010/219583

Geserick M, Vogel M, Gausche R, Lipek T, Spielau U, Keller E, Pfaffle R, Kiess W, Korner A (2018) Acceleration of BMI in early childhood and risk of sustained obesity. N Engl J Med 379(14):1303–1312. https://doi.org/10.1056/NEJMoa1803527

Giordano A, Murano I, Mondini E, Perugini J, Smorlesi A, Severi I, Barazzoni R, Scherer PE, Cinti S (2013) Obese adipocytes show ultrastructural features of stressed cells and die of pyroptosis. J Lipid Res 54(9):2423–2436. https://doi.org/10.1194/jlr.M038638

Glass CK, Olefsky JM (2012) Inflammation and lipid signaling in the etiology of insulin resistance. Cell Metab 15(5):635–645. https://doi.org/10.1016/j.cmet.2012.04.001

Guo W, Pirtskhalava T, Tchkonia T, Xie W, Thomou T, Han J, Wang T, Wong S, Cartwright A, Hegardt FG, Corkey BE, Kirkland JL (2007) Aging results in paradoxical susceptibility of fat cell progenitors to lipotoxicity. Am J Physiol Endocrinol Metab 292(4):E1041–E1051. https://doi.org/10.1152/ajpendo.00557.2006

Gyurina K, Yarmak M, Sasi-Szabó L, Molnár S, Méhes G, Röszer T (2023) Loss of uncoupling protein 1 expression in the subcutaneous adipose tissue predicts childhood obesity. Int J Mol Sci 24(23):16706

Haase J, Weyer U, Immig K, Kloting N, Bluher M, Eilers J, Bechmann I, Gericke M (2014) Local proliferation of macrophages in adipose tissue during obesity-induced inflammation. Diabetologia 57(3):562–571. https://doi.org/10.1007/s00125-013-3139-y

Hausberger FX (1966) Pathological changes in adipose tissue of obese mice. Anat Rec 154(3):651–660. https://doi.org/10.1002/ar.1091540311

Hill AA, Reid Bolus W, Hasty AH (2014) A decade of progress in adipose tissue macrophage biology. Immunol Rev 262(1):134–152. https://doi.org/10.1111/imr.12216

Hoang AC, Yu H, Röszer T (2021) Transcriptional landscaping identifies a beige adipocyte depot in the newborn mouse. Cells 10(9):2368

Hoang AC, Sasi-Szabó L, Pál T, Szabó T, Diedrich V, Herwig A, Landgraf K, Körner A, Röszer T (2022) Mitochondrial RNA stimulates beige adipocyte development in young mice. Nat Metab 4(12):1684–1696. https://doi.org/10.1038/s42255-022-00683-w

Honecker J, Ruschke S, Seeliger C, Laber S, Strobel S, Pröll P, Nellaker C, Lindgren CM, Kulozik U, Ecker J, Karampinos DC, Claussnitzer M, Hauner H (2022) Transcriptome and fatty-acid signatures of adipocyte hypertrophy and its non-invasive MR-based characterization in human adipose tissue. EBioMedicine 79:104020. https://doi.org/10.1016/j.ebiom.2022.104020

Huang Stanley C-C, Smith Amber M, Everts B, Colonna M, Pearce Erika L, Schilling Joel D, Pearce Edward J (2016) Metabolic reprogramming mediated by the mTORC2-IRF4 signaling axis is essential for macrophage alternative activation. Immunity 45(4):817–830. https://doi.org/10.1016/j.immuni.2016.09.016

Huang Y, Qi H, Zhang Z, Wang E, Yun H, Yan H, Su X, Liu Y, Tang Z, Gao Y, Shang W, Zhou J, Wang T, Che Y, Zhang Y, Yang R (2017) Gut REG3γ-associated lactobacillus induces anti-inflammatory macrophages to maintain adipose tissue homeostasis. Front Immunol 8:1063. https://doi.org/10.3389/fimmu.2017.01063

Hui X, Gu P, Zhang J, Nie T, Pan Y, Wu D, Feng T, Zhong C, Wang Y, Lam KS, Xu A (2015) Adiponectin enhances cold-induced browning of subcutaneous adipose tissue via promoting M2 macrophage proliferation. Cell Metab 22(2):279–290. https://doi.org/10.1016/j.cmet.2015.06.004

Jacks RD, Lumeng CN (2024) Macrophage and T cell networks in adipose tissue. Nat Rev Endocrinol 20(1):50–61. https://doi.org/10.1038/s41574-023-00908-2

Kopsch F (1914) Die Gewebe. In: Kopsch F (ed) Rauber's Lehrbuch der Anatomie des Menschen. Georg Thieme, Leipzig, pp 95–99

Kosteli A, Sugaru E, Haemmerle G, Martin JF, Lei J, Zechner R, Ferrante AW Jr (2010) Weight loss and lipolysis promote a dynamic immune response in murine adipose tissue. J Clin Invest 120(10):3466–3479. https://doi.org/10.1172/jci42845

Kováčiková M, Sengenes C, Kováčová Z, Šiklová-Vítková M, Klimčáková E, Polák J, Rossmeislová L, Bajzová M, Hejnová J, Hnĕvkovská Z, Bouloumié A, Langin D, Štich V (2011) Dietary intervention-induced weight loss decreases macrophage content in adipose tissue of obese women. Int J Obes 35(1):91–98. https://doi.org/10.1038/ijo.2010.112

Kuziel G, Moore BN, Arendt LM (2023) Obesity and fibrosis: setting the stage for breast cancer. Cancers (Basel) 15(11):2929. https://doi.org/10.3390/cancers15112929

Lackey DE, Olefsky JM (2016) Regulation of metabolism by the innate immune system. Nat Rev Endocrinol 12(1):15–28. https://doi.org/10.1038/nrendo.2015.189

Landgraf K, Rockstroh D, Wagner IV, Weise S, Tauscher R, Schwartze JT, Löffler D, Bühligen U, Wojan M, Till H, Kratzsch J, Kiess W, Blüher M, Körner A (2015) Evidence of early alterations in adipose tissue biology and function and its association with obesity-related inflammation and insulin resistance in children. Diabetes 64(4):1249–1261. https://doi.org/10.2337/db14-0744

Lee BC, Lee J (2014) Cellular and molecular players in adipose tissue inflammation in the development of obesity-induced insulin resistance. Biochim Biophys Acta 1842(3):446–462. https://doi.org/10.1016/j.bbadis.2013.05.017

Li P, Spann NJ, Kaikkonen MU, Lu M, Oh da Y, Fox JN, Bandyopadhyay G, Talukdar S, Xu J, Lagakos WS, Patsouris D, Armando A, Quehenberger O, Dennis EA, Watkins SM, Auwerx J, Glass CK, Olefsky JM (2013) NCoR repression of LXRs restricts macrophage biosynthesis of insulin-sensitizing omega 3 fatty acids. Cell 155(1):200–214. https://doi.org/10.1016/j.cell.2013.08.054

Li C, Menoret A, Farragher C, Ouyang Z, Bonin C, Holvoet P, Vella AT, Zhou B (2019) Single cell transcriptomics based-MacSpectrum reveals novel macrophage activation signatures in diseases. JCI Insight 5(10):e126453. https://doi.org/10.1172/jci.insight.126453

Li B, Liu S, Yang Q, Li Z, Li J, Wu J, Sun S, Xu Z, Sun S, Wu Q (2023) Macrophages in tumor-associated adipose microenvironment accelerate tumor progression. Adv Biol 7(1):2200161. https://doi.org/10.1002/adbi.202200161

Lin L, Kuhn C, Ditsch N, Kolben T, Czogalla B, Beyer S, Trillsch F, Schmoeckel E, Mayr D, Mahner S, Jeschke U, Hester A (2021) Breast adipose tissue macrophages (BATMs) have a stronger correlation with breast cancer survival than breast tumor stroma macrophages (BTSMs). Breast Cancer Res 23(1):45. https://doi.org/10.1186/s13058-021-01422-x

Linde N, Casanova-Acebes M, Sosa MS, Mortha A, Rahman A, Farias E, Harper K, Tardio E, Reyes Torres I, Jones J, Condeelis J, Merad M, Aguirre-Ghiso JA (2018) Macrophages orchestrate breast cancer early dissemination and metastasis. Nat Commun 9(1):21. https://doi.org/10.1038/s41467-017-02481-5

Lindhorst A, Raulien N, Wieghofer P, Eilers J, Rossi FMV, Bechmann I, Gericke M (2021) Adipocyte death triggers a pro-inflammatory response and induces metabolic activation of resident macrophages. Cell Death Dis 12(6):579. https://doi.org/10.1038/s41419-021-03872-9

Liu R, Nikolajczyk BS (2019) Tissue immune cells fuel obesity-associated inflammation in adipose tissue and beyond. Front Immunol 10:1587. https://doi.org/10.3389/fimmu.2019.01587

Lumeng CN, Bodzin JL, Saltiel AR (2007) Obesity induces a phenotypic switch in adipose tissue macrophage polarization. J Clin Invest 117(1):175–184. https://doi.org/10.1172/JCI29881

Lumeng CN, DelProposto JB, Westcott DJ, Saltiel AR (2008) Phenotypic switching of adipose tissue macrophages with obesity is generated by spatiotemporal differences in macrophage subtypes. Diabetes 57(12):3239–3246. https://doi.org/10.2337/db08-0872

Luo B, Wang Z, Zhang Z, Shen Z, Zhang Z (2019) The deficiency of macrophage erythropoietin signaling contributes to delayed acute inflammation resolution in diet-induced obese mice. Biochim Biophys Acta Mol basis Dis 1865(2):339–349. https://doi.org/10.1016/j.bbadis.2018.10.005

Morinaga H, Mayoral R, Heinrichsdorff J, Osborn O, Franck N, Hah N, Walenta E, Bandyopadhyay G, Pessentheiner AR, Chi TJ, Chung H, Bogner-Strauss JG, Evans RM, Olefsky JM, Oh DY (2015) Characterization of distinct subpopulations of hepatic macrophages in HFD/obese mice. Diabetes 64(4):1120–1130. https://doi.org/10.2337/db14-1238

Morris DL, Singer K, Lumeng CN (2011) Adipose tissue macrophages: phenotypic plasticity and diversity in lean and obese states. Curr Opin Clin Nutr Metab Care 14(4):341–346. https://doi.org/10.1097/MCO.0b013e328347970b

Nagareddy PR, Kraakman M, Masters SL, Stirzaker RA, Gorman DJ, Grant RW, Dragoljevic D, Hong ES, Abdel-Latif A, Smyth SS, Choi SH, Korner J, Bornfeldt KE, Fisher EA, Dixit VD, Tall AR, Goldberg IJ, Murphy AJ (2014) Adipose tissue macrophages promote myelopoiesis and monocytosis in obesity. Cell Metab 19(5):821–835. https://doi.org/10.1016/j.cmet.2014.03.029

Nickl B, Qadri F, Bader M (2021) Anti-inflammatory role of Gpnmb in adipose tissue of mice. Sci Rep 11(1):19614. https://doi.org/10.1038/s41598-021-99090-6

Nisoli E, Cardile A, Bulbarelli A, Tedesco L, Bracale R, Cozzi V, Morroni M, Cinti S, Valerio A, Carruba MO (2006) White adipocytes are less prone to apoptotic stimuli than brown adipocytes in rodent. Cell Death Differ 13(12):2154–2156. https://doi.org/10.1038/sj.cdd.4401956

Osborn O, Olefsky JM (2012) The cellular and signaling networks linking the immune system and metabolism in disease. Nat Med 18(3):363–374. https://doi.org/10.1038/nm.2627

Osuga J, Ishibashi S, Oka T, Yagyu H, Tozawa R, Fujimoto A, Shionoiri F, Yahagi N, Kraemer FB, Tsutsumi O, Yamada N (2000) Targeted disruption of hormone-sensitive lipase results in male sterility and adipocyte hypertrophy, but not in obesity. Proc Natl Acad Sci USA 97(2):787–792. https://doi.org/10.1073/pnas.97.2.787

Pietrobelli A, Agosti M, MeNu Group (2017) Nutrition in the first 1000 days: ten practices to minimize obesity emerging from published science. Int J Environ Res Public Health 14(12):1491. https://doi.org/10.3390/ijerph14121491

Pirzgalska RM, Seixas E, Seidman JS, Link VM, Sánchez NM, Mahú I, Mendes R, Gres V, Kubasova N, Morris I, Arús BA, Larabee CM, Vasques M, Tortosa F, Sousa AL, Anandan S, Tranfield E, Hahn MK, Iannacone M, Spann NJ, Glass CK, Domingos AI (2017) Sympathetic neuron–associated macrophages contribute to obesity by importing and metabolizing norepinephrine. Nat Med 23:1309. https://doi.org/10.1038/nm.4422

Prieur X, Röszer T, Ricote M (2010) Lipotoxicity in macrophages: evidence from diseases associated with the metabolic syndrome. Biochim Biophys Acta 1801(3):327–337. https://doi.org/10.1016/j.bbalip.2009.09.017

Qian S, Pan J, Su Y, Tang Y, Wang Y, Zou Y, Zhao Y, Ma H, Zhang Y, Liu Y, Guo L, Tang Q-q (2020) BMPR2 promotes fatty acid oxidation and protects white adipocytes from cell death in mice. Commun Biol 3(1):200. https://doi.org/10.1038/s42003-020-0928-y

Qiu Y, Nguyen KD, Odegaard JI, Cui X, Tian X, Locksley RM, Palmiter RD, Chawla A (2014) Eosinophils and type 2 cytokine signaling in macrophages orchestrate development of functional beige fat. Cell 157(6):1292–1308. https://doi.org/10.1016/j.cell.2014.03.066

Rolland-Cachera MF, Deheeger M, Bellisle F, Sempé M, Guilloud-Bataille M, Patois E (1984) Adiposity rebound in children: a simple indicator for predicting obesity. Am J Clin Nutr 39(1):129–135. https://doi.org/10.1093/ajcn/39.1.129

Rosen ED, Spiegelman BM (2014) What we talk about when we talk about fat. Cell 156(1-2):20–44. https://doi.org/10.1016/j.cell.2013.12.012

Röszer T (2017) Transcriptional control of apoptotic cell clearance by macrophage nuclear receptors. Apoptosis 22(2):284–294. https://doi.org/10.1007/s10495-016-1310-x

Röszer T (2018) Understanding the biology of self-renewing macrophages. Cells 7(8):103

Röszer T (2020a) Mechanisms which control the size of M2 macrophage-dominated tissue macrophage niches. In: Röszer T (ed) The M2 macrophage. Springer, pp 99–111

Röszer T (2020b) Signal mechanisms of M2 macrophage activation. The M2 macrophage. Springer, Cham, pp 73–97

Röszer T (2020c) What is an M2 macrophage? Historical overview of the macrophage polarization model. The Th1/Th2 and M1/M2 paradigm, the arginine fork. In: Röszer T (ed) The M2 macrophage. Springer, pp 3–25
Röszer T (2020d) Practical approaches in M2 macrophage biology: analysis, pharmacology, and didactical interpretation of M2 macrophage functions. In: Röszer T (ed) The M2 macrophage. Springer, Cham, pp 189–224
Röszer T (2020e) M2 macrophages in the metabolic organs and in the neuroendocrine system. In: Röszer T (ed) The M2 macrophage. Springer, pp 171–187
Röszer T (2021) Adipose tissue immunometabolism and apoptotic cell clearance. Cells 10(9):2288
Röszer T (2022) Metabolic impact of adipose tissue macrophages in the early postnatal life. J Leukoc Biol 112(6):1515–1524. https://doi.org/10.1002/jlb.3mr0722-201r
Russo L, Lumeng CN (2018) Properties and functions of adipose tissue macrophages in obesity. Immunology 155(4):10. https://doi.org/10.1111/imm.13002
Sacks H, Symonds ME (2013) Anatomical locations of human brown adipose tissue: functional relevance and implications in obesity and type 2 diabetes. Diabetes 62(6):1783–1790. https://doi.org/10.2337/db12-1430
Shapiro H, Lutaty A, Ariel A (2011) Macrophages, meta-inflammation, and immuno-metabolism. ScientificWorldJournal 11:2509–2529. https://doi.org/10.1100/2011/397971
Shaul ME, Bennett G, Strissel KJ, Greenberg AS, Obin MS (2010) Dynamic, M2-like remodeling phenotypes of CD11c+ adipose tissue macrophages during high-fat diet--induced obesity in mice. Diabetes 59(5):1171–1181. https://doi.org/10.2337/db09-1402
Siervogel RM, Roche AF, Guo SM, Mukherjee D, Chumlea WC (1991) Patterns of change in weight/stature2 from 2 to 18 years: findings from long-term serial data for children in the Fels longitudinal growth study. Int J Obes 15(7):479–485
Singer K, Morris DL, Oatmen KE, Wang T, DelProposto J, Mergian T, Cho KW, Lumeng CN (2013) Neuropeptide Y is produced by adipose tissue macrophages and regulates obesity-induced inflammation. PLoS One 8(3):e57929
Singh A, Mayengbam SS, Yaduvanshi H, Wani MR, Bhat MK (2022) Obesity programs macrophages to support cancer progression. Cancer Res 82(23):4303–4312. https://doi.org/10.1158/0008-5472.can-22-1257
Sorisky A, Magun R, Gagnon AM (2000) Adipose cell apoptosis: death in the energy depot. Int J Obes Relat Metab Disord 24(Suppl 4):S3–S7. https://doi.org/10.1038/sj.ijo.0801491
Spencer M, Yao-Borengasser A, Unal R, Rasouli N, Gurley CM, Zhu B, Peterson CA, Kern PA (2010) Adipose tissue macrophages in insulin-resistant subjects are associated with collagen VI and fibrosis and demonstrate alternative activation. Am J Physiol Endocrinol Metab 299(6):E1016–E1027. https://doi.org/10.1152/ajpendo.00329.2010
Sun K, Gao Z, Kolonin MG (2018) Transient inflammatory signaling promotes beige adipogenesis. Sci Signal 11(527):eaat3192. https://doi.org/10.1126/scisignal.aat3192
Tchkonia T, Tchoukalova YD, Giorgadze N, Pirtskhalava T, Karagiannides I, Forse RA, Koo A, Stevenson M, Chinnappan D, Cartwright A, Jensen MD, Kirkland JL (2005) Abundance of two human preadipocyte subtypes with distinct capacities for replication, adipogenesis, and apoptosis varies among fat depots. Am J Physiol Endocrinol Metab 288(1):E267–E277. https://doi.org/10.1152/ajpendo.00265.2004
Vandanmagsar B, Youm YH, Ravussin A, Galgani JE, Stadler K, Mynatt RL, Ravussin E, Stephens JM, Dixit VD (2011) The NLRP3 inflammasome instigates obesity-induced inflammation and insulin resistance. Nat Med 17(2):179–188. https://doi.org/10.1038/nm.2279
Vogel P, Read R, Hansen G, Wingert J, Dacosta CM, Buhring LM, Shadoan M (2011) Pathology of congenital generalized lipodystrophy in Agpat2−/− mice. Vet Pathol 48(3):642–654. https://doi.org/10.1177/0300985810383870
Waqas SFH, Hoang A, Lin Y, Ampem G et al (2017a) Neuropeptide FF increases M2 activation and self-renewal of adipose tissue macrophages. J Clin Invest 127(7):2842–2854

Waqas SFH, Noble A, Hoang A, Ampem G, Popp M, Strauß S, Guille M, Röszer T (2017b) Adipose tissue macrophages develop from bone marrow-independent progenitors in *Xenopus laevis* and mouse. J Leukoc Biol 102(3):845–855

Weisberg SP, McCann D, Desai M, Rosenbaum M, Leibel RL, Ferrante AW (2003) Obesity is associated with macrophage accumulation in adipose tissue. J Clin Invest 112(12):1796–1808. https://doi.org/10.1172/jci200319246

West M (2009) Dead adipocytes and metabolic dysfunction: recent progress. Curr Opin Endocrinol Diabetes Obes 16(2):178–182

Xu H, Barnes GT, Yang Q, Tan G, Yang D, Chou CJ, Sole J, Nichols A, Ross JS, Tartaglia LA, Chen H (2003) Chronic inflammation in fat plays a crucial role in the development of obesity-related insulin resistance. J Clin Invest 112(12):1821–1830. https://doi.org/10.1172/JCI19451

Yu H, Dilbaz S, Coßmann J, Hoang AC, Diedrich V, Herwig A, Harauma A, Hoshi Y, Moriguchi T, Landgraf K, Körner A, Lucas C, Brodesser S, Balogh L, Thuróczy J, Karemore G, Kuefner MS, Park EA, Rapp C, Travers JB, Röszer T (2019) Breast milk alkylglycerols sustain beige adipocytes through adipose tissue macrophages. J Clin Invest 129(6):2485–2499. https://doi.org/10.1172/JCI125646

Zeyda M, Farmer D, Todoric J, Aszmann O, Speiser M, Gyori G, Zlabinger GJ, Stulnig TM (2007) Human adipose tissue macrophages are of an anti-inflammatory phenotype but capable of excessive pro-inflammatory mediator production. Int J Obes 31(9):1420–1428. https://doi.org/10.1038/sj.ijo.0803632

Zheng C, Yang Q, Xu C, Shou P, Cao J, Jiang M, Chen Q, Cao G, Han Y, Li F, Cao W, Zhang L, Zhang L, Shi Y, Wang Y (2015) CD11b regulates obesity-induced insulin resistance via limiting alternative activation and proliferation of adipose tissue macrophages. Proc Natl Acad Sci USA 112(52):E7239–E7248. https://doi.org/10.1073/pnas.1500396113

Chapter 7
Liver Macrophage Diversity in Health and Disease

Paul Horn and Frank Tacke

Abstract The liver fulfils a plethora of metabolic and immunological functions. Liver macrophages are a heterogeneous immune cell population with high plasticity and are important for maintaining normal liver function but are also critically involved in disease processes. In this chapter, we review the heterogeneity and multifaceted functions of hepatic macrophages in liver health and in disease conditions, including acute liver injury, chronic liver diseases, and hepatocellular carcinoma. Under homeostatic conditions, the tissue resident Kupffer cells are phagocytic cells that have important functions in immune surveillance, antigen presentation, and metabolic regulation while the roles of other populations such as capsular, peritoneal, or monocyte-derived macrophages in liver health are less clearly defined. Upon liver injury, Kupffer cell numbers are markedly reduced while monocyte-derived macrophages significantly expand and take critical roles in driving and resolving liver injury, including important pathogenic involvements in inflammation, fibrosis, and regeneration. They also create and maintain an immunosuppressive and immune-excluded microenvironment in hepatocellular carcinoma. Single-cell and spatial omics technologies are significantly expanding our understanding of the diversity and plasticity of macrophage populations under different conditions and enable the reliable identification of specific hepatic macrophage subsets. This knowledge can now be applied to dissect the exact contributions of distinct macrophage populations to disease processes and hopefully will pave the way for new therapeutic interventions.

P. Horn
Department of Hepatology and Gastroenterology, Charité – Universitätsmedizin Berlin, Campus Virchow-Klinikum and Campus Charité Mitte, Berlin, Germany

Berlin Institute of Health at Charité – Universitätsmedizin Berlin, BIH Biomedical Innovation Academy, BIH Charité Digital Clinician Scientist Program, Berlin, Germany

F. Tacke (✉)
Department of Hepatology and Gastroenterology, Charité – Universitätsmedizin Berlin, Campus Virchow-Klinikum and Campus Charité Mitte, Berlin, Germany
e-mail: frank.tacke@charite.de

M. Kloc et al. (eds.), *Monocytes and Macrophages in Development, Regeneration, and Disease*, Results and Problems in Cell Differentiation 74,
https://doi.org/10.1007/978-3-031-65944-7_7

7.1 Introduction

The human liver is a remarkable parenchymal organ that fulfils a plethora of essential metabolic and immunologic functions. These include carbohydrate, lipid, and amino acid metabolism, regulation of trace element metabolism (e.g. iron and copper), detoxification, drug metabolism, bile formation and fat digestion, production of plasma proteins including albumin, lipoproteins and coagulation factors as well as immune surveillance (Renato et al. 2007). The liver is rich in tissue-resident macrophages, so-called Kupffer cells (KCs), which have important roles in regulating and maintaining normal function and microarchitecture of the liver (Jaeschke 2007). Recent research highlighted the dual role of hepatic macrophages under disease conditions. While the infiltration and inflammatory activation of macrophages is required for inducing immunological pathways that drive acute and chronic liver injury, as well as hepatocellular carcinogenesis, they are also essential players in the resolution of injury and restitution of liver function (Guillot and Tacke 2019; Hammerich and Tacke 2023).

Largely owing to the rapid developments in single-cell and spatial omics technologies, our understanding of the complexity and diversity of macrophage phenotypes is currently dramatically evolving (Guilliams and Scott 2022). Therefore, the previously suggested dichotomous distinction between inflammatory M1 and anti-inflammatory M2 liver macrophages needs to be replaced by a more nuanced description of macrophage phenotypes and functions (Guillot and Tacke 2019). Still, we are only beginning to understand the contribution of different macrophage populations to physiological and disease processes as well as how these subsets are shaped by their origin, spatial localisation as well as metabolic and immunologic microenvironmental cues. In this chapter we aim to give a comprehensive overview of our current understanding and ongoing developments relating to liver macrophage biology and their heterogeneity under homeostatic and disease conditions. We will cover the basic processes involved but will also highlight the most recent advances in our understanding of macrophage diversity and function to finally touch on promising targets for treating liver disease.

7.2 The Hepatic Microarchitecture

The liver lobule is the basic functional unit of the liver. The main metabolic liver functions are provided by hepatocytes, which absorb nutrients and xenobiotic substances on their basolateral, sinusoidal site and on their apical site secrete bile fluid into the biliary canaliculi, which collect and transport the bile towards the portal bile ducts. Blood flows from the portal branches of the portal vein and the hepatic artery through the hepatic sinusoids into the central vein. These liver sinusoids are lined by specialised, fenestrated endothelial cells (Semela and Shah 2007) and contain the tissue-resident macrophages of the liver, the so-called Kupffer cells (Jaeschke

2007). The subendothelial space (space of Disse), is built of loose connective tissue and a basement membrane-like matrix. This region contains the hepatic stellate cells, which function as vitamin A storing pericytes of the hepatic sinusoids and produce the fine connective tissue of the Disse space (Pinzani 2007). The localization in the subendothelial space allows hepatic stellate cells to communicate with neighbouring cells, including the hepatocytes, endothelial cells but also KCs (Tsuchida and Friedman 2017). In general, the major players of the hepatic inflammatory and fibrotic response are in intimate proximity in the space of Disse, allowing for close intercellular communication.

7.3 Macrophages in Liver Homeostasis

7.3.1 Kupffer Cells

Under homeostatic conditions, KCs are the major tissue-resident hepatic macrophage population and fulfil a plethora of functions. It has been thought for many decades that KCs originate and replenish from bone marrow precursors but more recent lineage tracing studies revealed that they are embryologically derived and originate from CSF1R^{+} yolk sac progenitors (Fig. 7.1) (Gomez Perdiguero et al. 2015). Under homeostatic conditions, it has been well established that they self-renew independent of bone marrow precursors in mice (Yona et al. 2013; Mass et al. 2016) but data are less clear in humans, although donor liver KCs can persist for many years after liver transplantation (Pallett et al. 2020). In response to liver injury, Kupffer cell numbers often decrease and during liver regeneration, the Kupffer cell pool can be replenished from self-renewing embryonically-derived KCs (emKCs) or from monocytes acquiring a Kupffer cell phenotype (moKCs) (Scott et al. 2016).

KCs reside on the luminal side of the sinusoidal endothelium where they sense their microenvironment through cytoplasmic extensions, giving them a star-shaped appearance (in contrast to the more round-shaped monocyte-derived macrophages). Related to their localization at the interface of hepatic blood flow and the sub-sinusoidal space of Disse, their high sensing capacity for microenvironmental factors as well as their high phagocytic activity, KCs exert diverse functions in immunity, defence against infections and regulation of metabolism.

The liver is exposed to a constant influx of antigens and pathogens from systemic circulation and the gut. The reticuloendothelial system, mainly composed of liver-resident macrophages, has a huge capacity for clearing circulating bacteria (Benacerraf et al. 1959) and thereby exerts important functions in antimicrobial defence. KCs express the complement receptor CRIg, which enables the detection and capture of opsonised (Helmy et al. 2006) and non-opsonised bacteria (Zeng et al. 2016). Consequently, chronic liver injury, which leads to reduced expression of CRIg and reductions in Kupffer cell numbers, results in increased susceptibility to bacterial infection (Duan et al. 2019; Peiseler et al. 2023). While CRIg on KCs is

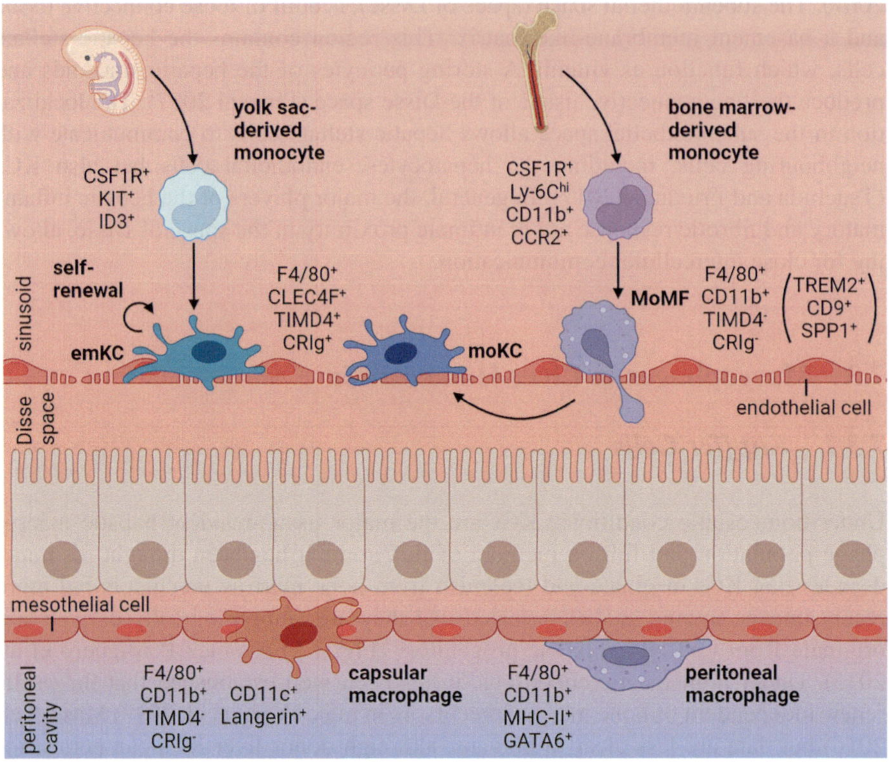

Fig. 7.1 Principal macrophage populations in healthy and diseased liver. Liver macrophages are a heterogeneous cell population with different ontogenetic origins. Kupffer cells are the principal liver-resident, self-renewing macrophage population and reside in the sinusoids. Homeostatic, embryonic KCs (moKC) originate from yolk sac-derived monocytes. Monocyte-derived macrophages (MoMF) however, differentiate from bone marrow-derived circulating monocytes, are lowly abundant in healthy liver but upon injury undergo pronounced proliferation and differentiation into diverse subpopulations, including monocyte-derived Kupffer cells (moKC). Capsular and peritoneal macrophages represent minor populations with to date insufficiently understood roles in liver homeostasis and disease. The depicted markers for liver macrophage subsets were identified in mice (created with BioRender.com)

crucial for preventing the spreading of gram-positive bacteria, KCs (and other liver macrophages) are also equipped with numerous "danger sensing receptors" such as toll-like receptors (TLRs), allowing them to sense infectious or non-infectious threats (e.g. LPS from gram-negative bacteria via TLR4) (Wen et al. 2021).

While KCs are important for removing blood-borne pathogens to prevent infection, they also suppress immune responses to harmless antigens, primarily those delivered to the liver via the portal vein. As one of the main antigen-presenting cell populations in liver homeostasis, KCs maintain immunological tolerance by promoting the activation and expansion of IL-10-producing regulatory T cells and suppress T cell activation (You et al. 2008; Heymann et al. 2015), thereby promoting immune tolerance to harmless self- or food-antigens.

KCs also have an important role in clearing aged circulating red blood cells and platelets. Iron homeostasis is tightly regulated and depends on the effective recycling of iron from red blood cells (Scott and Guilliams 2018). KCs are responsible for about half the red blood cell clearing capacity, as demonstrated by experiments using $Na_2{}^{51}CrO_4$-labeled oxidised red blood cells and haemoglobin-containing vesicles, a process dependent on the expression of scavenger receptors for polyinosinic acid and phosphatidylserine (Terpstra and van Berkel 2000; Willekens et al. 2005). Furthermore, KCs partake in the direct transport and storage of iron by hepatocytes through transferrin receptors and ferroportin (Scott and Guilliams 2018), thereby preventing iron toxicity from available free iron ions. The removal of aged platelets from circulation by KCs is dependent on the expression of galactose lectin and CLEC4F in mice, and maintains primary haemostasis function (Deppermann et al. 2020; Jiang et al. 2021).

Furthermore, KCs have been identified as players in hepatic cholesterol and lipid metabolism. They produce *cholesteryl ester transfer protein* (CETP), which regulates high-density lipoprotein (HDL) and very low-density lipoprotein (VLDL) levels, thereby controlling whole-body cholesterol metabolism (Wang et al. 2015). Using single-cell sequencing approaches, others have identified a subset of murine KCs characterised by the expression of genes related to lipid metabolism and this subset seems to contribute to glucose homeostasis and lipid processing, dependent on CD36 expression (Blériot et al. 2021).

In mice, KCs can be identified as CD11b^{low} F4/80^{high} macrophages that express *C-type lectin domain family 4 member F* (Clec4f), *T cell immunoglobulin (Ig) and mucin domain containing 4* (Timd4) and CRIg (also called Vsig4) (Helmy et al. 2006; Scott et al. 2016). The advent of single-cell transcriptomics approaches has enabled the identification of different murine subsets of KCs with proposed different functions but without obvious differences in spatial distribution (De Simone et al. 2021), including the metabolically enriched KC population discussed above (Blériot et al. 2021). The specific roles of different KC populations, however, still need to be defined.

Single-cell technologies have also significantly helped to characterise human Kupffer cell populations. Based on three independent transcriptomic studies from human liver, human KCs can be defined as CD163$^+$ MARCO$^+$ CD5L$^+$ TIMD4$^+$ macrophages (MacParland et al. 2018; Ramachandran et al. 2019; Aizarani et al. 2019). These newly identified markers can now be used to specifically isolate and further characterise these cells. Unlike in mice, MARCO$^+$ KCs show a preferential periportal zonal distribution (MacParland et al. 2018). MARCO$^+$ KCs are also less responsive to LPS and IFN-γ stimulation, which indicates a tolerogenic role in humans (MacParland et al. 2018). Similar to studies in mice, single-cell RNA sequencing studies also revealed heterogeneity of the human Kupffer cell population, with populations positive or negative for TIMD4 expression (Ramachandran et al. 2019) but their specific roles still need to be defined.

7.3.2 Monocyte-Derived Macrophages

While KCs represent the predominant macrophage population in the healthy liver, a small population of circulating monocyte-derived macrophages (MoMF) can be found under homeostasis as well. Originating from bone-marrow precursors, the recruitment of circulating CSFR1R$^+$CD11b$^+$Ly-6C$^+$ monocytes to the liver is mediated primarily by chemokines, such as C-C motif ligand 2 (CCL2), CCL3, CCL5, CCL9 and CCL10, and their receptors C-C chemokine receptor type 2 (CCR2) and CCR5 in mice (Krenkel and Tacke 2017). These monocytes eventually differentiate into macrophages and subsequently, depending on microenvironmental cues, can undergo pronounced dynamic phenotypical and functional changes, which have recently been illuminated by single-cell and spatial omics technologies (Peiseler et al. 2022). While MoMF mainly derive from inflammatory, immature Ly-6C^{high} monocytes in mice (Fig. 7.1), patrolling or mature Ly-6C^{low} hepatic monocytes have also been described and are associated with tissue repair processes and can differentiate into dendritic cells (Guilliams et al. 2018). In humans, CD14^{high}CD16$^-$ classical monocytes and CD14^{low}CD16$^+$ monocytes correspond to Ly-6C^{high} and Ly-6C^{low} murine monocytes, respectively. Similarly, freshly infiltrating liver macrophages are mainly Ly-6C^{high}/CD14^{high}CD16$^-$ and promote inflammatory processes while CD14$^+$ CD16$^+$ subsets that differentiate into CD14$^+$ CD16$^-$ monocytes upon liver tissue infiltration have also been described (Zimmermann et al. 2010; Liaskou et al. 2013).

Under homeostatic conditions, MoMF are lowly abundant but contribute to iron and cholesterol metabolism (Theurl et al. 2016). A population of portal macrophages closely localised to the biliary epithelium has been characterised as resembling lipid-associated macrophages (LAM), that are found in steatotic liver disease or obese adipose tissue (Guilliams et al. 2022). While LAM have a clear role in liver inflammation and fibrogenesis under diseased conditions, their role in the healthy liver has not been fully characterised yet. Taken together, MoMF are clearly present in the liver under homeostatic conditions, but their significance and regulation in liver health are not well studied. Their roles are best described in disease conditions, where they show striking plasticity and exert diverse functions relating to immune regulation, inflammation, and extracellular matrix organization. Therefore, the heterogeneity of MoMF will be discussed in more detail in the respective sections of this chapter.

7.3.3 Other Sources of Hepatic Macrophages

Capsular macrophages are another population of hepatic macrophages that only recently have been described in more detail. They are distinct from hepatic MoMF and KC, and can be identified in mice by the expression of CD11b, F4/80, CD11c, CD64 and Langerin and the absence of Ly-6C, Clec4F and Tim4 (Sierro et al. 2017;

Guilliams et al. 2022). They reside in the liver capsule at the interface of liver parenchyma and the abdominal cavity and extend their cytoplasmic protrusions into both compartments (Sierro et al. 2017). Here, they form a barrier to bacterial infiltration from the peritoneum and aid in local neutrophil recruitment in response to infection (Sierro et al. 2017). Apart from antimicrobial defence, the role of capsular macrophages in liver homeostasis and disease needs to be defined.

Peritoneal macrophages are a less well studied population of macrophages distinct from KC and MoMF. They reside in the peritoneal cavity and some have reported infiltration of peritoneal macrophages *per continuitatem* through the liver capsule into the parenchyma in response to thermal liver injury and also suggested a role of these cells in liver cancer (Wang and Kubes 2016). However, recent reports suggest that peritoneal macrophages identified by single-cell technologies may be originating from contamination with cells on the organ surface but not tissue-infiltrating peritoneal macrophages (Jin et al. 2021; Guilliams et al. 2022). Overall, it still needs to be demonstrated whether peritoneal macrophages have a role in liver homeostasis and disease.

7.4 Liver Macrophages in Acute Hepatic Injury

Acute liver injury (ALI) is defined as the sudden deterioration of liver function in the absence of prior chronic liver disease. A plethora of factors, including drugs, hepatotoxins, viruses (mainly hepatitis virus A, B or E) or autoimmune conditions can induce ALI, which in the absence of adequate regenerative capacity or upon overwhelming injury can progress to acute liver failure (ALF) with jaundice, coagulation failure and hepatic encephalopathy. Paracetamol overdose is the most prevalent cause of acute liver failure in the Western hemisphere while viral hepatitis is the dominant cause in developing countries (Bernal and Wendon 2013).

In the early necroinflammatory phase of ALI, hepatocellular necrosis leads to the release of cell constituents, so-called damage-associated molecular pattern (DAMP) molecules, including HMGB1, nucleotides or mitochondrial DNA (Bernal and Wendon 2013). KCs, which are highly sensitive in detecting danger-associated signals, subsequently get activated and release many chemokines and cytokines, including CXC-chemokine ligand 1 (CXCL1), CXCL2, and CXCL8, as well as CCl2, CCL25 and CXCL1, thereby attracting circulating immune cells, particularly monocytes and neutrophils (Heymann and Tacke 2016). Furthermore, activated KCs release TNFα and IL-1β and thereby further contribute to liver inflammation in early ALI (Fisher et al. 2013). Still, early depletion of KCs results in severely aggravated liver injury in murine models of acetaminophen (APAP) induced liver damage (Ju et al. 2002; Holt et al. 2010), which suggests a dual role for KCs. Indeed KC numbers significantly decrease early in liver injury, vacating their niche which can subsequently be repopulated by MoMF (Triantafyllou et al. 2018). The reduction in KC numbers is accompanied by a commensurately reduced bacterial clearing capacity of the remaining Kupffer cell population with increased expression of the checkpoint ligand PD-1 on KCs (Triantafyllou et al. 2021; Lopes et al. 2022). Conversely, PD-1 blockade improves bacterial clearing and protects from sepsis in murine APAP-induced ALF (Triantafyllou et al. 2021). A recent study showed that treatment with bone marrow-derived macrophages partly recovered bacterial clearing capacity, demonstrating that MoMF are capable of partly taking up KC functions (Lopes et al. 2022).

The early inflammatory response is characterised by the infiltration of inflammatory Ly-6C^{high} CCR2^{+} CCR5^{+} monocytes in mice and genetic depletion or pharmacologic blockage of CCR2 or CCR5 results in attenuation of injury (Choi et al. 2015; Mossanen et al. 2016) which highlights the important role of infiltrating monocytes and MoMF in driving early ALI. However, macrophages are also increasingly recognised for their adaptive functions in ALI. Mediated through endothelial IL-33, IL-4 and macrophage CCL24 release, they attract eosinophil granulocytes which limit ALI (Xu et al. 2022). Infiltrating inflammatory monocytes also readily differentiate into MoMF and, dependent on IL-4, IL-10, CX3CL1, neutrophil-derived reactive oxygen species (ROS), and phagocytosis of dead neutrophils (Yang et al. 2019b), shift towards an early repair phenotype which is vital for the initiation of the following tissue repair phase (Fig. 7.2).

Following the early inflammatory phase of ALI, repair mechanisms are essential to limit inflammation and allow for the restitution of liver function. This tissue repair phase is characterised by a reduction in necrotic area and decreased proinflammatory responses, which if successful eventually results in the full restoration of structural and functional integrity. KCs repopulate the liver but their contribution to tissue repair remains unclear. Although repopulation likely takes place mainly through self-renewal of the remaining emKC population (Zigmond et al. 2014), stringent lineage tracing studies are still needed to define sources of repopulating KCs in ALI. Macrophages fulfil an important role in this process and blocking monocyte infiltration by genetic deletion or pharmacological blockade of CCR2 in mice results in a delayed resolution of ALI (Holt et al. 2008; Zigmond et al. 2014). Clearance of dead and dying cells is an important function of macrophages in general, and indeed in ALI, apoptotic cells produce CXCL1 to attract restorative macrophages (Dal-Secco et al. 2015). Besides their phagocytic function that partakes in clearing necrotic areas and apoptotic cells, this resolution phenotype is characterised by a gene expression signature enriched in angiogenic and anti-inflammatory signalling (Zigmond et al. 2014). Furthermore, single-cell transcriptome studies showed that these resolution macrophages are *Trem2*$^{+}$ (Song et al. 2023), highly express *Mmp12* and lack capacity for antigen presentation and lymphocyte recruitment but upregulate chemokines *Ccl2*, *Ccl7* and *Ccl9*, indicating that they also contribute to the recruitment of new macrophages to the site of injury (Ben-Moshe et al. 2022). Neutrophils play an important role in the maturation of inflammatory MoMF towards a restorative phenotype. They mediate the differentiation of the restorative macrophage phenotype through ROS-dependent AMP-dependent protein kinase (AMPK) activation (Yang et al. 2019b) and relay early macrophage type-I interferon signals to induce maturation of restorative macrophages through CSF-1 signalling (Song et al. 2023). Interestingly, repair macrophages were enriched in the

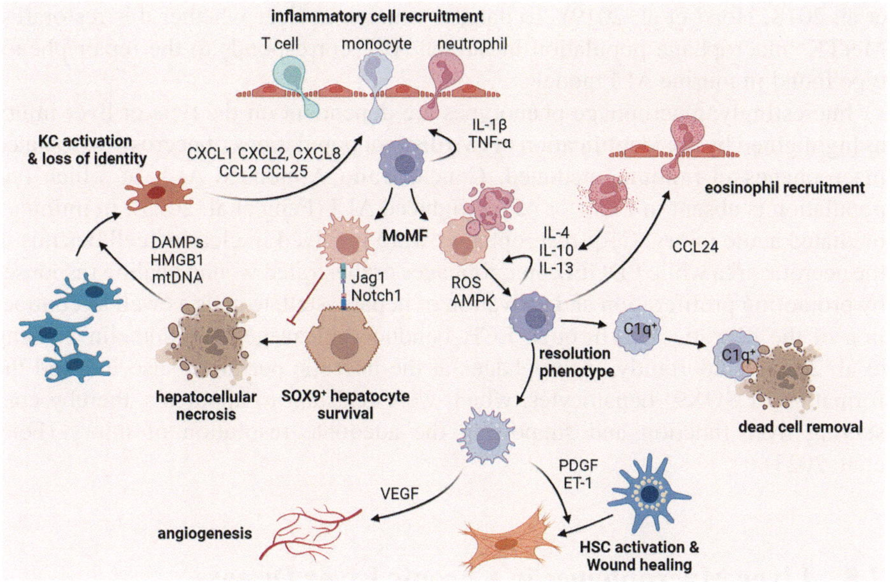

Fig. 7.2 Macrophages in acute necroinflammation and resolution of acute liver injury. Hepatocellular necrosis leads to the release of soluble signalling molecules (danger-associated molecular patterns, DAMPs) that trigger Kupffer cell (KC) activation and loss of KC identity. KCs secrete inflammatory cytokines and chemokines which leads to the recruitment of inflammatory cells and subsequent liver inflammation. Infiltrating monocytes differentiate into early inflammatory monocyte-derived macrophages (MoMF), which further maintain the inflammatory microenvironment through the release of inflammatory mediators. Different signals, including phagocytosis of dead neutrophils, neutrophil-derived reactive oxygen species (ROS) and anti-inflammatory cytokines (e.g. IL-4, IL-10) as well as type I interferon signaling induce a switch in macrophage phenotype towards a pro-resolution function. Pro-resolution macrophages are involved in hepatocyte survival, recruitment of eosinophils that promote liver regeneration, removal of dead cells, induction of wound-healing responses and angiogenesis. Abbreviations: CCL C–C motif ligand, CXCL C–X–C motif ligand, AMPK 5′ AMP-activated protein kinase, ET-1 endothelin 1, HMGB1 high mobility group box 1, IL interleukin, mtDNA mitochondrial DNA, PDGF platelet-derived growth factor, SOX9 SRY-box transcription factor 9, TNFα tumour necrosis factor alpha, VEGF vascular endothelial growth factor (created with BioRender.com)

livers of mice pre-treated with a low dose of APAP, which was associated with resistance to APAP injury, further underpinning the role of restorative macrophages in the resolution of APAP injury (Yu et al. 2023).

In human ALF, the role of macrophages is not as well described, but MerTK$^+$ macrophages likely play an important role in the resolution of injury. MerTK is critically involved in the phagocytosis of apoptotic cells and acts as an important modulator of macrophage function required for adequate tissue repair (Doran et al. 2020). The number of MerTK$^+$ monocytes is significantly increased in patients with ALF (Bernsmeier et al. 2015), and MerTK deficiency in mice leads to more severe liver injury in response to APAP, lower numbers of MHC-II^{high}/Ly6C^{low} restorative macrophages and increased infiltration of inflammatory neutrophils (Triantafyllou et al. 2018; Horst et al. 2019). To date, it remains unclear whether this restorative MerTK$^+$ macrophage population in human ALF corresponds to the repair phenotype found in murine ALI models.

Interestingly, macrophage phenotypes are dependent on the type of liver injury as highlighted by the identification of two distinct populations of necrosis-associated macrophages in immune-mediated, Concanavalin A-induced ALI, of which one population is absent in CCl_4- or APAP-induced ALI (Feng et al. 2023). In immune-mediated acute injury, C1q$^+$ macrophages were involved in clearing cell detritus in the necrotic area while PDGFB$^+$ macrophages orchestrated wound healing responses by promoting proliferation and activation of hepatic stellate cells as well as contraction of the necrotic area through ECE-1-induced cleavage of endothelin 1 (Feng et al. 2023). Importantly, macrophages at the necrotic perimeter also induced the formation of SOX9$^+$ hepatocytes, which were resistant to apoptosis, thereby conserving liver function and supporting the adequate resolution of injury (Feng et al. 2023).

7.5 Liver Macrophages in Chronic Liver Disease

Chronic liver disease (CLD) is characterised by ongoing hepatocellular injury, which over time leads to chronic inflammation and initiation of wound healing responses, which ultimately culminate into the excess deposition of extracellular matrix, a process called liver fibrosis (Hammerich and Tacke 2023). Liver fibrosis and cirrhosis are the main determinants of patient prognosis in CLD and can lead to liver failure and hepatocellular carcinoma, in sum accounting for 2–4% of deaths globally (Sepanlou et al. 2020). Underlying aetiologies of liver disease are diverse and include chronic viral infections (e.g., hepatitis B [HBV] and hepatitis C [HCV] virus), metabolic dysfunction associated steatotic liver disease (MASLD), alcohol-related liver disease (ALD), drug toxicity, autoimmune and biliary diseases as well as genetic disorders.

7.5.1 Liver Fibrosis

Hepatic fibrogenesis is a complex and dynamic process involving all hepatic cell populations but also extrahepatic signals such as from the intestine or adipose tissue (Marra and Svegliati-Baroni 2018; De Muynck et al. 2021). Long-lasting liver injury leads to a persistent inflammatory response and to the release of cell damage mediators, causing activation of KCs, infiltration of inflammatory leukocytes and activation of matrix-producing cells, namely hepatic stellate cells (HSCs) that secrete extracellular matrix (ECM) proteins (Fig. 7.3). A plethora of signals are involved in this process, including damage-associated molecular patterns (DAMPs), pathogen-associated molecular patterns (PAMPs), cytokines, growth factors,

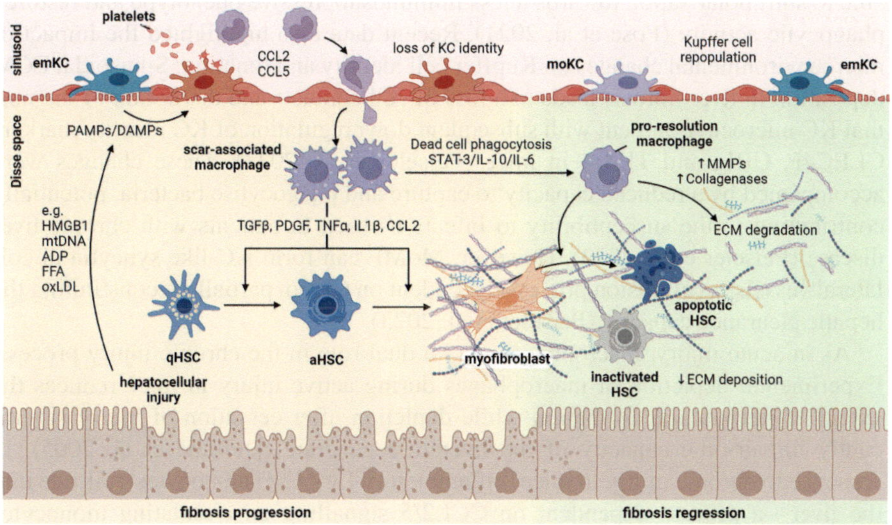

Fig. 7.3 Macrophages in liver fibrosis. Hepatic macrophages act as central regulators in the progression and regression of liver fibrosis. Hepatocellular injury drives Kupffer cell (KC) activation and loss of KC identity through the release of damage-associated mediators, including high mobility group box 1 (HMGB1), mitochondrial DNA (mtDNA), free fatty acids (FFA) and modified lipoproteins (e.g. oxidised LDL). These signals, as well as cytokines and chemokines released by KCs prompt the infiltration of circulating monocytes, which differentiate into scar-associated macrophages (SAMs). SAMs secrete several factors, including transforming growth factor beta (TGFβ), platelet-derived growth factor (PDGF), tumour necrosis factor alpha (TNFα), IL-1β and CCL2, which drive activation, proliferation and migration of hepatic stellate cells (HSCs) and subsequent deposition of extracellular matrix. Parallel to ongoing injury, phagocytosis and anti-inflammatory cytokine signalling differentiate macrophages towards a pro-resolution phenotype, which promotes inactivation and apoptosis of HSCs, as well as extracellular matrix (ECM) degradation. Abbreviations: ADP adenosine diphosphate, CCL C–C motif ligand, em/moKC embryonic/monocyte-derived Kupffer cell, (q/a)HSC (quiescent/activated) hepatic stellate cell, IL interleukin, MMP matrix metalloproteinase, STAT-3 signal transducer and activator of transcription 3 (created with BioRender.com)

hypoxia, changes in tissue biomechanics and others (Schuppan et al. 2001; Hammerich and Tacke 2023). Inflammation and the deposition of ECM eventually result in a disruption of the sinusoidal microarchitecture and over the course of years and decades can ultimately progress to liver cirrhosis.

In this process, liver macrophages have received increasing attention due to their important functions in modulating disease processes and due to their phenotypical and functional diversity. Like in ALI, KCs act as early responders to PAMPs and DAMPs and release inflammatory mediators that drive the recruitment of monocytes and their differentiation into inflammatory, pathogenic MoMF (Heymann and Tacke 2016). However, KC numbers are reduced in fibrotic livers and are not found in areas of increased ECM deposition (Ramachandran et al. 2019). Interestingly, fibrotic KCs express high levels of PD-L1 and show a more immunosuppressive phenotype compared to healthy KCs (Pose et al. 2021). Blockade of PD-L1 was able to shift polarisation towards a less immunosuppressive phenotype and restored phagocytic activity (Pose et al. 2021). Recent data also highlighted the impact of microenvironmental changes on Kupffer cell identity and function: Sinusoidal ECM deposition in liver fibrosis leads to a loss of cell contacts and disruption of the normal KC microenvironment with subsequent downregulation of KC identity markers CLEC4F, CRIg and TIM-4 in mice (Peiseler et al. 2023). These changes were accompanied by a reduced capacity to capture and phagocytise bacteria, potentially contributing to the susceptibility to infections seen in patients with chronic liver disease (Peiseler et al. 2023). However, MoMF can form KC-like syncytia in collateral vessels, a cell fusion process dependent on CD36, partially reconstituting the hepatic clearance capacity (Peiseler et al. 2023).

As in acute injury, macrophages play a dual role in the chronic injury process. Experimental depletion of macrophages during active injury in mice reduces the extent of fibrous scar formation while depletion after cessation of injury significantly impairs the capacity of the liver to regenerate (Duffield et al. 2005). In response to chronic injury in mice, inflammatory Ly-6C^{high} monocytes infiltrate into the liver—a process dependent on CCL2/5 signalling on circulating monocytes (Karlmark et al. 2009; Mitchell et al. 2009; Miura et al. 2012)—and differentiate into inflammatory macrophages. These activated macrophages maintain the fibrogenic microenvironment through the secretion of several inflammatory and growth mediators that are significantly involved in stellate cell activation. In particular, transforming growth factor-β (TGFβ), platelet-derived growth factor (PDGF), tumour necrosis factor alpha (TNFα), interleukin 1β (IL-1β) and CCL2 contribute to stellate cell activation through direct binding of their receptors on HSCs (Tsuchida and Friedman 2017). Macrophages also contribute to liver fibrogenesis through interactions with other cell types. IL-1β produced by MoMF, for example, can contribute to hepatocellular fatty degeneration and hepatocellular injury (Wang et al. 2022). Similarly, reactive oxygen species, TNFα and PDGF secreted by activated KCs initiate angiogenic processes in liver endothelial cells to further maintain fibrogenic processes (Ramirez-Pedraza and Fernández 2019). Furthermore, KCs recruit platelets through hyaluronan-CD44 binding in steatohepatitis, a process involved in the recruitment of immune cells and liver fibrosis progression (Malehmir et al. 2019).

Parallel to ongoing injury, these proinflammatory Ly-6C^{high} MoMF can differentiate into a restorative Ly-6C^{low} restorative macrophage phenotype in mice (Ramachandran et al. 2012). These restorative macrophages are important actors balancing wound healing responses and are involved in the process of fibrosis regression (Fig. 7.3). These restorative macrophages are characterised by high expression of matrix metalloproteinases (MMP) that degrade the extracellular matrix (Roeb 2018), and by expression of factors like TRAIL and MMP9 which induce HSC and myofibroblast apoptosis, thereby reducing the pool of ECM-producing cells (Ramachandran et al. 2012). Furthermore, these regenerative, pro-resolution macrophages shape and promote an anti-inflammatory microenvironment through the release of anti-inflammatory, angiogenic and growth mediators such as IL-10, VEGF and HGF, thereby dampening inflammation and promoting healing responses (Ramachandran et al. 2012; Yang et al. 2014; Tacke 2017). The signals promoting this phenotypic switch are currently under investigation but include phagocytosis of dead and dying cells (Ramachandran et al. 2012), STAT-3/IL-10/IL-6 signalling (Campana et al. 2018) and Nur77 signalling (Hu et al. 2021). The biology of restorative liver macrophages in fibrosis regression has mainly been studied in murine models of liver fibrosis, and to date the human equivalent of this macrophage population has not been identified.

Current single-cell sequencing data from human healthy and fibrotic diseased liver provide a new view on the heterogeneity of hepatic macrophage populations. In particular, the classic dichotomous division into classically activated, pro-inflammatory, fibrogenic (M1-like), and alternatively activated, anti-inflammatory, regenerative macrophages (M2-like) can no longer be maintained (Guillot and Tacke 2019). A key finding is the identification of specialised scar-associated/lipid-associated macrophages (SAMs/LAMs), characterised by their expression of CD9 and TREM2 in humans and mice, that are instrumental in shaping the fibrogenic microenvironment in chronic liver injury (Ramachandran et al. 2019). These SAMs localise to the fibrotic scar and can activate hepatic stellate cells to proliferate and produce collagen (Ramachandran et al. 2019). Similar TREM2^{+} CD9^{+} SAM/LAM subsets have been identified in mouse models of CCl_4-induced liver fibrosis (Ramachandran et al. 2019) as well as metabolic dysfunction-associated steatohepatitis (MASH) (Xiong et al. 2019; Seidman et al. 2020; Remmerie et al. 2020) and are conserved across species (Guilliams et al. 2022; Fabre et al. 2023). Recent integrated single-cell sequencing studies on human liver and lung fibrosis also suggest that these highly fibrogenic macrophages are shared across organs and represent a common fibrogenic macrophage phenotype induced by G-CSF, IL-17A and TGF-β signalling (Fabre et al. 2023). Strikingly, soluble TREM2 (sTREM2) is significantly increased under conditions of steatohepatitis and fibrosis, paralleling the infiltration of TREM2 macrophages into the liver, and sTREM2 is currently under investigation as a biomarker for activity and stage of disease (Hendrikx et al. 2022). Still, TREM2^{+} macrophages likely represent a very heterogeneous population with diverse functions. Besides their fibrogenic role, TREM2^{+} macrophages were also found to express high levels of MMPs, thereby potentially limiting fibrosis (Lee et al. 2023), and TREM2 has an important role in the phenotypic switch from inflammatory to restorative macrophages (Coelho et al. 2021). In cell type specific mouse models, TREM2 expression by MoMF (or bone marrow derived myeloid cells) limits steatohepatitis and fibrosis (Hendrikx et al. 2022). Disease-specific functions of TREM2^{+} macrophages are currently under investigation, particularly in metabolic dysfunction associated steatotic liver disease (MASLD, formerly known as non-alcoholic fatty liver disease, NAFLD), as outlined in the following section.

7.5.2 Metabolic Dysfunction-Associated Steatohepatitis

The global increase in overnutrition, obesity and a sedentary lifestyle is associated with a commensurate increase in the prevalence of MASLD (Younossi et al. 2018). Tightly linked to insulin resistance and the metabolic syndrome, hepatocellular lipid overload can lead to lipotoxicity, oxidative stress, mitochondrial dysfunction, unfolded protein response and endoplasmic reticulum stress, which ultimately result in hepatocellular injury and apoptosis (Marra and Svegliati-Baroni 2018). Thus, dysregulated hepatic lipid accumulation prompts inflammatory cascades and immune cell infiltration, resulting in a condition termed metabolic dysfunction-associated steatohepatitis (MASH), which in turn can progress to liver fibrosis, cirrhosis, and hepatocellular carcinoma.

The general mechanisms how macrophages contribute to MASLD progression are very similar to those outlined for liver fibrosis. Kupffer cell numbers are reduced in murine MASH (Devisscher et al. 2017), and the inflammatory process following hepatocellular injury in MASH prompts the infiltration of Ly6C^{high} inflammatory monocytes which fill the vacated Kupffer cell niche to form lipid- and scar associated macrophages (Daemen et al. 2021), as well as take on a KC-like phenotype (Scott et al. 2016; Daemen et al. 2021). KCs are important drivers of early disease processes, and clodronate or gadolinium chloride-induced experimental KC depletion protects from steatohepatitis and fibrosis in mice (Huang et al. 2010; Miura et al. 2012; Tosello-Trampont et al. 2012). Injured and dying lipid-laden hepatocytes release several damage-associated molecules (damage-associated molecular patterns) including mitochondrial DNA, adenosine-diphosphate (ADP), free cholesterol and HMGB1, which are sensed by KCs and drive their inflammatory activation through pattern recognition receptors (PRRs) (Canbay et al. 2003; Garcia-Martinez et al. 2016; Horn et al. 2022). Similarly, MASH is associated with intestinal barrier dysfunction and microbial dysbiosis, which partakes in early hepatic macrophage activation through the release of pathogen-associated molecular patterns (PAMPs), including LPS and other bacterial components, into the portal circulation, which stimulate macrophage inflammatory activation through TLR4 and TLR9 (Leroux et al. 2012; Garcia-Martinez et al. 2016; Mridha et al. 2017a; Marra and Svegliati-Baroni 2018). Furthermore, activated hepatic macrophages in MASH are critically involved in the recruitment of immune cells to the liver (Vonderlin et al. 2023), including monocytes, maintaining the inflammatory microenvironment in MASH. Not least, inflammatory MASH-associated MoMF promote survival, activation and tissue inhibitor of metalloproteinase (TIMP) expression in HSC through NFkB-dependent signalling (Tomita et al. 2006; Pradere et al. 2013; Tsuchida and Friedman 2017), and MerTK^{+} macrophages contribute to HSC activation through the ERK-TGFβ1 pathway (Cai et al. 2020), thereby driving disease progression to MASH fibrosis and cirrhosis.

Recent research highlights how liver macrophages interact with steatotic hepatocytes and how they balance the lipotoxic microenvironment to either limit or perturb inflammatory processes. Macrophages can limit the detrimental accumulation of free lipids and thereby help reduce lipotoxic stress but the inflammatory activation of macrophages in response to hepatocyte-derived lipids and apoptotic bodies has also been identified as important cues in the initiation of hepatic inflammation in MASH. The morphological representation of this interaction between fatty hepatocytes and macrophages are hepatic crown-like structures (hCLS), i.e. macrophages accumulating around fatty degenerated hepatocytes (Ioannou et al. 2013), a phenomenon well described for adipose tissue inflammation in obesity and type 2 diabetes (Weisberg et al. 2003; Xu et al. 2003). Although the specific role of hCLS in MASLD still needs to be defined, their presence is clearly associated with disease progression to steatohepatitis (Kleiner et al. 2005; Ioannou et al. 2013, 2015; Itoh et al. 2013, 2017). Lineage tracing studies in mice revealed that hCLS macrophages are derived from infiltrating monocytes and not from *Csfr1*-expressing resident emKC (Daemen et al. 2021). Using single-cell RNA sequencing on murine hepatic myeloid cells in MASH, Daemen et al. identified two clusters of *Timd4*-negative macrophages with increasing expression of LAM markers *Trem2*, *Cd9*, *CD63* and *Gpmnb* and decreasing expression of monocyte markers *Cx3cr1* and *Ccr2*, representing the transition from infiltrating monocytes to hCLS-associated LAMs (Daemen et al. 2021). Genetic deletion of functional *Ccr2* abrogates hCLS formation but strikingly leads to increased liver fibrosis in murine high fat diet (HFD) or fructose, palmitate, cholesterol (FPC) diet models of MASLD and MASH (Daemen et al. 2021). The protective effects of hCLS formation might be explained by their efferocytic capacity, clearing apoptotic cells and preventing the accumulation of free lipids, thereby limiting the release of DAMPs and subsequent inflammatory signalling (Wang et al. 2023). MASH-associated MoMF upregulate TREM2, an important player in efferocytosis, which is shed upon prolonged liver injury, leading to inadequate efferocytosis and the subsequent aggravation of inflammatory processes in the liver (Wang et al. 2023). Conversely, genetic knockout of macrophage scavenger receptor 1 (*Msr1*), a phagocytic receptor promoting macrophage activation through JNK upon recognizing modified lipoproteins (Guo et al. 2019), reduced the formation of foamy macrophage in hCLS in murine HFD-induced MASH (Govaere et al. 2022). This reduction in hCLS translated into reduced hepatic and adipose tissue inflammation and protection from liver fibrosis (Govaere et al. 2022), highlighting the role of hCLS in promoting steatohepatitis. Therefore, hCLS are a complex entity with dual roles in the initiation and progression of MASH (Fig. 7.4). Future research will need to address the specific mechanisms involved in driving these diverse functions to enable the development of therapeutic approaches targeting hCLS formation in human MASH.

Vice versa, inflammatory macrophages also promote hepatocellular lipid accumulation through different mechanisms. Activated macrophages release IL-1β and TNFα which induce insulin resistance and suppress fatty acid oxidation, thereby exacerbating lipotoxic cell injury (Stienstra et al. 2010; Huang et al. 2010; Tosello-Trampont et al. 2012). Endoplasmic reticulum (ER) stress in macrophages appears to play a major role in this process, as myeloid-specific deletion of the ER-stress effector *Xbp1* abrogates macrophage-induced hepatocellular steatosis in mouse models of MASLD and MASH (Wang et al. 2022). Furthermore, single-cell

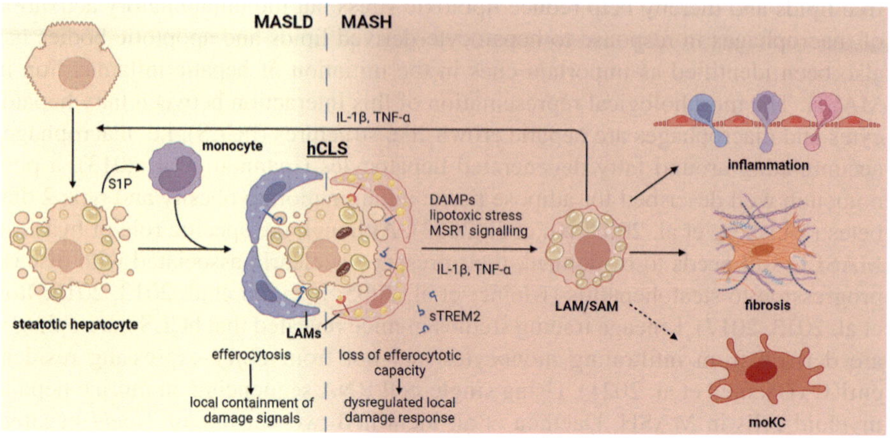

Fig. 7.4 The dual role of macrophages in the initiation and propagation of MASH. Macrophages have an important role in the initiation and progression of MASLD and MASH. Steatotic hepatocytes release factors (e.g. sphingosine-1-phosphate, S1P) that drive Kupffer cell activation and monocyte recruitment. The lipotoxic microenvironment drives differentiation of monocytes towards lipid-associated macrophages (LAMs) which aggregate surrounding lipid-laden hepatocytes to form hepatic crown-like structures (hCLS). Early in MASLD, TREM2$^+$ LAMs are characterised by efficient efferocytic capacity which helps in containing local damage signals and limits inflammation. Upon prolonged lipotoxic injury, however, LAMs shed TREM2, a process associated with loss of efferocytic capacity, which together with non-contained lipotoxic signals, e.g. via macrophage scavenger receptor 1 (MSR1) signalling, leads to a pronounced inflammatory activation of LAMs, which drive liver inflammation and fibrosis. Furthermore, inflammatory LAMs promote hepatocellular steatosis through the secretion of inflammatory mediators (e.g. IL-1β, TNF-α). Abbreviations: IL interleukin, moKC monocyte-derived Kupffer cell, SAM scar-associated macrophage, (s)TREM (soluble) triggering receptor expressed on myeloid cells 2, TNFα tumour necrosis factor alpha (created with BioRender.com)

sequencing studies recently identified a minor lipogenic *CD206^{high}ESAM$^+$CD36$^+$* KC population with transcriptional enrichment of metabolic pathways (Blériot et al. 2021). Specific deletion of this KC subset or knockout of CD36 significantly improved metabolic function and steatosis in murine HFD (Blériot et al. 2021) which underpins the significant contribution of macrophages to metabolic dysregulation in MASH.

7.5.3 Alcohol-Related Liver Disease

Alcohol-related liver disease (ALD) is one of the main causes of chronic liver disease across the globe. The pathophysiology and histological presentation highly resembles that of MASLD and MASH, and therefore, ALD has been integrated into the newly defined spectrum of steatotic liver diseases (Rinella et al. 2023). Nevertheless, liver damage in ALD is mainly related to direct hepatotoxic effects of alcohol and its metabolite acetaldehyde, which trigger oxidative stress, mitochondrial damage, *de novo* lipogenesis and liver steatosis (Vonghia et al. 2019). Following alcohol-induced hepatocellular injury, multiple mechanisms trigger the hepatic immune response (Wang et al. 2021): Alcohol induces intestinal barrier dysfunction which leads to translocation of bacterial products, transport to the liver via the portal vein and subsequent macrophage activation. Furthermore, ethanol and acetaldehyde directly trigger hepatocyte death with the subsequent release of inflammatory mediators and cell death signals, which modulate macrophage function. Not least, alcohol consumption causes adipose tissue inflammation and dysfunction with the release of free fatty acids, cytokine, chemokines and adipokines.

Overall, according to current knowledge, macrophages show similar responses in ALD compared to MASH. Monocyte and macrophage recruitment is highly dependent on CCL2 and CCR2/CCR5 blockade effectively inhibits expansion of MoMF in ALD and reduces hepatic injury, steatosis and fibrosis (Ambade et al. 2019). Relating to the underlying mechanisms of alcoholic hepatitis, IL1β produced through necroptotic pathways plays a particularly important role in driving ALD pathogenesis (Mathews and Gao 2013) and direct blockade of IL1β receptor or genetic models of inflammasome deletion have proven effective in reducing disease phenotypes in mouse models of ALD (Petrasek et al. 2012). Still, the heterogeneity of hepatic macrophage subpopulations and their phenotypes in ALD have not been studied to the same extent as for MASLD/MASH (Ait Ahmed et al. 2023), and future research will need to delineate the specific kinetics, subsets and roles of macrophages in ALD.

7.5.4 *Viral Hepatitis*

Chronic viral hepatitis, mainly caused by infection with the hepatitis B or hepatitis C virus, is still one of the leading causes of chronic liver disease and hepatocellular carcinoma, affecting around 330 million people or 4.5% of the world population (World Health Organization 2017). T cell responses tend to be in the focus of research on immune mechanisms against viral hepatitis and adequate $CD8^+$ and $CD4^+$ T cell responses are required for the resolution of infection, while T cell exhaustion and immune evasion result in persisting viral replication with subsequent liver inflammation and its complications liver fibrosis, cirrhosis and hepatocellular carcinoma (Shin et al. 2016). Macrophages are recognised as important modulators of the T cell response and recent research is shedding more light on the heterogeneity of macrophage populations in chronic viral hepatitis.

Virus particles induce macrophage activation and subsequent release of inflammatory mediators. In KCs, phagocytosis of viral particles stimulates the production of IL-10 and activation of NK cells (Wu et al. 2019) while the production of IL-18 and IL-1β in response to TLR2 signalling induces inflammasome activation in HCV infection (Chang et al. 2007; Huang et al. 2013). Similar to other chronic liver diseases, macrophage numbers are increased in viral hepatitis and correlate with disease activity and expression of inflammatory mediators IL-1β, IL-6, IL-18, TNFα and IFN-γ (Khakoo et al. 1997; McGuinness et al. 2000) while depletion of KCs in murine models increases monocyte infiltration and promotes clearance of HBV, again demonstrating the dual roles of macrophages in liver disease (Wu et al. 2019). In chronic HBV infection, macrophages promote impaired immune responses, T cell exhaustion and subsequent immune cell evasion and inflammatory responses. These responses are mediated by release of IL-10, increased TLR2 signalling on KCs and upregulation of PD-L1 on macrophages (Nebbia et al. 2012; Li et al. 2015; Tian et al. 2016; Liu et al. 2018). Conversely, in HCV infection, macrophages can inhibit viral replication through IL-6, IFN-γ, and induction of hepatocyte apoptosis (Boltjes et al. 2014).

Single-cell sequencing approaches begin to reveal the heterogeneity of macrophage populations in chronic viral hepatitis. One study comparing healthy liver and HBV cirrhosis identified seven distinct macrophage populations, including two subsets of *CD9*$^+$ macrophages as well as populations characterised by *C1QA*, *CCL5*, *NAMPT*, *PLCG2* or *THBS1* expression (Bai et al. 2023). HBV cirrhosis was associated with a significant increase in the proportion of *CD9*$^+$, *IL-18*$^+$ and *C1QA*$^+$ macrophages, which showed transcriptional enrichment relating to MHC class II signalling, antigen processing and presentation (Bai et al. 2023). Still, functional characterisation of these populations is needed to pin down their role in HBV infection and no comparison to non-HBV cirrhosis is available, making it difficult to attribute changes to HBV-specific effects. Others found increased inflammatory monocyte numbers in peripheral blood of HBV-infected patients compared to healthy controls, which is reversed upon treatment with pegylated IFN-α (Jiang et al. 2024). Similar observations have been made for intrahepatic myeloid populations in patients with sustained virological response after HCV treatment, where monocytes and macrophages shifted towards increased anti-inflammatory and pro-resolution responses (Cui et al. 2023).

7.5.5 Cholestatic Liver Disease

Cholestatic liver diseases or cholangiopathies are a heterogeneous group of conditions that are characterised by injury to the bile ducts, leading to inflammation and fibrosis that are predominantly localised in the portal area (Bruneau et al. 2022). The term covers autoimmune(-like) cholangiopathies, e.g. primary sclerosing cholangitis (PSC), primary biliary cholangitis (PBC) and IgG4-associated autoimmune cholangitis, idiopathic cholangiopathies, e.g. biliary atresia (BA), as well as genetic disorders including cystic fibrosis-associated liver disease, progressive familial intrahepatic cholestasis (PFIC), Caroli syndrome as well as secondary sclerosing cholangitis of different aetiologies. We are now becoming increasingly aware that the interaction of macrophages with different liver cell populations, including cholangiocytes, hepatocytes, liver immune cells and hepatic stellate cells, is an important driver of disease progression in cholestatic liver diseases (Bruneau et al. 2022). In fact, macrophage accumulation around bile ducts ("ductular reaction") is a hallmark of progressive fibrotic disease in cholangiopathies such as PSC as well as in MASLD/MASH (Guillot et al. 2023).

In cholestatic liver disease, macrophages mainly infiltrate into the portal and periportal area as shown in PSC (Cameron et al. 2001; Guicciardi et al. 2018; Chen et al. 2019) and BA (Taylor et al. 2021) in humans and in different murine models of cholestatic injury, including *Abcb4* knockout, bile duct ligation (BDL), and hCD59-transgenic mice (Guillot et al. 2021, 2023; Tian et al. 2021). Infiltration of macrophages is induced by specific cholangiocellular injury and senescence (Ferreira-Gonzalez et al. 2018; Guillot et al. 2021), and like in other liver disease conditions, CCL2 plays an important role in the recruitment of macrophages to the site of biliary injury. CCL2 is upregulated in acutely injured cholangiocytes in mice (Guillot et al. 2021), and pharmacological dual CCR2/CCR5 blockade with the inhibitor cenicriviroc effectively reduces macrophage infiltration in different murine models of sclerosing cholangitis (Guicciardi et al. 2018). In a small clinical trial involving adults with PSC, cenicriviroc showed promising effects on cholestatic enzymes (Eksteen et al. 2021), warranting further investigation in larger trials with longer follow-up. In addition, IFN-γ plays an important role in shaping macrophage responses in cholestatic liver diseases. $CD8^+$ T cells and NK cells in *Abcb4*$^{-/-}$ mice release high levels of IFN-γ, which shifts the balance towards inflammatory macrophage responses (Ravichandran et al. 2019). Accordingly, *Ifng* knockout and antibody-mediated neutralization reduced liver inflammation and fibrosis (Ravichandran et al. 2019).

The roles of hepatic macrophages in chronic cholestatic liver injury, however, still need to be defined. The mechanisms discussed above and reduced inflammatory and fibrotic responses after Kupffer cell depletion in murine models of cholestasis clearly show their contribution to liver injury (Duwaerts et al. 2013; Guillot et al. 2021; Tian et al. 2021). Still, infiltrating monocytes quickly acquire a pro-resolution phenotype that promotes cholangiocyte proliferation and biliary regeneration (Guillot et al. 2021). Furthermore, hepatic macrophages likely act as sensors and modulators of bile acid metabolism. One study showed high expression of the bile acid receptor TGR5 on $CD206^+$ macrophages in PSC with inhibitory functions on inflammatory macrophage activation in response to bile acid stimulation (Keitel et al. 2008). Still, TGR5 activation has also been associated with macrophage-mediated inflammation in other contexts (Haselow et al. 2013; Shi et al. 2021).

Currently published single-cell studies were focusing on T cells and did not include any data on macrophages (Poch et al. 2021; Han et al. 2022); however, a recent report shed more light on macrophage populations in PSC (Andrews et al. 2024). Macrophage populations mostly aligned with published data on other liver disease conditions, including KCs, inflammatory monocyte-like macrophages and *CD9*$^+$*TREM2*$^+$ scar-associated macrophages. However, scar-associated macrophages showed lower enrichment of lipid-associated genes and higher expression of pathways related to phagocytosis (Andrews et al. 2024). Strikingly, they identified a novel macrophage subset with high expression of genes involved in the formation of immune synapses and interactions with T cells, suggesting that these macrophages regulate inflammatory T cell responses in PSC (Andrews et al. 2024). More mechanistic work along these lines in the future will help to better define the roles of macrophages and their interactions with other cell types in cholestatic liver injury and regeneration.

7.6 Liver Macrophages in Hepatic Carcinogenesis

Hepatocellular carcinoma (HCC) is the most common type of primary liver cancer and accounts for a significant proportion of global cancer-related deaths (Yang et al. 2019a). HCC typically develops in chronically diseased liver, where a disrupted tissue microarchitecture, infiltrating immune cells, inflammatory cytokines and growth factors provide a niche for cancer cell growth and survival (Donne and Lujambio 2023). KCs and tumour-associated macrophages (TAMs) have important divergent roles as they generate a pro-tumorigenic microenvironment, remodel the cancer microarchitecture but are also involved in clearing tumour cells and contribute to the adaptive immune response (Fig. 7.5). Current evidence is also unravelling the role of TAMs in modulating the susceptibility to different treatment modalities and a better understanding of the role of TAMs in HCC will help to optimise and individualise treatment.

Early in cancer development, pre-malignant hepatocytes are characterised by cellular senescence, a stress response program characterised by cell cycle arrest and the release of various cytokines, including IL-1 and IL-6 (Kuilman et al. 2010). Macrophages, in lieu with $CD4^+$ T cells, are required for clearing senescent pre-malignant hepatocytes (Kang et al. 2011) and CCR2 positive monocytes play an important role in cancer cell surveillance. Reducing the influx of $CCR2^+$ cells through pharmacologic inhibition or genetic deletion of CCR2 results in a more aggressive HCC phenotype in a mouse model of tumour cell transfer (Eggert et al. 2016).

However, in our current understanding, TAMs mainly fulfil pro-tumorigenic functions in HCC. Upon tumour manifestation, infiltration of macrophages is driven by a range of different tumour-derived factors including CCL2, CSF1, IL-34, mitochondrial DNA, osteopontin and HCC-derived exosomes (Donne and Lujambio 2023). Additional tumour-derived mediators also shift macrophage polarisation towards a “M2-like” phenotype which is critically involved in maintaining the characteristic immunosuppressive microenvironment in HCC (Donne and Lujambio 2023). TAMs characteristically express high levels of the immune-checkpoint molecule PD-L1 (Kuang et al. 2009) and release several immunosuppressive cytokines, including IL-10 and TGF-β (Qian and Pollard 2010), which promote effector T cell exhaustion and the infiltration and differentiation of Th2 and regulatory T cells, the latter further contributing to an inadequate effector T cell response (Sica and Bronte 2007). The overall pro-tumorigenic role of TAMs in HCC development is

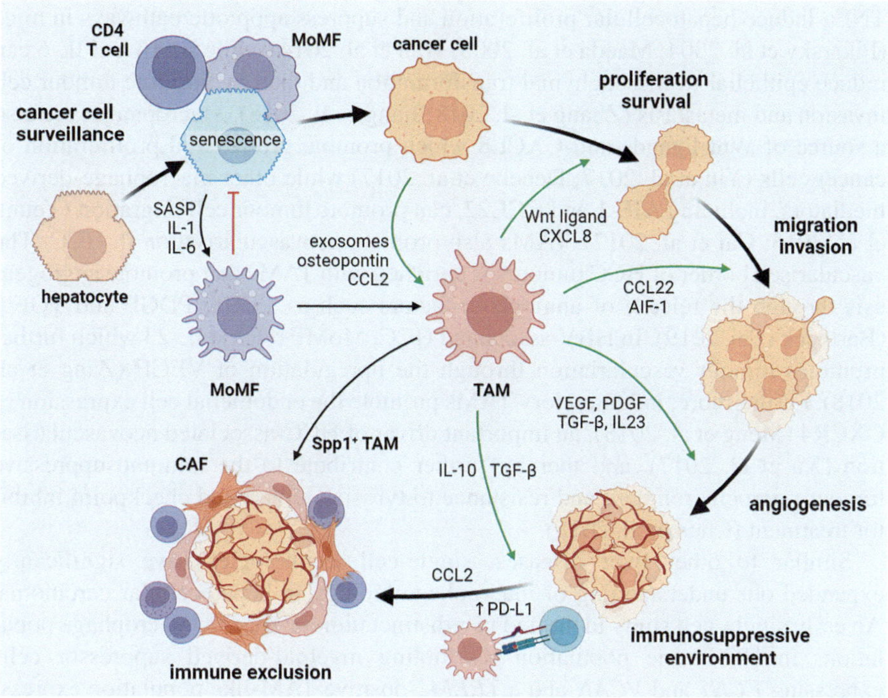

Fig. 7.5 Tumour-associated macrophages (TAMs) in hepatic carcinogenesis. Hepatic macrophages fulfil a plethora of functions in the development and progression of hepatocellular carcinoma, as well as in the resistance to treatment. The carcinogenic microenvironment in chronic liver diseases is characterised by an increased abundance of senescent hepatocytes and monocyte-derived macrophages (MoMF) have an important role in clearing these cells, thereby inhibiting cancer cell development. However, inflammatory factors released by senescent cells (senescence-associated secretory phenotype—SASP) and—upon cancer cell development—cancer cell-derived factors (e.g. exosomes, osteopontin, CCL2) promote the differentiation of monocytes and macrophages towards distinct tumour-associated macrophage (TAM) phenotypes. TAMs promote cancer cell growth through proliferative and pro-survival factors and contribute to migration and invasion of cancer cells. Furthermore, distinct angiogenic TAM populations secrete angiogenic factors (e.g. VEGF, PDGF, TGF-β and IL-23) that drive cancer neovascularisation. In addition, TAMs secrete anti-inflammatory cytokines and growth factors that maintain an immunosuppressive microenvironment (e.g. PD-L1 expression and stimulation of PD-L1 expression on cancer cells through IL-10 and TGF-β signalling). Finally, *Spp1*⁺ (osteopontin-expressing) TAMs together with cancer-associated fibroblasts (CAFs) form an immune barrier that excludes anti-tumour immune cells from entering the tumour. Abbreviations: AIF-1 allograft inflammatory factor 1, CCL C–C motif ligand, CXCL C–X–C motif ligand, IL interleukin, PDGF platelet-derived growth factor, PD-L1 programmed death-ligand 1, TGFβ transforming growth factor beta, VEGF vascular endothelial growth factor (created with BioRender.com)

underpinned by findings that blocking monocyte infiltration by targeting CCR2 inhibits tumour growth (Li et al. 2017).

Additional inflammatory mediators released by macrophages directly impact on tumour cell function. Sustained inflammation and macrophage release of IL-6 and

TNFα induce hepatocellular proliferation and suppress apoptotic pathways in mice (Pikarsky et al. 2004; Maeda et al. 2005; Wan et al. 2014) while IL-1β and IL-6 can induce epithelial-to-mesenchymal transformation and thereby promote tumour cell invasion and metastasis (Zhang et al. 2018; Jiang et al. 2018). Macrophages are also a source of Wnt ligands and CXCL8 which promote growth and proliferation of cancer cells (Yin et al. 2017; Debebe et al. 2017) while other macrophage-derived mediators, including AIF-1 and CCL22, can promote tumour cell migration (Yeung et al. 2015; Cai et al. 2017). TAMs also promote neovascularisation in HCC. The vascularised border of HCC tumours is enriched with TAMs that promote angiogenesis through the release of angiogenic factors such as VEGF, PDGF and TGF-β (Bartneck et al. 2019). In HBV-associated HCC, MoMF release IL-23 which further promotes tumour vasculariation through the upregulation of VEGF (Zang et al. 2018). Furthermore, inflammatory TAMs promote the endothelial cell expression of CXCR4 (Meng et al. 2018), an important driver of HCC-associated neovascularisation (Xu et al. 2017), and thereby further contribute to the immunosuppressive tumour microenvironment and resistance to tyrosine kinase and checkpoint inhibitor treatment (Chen et al. 2015).

Similar to other liver diseases, single-cell technologies have significantly expanded our understanding of macrophage biology in hepatocellular carcinoma. An early single-cell study identified two distinct tumour-specific macrophage populations in HCC: one population resembling myeloid-derived suppressor cells expressing *FCN1* and *VCAN* and a *TREM2*-positive TAM-like population expressing *APOE*, *C1QA* and *C1QB* (Zhang et al. 2019). Interestingly, only the *TREM2*$^+$ macrophage gene signature was associated with poorer prognosis, implicating them as drivers in HCC progression (Zhang et al. 2019). Later research added to this heterogeneity by identifying two separate and spatially distinct subpopulations of TAMs, namely a more immunosuppressive *C1QC*$^+$/*FOLR2*$^+$ population resembling fetal liver macrophages (Sharma et al. 2020; Cheng et al. 2021) and a *SPP1*$^+$ (osteopontin-expressing) population resembling MoMF (Sharma et al. 2020). In contrast to the immunosuppressive FOLR2$^+$ population, SPP1$^+$ cells seem to have a pivotal role in generating an immune excluded environment. These metabolically active, hypoxia-associated *SPP1*$^+$ macrophages were enriched in HCC tissue of non-responders to checkpoint inhibitor treatment and were associated with poor survival (Liu et al. 2023). *SPP1*$^+$ macrophages closely interacted with cancer-associated fibroblasts to promote the formation of a tumour immune barrier surrounding the cancer cells, preventing cancer-directed immune cell infiltration (Liu et al. 2023). The idea that TAMs are important for preventing infiltration and activity of anti-tumour immune cells is further supported by findings of progressively immunosuppressive TAM phenotypes in HCC and inverse correlations with T cell numbers and patient prognosis (Ho et al. 2021). Genetic silencing of *Spp1* in a murine HCC model prevented barrier formation, enabled T cell infiltration and thereby sensitised tumours to anti-PD-1 treatment (Liu et al. 2023). A recent comprehensive single-cell study based on 10 patients with HCC revealed even further TAM heterogeneity (Lu et al. 2022). Alongside *MARCO*$^+$ KCs, they identified five distinct macrophage populations shared across patients. They were able to identify *TREM2*$^+$ TAMs which resembled the immunosuppressive *C1QC*$^+$/*FOLR2*$^+$ hepatic TAMs described before (Sharma et al. 2020; Cheng et al. 2021). Besides two populations of classical monocyte-derived macrophages, they identified a *VEGFA*$^+$ subset enriched in oxidative stress response genes and a *MMP9*$^+$ *SPP1*$^+$ population (Lu et al. 2022). These *MMP9*$^+$ macrophages showed high tumorigenic potential and likely differentiate from both monocyte-derived macrophages and *TREM2*$^+$ TAMs in a *PPARG*-dependent manner (Lu et al. 2022). Strikingly, across different single-cell transcriptome studies, authors found macrophage clusters specifically associated with individual patients (Sharma et al. 2020; Cheng et al. 2021; Lu et al. 2022), which further highlights the heterogeneity of hepatocellular carcinoma and the need to investigate macrophage populations in larger patient populations. Already, immune cell signatures derived from single-cell sequencing studies can be used to define immunophenotypes in HCC (Lu et al. 2022). Hopefully, in the future this will help to stratify patients with HCC to find individualised effective treatment options.

7.7 Targeting Hepatic Macrophages as a Therapeutic Opportunity

Macrophages have a key role in regulating liver injury, liver inflammation and fibrosis in a range of different liver diseases. Accelerated by recent developments in understanding the underlying heterogeneity and biology, macrophages are more and more in the focus of therapeutic strategies targeting inflammatory and detrimental macrophage activation. At the same time, given our increasing understanding of the restorative potential of macrophages, therapeutic strategies ideally should be selective for those pathogenic macrophage subsets while promoting restorative and regenerative macrophage differentiation.

While KCs play important roles in liver homeostasis and in the initiation of injury, pathogenic liver macrophages mostly arise from infiltrating monocytes and therefore, inhibition of monocyte recruitment has been extensively studied as a therapeutic approach to limit liver inflammation. Chemokine receptors are particularly involved in this process, and CCR2 and CCR5 received notable attention. The dual CCR2/CCR5 inhibitor cenicriviroc efficiently inhibits monocyte recruitment, liver inflammation and fibrosis in mouse models of chronic liver disease (Lefebvre et al. 2016; Mossanen et al. 2016; Krenkel et al. 2018). In clinical trials however, despite promising data from a phase 2b trial (Friedman et al. 2018; Ratziu et al. 2020), cenicriviroc did not reach the primary endpoint in a subsequent phase 3 trial in patients with MASH (Anstee et al. 2024). Other promising compounds targeting monocyte recruitment are currently in development, including CCR2, CCL2 and CCL5 inhibitors (Baeck et al. 2014; Pérez-Martínez et al. 2014; Mulder et al. 2017), as well VEGF neutralizing antibodies (Yang et al. 2014). While preclinical data are promising, they still need to prove efficacy in clinical trials. Furthermore, accumulating data on the multifaceted roles of macrophages in liver disease should make us cautious, as blocking monocyte inflammation may also limit the formation of pro-resolution macrophages.

Another approach for targeting liver macrophages is modulating their activation status and phenotype. Promising approaches include inhibition of macrophage responses to PAMPs and DAMPs, for example by blocking toll-like receptors, other pattern recognition receptors, or their downstream signalling pathways (Tacke et al. 2023). Downstream signalling pathways that might be amenable to therapeutic targeting include NLRP3 inflammasome activation (Mridha et al. 2017b) or apoptosis signal-regulating kinase 1 (ASK-1) inhibition (Schuster-Gaul et al. 2020). Unfortunately, the ASK-1 inhibitor selonsertib did not show sufficient anti-fibrotic efficacy (Harrison et al. 2020), despite promising signals in a phase 2 trials in patients with MASH (Loomba et al. 2018). Another group of potential drugs is targeting nuclear receptors, including farnesoid X-receptor (FXR), liver X-receptor (LXR) or peroxisome-proliferator-activator receptors (PPARs), which have important functions in regulating macrophage substrate metabolism and activation (Puengel et al. 2022), but to date none of these drugs has shown sufficient effectiveness and a favourable risk-benefit-ratio in phase 3 clinical trials. In addition, disease-specific mechanisms are currently being unravelled, e.g. the role of scavenger receptors like MSR1 in the formation of inflammatory hCLS (Govaere et al. 2022) or the inflammation-limiting role of macrophage TREM2 in MASH (Wang et al. 2023), which will provide new avenues for drug development. Hopefully in the future, developments in our understanding of liver macrophage biology and their regulation in disease will enable the development of novel therapeutic approaches.

7.8 Concluding Remarks

In this chapter, we have characterised hepatic macrophages as a heterogeneous, versatile, and highly dynamic cell population that fulfils important functions in liver health. Dysregulation of normal macrophage function can induce inflammatory and fibrogenic cascades, which upon severe acute or prolonged chronic injury leads to loss of liver function, chronic liver failure or hepatocellular carcinoma. However, macrophages are also critically involved in regenerative processes by limiting inflammation, promoting wound healing responses and the proliferation of parenchymal cells. We currently only started to uncover the specific subsets involved, their regulation and specific functions. In the future, we will need to further delineate the high granularity of liver macrophages and uncover the underlying signalling circuits. Eventually, we can expect that this will pave the road to the development of novel therapeutic strategies tailored to the specific underlying disease condition and other patient-related factors.

References

Ait Ahmed Y, Lafdil F, Tacke F (2023) Ambiguous pathogenic roles of macrophages in alcohol-associated liver diseases. Hepatic Med 15:113–127. https://doi.org/10.2147/HMER.S326468

Aizarani N, Saviano A, Sagar et al (2019) A human liver cell atlas reveals heterogeneity and epithelial progenitors. Nature 572:199–204. https://doi.org/10.1038/s41586-019-1373-2

Ambade A, Lowe P, Kodys K et al (2019) Pharmacological Inhibition of CCR2/5 Signaling Prevents and Reverses Alcohol-Induced Liver Damage, Steatosis, and Inflammation in Mice. Hepatology 69:1105. https://doi.org/10.1002/hep.30249

Andrews TS, Nakib D, Perciani CT et al (2024) Single-cell and spatial transcriptomics characterisation of the immunological landscape in the healthy and PSC human liver. J Hepatol 80:730–743. https://doi.org/10.1016/j.jhep.2023.12.023

Anstee QM, Neuschwander-Tetri BA, Wong VW-S et al (2024) Cenicriviroc lacked efficacy to treat liver fibrosis in nonalcoholic steatohepatitis: AURORA phase III randomized study. Clin Gastroenterol Hepatol 22:124–134.e1. https://doi.org/10.1016/j.cgh.2023.04.003

Baeck C, Wei X, Bartneck M et al (2014) Pharmacological inhibition of the chemokine C-C motif chemokine ligand 2 (monocyte chemoattractant protein 1) accelerates liver fibrosis regression by suppressing Ly-6C+ macrophage infiltration in mice. Hepatology 59:1060–1072. https://doi.org/10.1002/hep.26783

Bai Q, Hong X, Lin H et al (2023) Single-cell landscape of immune cells in human livers affected by HBV-related cirrhosis. JHEP Rep 5:100883. https://doi.org/10.1016/j.jhepr.2023.100883

Bartneck M, Schrammen PL, Möckel D et al (2019) The CCR2+ macrophage subset promotes pathogenic angiogenesis for tumor vascularization in fibrotic livers. Cell Mol Gastroenterol Hepatol 7:371–390. https://doi.org/10.1016/j.jcmgh.2018.10.007

Benacerraf B, Sebestyen MM, Schlossman S (1959) A quantitative study of the kinetics of blood clearance of P32-labelled Escherichia coli and Staphylococci by the reticulo-endothelial system. J Exp Med 110:27–48. https://doi.org/10.1084/jem.110.1.27

Ben-Moshe S, Veg T, Manco R et al (2022) The spatiotemporal program of zonal liver regeneration following acute injury. Cell Stem Cell 29:973–989.e10. https://doi.org/10.1016/j.stem.2022.04.008

Bernal W, Wendon J (2013) Acute liver failure. N Engl J Med 369:2525–2534. https://doi.org/10.1056/NEJMra1208937

Bernsmeier C, Pop OT, Singanayagam A et al (2015) Patients with acute-on-chronic liver failure have increased numbers of regulatory immune cells expressing the receptor tyrosine kinase MERTK. Gastroenterology 148:603–615.e14. https://doi.org/10.1053/j.gastro.2014.11.045

Blériot C, Barreby E, Dunsmore G et al (2021) A subset of Kupffer cells regulates metabolism through the expression of CD36. Immunity 54:2101–2116.e6. https://doi.org/10.1016/j.immuni.2021.08.006

Boltjes A, Movita D, Boonstra A, Woltman AM (2014) The role of Kupffer cells in hepatitis B and hepatitis C virus infections. J Hepatol 61:660–671. https://doi.org/10.1016/j.jhep.2014.04.026

Bruneau A, Guillot A, Tacke F (2022) Macrophages in cholangiopathies. Curr Opin Gastroenterol 38:114. https://doi.org/10.1097/MOG.0000000000000814

Cai H, Zhu X-D, Ao J-Y et al (2017) Colony-stimulating factor-1-induced AIF1 expression in tumor-associated macrophages enhances the progression of hepatocellular carcinoma. Oncoimmunology 6:e1333213. https://doi.org/10.1080/2162402X.2017.1333213

Cai B, Dongiovanni P, Corey KE et al (2020) Macrophage MerTK promotes liver fibrosis in nonalcoholic steatohepatitis. Cell Metab 31:406–421.e7. https://doi.org/10.1016/j.cmet.2019.11.013

Cameron RG, Blendis LM, Neuman MG (2001) Accumulation of macrophages in primary sclerosing cholangitis. Clin Biochem 34:195–201. https://doi.org/10.1016/S0009-9120(01)00215-6

Campana L, Starkey Lewis PJ, Pellicoro A et al (2018) The STAT3–IL-10–IL-6 pathway is a novel regulator of macrophage efferocytosis and phenotypic conversion in sterile liver injury. J Immunol 200:1169–1187. https://doi.org/10.4049/jimmunol.1701247

Canbay A, Feldstein AE, Higuchi H et al (2003) Kupffer cell engulfment of apoptotic bodies stimulates death ligand and cytokine expression. Hepatology 38:1188–1198. https://doi.org/10.1053/jhep.2003.50472

Chang S, Dolganiuc A, Szabo G (2007) Toll-like receptors 1 and 6 are involved in TLR2-mediated macrophage activation by hepatitis C virus core and NS3 proteins. J Leukoc Biol 82:479–487. https://doi.org/10.1189/jlb.0207128

Chen Y, Ramjiawan RR, Reiberger T et al (2015) CXCR4 inhibition in tumor microenvironment facilitates anti-programmed death receptor-1 immunotherapy in sorafenib-treated hepatocellular carcinoma in mice. Hepatology 61:1591. https://doi.org/10.1002/hep.27665

Chen Y-Y, Arndtz K, Webb G et al (2019) Intrahepatic macrophage populations in the pathophysiology of primary sclerosing cholangitis. JHEP Rep 1:369–376. https://doi.org/10.1016/j.jhepr.2019.10.003

Cheng S, Li Z, Gao R et al (2021) A pan-cancer single-cell transcriptional atlas of tumor infiltrating myeloid cells. Cell 184:792–809.e23. https://doi.org/10.1016/j.cell.2021.01.010

Choi D-Y, Ban J-O, Kim S-C, Hong JT (2015) CCR5 knockout mice with C57BL6 background are resistant to acetaminophen-mediated hepatotoxicity due to decreased macrophages migration into the liver. Arch Toxicol 89:211–220. https://doi.org/10.1007/s00204-014-1253-3

Coelho I, Duarte N, Barros A et al (2021) Trem-2 promotes emergence of restorative macrophages and endothelial cells during recovery from hepatic tissue damage. Front Immunol 11:616044

Cui A, Li B, Wallace MS et al (2023) Single-cell atlas of the liver myeloid compartment before and after cure of chronic viral hepatitis. J Hepatol 80:251–267. https://doi.org/10.1016/j.jhep.2023.02.040

Daemen S, Gainullina A, Kalugotla G et al (2021) Dynamic shifts in the composition of resident and recruited macrophages influence tissue remodeling in NASH. Cell Rep 34:108626. https://doi.org/10.1016/j.celrep.2020.108626

Dal-Secco D, Wang J, Zeng Z et al (2015) A dynamic spectrum of monocytes arising from the in situ reprogramming of CCR2+ monocytes at a site of sterile injury. J Exp Med 212:447–456. https://doi.org/10.1084/jem.20141539

De Muynck K, Vanderborght B, Van Vlierberghe H, Devisscher L (2021) The gut–liver axis in chronic liver disease: a macrophage perspective. Cells 10:2959. https://doi.org/10.3390/cells10112959

De Simone G, Andreata F, Bleriot C et al (2021) Identification of a Kupffer cell subset capable of reverting the T cell dysfunction induced by hepatocellular priming. Immunity 54:2089–2100.e8. https://doi.org/10.1016/j.immuni.2021.05.005

Debebe A, Medina V, Chen C-Y et al (2017) Wnt/β-catenin activation and macrophage induction during liver cancer development following steatosis. Oncogene 36:6020–6029. https://doi.org/10.1038/onc.2017.207

Deppermann C, Kratofil RM, Peiseler M et al (2020) Macrophage galactose lectin is critical for Kupffer cells to clear aged platelets. J Exp Med 217:e20190723. https://doi.org/10.1084/jem.20190723

Devisscher L, Scott CL, Lefere S et al (2017) Non-alcoholic steatohepatitis induces transient changes within the liver macrophage pool. Cell Immunol 322:74–83. https://doi.org/10.1016/j.cellimm.2017.10.006

Donne R, Lujambio A (2023) The liver cancer immune microenvironment: therapeutic implications for hepatocellular carcinoma. Hepatology 77:1773. https://doi.org/10.1002/hep.32740

Doran AC, Yurdagul A, Tabas I (2020) Efferocytosis in health and disease. Nat Rev Immunol 20:254–267. https://doi.org/10.1038/s41577-019-0240-6

Duan Y, Llorente C, Lang S et al (2019) Bacteriophage targeting of gut bacterium attenuates alcoholic liver disease. Nature 575:505–511. https://doi.org/10.1038/s41586-019-1742-x

Duffield JS, Forbes SJ, Constandinou CM et al (2005) Selective depletion of macrophages reveals distinct, opposing roles during liver injury and repair. J Clin Invest 115:56–65. https://doi.org/10.1172/JCI200522675

Duwaerts CC, Gehring S, Cheng C-W et al (2013) Contrasting responses of Kupffer cells and inflammatory mononuclear phagocytes to biliary obstruction in a mouse model of cholestatic liver injury. Liver Int 33:255–265. https://doi.org/10.1111/liv.12048

Eggert T, Wolter K, Ji J et al (2016) Distinct functions of senescence-associated immune responses in liver tumor surveillance and tumor progression. Cancer Cell 30:533–547. https://doi.org/10.1016/j.ccell.2016.09.003

Eksteen B, Bowlus CL, Montano-Loza AJ et al (2021) Efficacy and safety of cenicriviroc in patients with primary sclerosing cholangitis: PERSEUS study. Hepatol Commun 5:478. https://doi.org/10.1002/hep4.1619

Fabre T, Barron AMS, Christensen SM et al (2023) Identification of a broadly fibrogenic macrophage subset induced by type 3 inflammation. Sci Immunol 8:eadd8945. https://doi.org/10.1126/sciimmunol.add8945

Feng D, Xiang X, Guan Y et al (2023) Monocyte-derived macrophages orchestrate multiple cell-type interactions to repair necrotic liver lesions in disease models. J Clin Invest 133:e166954. https://doi.org/10.1172/JCI166954

Ferreira-Gonzalez S, Lu W-Y, Raven A et al (2018) Paracrine cellular senescence exacerbates biliary injury and impairs regeneration. Nat Commun 9:1020. https://doi.org/10.1038/s41467-018-03299-5

Fisher JE, Mckenzie TJ, Lillegard JB et al (2013) Role of Kupffer cells and toll-like receptor 4 in acetaminophen-induced acute liver failure. J Surg Res 180:147–155. https://doi.org/10.1016/j.jss.2012.11.051

Friedman SL, Ratziu V, Harrison SA et al (2018) A randomized, placebo-controlled trial of cenicriviroc for treatment of nonalcoholic steatohepatitis with fibrosis. Hepatology 67:1754. https://doi.org/10.1002/hep.29477

Garcia-Martinez I, Santoro N, Chen Y et al (2016) Hepatocyte mitochondrial DNA drives nonalcoholic steatohepatitis by activation of TLR9. J Clin Invest 126:859–864. https://doi.org/10.1172/JCI83885

Gomez Perdiguero E, Klapproth K, Schulz C et al (2015) Tissue-resident macrophages originate from yolk-sac-derived erythro-myeloid progenitors. Nature 518:547–551. https://doi.org/10.1038/nature13989

Govaere O, Petersen SK, Martinez-Lopez N et al (2022) Macrophage scavenger receptor 1 mediates lipid-induced inflammation in non-alcoholic fatty liver disease. J Hepatol 76:1001–1012. https://doi.org/10.1016/j.jhep.2021.12.012

Guicciardi ME, Trussoni CE, Krishnan A et al (2018) Macrophages contribute to the pathogenesis of sclerosing cholangitis in mice. J Hepatol 69:676–686. https://doi.org/10.1016/j.jhep.2018.05.018

Guilliams M, Scott CL (2022) Liver macrophages in health and disease. Immunity 55:1515–1529. https://doi.org/10.1016/j.immuni.2022.08.002

Guilliams M, Mildner A, Yona S (2018) Developmental and functional heterogeneity of monocytes. Immunity 49:595–613. https://doi.org/10.1016/j.immuni.2018.10.005

Guilliams M, Bonnardel J, Haest B et al (2022) Spatial proteogenomics reveals distinct and evolutionarily conserved hepatic macrophage niches. Cell 185:379–396.e38. https://doi.org/10.1016/j.cell.2021.12.018

Guillot A, Tacke F (2019) Liver macrophages: old dogmas and new insights. Hepatol Commun 3:730–743. https://doi.org/10.1002/hep4.1356

Guillot A, Guerri L, Feng D et al (2021) Bile acid–activated macrophages promote biliary epithelial cell proliferation through integrin αvβ6 upregulation following liver injury. J Clin Invest 131:e132305. https://doi.org/10.1172/JCI132305

Guillot A, Winkler M, Silva Afonso M et al (2023) Mapping the hepatic immune landscape identifies monocytic macrophages as key drivers of steatohepatitis and cholangiopathy progression. Hepatology 78:150. https://doi.org/10.1097/HEP.0000000000000270

Guo M, Härtlova A, Gierliński M et al (2019) Triggering MSR1 promotes JNK-mediated inflammation in IL-4-activated macrophages. EMBO J 38:e100299. https://doi.org/10.15252/embj.2018100299
Hammerich L, Tacke F (2023) Hepatic inflammatory responses in liver fibrosis. Nat Rev Gastroenterol Hepatol 20:633–646. https://doi.org/10.1038/s41575-023-00807-x
Han Y, Bian Z-H, Yang S-Y et al (2022) Single-cell characterization of hepatic CD8+ T cells in a murine model of primary biliary cholangitis. Front Immunol 13:860311
Harrison SA, Wong VWS, Okanoue T et al (2020) Selonsertib for patients with bridging fibrosis or compensated cirrhosis due to NASH: results from randomized phase III STELLAR trials. J Hepatol 73:26. https://doi.org/10.1016/j.jhep.2020.02.027
Haselow K, Bode JG, Wammers M et al (2013) Bile acids PKA-dependently induce a switch of the IL-10/IL-12 ratio and reduce proinflammatory capability of human macrophages. J Leukoc Biol 94:1253–1264. https://doi.org/10.1189/jlb.0812396
Helmy KY, Katschke KJ, Gorgani NN et al (2006) CRIg: a macrophage complement receptor required for phagocytosis of circulating pathogens. Cell 124:915–927. https://doi.org/10.1016/j.cell.2005.12.039
Hendrikx T, Porsch F, Kiss MG et al (2022) Soluble TREM2 levels reflect the recruitment and expansion of TREM2+ macrophages that localize to fibrotic areas and limit NASH. J Hepatol 77:1373–1385. https://doi.org/10.1016/j.jhep.2022.06.004
Heymann F, Tacke F (2016) Immunology in the liver — from homeostasis to disease. Nat Rev Gastroenterol Hepatol 13:88–110. https://doi.org/10.1038/nrgastro.2015.200
Heymann F, Peusquens J, Ludwig-Portugall I et al (2015) Liver inflammation abrogates immunological tolerance induced by Kupffer cells. Hepatology 62:279. https://doi.org/10.1002/hep.27793
Ho DW-H, Tsui Y-M, Chan L-K et al (2021) Single-cell RNA sequencing shows the immunosuppressive landscape and tumor heterogeneity of HBV-associated hepatocellular carcinoma. Nat Commun 12:3684. https://doi.org/10.1038/s41467-021-24010-1
Holt MP, Cheng L, Ju C (2008) Identification and characterization of infiltrating macrophages in acetaminophen-induced liver injury. J Leukoc Biol 84:1410–1421. https://doi.org/10.1189/jlb.0308173
Holt MP, Yin H, Ju C (2010) Exacerbation of acetaminophen-induced disturbances of liver sinusoidal endothelial cells in the absence of Kupffer cells in mice. Toxicol Lett 194:34–41. https://doi.org/10.1016/j.toxlet.2010.01.020
Horn CL, Morales AL, Savard C et al (2022) Role of cholesterol-associated steatohepatitis in the development of NASH. Hepatol Commun 6:12. https://doi.org/10.1002/hep4.1801
Horst AK, Tiegs G, Diehl L (2019) Contribution of macrophage efferocytosis to liver homeostasis and disease. Front Immunol 10:2670
Hu M, Wang Y, Liu Z et al (2021) Hepatic macrophages act as a central hub for relaxin-mediated alleviation of liver fibrosis. Nat Nanotechnol 16:466–477. https://doi.org/10.1038/s41565-020-00836-6
Huang W, Metlakunta A, Dedousis N et al (2010) Depletion of liver Kupffer cells prevents the development of diet-induced hepatic steatosis and insulin resistance. Diabetes 59:347–357. https://doi.org/10.2337/db09-0016
Huang L-R, Wohlleber D, Reisinger F et al (2013) Intrahepatic myeloid-cell aggregates enable local proliferation of CD8+ T cells and successful immunotherapy against chronic viral liver infection. Nat Immunol 14:574–583. https://doi.org/10.1038/ni.2573
Ioannou GN, Haigh WG, Thorning D, Savard C (2013) Hepatic cholesterol crystals and crown-like structures distinguish NASH from simple steatosis[S]. J Lipid Res 54:1326–1334. https://doi.org/10.1194/jlr.M034876
Ioannou GN, Van Rooyen DM, Savard C et al (2015) Cholesterol-lowering drugs cause dissolution of cholesterol crystals and disperse Kupffer cell crown-like structures during resolution of NASH. J Lipid Res 56:277–285. https://doi.org/10.1194/jlr.M053785

Itoh M, Kato H, Suganami T et al (2013) Hepatic crown-like structure: a unique histological feature in non-alcoholic steatohepatitis in mice and humans. PLoS One 8:e82163. https://doi.org/10.1371/journal.pone.0082163

Itoh M, Suganami T, Kato H et al (2017) CD11c+ resident macrophages drive hepatocyte death-triggered liver fibrosis in a murine model of nonalcoholic steatohepatitis. JCI Insight 2:e92902. https://doi.org/10.1172/jci.insight.92902

Jaeschke H (2007) Kupffer cells. In: Textbook of hepatology, Wiley, pp 36–42

Jiang J, Wang G-Z, Wang Y et al (2018) Hypoxia-induced HMGB1 expression of HCC promotes tumor invasiveness and metastasis via regulating macrophage-derived IL-6. Exp Cell Res 367:81–88. https://doi.org/10.1016/j.yexcr.2018.03.025

Jiang Y, Tang Y, Hoover C et al (2021) Kupffer cell receptor CLEC4F is important for the destruction of desialylated platelets in mice. Cell Death Differ 28:3009–3021. https://doi.org/10.1038/s41418-021-00797-w

Jiang P, Jia H, Qian X et al (2024) Single-cell RNA sequencing reveals the immunoregulatory roles of PegIFN-α in patients with chronic hepatitis B. Hepatology 79:167–182. https://doi.org/10.1097/HEP.0000000000000524

Jin H, Liu K, Tang J et al (2021) Genetic fate-mapping reveals surface accumulation but not deep organ invasion of pleural and peritoneal cavity macrophages following injury. Nat Commun 12:2863. https://doi.org/10.1038/s41467-021-23197-7

Ju C, Reilly TP, Bourdi M et al (2002) Protective role of Kupffer cells in acetaminophen-induced hepatic injury in mice. Chem Res Toxicol 15:1504–1513. https://doi.org/10.1021/tx0255976

Kang T-W, Yevsa T, Woller N et al (2011) Senescence surveillance of pre-malignant hepatocytes limits liver cancer development. Nature 479:547–551. https://doi.org/10.1038/nature10599

Karlmark KR, Weiskirchen R, Zimmermann HW et al (2009) Hepatic recruitment of the inflammatory Gr1+ monocyte subset upon liver injury promotes hepatic fibrosis. Hepatology 50:261–274. https://doi.org/10.1002/hep.22950

Keitel V, Donner M, Winandy S et al (2008) Expression and function of the bile acid receptor TGR5 in Kupffer cells. Biochem Biophys Res Commun 372:78–84. https://doi.org/10.1016/j.bbrc.2008.04.171

Khakoo SI, Soni PN, Savage K et al (1997) Lymphocyte and macrophage phenotypes in chronic hepatitis C infection. Correlation with disease activity. Am J Pathol 150:963–970

Kleiner DE, Brunt EM, Van Natta M et al (2005) Design and validation of a histological scoring system for nonalcoholic fatty liver disease. Hepatology 41:1313–1321. https://doi.org/10.1002/hep.20701

Krenkel O, Tacke F (2017) Liver macrophages in tissue homeostasis and disease. Nat Rev Immunol 17:306–321. https://doi.org/10.1038/nri.2017.11

Krenkel O, Puengel T, Govaere O et al (2018) Therapeutic inhibition of inflammatory monocyte recruitment reduces steatohepatitis and liver fibrosis. Hepatology 67:1270–1283. https://doi.org/10.1002/hep.29544

Kuang D-M, Zhao Q, Peng C et al (2009) Activated monocytes in peritumoral stroma of hepatocellular carcinoma foster immune privilege and disease progression through PD-L1. J Exp Med 206:1327–1337. https://doi.org/10.1084/jem.20082173

Kuilman T, Michaloglou C, Mooi WJ, Peeper DS (2010) The essence of senescence. Genes Dev 24:2463–2479. https://doi.org/10.1101/gad.1971610

Lee K-J, An S, Kim M-Y et al (2023) Hepatic TREM2+ macrophages express matrix metalloproteinases to control fibrotic scar formation. Immunol Cell Biol 101:216–230. https://doi.org/10.1111/imcb.12616

Lefebvre E, Moyle G, Reshef R et al (2016) Antifibrotic effects of the dual CCR2/CCR5 antagonist cenicriviroc in animal models of liver and kidney fibrosis. PLoS One 11:e0158156. https://doi.org/10.1371/journal.pone.0158156

Leroux A, Ferrere G, Godie V et al (2012) Toxic lipids stored by Kupffer cells correlates with their pro-inflammatory phenotype at an early stage of steatohepatitis. J Hepatol 57:141–149. https://doi.org/10.1016/j.jhep.2012.02.028

Li M, Sun R, Xu L et al (2015) Kupffer cells support hepatitis B virus–mediated CD8+ T cell exhaustion via hepatitis B core antigen–TLR2 interactions in mice. J Immunol 195:3100–3109. https://doi.org/10.4049/jimmunol.1500839

Li X, Yao W, Yuan Y et al (2017) Targeting of tumour-infiltrating macrophages via CCL2/CCR2 signalling as a therapeutic strategy against hepatocellular carcinoma. Gut 66:157–167. https://doi.org/10.1136/gutjnl-2015-310514

Liaskou E, Zimmermann HW, Li K-K et al (2013) Monocyte subsets in human liver disease show distinct phenotypic and functional characteristics. Hepatology 57:385–398. https://doi.org/10.1002/hep.26016

Liu J, Yu Q, Wu W et al (2018) TLR2 stimulation strengthens intrahepatic myeloid-derived cell-mediated T cell tolerance through inducing Kupffer cell expansion and IL-10 production. J Immunol 200:2341–2351. https://doi.org/10.4049/jimmunol.1700540

Liu Y, Xun Z, Ma K et al (2023) Identification of a tumour immune barrier in the HCC microenvironment that determines the efficacy of immunotherapy. J Hepatol 78:770–782. https://doi.org/10.1016/j.jhep.2023.01.011

Loomba R, Lawitz E, Mantry PS et al (2018) The ASK1 inhibitor selonsertib in patients with nonalcoholic steatohepatitis: a randomized, phase 2 trial. Hepatology 67:549. https://doi.org/10.1002/hep.29514

Lopes ME, Nakagaki BN, Mattos MS et al (2022) Susceptibility to infections during acute liver injury depends on transient disruption of liver macrophage niche. Front Immunol 13:892114

Lu Y, Yang A, Quan C et al (2022) A single-cell atlas of the multicellular ecosystem of primary and metastatic hepatocellular carcinoma. Nat Commun 13:4594. https://doi.org/10.1038/s41467-022-32283-3

MacParland SA, Liu JC, Ma X-Z et al (2018) Single cell RNA sequencing of human liver reveals distinct intrahepatic macrophage populations. Nat Commun 9:4383. https://doi.org/10.1038/s41467-018-06318-7

Maeda S, Kamata H, Luo J-L et al (2005) IKKβ couples hepatocyte death to cytokine-driven compensatory proliferation that promotes chemical hepatocarcinogenesis. Cell 121:977–990. https://doi.org/10.1016/j.cell.2005.04.014

Malehmir M, Pfister D, Gallage S et al (2019) Platelet GPIbα is a mediator and potential interventional target for NASH and subsequent liver cancer. Nat Med 25:641–655. https://doi.org/10.1038/s41591-019-0379-5

Marra F, Svegliati-Baroni G (2018) Lipotoxicity and the gut-liver axis in NASH pathogenesis. J Hepatol 68:280–295. https://doi.org/10.1016/j.jhep.2017.11.014

Mass E, Ballesteros I, Farlik M et al (2016) Specification of tissue-resident macrophages during organogenesis. Science 353:aaf4238. https://doi.org/10.1126/science.aaf4238

Mathews S, Gao B (2013) Therapeutic potential of interleukin 1 inhibitors in the treatment of alcoholic liver disease. Hepatology 57:2078–2080. https://doi.org/10.1002/hep.26336

McGuinness P, Painter D, Davies S, McCaughan G (2000) Increases in intrahepatic CD68 positive cells, MAC387 positive cells, and proinflammatory cytokines (particularly interleukin 18) in chronic hepatitis C infection. Gut 46:260–269. https://doi.org/10.1136/gut.46.2.260

Meng Y-M, Liang J, Wu C et al (2018) Monocytes/macrophages promote vascular CXCR4 expression via the ERK pathway in hepatocellular carcinoma. Onco Targets Ther 7:e1408745. https://doi.org/10.1080/2162402X.2017.1408745

Mitchell C, Couton D, Couty J-P et al (2009) Dual role of CCR2 in the constitution and the resolution of liver fibrosis in mice. Am J Pathol 174:1766–1775. https://doi.org/10.2353/ajpath.2009.080632

Miura K, Yang L, van Rooijen N et al (2012) Hepatic recruitment of macrophages promotes nonalcoholic steatohepatitis through CCR2. Am J Physiol Gastrointest Liver Physiol 302:G1310–G1321. https://doi.org/10.1152/ajpgi.00365.2011

Mossanen JC, Krenkel O, Ergen C et al (2016) Chemokine (C-C motif) receptor 2–positive monocytes aggravate the early phase of acetaminophen-induced acute liver injury. Hepatology 64:1667. https://doi.org/10.1002/hep.28682

Mridha AR, Haczeyni F, Yeh MM et al (2017a) TLR9 is up-regulated in human and murine NASH: pivotal role in inflammatory recruitment and cell survival. Clin Sci 131:2145–2159. https://doi.org/10.1042/CS20160838

Mridha AR, Wree A, Robertson AAB et al (2017b) NLRP3 inflammasome blockade reduces liver inflammation and fibrosis in experimental NASH in mice. J Hepatol 66:1037–1046. https://doi.org/10.1016/j.jhep.2017.01.022

Mulder P, van den Hoek AM, Kleemann R (2017) The CCR2 inhibitor propagermanium attenuates diet-induced insulin resistance, adipose tissue inflammation and non-alcoholic steatohepatitis. PLoS One 12:e0169740. https://doi.org/10.1371/journal.pone.0169740

Nebbia G, Peppa D, Schurich A et al (2012) Upregulation of the Tim-3/Galectin-9 pathway of T cell exhaustion in chronic hepatitis B virus infection. PLoS One 7:e47648. https://doi.org/10.1371/journal.pone.0047648

Pallett LJ, Burton AR, Amin OE et al (2020) Longevity and replenishment of human liver-resident memory T cells and mononuclear phagocytes. J Exp Med 217:e20200050. https://doi.org/10.1084/jem.20200050

Peiseler M, Schwabe R, Hampe J et al (2022) Immune mechanisms linking metabolic injury to inflammation and fibrosis in fatty liver disease - novel insights into cellular communication circuits. J Hepatol 77:1136–1160. https://doi.org/10.1016/j.jhep.2022.06.012

Peiseler M, Araujo David B, Zindel J et al (2023) Kupffer cell–like syncytia replenish resident macrophage function in the fibrotic liver. Science 381:eabq5202. https://doi.org/10.1126/science.abq5202

Pérez-Martínez L, Pérez-Matute P, Aguilera-Lizarraga J et al (2014) Maraviroc, a CCR5 antagonist, ameliorates the development of hepatic steatosis in a mouse model of non-alcoholic fatty liver disease (NAFLD). J Antimicrob Chemother 69:1903–1910. https://doi.org/10.1093/jac/dku071

Petrasek J, Bala S, Csak T et al (2012) IL-1 receptor antagonist ameliorates inflammasome-dependent alcoholic steatohepatitis in mice. J Clin Invest 122:3476–3489. https://doi.org/10.1172/JCI60777

Pikarsky E, Porat RM, Stein I et al (2004) NF-κB functions as a tumour promoter in inflammation-associated cancer. Nature 431:461–466. https://doi.org/10.1038/nature02924

Pinzani M (2007) The hepatic stellate cell. In: Textbook of hepatology. Wiley, pp 43–51

Poch T, Krause J, Casar C et al (2021) Single-cell atlas of hepatic T cells reveals expansion of liver-resident naive-like CD4+ T cells in primary sclerosing cholangitis. J Hepatol 75:414–423. https://doi.org/10.1016/j.jhep.2021.03.016

Pose E, Coll M, Martínez-Sánchez C et al (2021) Programmed death ligand 1 is overexpressed in liver macrophages in chronic liver diseases, and its blockade improves the antibacterial activity against infections. Hepatology 74:296. https://doi.org/10.1002/hep.31644

Pradere J-P, Kluwe J, De Minicis S et al (2013) Hepatic macrophages but not dendritic cells contribute to liver fibrosis by promoting the survival of activated hepatic stellate cells in mice. Hepatology 58:1461–1473. https://doi.org/10.1002/hep.26429

Puengel T, Liu H, Guillot A et al (2022) Nuclear receptors linking metabolism, inflammation, and fibrosis in nonalcoholic fatty liver disease. Int J Mol Sci 23:2668. https://doi.org/10.3390/ijms23052668

Qian B-Z, Pollard JW (2010) Macrophage diversity enhances tumor progression and metastasis. Cell 141:39–51. https://doi.org/10.1016/j.cell.2010.03.014

Ramachandran P, Pellicoro A, Vernon MA et al (2012) Differential Ly-6C expression identifies the recruited macrophage phenotype, which orchestrates the regression of murine liver fibrosis. Proc Natl Acad Sci 109:E3186–E3195. https://doi.org/10.1073/pnas.1119964109

Ramachandran P, Dobie R, Wilson-Kanamori JR et al (2019) Resolving the fibrotic niche of human liver cirrhosis at single-cell level. Nature 575:512–518. https://doi.org/10.1038/s41586-019-1631-3

Ramirez-Pedraza M, Fernández M (2019) Interplay between macrophages and angiogenesis: a double-edged sword in liver disease. Front Immunol 10:2882

Ratziu V, Sanyal A, Harrison SA et al (2020) Cenicriviroc treatment for adults with nonalcoholic steatohepatitis and fibrosis: final analysis of the phase 2b CENTAUR study. Hepatology 72:892. https://doi.org/10.1002/hep.31108

Ravichandran G, Neumann K, Berkhout LK et al (2019) Interferon-γ-dependent immune responses contribute to the pathogenesis of sclerosing cholangitis in mice. J Hepatol 71:773–782. https://doi.org/10.1016/j.jhep.2019.05.023

Remmerie A, Martens L, Thoné T et al (2020) Osteopontin expression identifies a subset of recruited macrophages distinct from Kupffer cells in the fatty liver. Immunity 53:641–657.e14. https://doi.org/10.1016/j.immuni.2020.08.004

Renato P, Correa AV, Nathanson MH et al (2007) Functions of the liver. In: Textbook of hepatology. Wiley, pp 89–128

Rinella ME, Lazarus JV, Ratziu V et al (2023) A multisociety Delphi consensus statement on new fatty liver disease nomenclature. J Hepatol 79:1542–1556. https://doi.org/10.1016/j.jhep.2023.06.003

Roeb E (2018) Matrix metalloproteinases and liver fibrosis (translational aspects). Matrix Biol 68–69:463–473. https://doi.org/10.1016/j.matbio.2017.12.012

Schuppan D, Ruehl M, Somasundaram R, Hahn EG (2001) Matrix as a modulator of hepatic fibrogenesis. Semin Liver Dis 21:351–372. https://doi.org/10.1055/s-2001-17556

Schuster-Gaul S, Geisler LJ, McGeough MD et al (2020) ASK1 inhibition reduces cell death and hepatic fibrosis in an Nlrp3 mutant liver injury model. JCI Insight 5:e123294. https://doi.org/10.1172/jci.insight.123294

Scott CL, Guilliams M (2018) The role of Kupffer cells in hepatic iron and lipid metabolism. J Hepatol 69:1197–1199. https://doi.org/10.1016/j.jhep.2018.02.013

Scott CL, Zheng F, De Baetselier P et al (2016) Bone marrow-derived monocytes give rise to self-renewing and fully differentiated Kupffer cells. Nat Commun 7:10321. https://doi.org/10.1038/ncomms10321

Seidman JS, Troutman TD, Sakai M et al (2020) Niche-specific reprogramming of epigenetic landscapes drives myeloid cell diversity in nonalcoholic steatohepatitis. Immunity 52:1057–1074.e7. https://doi.org/10.1016/j.immuni.2020.04.001

Semela D, Shah VH (2007) Liver sinusoidal endothelial cells. In: Textbook of hepatology. Wiley, pp 29–35

Sepanlou SG, Safiri S, Bisignano C et al (2020) The global, regional, and national burden of cirrhosis by cause in 195 countries and territories, 1990–2017: a systematic analysis for the Global Burden of Disease Study 2017. Lancet Gastroenterol Hepatol 5:245–266. https://doi.org/10.1016/S2468-1253(19)30349-8

Sharma A, Seow JJW, Dutertre C-A et al (2020) Onco-fetal reprogramming of endothelial cells drives immunosuppressive macrophages in hepatocellular carcinoma. Cell 183:377–394.e21. https://doi.org/10.1016/j.cell.2020.08.040

Shi Y, Su W, Zhang L et al (2021) TGR5 regulates macrophage inflammation in nonalcoholic steatohepatitis by modulating NLRP3 inflammasome activation. Front Immunol 11:609060

Shin E-C, Sung PS, Park S-H (2016) Immune responses and immunopathology in acute and chronic viral hepatitis. Nat Rev Immunol 16:509–523. https://doi.org/10.1038/nri.2016.69

Sica A, Bronte V (2007) Altered macrophage differentiation and immune dysfunction in tumor development. J Clin Invest 117:1155–1166. https://doi.org/10.1172/JCI31422

Sierro F, Evrard M, Rizzetto S et al (2017) A liver capsular network of monocyte-derived macrophages restricts hepatic dissemination of intraperitoneal bacteria by neutrophil recruitment. Immunity 47:374–388.e6. https://doi.org/10.1016/j.immuni.2017.07.018

Song Q, Datta S, Liang X et al (2023) Type I interferon signaling facilitates resolution of acute liver injury by priming macrophage polarization. Cell Mol Immunol 20:143–157. https://doi.org/10.1038/s41423-022-00966-y

Stienstra R, Saudale F, Duval C et al (2010) Kupffer cells promote hepatic steatosis via interleukin-1β–dependent suppression of peroxisome proliferator-activated receptor α activity. Hepatology 51:511–522. https://doi.org/10.1002/hep.23337

Tacke F (2017) Targeting hepatic macrophages to treat liver diseases. J Hepatol 66:1300–1312. https://doi.org/10.1016/j.jhep.2017.02.026

Tacke F, Puengel T, Loomba R, Friedman SL (2023) An integrated view of anti-inflammatory and antifibrotic targets for the treatment of NASH. J Hepatol 79:552–566. https://doi.org/10.1016/j.jhep.2023.03.038

Taylor SA, Chen S-Y, Gadhvi G et al (2021) Transcriptional profiling of pediatric cholestatic livers identifies three distinct macrophage populations. PLoS One 16:e0244743. https://doi.org/10.1371/journal.pone.0244743

Terpstra V, van Berkel TJC (2000) Scavenger receptors on liver Kupffer cells mediate the in vivo uptake of oxidatively damaged red blood cells in mice. Blood 95:2157–2163. https://doi.org/10.1182/blood.V95.6.2157

Theurl I, Hilgendorf I, Nairz M et al (2016) On-demand erythrocyte disposal and iron recycling requires transient macrophages in the liver. Nat Med 22:945–951. https://doi.org/10.1038/nm.4146

Tian Y, Kuo C, Akbari O, Ou JJ (2016) Maternal-derived hepatitis B virus e antigen alters macrophage function in offspring to drive viral persistence after vertical transmission. Immunity 44:1204–1214. https://doi.org/10.1016/j.immuni.2016.04.008

Tian X, Wang Y, Lu Y et al (2021) Conditional depletion of macrophages ameliorates cholestatic liver injury and fibrosis via lncRNA-H19. Cell Death Dis 12:646. https://doi.org/10.1038/s41419-021-03931-1

Tomita K, Tamiya G, Ando S et al (2006) Tumour necrosis factor α signalling through activation of Kupffer cells plays an essential role in liver fibrosis of non-alcoholic steatohepatitis in mice. Gut 55:415–424. https://doi.org/10.1136/gut.2005.071118

Tosello-Trampont A-C, Landes SG, Nguyen V et al (2012) Kuppfer cells trigger nonalcoholic steatohepatitis development in diet-induced mouse model through tumor necrosis factor-α production*. J Biol Chem 287:40161–40172. https://doi.org/10.1074/jbc.M112.417014

Triantafyllou E, Pop OT, Possamai LA et al (2018) MerTK expressing hepatic macrophages promote the resolution of inflammation in acute liver failure. Gut 67:333–347. https://doi.org/10.1136/gutjnl-2016-313615

Triantafyllou E, Gudd CLC, Mawhin M-A et al (2021) PD-1 blockade improves Kupffer cell bacterial clearance in acute liver injury. J Clin Invest 131:e140196. https://doi.org/10.1172/JCI140196

Tsuchida T, Friedman SL (2017) Mechanisms of hepatic stellate cell activation. Nat Rev Gastroenterol Hepatol 14:397–411. https://doi.org/10.1038/nrgastro.2017.38

Vonderlin J, Chavakis T, Sieweke M, Tacke F (2023) The multifaceted roles of macrophages in NAFLD pathogenesis. Cell Mol Gastroenterol Hepatol 15:1311–1324. https://doi.org/10.1016/j.jcmgh.2023.03.002

Vonghia L, Van Herck MA, Weyler J, Francque S (2019) Targeting myeloid-derived cells: new Frontiers in the treatment of non-alcoholic and alcoholic liver disease. Front Immunol 10:563

Wan S, Zhao E, Kryczek I et al (2014) Tumor-associated macrophages produce interleukin 6 and signal via STAT3 to promote expansion of human hepatocellular carcinoma stem cells. Gastroenterology 147:1393–1404. https://doi.org/10.1053/j.gastro.2014.08.039

Wang J, Kubes P (2016) A reservoir of mature cavity macrophages that can rapidly invade visceral organs to affect tissue repair. Cell 165:668–678. https://doi.org/10.1016/j.cell.2016.03.009

Wang Y, van der Tuin S, Tjeerdema N et al (2015) Plasma cholesteryl ester transfer protein is predominantly derived from Kupffer cells. Hepatology 62:1710. https://doi.org/10.1002/hep.27985

Wang H, Mehal W, Nagy LE, Rotman Y (2021) Immunological mechanisms and therapeutic targets of fatty liver diseases. Cell Mol Immunol 18:73–91. https://doi.org/10.1038/s41423-020-00579-3

Wang Q, Zhou H, Bu Q et al (2022) Role of XBP1 in regulating the progression of non-alcoholic steatohepatitis. J Hepatol 77:312–325. https://doi.org/10.1016/j.jhep.2022.02.031

Wang X, He Q, Zhou C et al (2023) Prolonged hypernutrition impairs TREM2-dependent efferocytosis to license chronic liver inflammation and NASH development. Immunity 56:58–77.e11. https://doi.org/10.1016/j.immuni.2022.11.013
Weisberg SP, McCann D, Desai M et al (2003) Obesity is associated with macrophage accumulation in adipose tissue. J Clin Invest 112:1796–1808. https://doi.org/10.1172/JCI19246
Wen Y, Lambrecht J, Ju C, Tacke F (2021) Hepatic macrophages in liver homeostasis and diseases-diversity, plasticity and therapeutic opportunities. Cell Mol Immunol 18:45–56. https://doi.org/10.1038/s41423-020-00558-8
Willekens FLA, Werre JM, Kruijt JK et al (2005) Liver Kupffer cells rapidly remove red blood cell–derived vesicles from the circulation by scavenger receptors. Blood 105:2141–2145. https://doi.org/10.1182/blood-2004-04-1578
World Health Organization (2017) Global hepatitis report 2017. World Health Organization, Geneva
Wu L, Peng W, Wu H et al (2019) Lymphocyte antigen 6 complex, locus C+ monocytes and Kupffer cells orchestrate liver immune responses against hepatitis B virus in mice. Hepatology 69:2364–2380. https://doi.org/10.1002/hep.30510
Xiong X, Kuang H, Ansari S et al (2019) Landscape of intercellular crosstalk in healthy and NASH liver revealed by single-cell secretome gene analysis. Mol Cell 75:644–660.e5. https://doi.org/10.1016/j.molcel.2019.07.028
Xu H, Barnes GT, Yang Q et al (2003) Chronic inflammation in fat plays a crucial role in the development of obesity-related insulin resistance. J Clin Invest 112:1821–1830. https://doi.org/10.1172/JCI19451
Xu J, Liang J, Meng Y-M et al (2017) Vascular CXCR4 expression promotes vessel sprouting and sensitivity to Sorafenib treatment in hepatocellular carcinoma. Clin Cancer Res 23:4482–4492. https://doi.org/10.1158/1078-0432.CCR-16-2131
Xu L, Yang Y, Wen Y et al (2022) Hepatic recruitment of eosinophils and their protective function during acute liver injury. J Hepatol 77:344–352. https://doi.org/10.1016/j.jhep.2022.02.024
Yang L, Kwon J, Popov Y et al (2014) Vascular endothelial growth factor promotes fibrosis resolution and repair in mice. Gastroenterology 146:1339–1350.e1. https://doi.org/10.1053/j.gastro.2014.01.061
Yang JD, Hainaut P, Gores GJ et al (2019a) A global view of hepatocellular carcinoma: trends, risk, prevention and management. Nat Rev Gastroenterol Hepatol 16:589–604. https://doi.org/10.1038/s41575-019-0186-y
Yang W, Tao Y, Wu Y et al (2019b) Neutrophils promote the development of reparative macrophages mediated by ROS to orchestrate liver repair. Nat Commun 10:1076. https://doi.org/10.1038/s41467-019-09046-8
Yeung OWH, Lo C-M, Ling C-C et al (2015) Alternatively activated (M2) macrophages promote tumour growth and invasiveness in hepatocellular carcinoma. J Hepatol 62:607–616. https://doi.org/10.1016/j.jhep.2014.10.029
Yin Z, Huang J, Ma T et al (2017) Macrophages activating chemokine (C-X-C motif) ligand 8/miR-17 cluster modulate hepatocellular carcinoma cell growth and metastasis. Am J Transl Res 9:2403–2411
Yona S, Kim K-W, Wolf Y et al (2013) Fate mapping reveals origins and dynamics of monocytes and tissue macrophages under homeostasis. Immunity 38:79–91. https://doi.org/10.1016/j.immuni.2012.12.001
You Q, Cheng L, Kedl RM, Ju C (2008) Mechanism of T cell tolerance induction by murine hepatic Kupffer cells. Hepatology 48:978–990. https://doi.org/10.1002/hep.22395
Younossi Z, Anstee QM, Marietti M et al (2018) Global burden of NAFLD and NASH: trends, predictions, risk factors and prevention. Nat Rev Gastroenterol Hepatol 15:11–20. https://doi.org/10.1038/nrgastro.2017.109
Yu L, Yan J, Zhan Y et al (2023) Single-cell RNA sequencing reveals the dynamics of hepatic non-parenchymal cells in autoprotection against acetaminophen-induced hepatotoxicity. J Pharm Anal 13:926–941. https://doi.org/10.1016/j.jpha.2023.05.004

Zang M, Li Y, He H et al (2018) IL-23 production of liver inflammatory macrophages to damaged hepatocytes promotes hepatocellular carcinoma development after chronic hepatitis B virus infection. Biochim Biophys Acta (BBA) - Mol Basis Dis 1864:3759–3770. https://doi.org/10.1016/j.bbadis.2018.10.004

Zeng Z, Surewaard BGJ, Wong CHY et al (2016) CRIg functions as a macrophage pattern recognition receptor to directly bind and capture blood-borne gram-positive bacteria. Cell Host Microbe 20:99–106. https://doi.org/10.1016/j.chom.2016.06.002

Zhang J, Zhang Q, Lou Y et al (2018) Hypoxia-inducible factor-1α/interleukin-1β signaling enhances hepatoma epithelial–mesenchymal transition through macrophages in a hypoxic-inflammatory microenvironment. Hepatology 67:1872. https://doi.org/10.1002/hep.29681

Zhang Q, He Y, Luo N et al (2019) Landscape and dynamics of single immune cells in hepatocellular carcinoma. Cell 179:829–845.e20. https://doi.org/10.1016/j.cell.2019.10.003

Zigmond E, Samia-Grinberg S, Pasmanik-Chor M et al (2014) Infiltrating monocyte-derived macrophages and resident Kupffer cells display different ontogeny and functions in acute liver injury. J Immunol 193:344–353. https://doi.org/10.4049/jimmunol.1400574

Zimmermann HW, Seidler S, Nattermann J et al (2010) Functional contribution of elevated circulating and hepatic non-classical CD14+CD16+ monocytes to inflammation and human liver fibrosis. PLoS One 5:e11049. https://doi.org/10.1371/journal.pone.0011049

Part III
Macrophage Functions and Dysfunction in Diseases

Chapter 8
Crosstalk Between Macrophages and Breast Cancer Cells: Networking Within Tumors

Pooja Kamal Melwani, Rahul Checker, Murali Mohan Sagar Balla, and Badri Narain Pandey

Abstract Tumor associated macrophages (TAMs) are one of the most prominent immune cells in the breast tumor microenvironment (TME). TAMs are categorised into classically activated anti-tumorigenic M1 and alternatively activated pro-tumorigenic M2 macrophages. TAMs are known to promote cancer pathogenesis by facilitating cancer cell and cancer stem cell growth, angiogenesis, immune evasion, invasion, and migration. Consequently, TAMs drive cancer progression towards metastasis. This chapter describes the role of TME in driving monocyte recruitment and polarization toward the M2 phenotype. We also illustrate the modalities of intercellular networking such as paracrine signaling, exosomes, and tunneling nanotubes (TNTs) that TAMs and cancer cells employ within TME to communicate with each other and with other cells of TME to facilitate the dynamic process of cancer progression. Finally, we discuss the clinical implications of TAMs in breast cancer and potential therapeutic strategies targeting TAM recruitment, polarization, and TAM-mediated immune evasion for effective cancer therapy.

Abbreviations

Ang	Angiopoietins
Arg-1	Arginase-1
CCL	Chemokine (C-C motif) ligands
CCR (n)	C-C motif chemokine receptor (n)
CD	Cluster of Differentiation

P. K. Melwani (✉) · R. Checker · M. M. S. Balla · B. N. Pandey
Radiation Biology & Health Sciences Division, Bhabha Atomic Research Centre, Mumbai, India

Homi Bhabha National Institute, Anushaktinagar, Mumbai, India
e-mail: pathejap@barc.gov.in

M. Kloc et al. (eds.), *Monocytes and Macrophages in Development, Regeneration, and Disease*, Results and Problems in Cell Differentiation 74,
https://doi.org/10.1007/978-3-031-65944-7_8

CHI3L1	Chitinase 3-like protein 1
CSCs	Cancer stem cells
CSF-1	Colony-stimulating factor-1
CTCs	Circulating tumor cells
CTLA4	Cytotoxic T-lymphocyte associated protein 4
CTLs	Cytotoxic T-lymphocytes
CXCL-(n)	Chemokine (C-X-C motif) ligand –(n)
CXCR (n)	CXC chemokine receptor-(n)
DC	Dendritic cells
ECM	Extra cellular matrix
EGF	Epidermal growth factor
EGFR	Epidermal growth factor receptor
EMT	Epithelial to mesenchymal transition
EVs	Extracellular vesicles
FGF (n)	Fibroblast growth factor
GM-CSF	Granulocyte-macrophage colony-stimulating factor
HGF	Hepatocyte growth factor
HLA	Human leukocyte antigen
IFN	Interferon
IKK	I kappaB kinase
IL-(n)	Interleukin
IL-(n)R	Interleukin (n) receptor
iNOS	Inducible nitric oxide synthase
IRF5	Interferon regulatory factor 5
JMJD3	Jumonji domain-containing protein 3
LIT-2	Leukocyte immunoglobulin-like receptor
LOX	Lysyl oxidase
MAPK	Mitogen Activated Protein Kinase
MCP-(n)	Monocyte chemoattractant protein (n)
MDSCs	Myeloid derived suppressor cells
miRNAs	microRNAs (n)
MMP (n)	Matrix metalloproteases (n)
MR	Mannose receptor
mRNA	Messenger ribonucleic acid
NF-κB	Nuclear factor kappa-light-chain-enhancer of activated B cells
NK	Natural killer cells
PDL1	Programmed death-ligand 1
PGE2	Prostaglandin E2
PI3K	Phosphoinositide 3-kinase
PMNs	Pre-metastatic niches
PRMT1	Protein arginine methyltransferase 1
RNA	Ribo nucleic acid
ROS	Reactive oxygen species
siRNA	Small interfering RNA
SMYD3	SET and MYND domain-containing protein 3

SR	Scavenger receptor
TAMs	Tumor associated macrophages
TCR	T-cell receptor
TGF	Transforming growth factor
Tie-2	Tyrosine-protein kinase receptor TEK
TIMP	Tissue inhibitor of metallopeptidase
TME	Tumor microenvironment
TNF-α/β	Tumor necrosis factor-alpha/beta
TNT	Tunneling nanotubes
TP	Thymidine phosphorylase
Tregs	Regulatory T cells
TSP-1	Thrombospondin-1
uPA	Urokinase-type plasminogen activator
VCAM-1	Vascular cell adhesion molecule-1
VEGF	Vascular endothelial growth factor
VEGFR	Vascular endothelial growth factor receptor
(n)	homolog or variant number

8.1 Introduction

Breast cancer is the most common cancer globally in terms of both incidence and prevalence as of data collected in 2020 (Ferlay et al. 2021). The estimated number of deaths due to breast cancer is highest in women (15.5%) globally, making it the fifth leading cause of death due to cancer worldwide (Ferlay et al. 2021). One of the major goals of cancer research is to decipher the mechanism of metastasis which is the primary cause of death due to cancer (Geiger and Peeper 2009). During the process of malignant transformation, cancer cells acquire distinctive capabilities referred to as "hallmarks of cancer" described by Douglas Hanahan and Robert Weinberg in their two classic reviews (Hanahan and Weinberg 2000, 2011). These capabilities comprise "sustaining proliferative signalling, evading growth suppressors, resisting cell death, enabling replicative immortality, inducing angiogenesis, deregulating energy metabolism, avoiding immune destruction, and activating invasion and metastasis" (Hanahan and Weinberg 2000, 2011). Together all these hallmarks enable cancer cells to survive, grow, proliferate, and disseminate to distant organs. Acquisition of these capabilities is driven by the underlying genetic diversity caused by genomic instability in the cancer cells (Hanahan and Weinberg 2000) and by tumor-promoting inflammation (Hanahan and Weinberg 2011).

Owing to chronic inflammation in tumor lesions and its similarity to the stroma of wounds, tumors have been described as "wounds that never heal" (Dvorak 1986; Schäfer and Werner 2008). Tumor-associated inflammation has rather a paradoxical effect on tumorigenesis and malignancy. Cancer lesions consist of a heterogenous population of cancer cells and cancer stem cells (CSCs) and are often infiltrated by cells of innate and adaptive arms of the immune system such as lymphocytes, macrophages, dendritic cells (Mantovani et al. 2008), stromal cells such as fibroblasts, myofibroblasts, endothelial cells, pericytes, adipocytes, and stellate cells, and non-cellular components such as the extracellular matrix (ECM) and extracellular vesicles (Hanahan and Weinberg 2011). Immune cells were initially thought to infiltrate the tumors as an attempt to recognise and eradicate tumor cells (Dunn et al. 2004), a phenomenon known as 'immune surveillance'. However, during this process some cancer cells appear to avoid detection by immune cells by a phenomenon referred to as 'immunoediting' (Dunn et al. 2004) and go on to develop into solid tumors and eventually metastasise, a phenomenon referred to as 'immune evasion'. Cancer cells can evade the immune system by selectively disabling its components such as suppression of cytotoxic T-lymphocyte associated protein 4 (CTLA4) and programmed death-ligand 1 (PDL1) on cytotoxic T-lymphocytes (CTLs), secretion of transforming growth factor-β (TGF-β) to inactivate CTLs and natural killer (NK) cells, and recruitment of myeloid-derived suppressor cells (MDSCs), macrophages, and regulatory T cells (T_{regs}) which suppress functions of CTLs (Yang et al. 2010; Shields et al. 2010; Mougiakakos et al. 2010; Ostrand-Rosenberg and Sinha 2009). Immune cells are also directed by cancer cells to support the process of carcinogenesis. Components of the immune system provide growth stimulatory signals, survival factors, pro-angiogenic factors, and ECM modifying enzymes to facilitate invasion and metastasis, and signals that lead to epithelial to mesenchymal transition (EMT) in cancer cells (Mantovani et al. 2008). Immune cells, through secreted cytokines and chemokines, also augment cancer-promoting inflammation. Inflammation also leads to the generation of highly mutagenic reactive oxygen species (ROS) that can lead to genetic instability in nearby cancer cells (Ziech et al. 2011; Liou and Storz 2010).

8.2 Tumor Associated Macrophages (TAMs)

Macrophages, one of most versatile cells of innate immune system, offer first line of defence against pathogens. Macrophages execute as antigen presenting cells and phagocytic cells and secrete inflammatory cytokines and chemokines. Macrophages also play an important role in wound healing and tissue homeostasis (Zhou et al. 2020). Macrophages infiltrating and populating the solid tumors are called tumor associated macrophages (TAMs) (Zhou et al. 2020). Tumor infiltrating macrophages can make up to 50% of tumor mass in breast cancer (Hambardzumyan et al. 2016; Condeelis and Pollard 2006). Such high density of macrophage infiltration in tumor microenvironment (TME) is linked to unfavourable prognosis in breast (Mahmoud et al. 2012; Zhao et al. 2017), lung (Wu et al. 2016; Hwang et al. 2020), and gastric cancer (Rihawi et al. 2021; Zhang et al. 2016; Yin et al. 2017) and with contrasting outcomes in colorectal cancer (Edin et al. 2013; Li et al. 2020a). Clinical studies and experimental xenograft models suggest that TAMs play a pro-tumoral role in several ways. Being a crucial component of TME, TAMs support several processes of carcinogenesis including tumor growth, angiogenesis, tumor cell invasion, migration, immune regulation, and chemo-resistance (Condeelis and Segall 2003; Pan et al. 2020).

8.2.1 Origin of TAMs

For a long time, it was believed that circulating monocytes derived from hematopoietic stem cells in the bone marrow are a major source of tissue macrophages, however, recent studies suggest that bulk of the tissue resident macrophages such as alveolar cells, microglial cells, and Kupffer cells are derived from tissue-specific yolk sac progenitor cells (Wynn et al. 2013). While monocytes recruited from the periphery contribute to the bulk of TAMs, embryonic-derived resident macrophages also contribute as a source of TAMs in certain tumors. However, the role of embryonic-derived TAMs in tumor growth remains to be elucidated (Mantovani et al. 2017). Circulating immune suppressive MDSCs also contribute to TAM populations. In mice, MDSCs expressing surface markers $Ly6C^{+}/Ly6G^{-}$ are classified as monocyte-related M-MDSCs while MDSCs expressing $Ly6C^{-}/Ly6G^{+}$ are classified as granulocyte-related G-MDSCs. It is shown that the M-MDSCs that are recruited and differentiated into TAMs in TME (Bronte et al. 2016; Kumar et al. 2016).

Circulating monocytes are recruited to the tumor site in response to hypoxia, necrosis, growth factors such as colony-stimulating factor-1 (CSF-1), vascular endothelial growth factor-A (VEGF-A), and platelet- derived growth factor (PDGF), cytokines such as interleukin-6 (IL-6), and chemokines such as CCL2, CXCR4, and CXCL12 secreted by tumor and other cell types in TME (Aras and Zaidi 2017). After infiltrating, monocytes differentiate into macrophages and polarize into different phenotypes based on the molecular milieu in TME.

8.2.2 Macrophage Polarization

Macrophages display a high level of plasticity and depending upon the extracellular milieu can be found in different, even opposing phenotypes such as pro-inflammatory classically activated M1 macrophages and anti-inflammatory alternatively activated M2 macrophages (Fig. 8.1). Cytokines such as interferon-γ(IFN-γ) and bacterial products such as LPS polarise macrophages towards M1 phenotype, while cytokines such as CSF-1, IL-10, IL-4, IL-13 polarise macrophages towards M2 phenotype (Zhou et al. 2020; Aras and Zaidi 2017). M1 macrophages participate in phagocytosis, and kill pathogens and tumor cells, while M2 macrophages play a crucial role in tissue homeostasis and wound healing, and promote tumorigenesis (Condeelis and Pollard 2006; Pan et al. 2020; Aras and Zaidi 2017). The two types of macrophage populations express distinct surface and molecular markers. M1 macrophages secrete pro-inflammatory cytokines such as tumor necrosis factor-α (TNF-α), IL-12, IFN-γ, and a high amount of nitric oxide synthase (NOS). M2 macrophages secrete anti-inflammatory cytokines such as IL-4, IL-13, and IL-10, and express high amounts of arginase-1 (Arg-1), mannose receptor (CD206, MR), and scavenger receptor (SR) (Condeelis and Pollard 2006; Pan et al. 2020; Aras and Zaidi 2017).

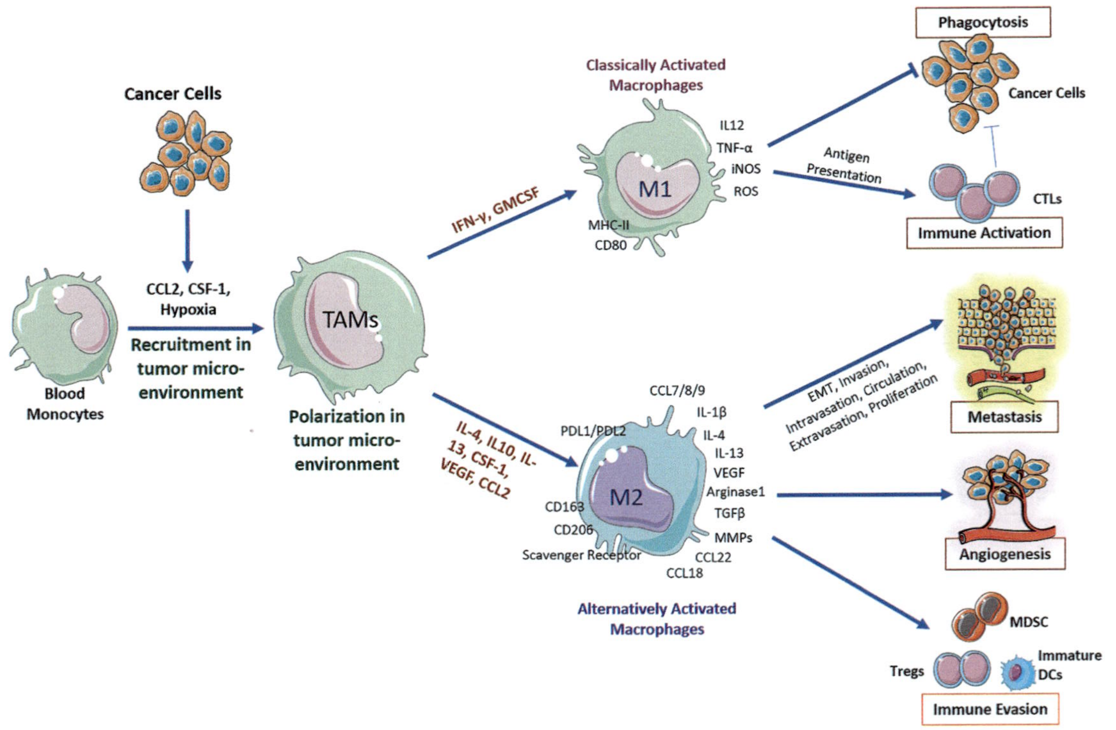

Fig. 8.1 Recruitment and polarization of tumor associated macrophages (TAMs) in tumor microenvironment. Macrophages are classified into two main sub-populations: classically activated macrophages (M1) and alternatively activated macrophages (M2). Macrophages are polarised into the M1 phenotype by IFN-γ, GM-CSF, and other stimulations, and into the M2 phenotype by IL-4, IL-13, IL-10, and other stimulation. M1 macrophages play an important role in immune function and phagocytosis and M2 macrophages participate in metastasis, angiogenesis, and immune cell evasion. The scheme was partly generated using Servier Medical Art (https://smart.servier.com) by Servier, licensed under a Creative Commons Attribution 3.0 unported license

8.2.3 Mechanisms Driving Macrophage Polarization in Breast Cancer

M1 macrophages can be polarised to M2 phenotype and vice versa depending upon the extracellular factors (Lin et al. 2019). Cytokines such as IL-4 and IL-10 in TME promote tumor-supportive M2 phenotype. Figure 8.1 shows the molecular signals in TME that induce polarization of macrophages into M1 and M2 TAMs. CCL2 and CSF-1 are the most common monocyte recruiters and M2 polarizing factors in breast TME. CCL2 signals M2 polarization through the CCR2 receptor expressed on the macrophage surface. Interestingly, overexpression of CCL2 and CSF-1 is associated with high TAM infiltration and metastasis and linked to poor clinical outcomes in breast cancer patients (Qian et al. 2011; Mizutani et al. 2009; Qian and Pollard 2010). In line with this, depletion of CSF-1 and CCL2/CCR2 interaction reduced TAM density in xenograft mouse models and delayed metastasis (Gazzaniga et al. 2007; Sierra-Filardi et al. 2014; Abraham et al. 2010; Hume and MacDonald 2012). VEGF-A also promotes M2 polarization in presence of IL-4 and IL-10 (Linde et al. 2012). Other factors that promote M2 polarization in TME include hypoxia, epidermal growth factor receptor (EGFR), prostaglandin E2 (PGE2), CCN3 (also known as NOV, nephroblastoma overexpressed), IL-13, IL-6, CCL7/8/9/18, and CXCL12. In addition to molecular signals, physiological signals of TME such as hypoxia also support monocyte recruitment in TME. Under hypoxic conditions, lactic acid is produced due to aerobic glycolysis. This lactic acid is sensed by G-protein coupled receptor 132 (Gpr132) and subsequent downstream signalling modulates polarization- associated genes in macrophages. Grp132 overexpression is also associated with poor outcomes in breast cancer patients (Lin et al. 2019). Furthermore, epigenetic reprogramming by enzymes such as protein arginine methyltransferase 1 (PRMT1), SET, and MYND domain-containing protein 3 (SMYD3), Jumonji domain-containing protein 3 (JMJD3), and microRNAs (miRNAs) such as miR-125, miR-155, miR-378, miR-9, miR-21, miR-146, miR-147, miR-187, miR-222, and miRlet7b have also been reported as modulators of M2 TAM phenotype (Lin et al. 2019) in the TME.

Though both M1 and M2 TAMs are found within TME, several intermediate states of macrophage activation can be found, and it is therefore suggested that TAMs should be classified based on their function in tumor progression instead of M1/M2 phenotype (Lin et al. 2019). Macrophages express CD14/CD68/CD204/HLA-DR, while TAMs are high in MMP2/9/B7-H4/STAT-3/CD206 (Lin et al. 2019). Nevertheless, M2 macrophages are often characterized as TAMs in TME due to their role in promoting cancer progression. For instance, like tumor cells, M2 TAMs also undergo metabolic reprogramming by activating glycolysis, fatty acid synthesis, modulation of TCA, and nitrogen cycle (Netea-Maier et al. 2018). In addition to this, tumor educated M2 TAMs acquire low MHC class II expression, mannose receptor, and stabilin, and do not function as antigen-presenting cells. TAMs in M2 state secrete cytokines, chemokines, growth factors, and enzymes such as IL10, TGF-β, hepatocyte growth factor (HGF), EGF, fibroblast growth factor (bFGF), VEGF, matrix metalloproteases (MMPs), arginase, adrenomedullin, urokinase-type plasminogen activator (uPA), thymidine phosphorylase (TP), PGE2, semaphoring 4D (Sema 4D), CCL2, CCL5, CCL3, CCL8, CCL22, etc (Pollard 2004; Mantovani et al. 2002; Riabov et al. 2014). These secreted molecules act on cancer cells, stromal cells, and other immune cells within TME in a paracrine manner and promote cancer progression. The major modalities of intercellular communication between macrophages and breast cancer cells are discussed in the next section.

8.3 Communication Pathways Between Macrophages and Breast Cancer Cells

Intercellular communication is crucial in cancer cell growth, proliferation, invasion, and metastasis. One of two enabling characteristics of hallmarks of cancer is inflammation and interaction between inflammatory cells and cancer cells relies on different modes of intercellular communication. Besides, the acquisition of all the hallmarks of cancer involves intercellular communication between cells in TME at one or the other stage. Studies have also shown the important role of intercellular communication in chemo- and radio-resistance in cancer cells.

Cancer cells within tumors are traditionally known to communicate with each other or with other cells in TME through the secretion of signaling molecules or by direct cell-to-cell contact such as gap junctions or cellular synapses. Another modality of intercellular communication between cancer cells includes communication through the secretion of membrane-bound small and large extracellular vesicles (EVs) such as exosomes, microvesicles, apoptotic bodies, oncosomes, and viable cytoplasmic fragments such as cytoplasts/microplasts (Chuo et al. 2018; Vader et al. 2014; Han et al. 2019; Bertolini et al. 2019; Di Vizio et al. 2009, 2012; Minciacchi et al. 2015; Yount et al. 2007; Headley et al. 2016). During the last decade, intercellular communication through the formation of intercellular bridges widely referred to as tunneling nanotubes (TNTs) (Roehlecke and Schmidt 2020; Pinto et al. 2020) in cancer progression has also gained attention. This section elaborates on these modes of communication between breast cancer cells and macrophages in the TME.

8.3.1 Secretion of Diffusible Molecules: Ligand Receptor Interaction

TAMs and breast cancer cells in TME communicate with each other through the secretion of diffusible soluble molecules. These secreted molecules are referred to as ligands. Ligands are synthesized and secreted by signaling cells and produce a specific response only in target cells that have receptors for the signaling molecules. The soluble factors include proteins such as growth factors, cytokines, chemokines; small molecules; and small peptides etc. Ligands after binding with the receptors on the target cells initiate a cascade of signaling pathways in the target cells which leads to the generation of an array of responses in the target cells. The responses are often manifested in terms of differential expression of genes, proteins, phenotype, functions of the target cells, alteration in metabolism and migration, among many others. Under the influence of the signaling cascade the target cells can also secrete soluble factors which can alter the response of other target cells. Such signalling is more relevant between macrophages and cancer cells where several positive or negative feedback signalling loops modulate phenotypic changes in macrophages as well as cancer cells.

Signalling through ligand-receptor interaction is involved in all the avenues of macrophage-cancer cell interaction (Fig. 8.1). For instance, monocyte recruitment relies on interactions between surface receptors of circulating monocytes such as $CCR2^+$ $Ly6C^{hi}$ $CX3CR1^{low}$ (Movahedi et al. 2010) and breast cancer cells derived chemoattractants such as CSF1, CCL2, VEGF-A etc. driving the accumulation of macrophages in tumors (Okabe and Medzhitov 2014; Strachan et al. 2013). CSF1R inhibition delays tumor growth in murine models of breast and cervical cancer by diminishing the turnover of TAMs (Strachan et al. 2013). After recruitment, monocytes are differentiated into TAMs and a subset of TAMs undergo self-renewal in CSF1-dependent autocrine and paracrine signaling. TAMs secrete EGF-1 which binds to EGF receptors on breast cancer cells, inducing breast cancer cells to secrete more CSF-1. The positive feedback loop between TAMs and breast cancer cells is essential for breast cancer progression and dissemination (Wyckoff et al. 2004). TAMs also secrete several cytokines such as IL-1, IL-8, IL-10, TNF-α, TGF-β, growth factors such as VEGF-A and FGF, and chemokines such as CCL22 and CCL20. These secreted molecules promote angiogenesis, EMT, and CSC phenotype and suppress immune response within the TME (Dallavalasa et al. 2021; Fu et al. 2020). The role of this signalling has been discussed in great detail in the forthcoming sections.

8.3.2 Extracellular Vesicles-Mediated Communication Between Breast Cancer Cells and Macrophages

Although extracellular vesicles (EVs) are released by almost all cell types, these are emerging as one of the important mediators of intercellular communication in cancer progression (Chuo et al. 2018; Vader et al. 2014; Han et al. 2019; György et al. 2011). The most common EVs studied in TAM-breast cancer interaction are exosomes. Exosomes (diameter: 30–150 nm), nanosized endosomal-derived membrane vesicles, are produced by most eukaryotic cells and are released into the extracellular environment. Exosomes carry bioactive materials including proteins, lipids, and nucleic acids between cells and across biological barriers and can effectively communicate with nearby cells. They can transfer biological information from secretory cells to recipient cells through the circulatory system, thereby influencing the progression of various human diseases, including cancer (Li et al. 2023). Recent studies suggest the role of TAM-derived exosomes in cancer cell proliferation, metastasis, angiogenesis, drug resistance, and metabolic reprogramming. This makes TAM-derived exosomes an ideal candidate which can be targeted for tumor therapy.

Initial studies using TAM-derived exosomes revealed that TAMs control the invasiveness of breast cancer cells through exosome-mediated delivery of oncogenic miRNAs (Yang et al. 2011). This study showed that miR-223 (a miRNA specific for IL-4-activated macrophages), was present in the TAMs-derived exosomes and was significantly elevated in the co-cultivated SKBR3 and MDA-MB-231 cells. Further, treatment with miR-223 antisense oligonucleotide led to a decrease in the invasiveness of co-cultivated breast cancer cells. Lee H D et al., reported that ADAM15 (A disintegrin and metalloproteinase 15) that regulates tumor progression and suppression is released as an exosomal component and exerts tumor suppressive effects in breast cancer cells. They observed that exosomes rich in ADAM15 display an enhanced binding affinity for integrin αvβ3 in an RGD-dependent manner. Further, these exosomes inhibit cell adhesion, growth, migration, and tumor growth *in vivo* (Lee et al. 2012). Xiuyan Yu X et al. studied the role of macrophage-derived exosomes in cancer recurrence and metastasis after chemotherapy. They observed that exposure of macrophages to breast cancer cells that have experienced chemotherapy increased the proliferation and metastasis of MCF-7 or MDA-MB-231 cells. These exosomes had more IL-6 which activated the STAT3 pathway and induced the transcription of STAT3-dependent genes such as CyclinD1, MMP2, and MMP9 (Yu et al. 2019). A recent study by Yue S., explored the effects of exosomes derived from Progranulin null (PGRN−/−) TAMs on invasion and migration of breast cancer cells. They reported a role of TAMs-derived exosomes in the observed inhibition of invasion, migration, and EMT of breast cancer cells in PGRN−/− mice. They also showed that PGRN−/− TAM-derived exosomes have higher levels of miR-5100 which modulated the expression of CXCL12 leading to suppression of invasion, migration, and EMT of breast cancer cells via inhibition of CXCL12/CXCR4 axis (Yue et al. 2021).

These studies clearly indicate that TAM-derived exosomes are an important component of the bidirectional communication between TAMs and cancer cells in the TME. These TAM-derived exosomes carry diverse bioactive constituents and can influence multiple aspects of tumor growth and progression such as malignant proliferation, metastasis, angiogenesis, immune evasion, metabolic reprogramming, and therapy resistance (Su et al. 2021). The important role of TAM-derived exosomes in tumor growth makes them attractive therapeutic targets and provides new avenues for the treatment of malignant tumors.

8.3.3 Macrophages Modulate Tunneling Nanotube Mediated Communication in Breast Cancer Cells

As described earlier, cross-talk between macrophages and tumor cells is mediated through the secretion of cytokines, chemokines, growth factors, and exosomes. However, cancer cells can also communicate with each other or with other cells in TME through the formation of long-range intercellular conduits referred to as tunneling nanotubes (TNTs) (Rustom et al. 2004). TNTs are membrane-bound intercellular bridges ranging in diameter from 50 nm to 700 nm which can mediate communication between cells that are few to few hundred microns apart (Pinto et al. 2020). TNTs more than 700 nm thick have also been reported in immune cells such as macrophages (Onfelt et al. 2006). A wide range of cargo including molecules such as proteins, miRNAs, and Ca2+ , and organelles such as mitochondria, lysosomes, and vesicles can be transported from one cell to another via TNTs (Pinto et al. 2020; Melwani and Pandey 2023). Intercellular communication through TNTs between cancer cells and other cells of TME has been shown to enhance cancer progression, metastasis, angiogenesis, as well as resistance to chemo- and radiotherapy (Melwani and Pandey 2023). Since macrophages are the most prominent immune cells in breast cancer TME, our lab for the first time investigated the effect of paracrine signaling between macrophages and breast cancer cells MCF-7 on the formation of TNTs between MCF-7 cells (Patheja and Sahu 2017). Our results showed that the conditioned media derived from macrophages significantly enhanced TNT formation in MCF-7 cells, leading to the formation of a physically connected network of cells. In this study, we also showed the release of large EVs referred to as microplasts from MCF-7 cells through the formation of TNTs and role of these microplasts in formation of complex cellular networks. While microplasts also contain protein and mRNA cargo similar to parent cells, their role in intercellular communication is not fully understood (Melwani et al. 2021). Later, Hanna et al., showed that EGF secreted by macrophages induces TNT formation between rat mammary adenocarcinoma MTLn3 cells in an M-Sec dependent manner and that the enhanced TNT formation between these cells is associated with increased migration in these cells (Hanna et al. 2019; Carter et al. 2019). The formation of heterotypic TNTs between rat RAW/LR5 macrophages and MTLn3 or human breast adenocarcinoma cell line MDA-MB-231 MTLn3 was also demonstrated (Hanna et al. 2019). The authors in this study showed that the TNT- based interaction between macrophages and breast cancer cells induced invasive phenotype in breast cancer cells in an EGF-EGFR-dependent manner. Chia-Wei Lee et al. also investigated the role of inflammatory stress induced by conditioned media derived from THP-1 cells in enhancing the TNT formation between pancreatic cells (Lee et al. 2022). Together, all these studies suggested that macrophages not only communicate with cancer cells and other cells of TME via the secretion of inflammatory molecules and exosomes but also modulate the formation of a physically connected network of cancer cells. The formation of such networks of cells has been shown to induce radiation resistance in glioblastoma cells (Osswald et al. 2015). Whether the formation of macrophage- mediated intercellular cellular networks in breast cancer cells aids in therapy resistance has not yet been investigated. Nevertheless, in addition to several other aspects of breast cancer pathogenesis mediated by macrophages as discussed in the next section, the novel role of macrophages in supporting breast cancer cell invasion via enhanced TNT formation further substantiates the need to develop macrophage-targeted therapy for effective management of the disease.

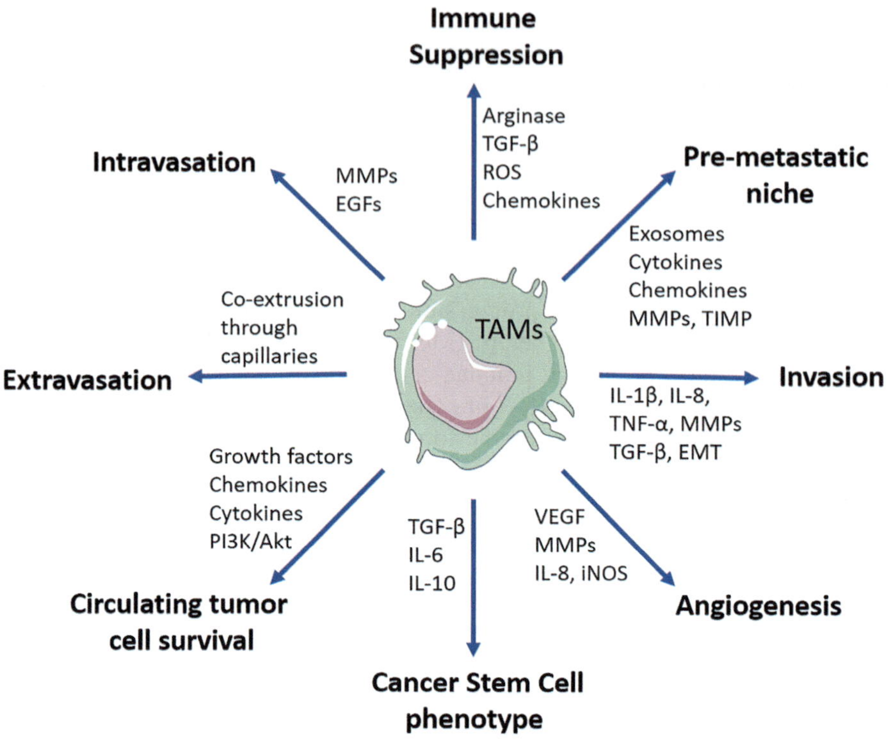

Fig. 8.2 Role of TAMs secreted cytokines, chemokines, growth factors, enzymes, and other signaling molecules in immune suppression, generation of cancer stem cells, and several steps of metastasis cascade. The scheme was partly generated using Servier Medical Art (https://smart.servier.com) by Servier, licensed under a Creative Commons Attribution 3.0 unported license

8.4 Functional Consequences of Macrophage-Breast Cancer Cell Crosstalk

Monocyte recruitment and their differentiation in TAMs/M2 phenotype are important characteristics of breast cancer TME. TAMs have a broad spectrum of actions from modulation of immune response to promoting tumor growth, angiogenesis, and metastasis (Figs. 8.1 and 8.2). The role of macrophages in breast cancer progression is discussed below.

8.4.1 Role of Macrophages in the Modulation of Immune Response

TAMs can modulate the cytotoxic role of CTLs and NK cells. Arg-1 secreted by M2 TAMs depletes L-arginine which leads to downregulation of T-cell receptor (TCR) ζ chain (Munder 2009). TAMs also downregulate CD3 ζ expression in T lymphocytes through oxidative stress rendering TCR ineffective (Otsuji et al. 1996). TAMs recruit $CCR4^+$ and $CCR6^+$ T_{regs} cells via secretion of CCL22 and CCL20, respectively (Noy and Pollard 2014). Soluble and membrane forms of HLA-C, -E, and -G secreted by TAMs can eliminate NKG2 and inhibitory leukocyte immunoglobulin-like receptor LIT-2 expressing NK cells leading to inhibition of IFN-γ secretion by these cells (Noy and Pollard 2014). In addition to this, TGF-β secreted by TAMs and tumor cells suppresses NK cells, CTLs, and migration of DC cells, and promotes differentiation of $CD4^+$ T cells into T_{regs} and Th_2 cells (Hao et al. 2012). M2 TAMs also upregulate PD-1 and PD-L1 expression which correlates with a decrease in phagocytosis by TAMs (Gordon et al. 2017).

8.4.2 *Role of TAMs in Neo-Vascularization*

Endothelial cells, the structural component of blood vessels, are crucial for neovascularization through angiogenesis (sprouting of new vessels from pre-existing ones). Angiogenesis is pivotal in sustaining the growth of tumor cells (Bergers and Benjamin 2003). During tumorigenesis, the otherwise quiescent vasculature is activated by turning on the "angiogenic switch". Regulation of proangiogenic proteins such as VEGF-A, PDGF, EGF, and FGF, and anti-angiogenic signals such as thrombospondin-1 (TSP-1), angiostatin, and endostatin in TME govern the process of angiogenesis by endothelial cells (Bergers and Benjamin 2003; Baeriswyl and Christofori 2009). TAMs are important players in regulating the angiogenic switch (Fu et al. 2020). TAMs induce angiogenesis by releasing growth factors, cytokines, and ECM remodelling enzymes such as EGF, VEGF-A, IL-1, IL-8, uPA, TNF-α, PGE2, adrenomedullin, Sema 4D, FGF, cyclooxygenase (COX)-2, nitric oxides (iNOS), MMP-9, MMP-7, and TP (Riabov et al. 2014; Fu et al. 2020; Lin et al. 2006). This is evident from the fact that in the absence of TAMs vessel density within tumors is reduced significantly (Lin et al. 2006). A significant positive correlation between vessel density, VEGF, and TAMs has been observed in mammary tumors (Valković et al. 2002). Similarly, expression of CCL2 (monocyte chemoattractant protein-1, MCP-1) is positively correlated with neovascularization, progression, and survival of breast cancer (Ueno et al. 2000; Saji et al. 2001). Apart from inducing angiogenesis, TAMs also induce remodelling of pre-existing vasculature to make them more aberrant and leaky (Lin et al. 2006). This drives tumor cell dissemination to distant sites. Also, the hypoxic TME not only recruits and polarizes macrophages into TAMs but also stimulates the expression of hypoxia-inducible factors (HIFs) in TAMs (Fang et al. 2009) and promotes angiogenesis in breast cancer (Giatromanolaki et al. 2006). In addition, a subset of TAM expressing angiopoietin-1 receptor Tie-2 affects the pro-angiogenic process since Tie-2 binds to all angiopoietins (Ang1, 2, 3, and 4) (Murdoch et al. 2007). Selective elimination of this subset of TAMs has been suggested as an effective targeted therapy approach (Venneri et al. 2007). TAMs also drive lymphangiogenesis, through which tumor cells escape to local lymph nodes and eventually distant organs. The process is driven through the VEGF-C/VEGFR-3 axis by either directly modulating lymphatic endothelial cells (LECs) or indirectly elevating the cathepsin-dependent heparanase secretion (Riabov et al. 2014; Cao et al. 2015). Targeting VEGF-C/VEGFR-3 signalling or genetic ablation of heparanase also resulted in altered lymphatic vessels and reduced tumor growth and metastasis (Riabov et al. 2014).

8.4.3 Role of TAMs in Breast Cancer Metastasis

Metastasis, one of the hallmarks of cancer, is a multi-step process involving tumor cells escaping from primary tumors into circulation and eventually disseminating to a distant site. The metastatic cascade involves invasion of cancer cells in the nearby tissue, intravasation in blood/lymph vessels, survival in circulation, extravasation out of blood vessels, and seeding, survival, and growth in distant sites. TAMs participate in all these steps of metastasis and promote cancer progression towards malignancy (Fu et al. 2020).

8.4.3.1 Role of TAMs in Invasion

Invasion involves the migration of cancer cells through the basement membrane into the stroma of surrounding tissue. Processes such as EMT and amoebiasis enable cancer cells to migrate through the ECM. EMT is characterised by loss of cell-to-cell junctions, cell polarity, and acquisition of motile mesenchymal morphology. EMT includes downregulation of epithelial markers such as E-cadherin and up-regulation of mesenchymal markers such as vimentin, N-cadherin, Snail etc. in tumor cells (Aras and Zaidi 2017). A large body of literature suggests an important role of TAM in the EMT process (Su et al. 2014; Zhang et al. 2015; Fu et al. 2015; Ravi et al. 2016). TAMs promote EMT and the CSC properties in triple-negative breast cancer through CCL2/Akt/β-catenin pathway (Chen et al. 2022). TAMs also mediate ECM remodelling, by secreting ECM-degrading enzymes such as cathepsins, MMP2, MMP7, MMP9, uPA, and serine proteases in breast and lung cancer (Kessenbrock et al. 2010; Gocheva et al. 2010; Vasiljeva et al. 2006). M2 macrophages have been shown to promote breast and gastric cancer cell invasion by secreting chitinase 3-like protein 1 (CHI3L1) which upregulates MMP expression via the MAPK pathway (Chen et al. 2017). After the acquisition of EMT phenotype and ECM degradation, the tumor cells move along the ECM fibres such as fibronectin and vitronectin (Condeelis and Segall 2003; Sangaletti et al. 2008). Moreover, TGF-β, IL-6, IL-10, and TNF-α derived from TAMs also enhance the CSC population by promoting EMT (Allavena et al. 2021). The formation of TNTs between macrophages and MTLn3 breast cancer cells has also been shown to enhance invasion in breast cancer cells (Hanna et al. 2019). Although, research on the role of TNTs in migration, invasion, and EMT is still in its infancy, and many such investigations are needed to validate the role of TNTs in this process.

8.4.3.2 Role of Macrophages in Intravasation

Tumor cells move through the small gaps between endothelial cells to come into circulation, a process known as intravasation. Direct visualization of macrophages accompanying tumor cells during the process using multi-photon microscopy suggests a role of TAMs in intravasation in mammary cancer (Wyckoff et al. 2004). In addition to this, a tripartite arrangement of TAMs, endothelial cells, and tumor cells, referred to as tumor microenvironment of metastasis (TMEM), has been observed in clinical samples and is linked to metastasis and poor prognosis in breast cancer patients (Robinson et al. 2009). Macrophages also facilitate intravasation by secreting ECM- degrading enzymes such as cathepsins, MMPs, and serine proteases around endothelium (Kessenbrock et al. 2010; Gocheva et al. 2010; Vasiljeva et al. 2006). Further, TAMs secrete EGF which functions in the positive feedback loop with breast cancer cell secreted CSF-1 to drive tumor cell migration towards endothelium (Wyckoff et al. 2004).

8.4.3.3 Role of Macrophages in the Survival of CTCs

Tumor cells entering circulation following intravasation are called circulating tumor cells (CTCs). CTCs must be primed for survival in circulation. CTCs can either exist as single cancer cells or in clusters of cancer cells (Amintas et al. 2020). CTCs associate with platelets and form clots, a strategy to escape targeting from NK cells in circulation (Palumbo et al. 2007; Nierodzik and Karpatkin 2006). Despite clot formation, genetic disruption of macrophages diminished CTC survival in lung capillaries and their metastatic seeding, suggesting the importance of macrophages in the process (Gil-Bernabé et al. 2012). In breast cancer cells, macrophages are shown to upregulate the PI3K/Akt survival pathway in vascular cell adhesion molecule-1 (VCAM-1) dependent manner via α4 integrins (Chen et al. 2011a; Lu et al. 2011), thereby promoting their survival in circulation. Alternatively, cytokines and chemokines secreted by macrophages promote the survival of CTCs (Gil-Bernabé et al. 2012).

8.4.3.4 Role of Macrophages in Extravasation

CTCs that manged to survive in circulation and settle in the capillaries of the target organs, attempt to attach, and extrude through the capillary walls. The close associations between macrophages and breast cancer cells in animal models of breast cancer metastasis during the process of extravasation were visualized and analysed quantitatively within the lungs (Qian et al. 2009). Genetic and chemical ablation of macrophages dramatically reduced the rate of metastatic seeding and subsequent growth (Qian et al. 2009).

8.4.3.5 Role of Macrophages in Metastatic Seeding: Pre-metastatic Niche

Metastasis is believed to be an event that can occur even in the early stages of tumor progression (Kaplan et al. 2005). The primary tumors are believed to "prime" the secondary sites and drive organ-specific dissemination long before the seeding of tumor cells. These primed sites harbour conditions for metastatic seeding and constitute pre-metastatic niches (PMNs) (Kaplan et al. 2005). Macrophages are thought to be one of the key immune cells involved in the formation of PMNs. Macrophages are mobilised and clustered in the pre-metastatic sites by a variety of tumor-secreted factors, such as CCL2, CSF-1, VEGF, PLGF, TNF-α, TGF-β, tissue inhibitor of metallopeptidase (TIMP)-1, and exosomes (Kaplan et al. 2005; Sceneay et al. 2013). Additionally, the tissue-resident macrophages are also involved in orchestrating PMN formation upon stimulation (Kaplan et al. 2006; Lu and Kang 2007). The presence of macrophages at these sites provides favourable conditions for homing of CTCs into the PMNs. Homing of CTCs is supported by enhanced expression of chemokines and remodelling of the ECM by secreting enzymes like MMPs and lysyl oxidase (LOX) (Sceneay et al. 2013; Kaplan et al. 2006; Lu and Kang 2007; Müller et al. 2001). Besides, macrophages also suppress the immune response mounted by Th1 cells and DCs at these sites ultimately promoting the growth of newly lodged tumor cells in the immune-suppressive environment (Sceneay et al. 2013; Morrissey et al. 2021).

8.4.4 Clinical Implications of Breast Cancer Cells and TAM Interactions

As described previously, TAMs play a pivotal role in the TME of breast cancer and their abundance correlates with poor prognosis and aggressive disease progression due to their role in promoting tumor growth, angiogenesis, and metastasis. Several clinical studies have demonstrated a link between the prognosis of breast cancer and the quantity of macrophages within the tumor mass. One of the earliest studies on the correlation between macrophage infiltration and the prognosis of invasive breast carcinoma was conducted by Leek R D et al., wherein they reported a strong relationship between increased macrophage counts, reduced relapse-free survival, and reduced overall survival (Leek et al. 1996). Lee AH et al., reported that diffuse inflammation, particularly macrophages, was associated with high tumour grade, tumour necrosis, large tumour size, and c-erbB-2 expression (Lee et al. 1997). Volodoko N et al., studied macrophage infiltration in 34 node-negative and 86 node-positive primary breast cancers and observed that macrophage infiltration was strongly associated with poor prognosis, high tumour grade, and high mitotic rate (Volodko et al. 1998). Goede V et al., reported that invasive ductal tumours with poor prognosis are associated with higher TAM counts compared to those with good prognosis (Goede et al. 1999). Another study suggested that the prognostic implications of TAM infiltration are due to the involvement of TAMs in tumor angiogenesis. They showed that TAM density had a significant correlation with VEGF expression and that patients with a high TAM density had a significantly worse disease-free survival prognosis than those with a low TAM density (Tsutsui et al. 2005). Recently, Allison E et al., performed a meta-analysis of markers for TAMs and breast cancer survival outcomes. They suggested that initial studies employed CD68 (a pan-macrophage marker CD68), whereas, an increasing number of recent studies have used CD163 (a marker expressed by alternatively activated M2 macrophages and TAM subsets) as a TAMs marker (Allison et al. 2023). Their meta-analysis included 32 studies with 8446 patients and was based on hazard ratios for survival outcomes of breast cancer patients with a high density of TAMs (CD68+ and/or CD163+) compared to a low density of TAMs. They observed that a high density of TAMs was associated with decreased overall survival and reduced progression-free survival. Further, they reported that subgrouping by CD marker type showed a lower overall survival for a high density of CD163+ TAMs compared to a high density of CD68+ TAMs. They also observed lower overall survival associated with a high density of CD163+ and/or CD68+ TAMs in TNBC compared to non-TNBC cases.

These studies clearly suggest that targeting TAMs in breast cancer therapy holds promise for improving patient outcomes by disrupting the tumor-promoting environment and enhancing immune responses against the tumor. Given these findings, there is increasing interest in defining new therapies aimed at reducing the persistent activation of macrophages and their release of proinflammatory and protumor mediators. Yondelis (Trabectedin), a novel antitumor agent of marine origin, was found to suppress functions of TAMs by inhibiting their ability to secrete major monocyte chemoattractants IL-6 and CCL2 (Torroella-Kouri et al. 2005). In another study, Nie X et al., studied the effect of Garcinone E (GE) on TAMs as a therapeutic target for the treatment of metastatic breast cancer. They showed that GE did not affect polarization to the M1-like macrophages but inhibited the expression of the M2 macrophage-specific genes in RAW 264.7 cells. GE also decreased M2 macrophage-induced breast cancer cell proliferation, migration, and angiogenesis. Further, GE suppressed tumor growth and lung metastasis in tumor-bearing mice and decreased the proportion of M2-like TAMs whilst simultaneously increasing M1-like TAMs via modulation of the STAT6 signaling pathway (Nie et al. 2023). Overall, the studies indicate that a high density of TAMs is associated with poor survival outcomes in breast cancer. Moreover, a high density of CD163+ TAMs is a better predictor of poor outcomes, suggesting that TAMs expressing a similar phenotype to alternatively activated macrophages are responsible for tumor progression and metastasis. Therefore, a better understanding of the clinical significance of TAMs may lead to the development of novel therapeutic strategies to effectively combat breast cancer.

8.5 Therapeutic Targeting of Macrophage-Breast Cancer Cell Interactions

A large body of literature demonstrates the pivotal role of macrophages in breast cancer progression and therefore suggests targeting TAMs for therapeutic interventions. In this regard, several studies have employed different strategies such as targeting macrophage polarization using drugs, inhibitors, or exosomes, interfering with macrophage recruitment to TME by blocking communication pathways, and blocking TAMs using immunotherapy approaches. These therapeutic strategies are described in the next section.

8.5.1 Targeting Macrophage Polarization

In mammary tumors, around 40% of CD45+ cells comprise macrophages suggesting their importance to the immunologic state of the tumor (Franklin et al. 2014). As described earlier, many tumor- secreted factors like IL-10, TGF-β, and IL-1R antagonist in TME participate in the polarization of macrophages leading to the acquisition of the M2 phenotype. Additionally, macrophages present in tumors down regulate NF-κB, STAT1, and IRF5 whereas up regulate IRF4, STST6, MYC, PPARγ, and KLF4 factors associated with tissue repair and remodelling (Murray et al. 2014; Krausgruber et al. 2011). Macrophage cytotoxicity was also reduced by downregulating NF-κB and CCAAT/enhancer binding protein (C/EBP) expression (Allavena et al. 2005). Furthermore, TAMs down regulate MHC class II mediated by migration inhibitory factor (MIF), preventing antigen presentation and adaptive immune induction. Alternatively, abnormal vasculature in tumor cells is perceived as one of the reasons that limits the delivery of drugs to tumor sites which in turn allows the cancer cells to escape for further disease progression. This abnormal microenvironment reduces the efficacy of cancer treatments and therefore, restoring the vascular function to normal status is one of the strategies that are emerging in cancer treatment. In this regard, Rolyny et al., showed that normalization of tumor vessels can be achieved by polarizing TAMs with M2-like phenotype to M1-like cells. TAMs with M2-like phenotype produce angiogenic factors like PIGF (placental growth factor), and cytokines such as IL-10 and CCL22 (Rolny et al. 2011). These cytokines also suppress immune functions of the cells. The authors demonstrated that elevated levels of histidine-rich glycoprotein which may binds to Fcγ receptors, partly down regulates PIGF and thereby polarize TAMs with M2-like phenotype to M1-like phenotype. This resulted in normalization of tumor vessels, decreased hypoxia, improved delivery of chemotherapeutic agents, and decreased metastasis. Exposure of TAMs to HRG also down regulated M2 markers such as MRC1, Arg1, IL-10, and CCL22 and increased expression of M1 markers such as IL-6, IFN-β, IL-12, and CXCL9. Furthermore, there was increased infiltration of CD8+ T cells, natural killer cells, and dendritic cells which resulted in the restoration of immune surveillance and reduced activity of the M2 phenotype. These studies suggest the important role of different phenotypes of macrophages in the TME and indicate that the strategy associated with macrophage plasticity may be exploited for cancer treatment. One such study employed nanoparticles to switch macrophage polarization from the M2 to the M1 state (Ramesh et al. 2019). The authors demonstrated that macrophage colony-stimulating factor (MCSF) and signal regulatory protein α (SIRPα) self-assembled dual inhibitor loaded nanoparticles (DNTs) target M2 macrophages and polarize them to M1 phenotype by inhibiting CSF1R and SHP2 pathways. This treatment also demonstrated enhanced anti-tumor activity on highly aggressive breast cancers (Ramesh et al. 2019). It was also demonstrated by Wanderley et al. that paclitaxel induced reprogramming of M2 macrophages to the M1 type in TLR4-dependent manner (Wanderley et al. 2018). Exosomes isolated from M1 macrophages when added to breast cancer cells, caspase-3 expression followed by increased production of pro-inflammatory cytokines was observed. Furthermore, it was demonstrated that treatment with nano-formulated paclitaxel containing exosomes exhibited enhanced anti-tumor effect in in vivo tumors (Wang et al. 2019). It was also demonstrated by Xuemei Jia et al., that in breast cancer cells emodin prevented macrophage polarization to M2 type by decreasing p-STAT6 levels and reduced C/EBPβ expression in macrophages (Jia et al. 2014). To target M2 macrophage-specific polarization, Hong Xiao et al., demonstrated that smart nanodrug containing IKKβ siRNA and STAT6 inhibitor AS1517499 repolarized M2 macrophages to M1 type in TME and not in healthy tissue (Xiao et al. 2020).

8.5.2 Inhibition of Communication Pathways

In breast cancer TAMs secrete VEGF and the interaction of VEGF with VEGF receptors triggers angiogenesis. Apart from VEGF secretion TAMs also express higher EGF which in turn activates EGFR receptors in breast cancer cells. This induces cancer cells to secrete CSF-1 which in turn establishes a positive feedback loop to attract TAMs. Furthermore, a hypoxic environment induces TAMs to secrete Hypoxia-inducible factors (HIF-1α and HIF-2α) which trigger the activation of VEGF secretion resulting in angiogenesis, thus promoting tumor growth. Inhibiting VEGF/VEGFR and CSF1 receptor communication has been demonstrated as a strategy to prevent TAM recruitment and thereby inhibit metastasis and angiogenesis (Dallavalasa et al. 2021; Fang et al. 2009; Farzaneh Behelgardi et al. 2020; Genard et al. 2017; LaGory and Giaccia 2016; Tamura et al. 2019).

Another communication pathway that helps breast cancer cells to metastasize in lungs and bone is interactions between breast cancer cells CCL2 ligand and macrophages CCR2 receptor (Lu and Kang 2009). Thus, inhibitors for CCL2/CCR2 communication also effectively inhibit breast cancer metastasis. Alternatively, interactions between the CCL18 ligand secreted by TAMs and PYK2 N-terminal domain interacting receptor 1 (PITPNM3) expressed on breast cancer cells reduce the invasiveness and metastatic effects of the CCL18 ligand. Thus, blocking both CCL5 and CCL18 ligands significantly suppressed tumor metastasis (Li et al. 2020b; An et al. 2019; Chen et al. 2011b; Nie et al. 2019).

8.5.3 *Immunotherapy Approaches*

Mainly these approaches involve enhancing the tumor-killing function of CD8+ T cells. TAMs modulate through various ways to suppress T cell function such as decrease of L-arginine, increase of production of NO, expression of PD-1, increase in expression of IFNγ, high expression of calcium/calmodulin-dependent protein kinase kinase (CaMKK2), enrichment of lymphocyte activating 3 (LAG 3), and T cell immunoglobin and mucin domain-containing protein 3 (TIM3) (de Boniface et al. 2012; Farhood et al. 2019; Hossain et al. 2021; Racioppi et al. 2019). Furthermore, TAMs stimulate myeloid-derived suppressor cells to secrete IL-12 which in turn mitigates T cell functions (Ugel et al. 2015). Therefore, enhancing CD8+ T cell function by employing strategies that interfere with the immune modulatory functions of TAMs in TME appears promising.

8.6 Conclusion

Currently, there are several challenges to the therapy of breast cancer metastasis. TAMs are recruited and educated by molecular cues in TME to promote breast cancer progression toward metastasis by supporting processes such as angiogenesis, migration, invasion, and immune evasion. TAMs communicate with cancer cells and other cells of TME by secretion of signalling molecules and exosomes. In this context, the novel role of macrophages in modulating direct cell-to-cell communication via TNTs between breast cancer cells (homologous TNTs) and between cancer cells and macrophages (heterologous TNTs) is also described by recent studies. TNT formation between cancer cells and other cells of TME under the effect of physiological conditions of TME has been implicated in cancer progression, angiogenesis, invasion, and therapy resistance via intercellular transfer of mitochondria, miRNAs etc. Therefore, inhibiting communication channels (including paracrine signaling and TNTs) that exist between cancer cells and TAMs appear as a promising strategy for cancer therapy. Furthermore, inhibitors targeting TAM recruitment and polarization in TME can offset their role in breast cancer progression. In addition to this, immunotherapies targeting TAMs in combination with current breast cancer therapies can offer new therapeutic approaches for cancer treatment. With the increasing interest in understanding the role of TAMs in cancer progression combined with technological advances to study the heterogeneity of TAMs in TME, the development of new therapeutic strategies for breast cancer will continue to grow.

References

Abraham D et al (2010) Stromal cell-derived CSF-1 blockade prolongs xenograft survival of CSF-1-negative neuroblastoma. Int J Cancer 126(6):1339–1352

Allavena P et al (2005) Anti-inflammatory properties of the novel antitumor agent yondelis (trabectedin): inhibition of macrophage differentiation and cytokine production. Cancer Res 65(7):2964–2971

Allavena P, Digifico E, Belgiovine C (2021) Macrophages and cancer stem cells: a malevolent alliance. Mol Med 27(1):121

Allison E et al (2023) Breast cancer survival outcomes and tumor-associated macrophage markers: a systematic review and meta-analysis. Oncol Ther 11(1):27–48

Amintas S et al (2020) Circulating tumor cell clusters: united we stand divided we fall. Int J Mol Sci 21(7)

An G et al (2019) Effects of CCL5 on the biological behavior of breast cancer and the mechanisms of its interaction with tumor-associated macrophages. Oncol Rep 42(6):2499–2511

Aras S, Zaidi MR (2017) TAMeless traitors: macrophages in cancer progression and metastasis. Br J Cancer 117(11):1583–1591

Baeriswyl V, Christofori G (2009) The angiogenic switch in carcinogenesis. Semin Cancer Biol 19(5):329–337

Bergers G, Benjamin LE (2003) Tumorigenesis and the angiogenic switch. Nat Rev Cancer 3(6):401–410

Bertolini I et al (2019) A GBM-like V-ATPase signature directs cell-cell tumor signaling and reprogramming via large oncosomes. EBioMedicine 41:225–235

Bronte V et al (2016) Recommendations for myeloid-derived suppressor cell nomenclature and characterization standards. Nat Commun 7:12150

Cao R et al (2015) Collaborative effects between the TNFα-TNFR1-macrophage axis and the VEGF-C-VEGFR3 signaling in lymphangiogenesis and metastasis. Oncoimmunology 4(3):e989777

Carter KP et al (2019) Macrophages enhance 3D invasion in a breast cancer cell line by induction of tumor cell tunneling nanotubes. Cancer Rep (Hoboken) 2(6):e1213

Chen Q, Zhang XH, Massagué J (2011a) Macrophage binding to receptor VCAM-1 transmits survival signals in breast cancer cells that invade the lungs. Cancer Cell 20(4):538–549

Chen J et al (2011b) CCL18 from tumor-associated macrophages promotes breast cancer metastasis via PITPNM3. Cancer Cell 19(4):541–555

Chen Y et al (2017) Tumor-recruited M2 macrophages promote gastric and breast cancer metastasis via M2 macrophage-secreted CHI3L1 protein. J Hematol Oncol 10(1):36

Chen X et al (2022) Tumor-associated macrophages promote epithelial-mesenchymal transition and the cancer stem cell properties in triple-negative breast cancer through CCL2/AKT/β--catenin signaling. Cell Commun Signal 20(1):92

Chuo ST, Chien JC, Lai CP (2018) Imaging extracellular vesicles: current and emerging methods. J Biomed Sci 25(1):91

Condeelis J, Pollard JW (2006) Macrophages: obligate partners for tumor cell migration, invasion, and metastasis. Cell 124(2):263–266

Condeelis J, Segall JE (2003) Intravital imaging of cell movement in tumours. Nat Rev Cancer 3(12):921–930

Dallavalasa S et al (2021) The role of tumor associated macrophages (TAMs) in cancer progression, chemoresistance, angiogenesis and metastasis - current status. Curr Med Chem 28(39):8203–8236

de Boniface J et al (2012) Expression patterns of the immunomodulatory enzyme arginase 1 in blood, lymph nodes and tumor tissue of early-stage breast cancer patients. Oncoimmunology 1(8):1305–1312

Di Vizio D et al (2009) Oncosome formation in prostate cancer: association with a region of frequent chromosomal deletion in metastatic disease. Cancer Res 69(13):5601–5609

Di Vizio D et al (2012) Large oncosomes in human prostate cancer tissues and in the circulation of mice with metastatic disease. Am J Pathol 181(5):1573–1584
Dunn GP, Old LJ, Schreiber RD (2004) The immunobiology of cancer immunosurveillance and immunoediting. Immunity 21(2):137–148
Dvorak HF (1986) Tumors: wounds that do not heal. Similarities between tumor stroma generation and wound healing. N Engl J Med 315(26):1650–1659
Edin S et al (2013) Macrophages: good guys in colorectal cancer. Oncoimmunology 2(2):e23038
Fang HY et al (2009) Hypoxia-inducible factors 1 and 2 are important transcriptional effectors in primary macrophages experiencing hypoxia. Blood 114(4):844–859
Farhood B, Najafi M, Mortezaee K (2019) CD8(+) cytotoxic T lymphocytes in cancer immunotherapy: a review. J Cell Physiol 234(6):8509–8521
Farzaneh Behelgardi M et al (2020) Targeting signaling pathways of VEGFR1 and VEGFR2 as a potential target in the treatment of breast cancer. Mol Biol Rep 47(3):2061–2071
Ferlay J et al (2021) Cancer statistics for the year 2020: an overview. Int J Cancer
Franklin RA et al (2014) The cellular and molecular origin of tumor-associated macrophages. Science 344(6186):921–925
Fu XT et al (2015) Macrophage-secreted IL-8 induces epithelial-mesenchymal transition in hepatocellular carcinoma cells by activating the JAK2/STAT3/Snail pathway. Int J Oncol 46(2):587–596
Fu LQ et al (2020) The roles of tumor-associated macrophages in tumor angiogenesis and metastasis. Cell Immunol 353:104119
Gazzaniga S et al (2007) Targeting tumor-associated macrophages and inhibition of MCP-1 reduce angiogenesis and tumor growth in a human melanoma xenograft. J Invest Dermatol 127(8):2031–2041
Geiger TR, Peeper DS (2009) Metastasis mechanisms. Biochim Biophys Acta 1796(2):293–308
Genard G, Lucas S, Michiels C (2017) Reprogramming of tumor-associated macrophages with anticancer therapies: radiotherapy versus chemo- and immunotherapies. Front Immunol 8:828
Giatromanolaki A et al (2006) Hypoxia-inducible factor-2 alpha (HIF-2 alpha) induces angiogenesis in breast carcinomas. Appl Immunohistochem Mol Morphol 14(1):78–82
Gil-Bernabé AM et al (2012) Recruitment of monocytes/macrophages by tissue factor-mediated coagulation is essential for metastatic cell survival and premetastatic niche establishment in mice. Blood 119(13):3164–3175
Gocheva V et al (2010) IL-4 induces cathepsin protease activity in tumor-associated macrophages to promote cancer growth and invasion. Genes Dev 24(3):241–255
Goede V et al (1999) Induction of inflammatory angiogenesis by monocyte chemoattractant protein-1. Int J Cancer 82(5):765–770
Gordon SR et al (2017) PD-1 expression by tumour-associated macrophages inhibits phagocytosis and tumour immunity. Nature 545(7655):495–499
György B et al (2011) Membrane vesicles, current state-of-the-art: emerging role of extracellular vesicles. Cell Mol Life Sci 68(16):2667–2688
Hambardzumyan D, Gutmann DH, Kettenmann H (2016) The role of microglia and macrophages in glioma maintenance and progression. Nat Neurosci 19(1):20–27
Han L, Lam EW, Sun Y (2019) Extracellular vesicles in the tumor microenvironment: old stories, but new tales. Mol Cancer 18(1):59
Hanahan D, Weinberg RA (2000) The hallmarks of cancer. Cell 100(1):57–70
Hanahan D, Weinberg RA (2011) Hallmarks of cancer: the next generation. Cell 144(5):646–674
Hanna SJ et al (2019) Tunneling nanotubes, a novel mode of tumor cell-macrophage communication in tumor cell invasion. J Cell Sci 132(3)
Hao NB et al (2012) Macrophages in tumor microenvironments and the progression of tumors. Clin Dev Immunol 2012:948098
Headley MB et al (2016) Visualization of immediate immune responses to pioneer metastatic cells in the lung. Nature 531(7595):513–517
Hossain MA et al (2021) Reinvigorating exhausted CD8(+) cytotoxic T lymphocytes in the tumor microenvironment and current strategies in cancer immunotherapy. Med Res Rev 41(1):156–201

Hume DA, MacDonald KP (2012) Therapeutic applications of macrophage colony-stimulating factor-1 (CSF-1) and antagonists of CSF-1 receptor (CSF-1R) signaling. Blood 119(8):1810–1820
Hwang I et al (2020) Tumor-associated macrophage, angiogenesis and lymphangiogenesis markers predict prognosis of non-small cell lung cancer patients. J Transl Med 18(1):443
Jia X et al (2014) Emodin suppresses pulmonary metastasis of breast cancer accompanied with decreased macrophage recruitment and M2 polarization in the lungs. Breast Cancer Res Treat 148(2):291–302
Kaplan RN et al (2005) VEGFR1-positive haematopoietic bone marrow progenitors initiate the pre-metastatic niche. Nature 438(7069):820–827
Kaplan RN, Psaila B, Lyden D (2006) Bone marrow cells in the 'pre-metastatic niche': within bone and beyond. Cancer Metastasis Rev 25(4):521–529
Kessenbrock K, Plaks V, Werb Z (2010) Matrix metalloproteinases: regulators of the tumor microenvironment. Cell 141(1):52–67
Krausgruber T et al (2011) IRF5 promotes inflammatory macrophage polarization and TH1-TH17 responses. Nat Immunol 12(3):231–238
Kumar V et al (2016) The nature of myeloid-derived suppressor cells in the tumor microenvironment. Trends Immunol 37(3):208–220
LaGory EL, Giaccia AJ (2016) The ever-expanding role of HIF in tumour and stromal biology. Nat Cell Biol 18(4):356–365
Lee AH et al (1997) Angiogenesis and inflammation in invasive carcinoma of the breast. J Clin Pathol 50(8):669–673
Lee HD et al (2012) Exosome release of ADAM15 and the functional implications of human macrophage-derived ADAM15 exosomes. FASEB J 26(7):3084–3095
Lee CW et al (2022) Effects of the media conditioned by various macrophage subtypes derived from THP-1 cells on tunneling nanotube formation in pancreatic cancer cells. BMC Mol Cell Biol 23(1):26
Leek RD et al (1996) Association of macrophage infiltration with angiogenesis and prognosis in invasive breast carcinoma. Cancer Res 56(20):4625–4629
Li J et al (2020a) Tumor-associated macrophage infiltration and prognosis in colorectal cancer: systematic review and meta-analysis. Int J Colorectal Dis 35(7):1203–1210
Li D et al (2020b) Tumor-associated macrophages secrete CC-chemokine ligand 2 and induce tamoxifen resistance by activating PI3K/Akt/mTOR in breast cancer. Cancer Sci 111(1):47–58
Li J et al (2023) A comprehensive review on the composition, biogenesis, purification, and multifunctional role of exosome as delivery vehicles for cancer therapy. Biomed Pharmacother 165:115087
Lin EY et al (2006) Macrophages regulate the angiogenic switch in a mouse model of breast cancer. Cancer Res 66(23):11238–11246
Lin Y, Xu J, Lan H (2019) Tumor-associated macrophages in tumor metastasis: biological roles and clinical therapeutic applications. J Hematol Oncol 12(1):76
Linde N et al (2012) Vascular endothelial growth factor-induced skin carcinogenesis depends on recruitment and alternative activation of macrophages. J Pathol 227(1):17–28
Liou GY, Storz P (2010) Reactive oxygen species in cancer. Free Radic Res 44(5):479–496
Lu X, Kang Y (2007) Organotropism of breast cancer metastasis. J Mammary Gland Biol Neoplasia 12(2-3):153–162
Lu X, Kang Y (2009) Chemokine (C-C motif) ligand 2 engages CCR2+ stromal cells of monocytic origin to promote breast cancer metastasis to lung and bone. J Biol Chem 284(42):29087–29096
Lu X et al (2011) VCAM-1 promotes osteolytic expansion of indolent bone micrometastasis of breast cancer by engaging α4β1-positive osteoclast progenitors. Cancer Cell 20(6):701–714
Mahmoud SM et al (2012) Tumour-infiltrating macrophages and clinical outcome in breast cancer. J Clin Pathol 65(2):159–163
Mantovani A et al (2002) Macrophage polarization: tumor-associated macrophages as a paradigm for polarized M2 mononuclear phagocytes. Trends Immunol 23(11):549–555
Mantovani A et al (2008) Cancer-related inflammation. Nature 454(7203):436–444
Mantovani A et al (2017) Tumour-associated macrophages as treatment targets in oncology. Nat Rev Clin Oncol 14(7):399–416

Melwani PK, Pandey BN (2023) Tunneling nanotubes: the intercellular conduits contributing to cancer pathogenesis and its therapy. Biochim Biophys Acta Rev Cancer 1878(6):189028
Melwani PK et al (2021) Integrated transcriptomic and proteomic analysis of microplasts derived from macrophage-conditioned medium-treated MCF-7 breast cancer cells. FEBS Lett 595(13):1844–1860
Minciacchi VR, Freeman MR, Di Vizio D (2015) Extracellular vesicles in cancer: exosomes, microvesicles and the emerging role of large oncosomes. Semin Cell Dev Biol 40:41–51
Mizutani K et al (2009) The chemokine CCL2 increases prostate tumor growth and bone metastasis through macrophage and osteoclast recruitment. Neoplasia 11(11):1235–1242
Morrissey SM et al (2021) Tumor-derived exosomes drive immunosuppressive macrophages in a pre-metastatic niche through glycolytic dominant metabolic reprogramming. Cell Metab 33(10):2040–2058.e10
Mougiakakos D et al (2010) Regulatory T cells in cancer. Adv Cancer Res 107:57–117
Movahedi K et al (2010) Different tumor microenvironments contain functionally distinct subsets of macrophages derived from Ly6C(high) monocytes. Cancer Res 70(14):5728–5739
Müller A et al (2001) Involvement of chemokine receptors in breast cancer metastasis. Nature 410(6824):50–56
Munder M (2009) Arginase: an emerging key player in the mammalian immune system. Br J Pharmacol 158(3):638–651
Murdoch C et al (2007) Expression of Tie-2 by human monocytes and their responses to angiopoietin-2. J Immunol 178(11):7405–7411
Murray PJ et al (2014) Macrophage activation and polarization: nomenclature and experimental guidelines. Immunity 41(1):14–20
Netea-Maier RT, Smit JWA, Netea MG (2018) Metabolic changes in tumor cells and tumor-associated macrophages: a mutual relationship. Cancer Lett 413:102–109
Nie Y et al (2019) Breast phyllodes tumors recruit and repolarize tumor-associated macrophages via secreting CCL5 to promote malignant progression, which can be inhibited by CCR5 inhibition therapy. Clin Cancer Res 25(13):3873–3886
Nie X et al (2023) Garcinone E suppresses breast cancer growth and metastasis by modulating tumor-associated macrophages polarization via STAT6 signaling. Phytother Res 37(10):4442–4456
Nierodzik ML, Karpatkin S (2006) Thrombin induces tumor growth, metastasis, and angiogenesis: evidence for a thrombin-regulated dormant tumor phenotype. Cancer Cell 10(5):355–362
Noy R, Pollard JW (2014) Tumor-associated macrophages: from mechanisms to therapy. Immunity 41(1):49–61
Okabe Y, Medzhitov R (2014) Tissue-specific signals control reversible program of localization and functional polarization of macrophages. Cell 157(4):832–844
Onfelt B et al (2006) Structurally distinct membrane nanotubes between human macrophages support long-distance vesicular traffic or surfing of bacteria. J Immunol 177(12):8476–8483
Osswald M et al (2015) Brain tumour cells interconnect to a functional and resistant network. Nature 528(7580):93–98
Ostrand-Rosenberg S, Sinha P (2009) Myeloid-derived suppressor cells: linking inflammation and cancer. J Immunol 182(8):4499–4506
Otsuji M et al (1996) Oxidative stress by tumor-derived macrophages suppresses the expression of CD3 zeta chain of T-cell receptor complex and antigen-specific T-cell responses. Proc Natl Acad Sci U S A 93(23):13119–13124
Palumbo JS et al (2007) Tumor cell-associated tissue factor and circulating hemostatic factors cooperate to increase metastatic potential through natural killer cell-dependent and-independent mechanisms. Blood 110(1):133–141
Pan Y et al (2020) Tumor-associated macrophages in tumor immunity. Front Immunol 11:583084
Patheja P, Sahu K (2017) Macrophage conditioned medium induced cellular network formation in MCF-7 cells through enhanced tunneling nanotube formation and tunneling nanotube mediated release of viable cytoplasmic fragments. Exp Cell Res 355(2):182–193
Pinto G, Brou C, Zurzolo C (2020) Tunneling nanotubes: the fuel of tumor progression? Trends Cancer 6(10):874–888

Pollard JW (2004) Tumour-educated macrophages promote tumour progression and metastasis. Nat Rev Cancer 4(1):71–78
Qian BZ, Pollard JW (2010) Macrophage diversity enhances tumor progression and metastasis. Cell 141(1):39–51
Qian B et al (2009) A distinct macrophage population mediates metastatic breast cancer cell extravasation, establishment and growth. PLoS One 4(8):e6562
Qian BZ et al (2011) CCL2 recruits inflammatory monocytes to facilitate breast-tumour metastasis. Nature 475(7355):222–225
Racioppi L et al (2019) CaMKK2 in myeloid cells is a key regulator of the immune-suppressive microenvironment in breast cancer. Nat Commun 10(1):2450
Ramesh A et al (2019) CSF1R- and SHP2-inhibitor-loaded nanoparticles enhance cytotoxic activity and phagocytosis in tumor-associated macrophages. Adv Mater 31(51):e1904364
Ravi J et al (2016) Cannabinoid receptor-2 agonist inhibits macrophage induced EMT in non-small cell lung cancer by downregulation of EGFR pathway. Mol Carcinog 55(12):2063–2076
Riabov V et al (2014) Role of tumor associated macrophages in tumor angiogenesis and lymphangiogenesis. Front Physiol 5:75
Rihawi K et al (2021) Tumor-associated macrophages and inflammatory microenvironment in gastric cancer: novel translational implications. Int J Mol Sci 22(8)
Robinson BD et al (2009) Tumor microenvironment of metastasis in human breast carcinoma: a potential prognostic marker linked to hematogenous dissemination. Clin Cancer Res 15(7):2433–2441
Roehlecke C, Schmidt MHH (2020) Tunneling nanotubes and tumor microtubes in cancer. Cancers (Basel) 12(4)
Rolny C et al (2011) HRG inhibits tumor growth and metastasis by inducing macrophage polarization and vessel normalization through downregulation of PlGF. Cancer Cell 19(1):31–44
Rustom A et al (2004) Nanotubular highways for intercellular organelle transport. Science 303(5660):1007–1010
Saji H et al (2001) Significant correlation of monocyte chemoattractant protein-1 expression with neovascularization and progression of breast carcinoma. Cancer 92(5):1085–1091
Sangaletti S et al (2008) Macrophage-derived SPARC bridges tumor cell-extracellular matrix interactions toward metastasis. Cancer Res 68(21):9050–9059
Sceneay J, Smyth MJ, Möller A (2013) The pre-metastatic niche: finding common ground. Cancer Metastasis Rev 32(3–4):449–464
Schäfer M, Werner S (2008) Cancer as an overhealing wound: an old hypothesis revisited. Nat Rev Mol Cell Biol 9(8):628–638
Shields JD et al (2010) Induction of lymphoidlike stroma and immune escape by tumors that express the chemokine CCL21. Science 328(5979):749–752
Sierra-Filardi E et al (2014) CCL2 shapes macrophage polarization by GM-CSF and M-CSF: identification of CCL2/CCR2-dependent gene expression profile. J Immunol 192(8):3858–3867
Strachan DC et al (2013) CSF1R inhibition delays cervical and mammary tumor growth in murine models by attenuating the turnover of tumor-associated macrophages and enhancing infiltration by CD8. Oncoimmunology 2(12):e26968
Su S et al (2014) A positive feedback loop between mesenchymal-like cancer cells and macrophages is essential to breast cancer metastasis. Cancer Cell 25(5):605–620
Su T et al (2021) Exosomal MicroRNAs mediating crosstalk between cancer cells with cancer-associated fibroblasts and tumor-associated macrophages in the tumor microenvironment. Front Oncol 11:631703
Tamura R et al (2019) The role of vascular endothelial growth factor in the hypoxic and immunosuppressive tumor microenvironment: perspectives for therapeutic implications. Med Oncol 37(1):2
Torroella-Kouri M et al (2005) Diminished expression of transcription factors nuclear factor kappaB and CCAAT/enhancer binding protein underlies a novel tumor evasion mechanism affecting macrophages of mammary tumor-bearing mice. Cancer Res 65(22):10578–10584

Tsutsui S et al (2005) Macrophage infiltration and its prognostic implications in breast cancer: the relationship with VEGF expression and microvessel density. Oncol Rep 14(2):425–431
Ueno T et al (2000) Significance of macrophage chemoattractant protein-1 in macrophage recruitment, angiogenesis, and survival in human breast cancer. Clin Cancer Res 6(8):3282–3289
Ugel S et al (2015) Tumor-induced myeloid deviation: when myeloid-derived suppressor cells meet tumor-associated macrophages. J Clin Invest 125(9):3365–3376
Vader P, Breakefield XO, Wood MJ (2014) Extracellular vesicles: emerging targets for cancer therapy. Trends Mol Med 20(7):385–393
Valković T et al (2002) Correlation between vascular endothelial growth factor, angiogenesis, and tumor-associated macrophages in invasive ductal breast carcinoma. Virchows Arch 440(6):583–588
Vasiljeva O et al (2006) Tumor cell-derived and macrophage-derived cathepsin B promotes progression and lung metastasis of mammary cancer. Cancer Res 66(10):5242–5250
Venneri MA et al (2007) Identification of proangiogenic TIE2-expressing monocytes (TEMs) in human peripheral blood and cancer. Blood 109(12):5276–5285
Volodko N et al (1998) Tumour-associated macrophages in breast cancer and their prognostic correlations. The Breast 7(2):99–105
Wanderley CW et al (2018) Paclitaxel reduces tumor growth by reprogramming tumor-associated macrophages to an M1 profile in a TLR4-dependent manner. Cancer Res 78(20):5891–5900
Wang P et al (2019) Exosomes from M1-polarized macrophages enhance paclitaxel antitumor activity by activating macrophages-mediated inflammation. Theranostics 9(6):1714–1727
Wu P et al (2016) Inverse role of distinct subsets and distribution of macrophage in lung cancer prognosis: a meta-analysis. Oncotarget 7(26):40451–40460
Wyckoff J et al (2004) A paracrine loop between tumor cells and macrophages is required for tumor cell migration in mammary tumors. Cancer Res 64(19):7022–7029
Wynn TA, Chawla A, Pollard JW (2013) Macrophage biology in development, homeostasis and disease. Nature 496(7446):445–455
Xiao H et al (2020) M2-like tumor-associated macrophage-targeted codelivery of STAT6 inhibitor and IKKβ siRNA induces M2-to-M1 repolarization for cancer immunotherapy with low immune side effects. ACS Cent Sci 6(7):1208–1222
Yang L, Pang Y, Moses HL (2010) TGF-beta and immune cells: an important regulatory axis in the tumor microenvironment and progression. Trends Immunol 31(6):220–227
Yang M et al (2011) Microvesicles secreted by macrophages shuttle invasion-potentiating microRNAs into breast cancer cells. Mol Cancer 10:117
Yin S et al (2017) The prognostic and clinicopathological significance of tumor-associated macrophages in patients with gastric cancer: a meta-analysis. PLoS One 12(1):e0170042
Yount G et al (2007) Independent motile microplast formation correlates with glioma cell invasiveness. J Neurooncol 81(2):113–121
Yu X et al (2019) Exosomes from macrophages exposed to apoptotic breast cancer cells promote breast cancer proliferation and metastasis. J Cancer 10(13):2892–2906
Yue S et al (2021) PGRN. Life Sci 264:118687
Zhang J et al (2015) Regulation of epithelial-mesenchymal transition by tumor-associated macrophages in cancer. Am J Transl Res 7(10):1699–1711
Zhang J et al (2016) High infiltration of tumor-associated macrophages influences poor prognosis in human gastric cancer patients, associates with the phenomenon of EMT. Medicine (Baltimore) 95(6):e2636
Zhao X et al (2017) Prognostic significance of tumor-associated macrophages in breast cancer: a meta-analysis of the literature. Oncotarget 8(18):30576–30586
Zhou J et al (2020) Tumor-associated macrophages: recent insights and therapies. Front Oncol 10:188
Ziech D et al (2011) Reactive oxygen species (ROS)--induced genetic and epigenetic alterations in human carcinogenesis. Mutat Res 711(1–2):167–173

Chapter 9
Macrophage Dysfunction in Respiratory Disease

Shyreen Hassibi and Louise E. Donnelly

Abstract Our understanding of the origin, phenotype, and function of pulmonary macrophages has evolved over recent years. The use of lineage tracing and single-cell RNA sequencing has led to a greater understanding of how these cells regulate homeostasis in the lung. The primary function of alveolar macrophages is to clear any inhaled particles or pathogens and they as well as tissue-resident cells also play a role in the clearance of apoptotic cells and the resolution of inflammation. Lung diseases affect over half a billion people globally and are attributable to 7% of all deaths each year. The common diseases are chronic obstructive pulmonary disease (COPD) and asthma but others that contribute to this statistic include cystic fibrosis and idiopathic pulmonary fibrosis (IPF). Macrophages are aberrant in all these diseases with a reduced phagocytic capacity and a high proinflammatory phenotype with changes to their capacity to resolve inflammation. The pathways leading to these macrophage dysfunctions differ with disease and may relate to the specific lung environment in each condition. However, there are clear changes in metabolic profiles and mitochondrial activity in many of these conditions that contribute to a change in macrophage phenotype towards a more proinflammatory, less homeostatic cell. Understanding the mechanisms that drive these changes will allow for more targeted therapies for the treatment of these long-term and debilitating conditions.

9.1 Introduction

Our understanding of alveolar macrophages has moved on since they were described as 'dust cells' of unknown origin, although it was speculated that they could derived from the circulation or the endothelium (Carleton 1927). Historically, alveolar

S. Hassibi · L. E. Donnelly (✉)
Airway Disease, National Heart and Lung Institute, Imperial College London, London, UK
e-mail: l.donnelly@imperial.ac.uk

M. Kloc et al. (eds.), *Monocytes and Macrophages in Development, Regeneration, and Disease*, Results and Problems in Cell Differentiation 74,
https://doi.org/10.1007/978-3-031-65944-7_9

macrophages were thought to derive from either blood monocytes or from interstitial macrophages that migrated out to the airspaces from the lung parenchyma (Morales-Nebreda et al. 2015; Huang et al. 2023). Over the past 15 years or so, we have a much clearer idea of the origin of pulmonary macrophages using murine models, and using lineage tracing, it appears that pulmonary macrophages derive from yolk sac progenitors (Guilliams et al. 2013) or the fetal liver and can self-renew (Tan and Krasnow 2016). However, under stressors, the replenishment of resident macrophages via proliferation may not be sufficient to maintain the population, thus the monocytes are recruited from the circulation and differentiate locally (Martin et al. 2021; Hashimoto et al. 2013). It has been suggested that newly recruited macrophages may be more activated compared to resident cells (Saeed et al. 2014). There appears to be epigenetic changes as a monocyte differentiates towards a macrophage *in vitro* (Saeed et al. 2014), but the extent of similar changes in the human lung is not clear. Nevertheless, it appears that newly recruited macrophages derived from monocytes are highly reactive but under conditions of health, could become less reactive like the resident population (Kulikauskaite and Wack 2020). More recently, more selective lineage tracing (Liu et al. 2019) has allowed further understanding as to the role of recruited monocyte-derived macrophages in the lung. There is a turnover of fetal-derived macrophages in the lung that are replenished from circulating cells (Zhang et al. 2024b). At 40 weeks (adulthood), 70% of the macrophages in the lung are monocyte derived. There are some functional differences in these cells in response to influenza virus but over time these are lost. This change appears to be subtle but when alveolar macrophages were depleted by chlodronate, the incoming monocyte-derived cells were functionally impaired (Zhang et al. 2024b). This scenario is unlikely to occur during disease in humans, but these authors suggest that the resident fetal-derived cells may train the incoming macrophages such that the result is identical function. This is substantiated by the observation that murine alveolar macrophage and blood-derived cells share ~98% transcriptomic similarity (Gibbings et al. 2015). This suggests the lung environment has a more important role in macrophage phenotype than the origin and is further substantiated by the transplantation of peritoneal macrophages to the lung that then show characteristics of the alveolar macrophages (Suzuki et al. 2014).

Pulmonary macrophage origin in humans is considered to follow rodent mechanisms but the lack of specific lineage markers has hampered further analysis. Studies following macrophages in the lungs of transplant recipients suggest that there is a small resident population but that most macrophages are the result of recruited monocytes differentiating locally (Byrne et al. 2020). These authors also showed that recruited monocytes are also the source of macrophages in the aged lung. This aligns with the scenario outlined by (Martin et al. 2021) where essentially in humans, the effect of the environment, such as smoking, pollution, and infection, can deplete the resident macrophage population and self-renewal is not sufficient to maintain a population of macrophages required for homeostasis throughout life. Therefore, a proportion of these recruited macrophages may become resident and contribute longer term to the pulmonary population. This observation has led to further work characterizing and using monocyte-derived macrophages as *in vitro* models to study macrophage function. We have known for some time that culture conditions during the differentiation of monocytes to macrophages can be used to manipulate macrophage phenotype (Lacey et al. 2012), but models can now be characterized further so that we can use these models to begin to understand function under specific conditions (Hickman et al. 2023).

There is no question that there is heterogeneity in the macrophage population in the lung (Chana et al. 2014; Arora et al. 2018; Hou et al. 2021). This may be related to the localization of the macrophage in the lung. Aside from alveolar macrophages that inhabit the alveoli, there are also macrophages within the airways and tissue. Each of these will experience a different local environment and it is considered that this may influence cell function and metabolism. Macrophages appear to utilize energy obtained from either glycolysis or oxidative phosphorylation that may align with macrophage polarization (Galván-Peña and O'Neill 2014). It has been suggested that M1-like macrophages are glycolytically active, whereas the more resolving M2-like macrophage tends towards oxidative phosphorylation. However, the use of M1 and M2 to describe macrophages is an oversimplification and the fluctuations in the availability of glucose in the lung can also affect macrophage metabolism (Baker and Baines 2018). Furthermore, tissue-resident alveolar macrophages tend to shift their energy from glycolysis even when stimulated with lipopolysaccharide (LPS) or viral challenge (Woods et al. 2020).

Irrespective of their origin, the role of alveolar macrophages remains one of surveillance and homeostasis. The role of these cells is to remove particles and pathogens that encounter the pulmonary surface (Franken et al. 2016; Belchamber and Donnelly 2020). In addition, macrophages have a role in the removal of apoptotic and senescent cells in the lungs (McCubbrey and Curtis 2013; Sagiv and Krizhanovsky 2013) and as such will have roles to play in the repair and remodeling of the lung. Macrophages recognize danger signals and foreign bodies via a plethora of cell surface receptors. These recognition signals include toll-like receptors (TLR), scavenger receptors such as CD206, and alveolar macrophages also express macrophage receptors with collagenous structure (MARCO) but can also recognize opsonized particles via complement receptors or Fcγ receptors such as CD16.

Macrophages are the most abundant leukocyte in the airways yet their homeostatic function may become dysregulated in chronic lung diseases, leading to changes in the lung microbiome (Han et al. 2012) leading to chronic colonization and susceptibility to exacerbation. The failure to remove apoptotic or senescent cells would contribute to remodeling and promotion of inflammation in the lung leading to the chronic inflammatory profile observed in many pulmonary conditions. The advent of novel analytic tools such as bulk and single-cell RNA sequencing (scRNA seq) has allowed for further characterization of macrophages both in health and disease (Aegerter et al. 2022). These markers will prove useful in the future to further understand the link between macrophage phenotype and disease, opening the possibility for further work into treatments for diseases that are common but for which few new treatments have come forward.

9.2 Chronic Lung Diseases

Chronic respiratory conditions affect over half a billion people globally and are the third leading cause of death worldwide (Collaborators 2020). The healthcare and societal costs of respiratory conditions are enormous but until recently there had been little advance in novel treatments for many of these conditions. Our understanding of many mechanisms driving the underlying pathophysiologies has aided the discovery of new therapeutics however many are sub-optimal and only target a subset of patients. Pulmonary macrophages are the most abundant leukocytes in the lung and exhibit plasticity depending on their environment. This can lead to dysfunction and changes in phenotype that are very specific to a disease. Understanding further macrophage changes during disease onset and progression might provide additional therapeutic targets.

9.2.1 Chronic Obstructive Pulmonary Disease

Chronic obstructive pulmonary disease (COPD) is a debilitating lung condition with increasing incidence and rising morbidity and mortality globally (Christenson et al. 2022). It is largely associated with tobacco smoke exposure but other environmental factors including exposure to particulate matter and biomass smoke contribute to risk (Salvi and Barnes 2010; Mah et al. 2024). COPD is heterogeneous and typically comprises chronic bronchitis, small airways disease, and emphysema although not all subjects will necessarily exhibit all three. Subjects with COPD will exhibit shortness of breath and airflow limitation due to the remodeling of the lung. Subjects with COPD often experience acute worsening of symptoms or exacerbations that can lead to hospital admission and worsening of disease progression (Wedzicha and Donaldson 2003). These may be driven by a bacterial or viral infection or even environmental stressors (Mah et al. 2024). More recently, the concept that COPD may be an example of accelerating aging brought on by exposure to exogenous oxidative stress has been proposed (Ito and Barnes 2009; Schuliga et al. 2021). The inflammatory profile observed in the lungs of patients with COPD closely resembles that of the senescence-associated secretory phenotype (SASP) with increased levels of interleukin (IL)-1β, CXCL8, CXCL1, CXCL9, matrix metalloproteinase (MMP)-2 and MMP-9 (Barnes et al. 2019; Barnes 2017). The SASP is released from senescent cells and both small airway epithelial cells and small airway fibroblasts show increased markers of senescence in COPD (Baker et al. 2016, 2018; Wrench et al. 2024).

Macrophages have long been considered important in the pathophysiology of COPD (Barnes 2004) (Fig. 9.1), however how they now fit within the paradigm of accelerated aging as a driver of COPD is not clear. Aging is associated with a decline in immune cell function known as immunosenescence (Murray and Chotirmall 2015). This inflammaging may also be considered sterile inflammation as there is a

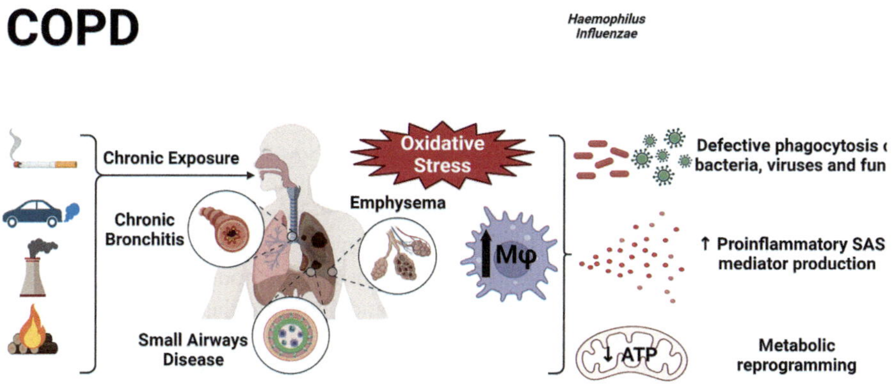

Fig. 9.1 Macrophages in chronic obstructive pulmonary disease (COPD). COPD is associated with chronic exposure to cigarette smoke, pollution or biomass smoke, all of which increase oxidative stress levels in the lung. COPD is comprised of three main pathologies: chronic bronchitis, emphysema and small airways disease. There are increased numbers of macrophages in the COPD lung and these cells have defective phagocytic function, are more proinflammatory and display metabolic reprogramming due to increased oxidative stress. (Figure created with Biorender.com)

low-level inflammation in the absence of any pathogens (Feldman et al. 2015; Lee et al. 2022). Most macrophages in the lung are terminally differentiated and as such could be described as senescent (Behmoaras and Gil 2021). Macrophages express high levels of checkpoint inhibitors including $p16^{INK4a}$ and lysosomal β-galactosidase activity (Hall et al. 2017). Furthermore, other markers of senescent cells are expressed in alveolar macrophages from smokers and COPD subjects including the anti-apoptotic B-cell lymphoma (BCL)-X_L and $p21^{CIP1}$ (Tomita et al. 2002). Aging is associated with a reduction in macrophage function including phagocytosis, efferocytosis, autophagy, and an aberrant inflammatory response (Sharma 2021). All of these features are seen in macrophages from COPD patients, including a reduced phagocytic response to bacteria (Berenson et al. 2006; Taylor et al. 2010) and efferocytosis (Hodge et al. 2007). The former can lead to impaired clearance of pathogens and bacterial colonization that is seen in the airways of patients with COPD (Singh et al. 2021).

There have been advances in understanding what is driving these aberrant macrophage functions in COPD, over and above those seen in 'healthy' aging. Oxidative stress is known to induce cellular senescence and exogenous oxidative stress has been shown to further impair phagocytosis of *Haemophilus influenzae* by COPD monocyte-derived and lung macrophages (Belchamber et al. 2019). Furthermore, this was associated with defective mitochondrial function (Belchamber et al. 2019; Bewley et al. 2017). Alveolar macrophages from COPD subjects appear to be metabolically reprogrammed with a reduction in ATP generation, suggestive of metabolic exhaustion with a depletion in both glycolytic and mitochondrial respiration (Ryan et al. 2023). This is associated with decreased expression of malic enzyme 1 (ME1) required for shuttling pyruvate into the TCA cycle. ME1 is regulated by the transcription factor Nrf-2 which is also downregulated in COPD macrophages

(Mercado et al. 2011). Activation of Nrf2 by KI-696, a molecule that disrupts the Keap-Nrf2 restored ME1 expression and improved phagocytosis (Ryan et al. 2023) demonstrating that changes in COPD macrophage phenotype are possible.

In COPD, oxidative stress is likely to be due to exposure to cigarette smoke. A recent study of a mouse model of COPD utilized scRNA seq to examine the changes in macrophages following 12-week exposure to cigarette smoke (Wohnhaas et al. 2024). Using this model, it appeared that monocyte-derived macrophages were actively recruited to the lung and differentiated giving a signature that was associated with emphysematous changes. Using data from this model, they were able to analyze a human data set (Baßler et al. 2022) and show similar changes in macrophage profile in the COPD lung.

Other environmental stressors also have been shown to affect macrophage function in COPD. Exposure of alveolar macrophages from COPD to human rhinovirus led to further impairment of their capacity to phagocytose *H. influenzae* and *Streptococcus pneumoniae* (Finney et al. 2019). This may, in part explain the emergence of bacterial infections after viral infection in COPD subjects undergoing exacerbations (George et al. 2014). There remains much to be understood as to the role of the macrophage in COPD pathophysiology but with the rapid onset of novel analytical tools, we are beginning to better understand their role and identify novel therapeutic targets that may be beneficial as disease-modifying therapies.

9.2.2 Asthma

Asthma is a also a global disease that affects both adults and children and is defined as reversible airflow limitation with persistent airway inflammation, bronchial hyper responsiveness and airway remodeling (GINA 2023). Between 5–10% of subjects with asthma are considered severe asthmatics and whilst the disease is considered primarily to be associated with epithelial dysfunction (Russell et al. 2024) and airway smooth muscle remodeling (Hough et al. 2020), there have been recent advances in biologic therapies which are transforming the lives of these patients. An example is benralizumab, an anti-IL-5 receptor monoclonal antibody that has proved to reduce exacerbations in previously uncontrolled asthma (Castro et al. 2014). Essentially, benralizumab reduces eosinophils in the lungs of these subjects (Pham et al. 2016), however, whether these treatments are changing the lung environment and as such macrophage phenotypes and function has not been well explored. However, depletion of eosinophils via antibody-dependent phagocytosis/efferocytosis has been proposed (Dagher et al. 2022) and would suggest that macrophages are involved in this process.

The role of the macrophage in asthma remains less well-defined (Fig. 9.2). There is evidence of an altered macrophage phenotype that appears to be specific to the asthmatic lung (Draijer et al. 2017). There is also evidence that associates a specific macrophage transcriptome with neutrophilic asthma (Fricker et al. 2022). However, unlike COPD, there is no increase in the numbers of macrophages in the lungs of

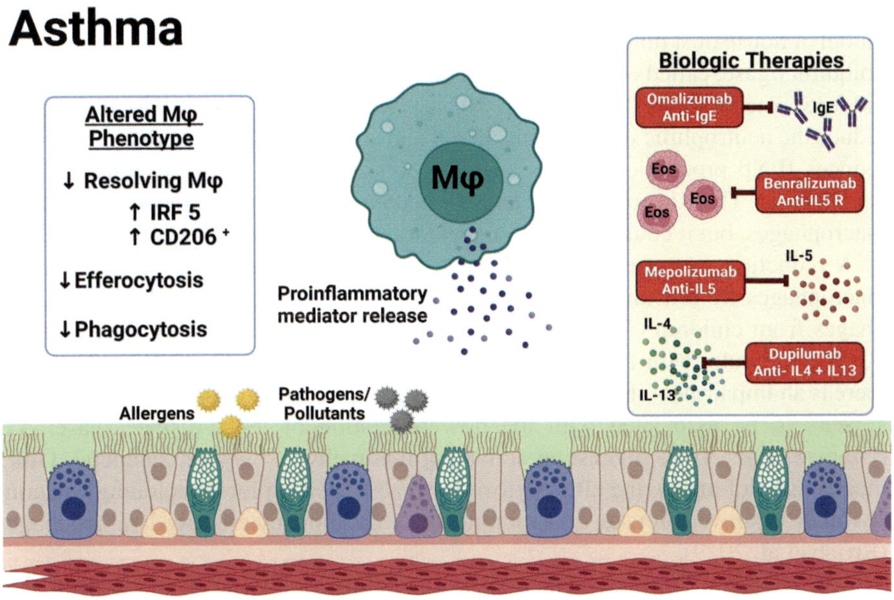

Fig. 9.2 Macrophages in asthma. In the asthmatic airways there is airway epithelial remodelling due to exposure to allergens, pathogens or pollutants. The effect of these on macrophage phenotype is less defined, but it appears that macrophages from asthmatic subjects have reduced phagocytic function and are shifted to a proinflammatory phenotype with increased expression of interferon regulatory factor (IRF) 5 and CD206⁺. Current therapies for asthma include biologic therapies, but the effect of these on macrophages is unknown. (Figure created with Biorender.com)

asthmatic subjects but macrophages do appear to exhibit increased expression of interferon regulatory factor (IRF)5+ and CD206+ macrophages (Draijer et al. 2017). This altered phenotype may relate to the production of proinflammatory mediators contributing to the inflammatory profile seen in asthma.

The role of macrophages in the resolution of inflammation is important as a failure of this function could lead to airway remodeling such as that seen in asthmatic airways. There has been interest in the role of specialized pro-resolving mediators (SPM) such as lipoxins, resolvins, and maresins amongst others (Barnig et al. 2018; Centanni et al. 2023). SPMs may have a role in improving macrophage phagocytosis (Rogerio et al. 2012; Serhan 2014) as well as shifting macrophages towards are more resolving phenotype (Dalli and Serhan 2012). A recent study has focused on the role of the alveolar macrophage in the production of mediators that could contribute to the resolution of inflammation in asthma. Townsend et al. (2023) found that activation of $P2X_7$ receptors on the surface of macrophages by exogenous ATP can induce the release of SPM in the short term, but longer-term exposure leads to the release of proinflammatory eicosanoids. Therefore, the balance between resolving and inflammatory mediators is carefully regulated, and more work is needed to understand how best to target macrophages towards a resolving phenotype.

Macrophages may have a role in exacerbations and a recent study using a mouse model of house dust mite-induced asthma showed that a component of CRL5, an E3 ubiquitin ligase, called cullin5, is upregulated in alveolar macrophages during viral-induced exacerbations (Zhang et al. 2024a). Knockout of this protein appeared to reduce the neutrophilic influx in this model and ameliorate the exacerbation. Cullin5 reduces IFNβ production and as such suppresses the anti-viral response (Zhang et al. 2024a). There is no evidence as yet whether cullin5 is upregulated in asthmatic macrophages, but it could provide a novel target for exacerbations in this condition.

A reduction in macrophage phagocytosis has also been reported for alveolar macrophages from asthmatic subjects. This has been observed in alveolar macrophages from children (Alexis et al. 2001; Fitzpatrick et al. 2008; Grunwell et al. 2019) and adults with neutrophilic, severe asthma (Liang et al. 2014). Similarly, there is an impairment in efferocytosis (Huynh et al. 2005; Simpson et al. 2013) that appears to be associated with the downregulation of the tyrosine kinase, Axl (Grabiec et al. 2017) suggesting activating this pathway may be beneficial. Other mechanisms of improving efferocytosis have also been explored including galectin-3 which improved uptake of apoptotic cells by monocyte-derived macrophages (Erriah et al. 2019).

9.2.3 *Idiopathic Pulmonary Fibrosis*

Idiopathic pulmonary fibrosis (IPF) is described as irreversible as lung tissue becomes thickened due to fibrosis leading to shortness of breath and has a poor prognosis with a survival of less than 5 years from diagnosis (Sharif 2017). It is relatively rare but incidence is increasing (Sharif 2017; Maher et al. 2021). The causes underlying this disease are not entirely clear but injury to the alveolar epithelium with the failure to repair leads to excessive deposition of extracellular matrix and fibrosis (Camelo et al. 2014). As such, this condition has been considered to be a disease of dysregulated mesenchymal cells such as fibroblasts and epithelial cells (Hewlett et al. 2018). However, the involvement of immune cells in the propagation of this disease has gained more interest (Mutsaers et al. 2023), in particular the role of the macrophage (Perrot et al. 2023).

Alveolar macrophages have a role to play in the resolution of inflammation, and if this becomes dysfunctional, aberrant repair may occur, and may lead to lung fibrosis (Fig. 9.3). Exactly how the macrophage may be involved is still under investigation but there is evidence of altered macrophage phenotypes in IPF (Morse et al. 2019; Reyfman et al. 2019) with reduced efferocytosis (Morimoto et al. 2012) and phagocytosis (Allden et al. 2019). There is also a shift in the macrophage population toward a lack of the transferrin receptor, CD71 (Allden et al. 2019), and as such these macrophages have a reduced capacity to sequester iron that represents a more immature phenotype (Morrell et al. 2023). Other changes in IPF macrophages have also been reported including upregulation of Mer tyrosine kinase (MERTK) (She et al. 2023). The relevance of this remains unclear as MERTK is important in

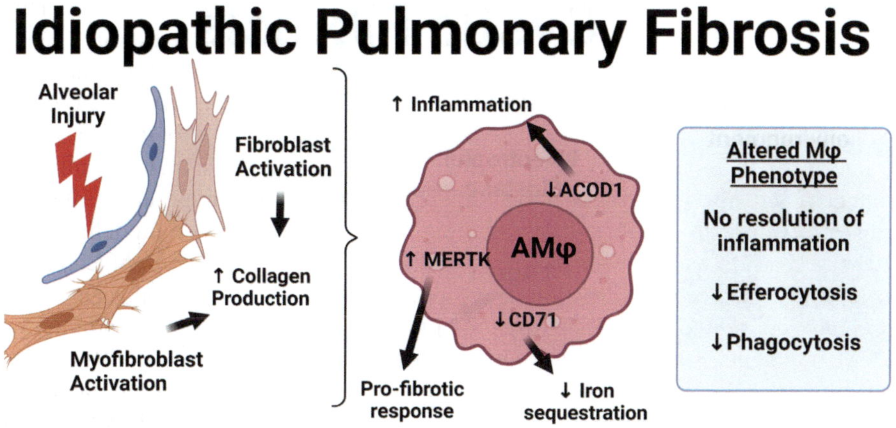

Fig. 9.3 Macrophages in idiopathic pulmonary fibrosis (IPF). IPF is characterized by injury to the alveolar epithelium with aberrant repair resulting in activation of fibroblasts and myofibroblasts, increased collagen production, fibrosis and inflammation. Alveolar macrophages in IPF are shifted towards a proinflammatory and profibrotic phenotype and display reduced phagocytic functions. In addition, these macrophages have reduced expression of transferrin receptor CD71 and ACOD1, and increased MERTK, which are involved in iron sequestration, increased inflammation and fibrosis development. (Figure created with Biorender.com). ACOD1: aconitate decarboxylase 1, AMφ: Alveolar macrophages, MERTK: Mer tyrosine kinase

regulating efferocytosis. In healthy macrophages, high expression of MERTK increases efferocytosis but this does not seem to be the case in macrophages from IPF subjects, but does appear to drive a pro-fibrotic response (She et al. 2023).

Macrophages from IPF patients also have decreased expression of aconitate decarboxylase 1 (ACOD1) which is associated with decreased levels of itaconate (Ogger et al. 2020). Itaconate is considered anti-inflammatory via the KEAP1-Nrf2 axis (Mills et al. 2018). Evidence for involvement in IPF has come from murine models where ACOD knockout mice show more severe disease (Ogger et al. 2020). Macrophages may contribute to disease via interactions with structural cells including lung fibroblasts. In health, it appears that interactions between alveolar macrophages and fibroblasts suppress the production of collagen but this is lost in IPF macrophages (Novak et al. 2023).

9.2.4 *Cystic Fibrosis*

Cystic fibrosis (CF) is a chronic progressive disease characterized by various mutations in the CF transmembrane regulator protein (CFTR) (Chmiel et al. 2002). Over the past few years, the advent of CFTR modulator therapies has led to a paradigm shift in how this disease is managed and this has begun to lead to investigations as to how these drugs may also be impacting macrophage function (Fig. 9.4). This is important as CFTR is expressed by macrophages and may have a role in the

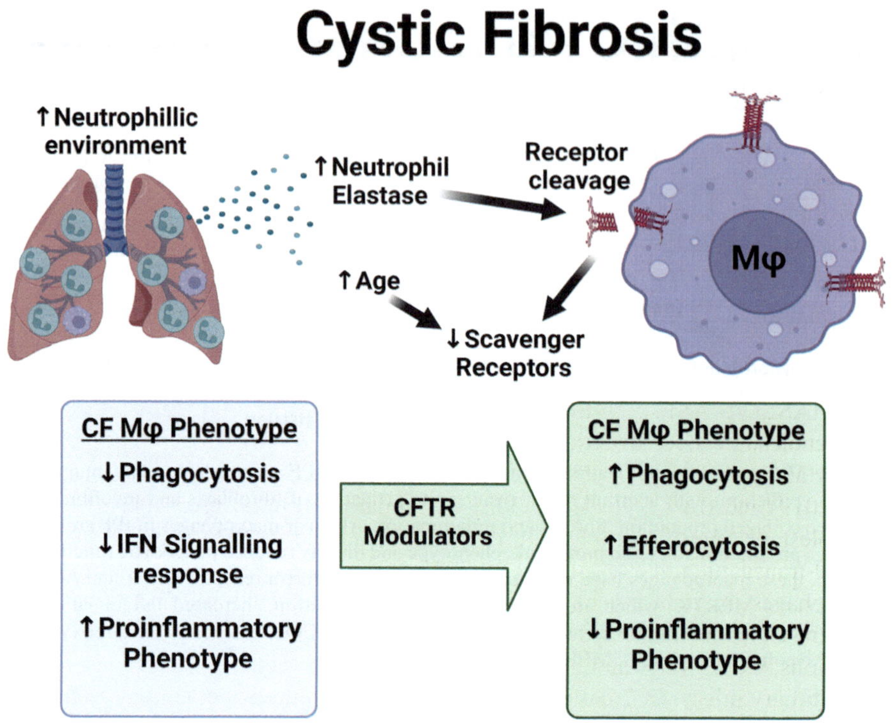

Fig. 9.4 Macrophages in cystic fibrosis. In the cystic fibrotic lung there are increased numbers of neutrophils, which secrete proinflammatory mediators and alter the lung environment. Alveolar macrophages from cystic fibrosis patients display reduced phagocytic function and an age dependent decrease in the expression of scavenger receptors, possibly due to cleavage of phagocytosis-associated proteins by mediators secreted by neutrophils, such as neutrophil elastase. Macrophages from cystic fibrosis patients also have reduced interferon (IFN) signalling response and a more proinflammatory phenotype. CF transmembrane regulator protein (CFTR) modulator treatments in macrophages from cystic fibrosis patients shifts their phenotype to increased phagocytic and efferocytosis function and reduced inflammation. (Figure created with Biorender.com)

acidification of the phagolysosome (Di et al. 2006) but there remains controversy as to whether CFTR is involved in bacterial killing with some suggesting not (Haggie and Verkman 2007) and more recently a more nuanced role (Das Gupta et al. 2024). Similar to other chronic lung diseases, macrophages from subjects with cystic fibrosis have decreased phagocytosis and increased proinflammatory phenotype (Simonin-Le Jeune et al. 2013). More recently, these cells also appear to have a reduced IFN signaling response (Gillan et al. 2023), again highlighting a role for macrophages in modulating the immune response in cystic fibrosis. In children with cystic fibrosis, there is an age-dependent decrease in the expression of scavenger receptors on alveolar macrophages (Slimmen et al. 2023). The mechanism underlying this observation is not clear but earlier studies have suggested that in a neutrophilic environment such as that in the cystic fibrosis lung, there are high levels of neutrophil elastase that cleave receptors such as those that have been shown for the phosphatidyl serine receptor (Alexis et al. 2006; Vandivier et al. 2002). It appears that this may be a more generalized effect on several proteins associated with the phagocytic process (Ma et al. 2021).

Studies have examined the effects of CFTR modulators *in vitro* using monocyte-derived macrophages and have shown that lumacaftor could improve phagocytosis and killing of *Pseudomonas aeruginosa* (Barnaby et al. 2018), but had little effect on cytokine expression. However, when used in combination with ivacaftor, there was a decrease in proinflammatory cytokines (Barnaby et al. 2018). Differential effects of these modulators on macrophage function have also been reported by others (Zhang et al. 2018). More recently a triple therapy of elexacaftor/tezacaftor/ivacaftor also shifts cystic fibrosis macrophages towards a more phagocytic but less inflammatory phenotype possibly due to altered mitochondrial activity (Aridgides et al. 2023). In a clinical study where triple therapy was given to patients, changes in the function of monocyte-derived macrophages increased over 12 months with similar changes in macrophage function including improved phagocytosis and efferocytosis (Zhang et al. 2023). Further work is ongoing to understand whether changes in the macrophage population in cystic fibrosis is due to the lung environment or the disease, but as new therapies emerge, it will be interesting to understand further how these will impact macrophages in the lung.

9.3 Conclusion

Recently, our understanding of the origin and function of macrophages in the lung has moved forward. However, it remains that there are considerable alterations in macrophage function associated with many chronic lung diseases. These effects on macrophage dysfunction may appear similar with differences in functions such as phagocytosis and efferocytosis, but the mechanisms underlying this appear to be different depending on disease. The impact of the lung environment is instrumental in driving macrophage phenotype, whether this is derived from resident macrophages or those that are newly recruited. Environmental challenges can not only include changes to lung architecture as seen in COPD and IPF but also external sources such as cigarette smoking and air pollution. The advent of vaping as a smoking cessation tool may also present new challenges in our understanding of macrophage function. Electronic (e)-cigarette exposure appears to alter macrophage functions *in vitro* (Scott et al. 2018; Jasper et al. 2021) and appears to lead to lipid-laden macrophages in the lungs of subjects using these devices (Maddock et al. 2019). The complex nature of these devices means elucidation of the mechanism of alteration in macrophage function is unclear as is how this will impact long-term users remains to be seen.

The advent of new technologies including spatial transcriptomics and scRNAseq that can detect granulocytes along with other techniques, will allow for mechanistic investigations into various aspects of macrophage biology. Further research into macrophage function and plasticity in the lung will no doubt lead to a better understanding of the underlying pathophysiologies of lung diseases with the hope of identifying novel therapeutic strategies.

References

Aegerter H, Lambrecht BN, Jakubzick CV (2022) Biology of lung macrophages in health and disease. Immunity 55:1564–1580

Alexis NE, Soukup J, Nierkens S, Becker S (2001) Association between airway hyperreactivity and bronchial macrophage dysfucntion in individuals with mild asthma. Am J Physiol - Lung Cell Mol Physiol 280:369–375

Alexis NE, Muhlebach MS, Peden DB, Noah TL (2006) Attenuation of host defense function of lung phagocytes in young cystic fibrosis patients. J Cyst Fibros 5:17–25

Allden SJ, Ogger PP, Ghai P, Mcerlean P, Hewitt R, Toshner R, Walker SA, Saunders P, Kingston S, Molyneaux PL, Maher TM, Lloyd CM, Byrne AJ (2019) The transferrin receptor CD71 delineates functionally distinct airway macrophage subsets during idiopathic pulmonary fibrosis. Am J Respir Crit Care Med 200:209–219

Aridgides DS, Mellinger DL, Gwilt LL, Hampton TH, Mould DL, Hogan DA, Ashare A (2023) Comparative effects of CFTR modulators on phagocytic, metabolic and inflammatory profiles of CF and nonCF macrophages. Sci Rep 13:11995

Arora S, Dev K, Agarwal B, Das P, Syed MA (2018) Macrophages: their role, activation and polarization in pulmonary diseases. Immunobiology 223:383–396

Baker EH, Baines DL (2018) Airway glucose homeostasis: a new target in the prevention and treatment of pulmonary infection. Chest 153:507–514

Baker JR, Vuppusetty C, Colley T, Papaioannou AI, Fenwick P, Donnelly L, Ito K, Barnes PJ (2016) Oxidative stress dependent microRNA-34a activation via PI3Kalpha reduces the expression of sirtuin-1 and sirtuin-6 in epithelial cells. Sci Rep 6:35871

Baker JR, Vuppusetty C, Colley T, Hassibi S, Fenwick PS, Donnelly LE, Ito K, Barnes PJ (2018) MicroRNA-570 is a novel regulator of cellular senescence and inflammaging. FASEB J: fj201800965R

Barnaby R, Koeppen K, Nymon A, Hampton TH, Berwin B, Ashare A, Stanton BA (2018) Lumacaftor (VX-809) restores the ability of CF macrophages to phagocytose and kill Pseudomonas aeruginosa. Am J Physiol Lung Cell Mol Physiol 314:L432–l438

Barnes PJ (2004) Macrophages as orchestrators of COPD. J COPD 1:50–70

Barnes PJ (2017) Senescence in COPD and its comorbidities. Annu Rev Physiol 79:517–539

Barnes PJ, Baker J, Donnelly LE (2019) Cellular senescence as a mechanism and target in chronic lung diseases. Am J Respir Crit Care Med 200:556–564

Barnig C, Frossard N, Levy BD (2018) Towards targeting resolution pathways of airway inflammation in asthma. Pharmacol Ther 186:98–113

Baßler K, Fujii W, Kapellos TS, Dudkin E, Reusch N, Horne A, Reiz B, Luecken MD, Osei-Sarpong C, Warnat-Herresthal S, Bonaguro L, Schulte-Schrepping J, Wagner A, Günther P, Pizarro C, Schreiber T, Knoll R, Holsten L, Kröger C, De Domenico E, Becker M, Händler K, Wohnhaas CT, Baumgartner F, Köhler M, Theis H, Kraut M, Wadsworth MH 2nd, Hughes TK, Ferreira HJ, Hinkley E, Kaltheuner IH, Geyer M, Thiele C, Shalek AK, Feißt A, Thomas D, Dickten H, Beyer M, Baum P, Yosef N, Aschenbrenner AC, Ulas T, Hasenauer J, Theis FJ, Skowasch D, Schultze JL (2022) Alveolar macrophages in early stage COPD show functional deviations with properties of impaired immune activation. Front Immunol 13:917232

Behmoaras J, Gil J (2021) Similarities and interplay between senescent cells and macrophages. J Cell Biol 220

Belchamber KBR, Donnelly LE (2020) Targeting defective pulmonary innate immunity - a new therapeutic option? Pharmacol Ther 209:107500

Belchamber KBR, Singh R, Batista CM, Whyte MK, Dockrell DH, Kilty I, Robinson MJ, Wedzicha JA, Barnes PJ, Donnelly LE (2019) Defective bacterial phagocytosis is associated with dysfunctional mitochondria in COPD macrophages. Eur Respir J

Berenson CS, Garlipp MA, Grove LJ, Maloney J, Sethi S (2006) Impaired phagocytosis of non-typeable *Haemophilus influenzae* by human alveolar macrophages in chronic obstructive pulmonary disease. J Infect Dis 194:1375–1384

Bewley MA, Preston JA, Mohasin M, Marriott HM, Budd RC, Swales J, Collini P, Greaves DR, Craig RW, Brightling CE, Donnelly LE, Barnes PJ, Singh D, Shapiro SD, Whyte MKB, Dockrell DH (2017) Impaired mitochondrial microbicidal responses in chronic obstructive pulmonary disease macrophages. Am J Respir Crit Care Med 196:845–855

Byrne AJ, Powell JE, O'sullivan BJ, Ogger PP, Hoffland A, Cook J, Bonner KL, Hewitt RJ, Wolf S, Ghai P, Walker SA, Lukowski SW, Molyneaux PL, Saglani S, Chambers DC, Maher TM, Lloyd CM (2020) Dynamics of human monocytes and airway macrophages during healthy aging and after transplant. J Exp Med 217

Camelo A, Dunmore R, Sleeman MA, Clarke DL (2014) The epithelium in idiopathic pulmonary fibrosis: breaking the barrier. Front Pharmacol 4

Carleton HM (1927) The origin of dust-cells in the lung. J Cell Sci s2-71:223–237

Castro M, Wenzel SE, Bleecker ER, Pizzichini E, Kuna P, Busse WW, Gossage DL, Ward CK, Wu Y, Wang B, Khatry DB, Van Der Merwe R, Kolbeck R, Molfino NA, Raible DG (2014) Benralizumab, an anti-interleukin 5 receptor α monoclonal antibody, versus placebo for uncontrolled eosinophilic asthma: a phase 2b randomised dose-ranging study. Lancet Respir Med 2:879–890

Centanni D, Henricks PAJ, Engels F (2023) The therapeutic potential of resolvins in pulmonary diseases. Eur J Pharmacol 958:176047

Chana KK, Fenwick PS, Nicholson AG, Barnes PJ, Donnelly LE (2014) Identification of a distinct glucocorticosteroid-insensitive pulmonary macrophage phenotype in patients with chronic obstructive pulmonary disease. J Allergy Clin Immunol 133:207-16 e1-11

Chmiel JF, Berger M, Konstan MW (2002) The role of inflammation in the pathophysiology of CF lung disease. Clin Rev Allergy Immunol 23:5–27

Christenson SA, Smith BM, Bafadhel M, Putcha N (2022) Chronic obstructive pulmonary disease. Lancet 399:2227–2242

Collaborators GBDCRD (2020) Prevalence and attributable health burden of chronic respiratory diseases, 1990-2017: a systematic analysis for the Global Burden of Disease Study 2017. Lancet Respir Med 8:585–596

Dagher R, Kumar V, Copenhaver AM, Gallagher S, Ghaedi M, Boyd J, Newbold P, Humbles AA, Kolbeck R (2022) Novel mechanisms of action contributing to benralizumab's potent anti-eosinophilic activity. Eur Respir J 59

Dalli J, Serhan CN (2012) Specific lipid mediator signatures of human phagocytes: microparticles stimulate macrophage efferocytosis and pro-resolving mediators. Blood 120:e60–e72

Das Gupta K, Curson JEB, Tarique AA, Kapetanovic R, Schembri MA, Fantino E, Sly PD, Sweet MJ (2024) CFTR is required for zinc-mediated antibacterial defense in human macrophages. Proc Natl Acad Sci U S A 121:e2315190121

Di A, Brown ME, Deriy LV, Li C, Szeto FL, Chen Y, Huang P, Tong J, Naren AP, Bindokas V, Palfrey HC, Nelson DJ (2006) CFTR regulates phagosome acidification in macrophages and alters bactericidal activity. Nat Cell Biol 8:933–944

Draijer C, Boorsma CE, Robbe P, Timens W, Hylkema MN, Ten Hacken NH, Van Den Berge M, Postma DS, Melgert BN (2017) Human asthma is characterized by more IRF5+ M1 and CD206+ M2 macrophages and less IL-10+ M2-like macrophages around airways compared with healthy airways. J Allergy Clin Immunol 140:280–283.e3

Erriah M, Pabreja K, Fricker M, Baines KJ, Donnelly LE, Bylund J, Karlsson A, Simpson JL (2019) Galectin-3 enhances monocyte-derived macrophage efferocytosis of apoptotic granulocytes in asthma. Respir Res 20:1

Feldman N, Rotter-Maskowitz A, Okun E (2015) DAMPs as mediators of sterile inflammation in aging-related pathologies. Ageing Res Rev 24:29–39

Finney LJ, Belchamber KBR, Fenwick PS, Kemp SV, Edwards MR, Mallia P, Donaldson G, Johnston SL, Donnelly LE, Wedzicha JA (2019) Human rhinovirus impairs the innate immune response to bacteria in alveolar macrophages in chronic obstructive pulmonary disease. Am J Respir Crit Care Med 199:1496–1507

Fitzpatrick AM, Holguin F, Teague WG, Brown LA (2008) Alveolar macrophage phagocytosis is impaired in children with poorly controlled asthma. J Allergy Clin Immunol 121:1372-8, 1378 e1-3

Franken L, Schiwon M, Kurts C (2016) Macrophages: sentinels and regulators of the immune system. Cell Microbiol 18:475–487

Fricker M, Qin L, Sánchez-Ovando S, Simpson JL, Baines KJ, Riveros C, Scott HA, Wood LG, Wark PA, Kermani NZ, Chung KF, Gibson PG (2022) An altered sputum macrophage transcriptome contributes to the neutrophilic asthma endotype. Allergy 77:1204–1215

Galván-Peña S, O'neill LA (2014) Metabolic reprograming in macrophage polarization. Front Immunol 5:420

George SN, Garcha DS, Mackay AJ, Patel AR, Singh R, Sapsford RJ, Donaldson GC, Wedzicha JA (2014) Human rhinovirus infection during naturally occurring COPD exacerbations. Eur Respir J 44:87–96

Gibbings SL, Goyal R, Desch AN, Leach SM, Prabagar M, Atif SM, Bratton DL, Janssen W, Jakubzick CV (2015) Transcriptome analysis highlights the conserved difference between embryonic and postnatal-derived alveolar macrophages. Blood 126:1357–1366

Gillan JL, Chokshi M, Hardisty GR, Clohisey Hendry S, Prasca-Chamorro D, Robinson NJ, Lasota B, Clark R, Murphy L, Whyte MKB, Baillie JK, Davidson DJ, Bao G, Gray RD (2023) CAGE sequencing reveals CFTR-dependent dysregulation of type I IFN signaling in activated cystic fibrosis macrophages. Sci Adv 9:eadg5128

Gina (2023) Global Initiative for Asthma (GINA) Main Report [Online]. Available: https://ginasthma.org/2023-gina-main-report/ [Accessed]

Grabiec AM, Denny N, Doherty JA, Happonen KE, Hankinson J, Connolly E, Fife ME, Fujimori T, Fujino N, Goenka A, Holden S, Tavernier G, Shah R, Cook PC, Macdonald AS, Niven RM, Dahlbäck B, Fowler SJ, Simpson A, Hussell T (2017) Diminished airway macrophage expression of the Axl receptor tyrosine kinase is associated with defective efferocytosis in asthma. J Allergy Clin Immunol 140:1144–1146.e4

Grunwell JR, Stephenson ST, Tirouvanziam R, Brown LAS, Brown MR, Fitzpatrick AM (2019) Children with neutrophil-predominant severe asthma have proinflammatory neutrophils with enhanced survival and impaired clearance. J Allergy Clin Immunol Pract 7:516–525.e6

Guilliams M, De Kleer I, Henri S, Post S, Vanhoutte L, De Prijck S, Deswarte K, Malissen B, Hammad H, Lambrecht BN (2013) Alveolar macrophages develop from fetal monocytes that differentiate into long-lived cells in the first week of life via GM-CSF. J Exp Med 210:1977–1992

Haggie PM, Verkman AS (2007) Cystic fibrosis transmembrane conductance regulator-independent phagosomal acidification in macrophages. J Biol Chem 282:31422–31428

Hall BM, Balan V, Gleiberman AS, Strom E, Krasnov P, Virtuoso LP, Rydkina E, Vujcic S, Balan K, Gitlin I, Leonova KI, Consiglio CR, Gollnick SO, Chernova OB, Gudkov AV (2017) p16(Ink4a) and senescence-associated beta-galactosidase can be induced in macrophages as part of a reversible response to physiological stimuli. Aging (Albany NY) 9:1867–1884

Han MK, Huang YJ, Lipuma JJ, Boushey HA, Boucher RC, Cookson WO, Curtis JL, Erb-Downward J, Lynch SV, Sethi S, Toews GB, Young VB, Wolfgang MC, Huffnagle GB, Martinez FJ (2012) Significance of the microbiome in obstructive lung disease. Thorax 67:456–463

Hashimoto D, Chow A, Noizat C, Teo P, Beasley MB, Leboeuf M, Becker CD, See P, Price J, Lucas D, Greter M, Mortha A, Boyer SW, Forsberg EC, Tanaka M, Van Rooijen N, García-Sastre A, Stanley ER, Ginhoux F, Frenette PS, Merad M (2013) Tissue-resident macrophages self-maintain locally throughout adult life with minimal contribution from circulating monocytes. Immunity 38:792–804

Hewlett JC, Kropski JA, Blackwell TS (2018) Idiopathic pulmonary fibrosis: epithelial-mesenchymal interactions and emerging therapeutic targets. Matrix Biol 71–72:112–127

Hickman E, Smyth T, Cobos-Uribe C, Immormino R, Rebuli ME, Moran T, Alexis NE, Jaspers I (2023) Expanded characterization of in vitro polarized M0, M1, and M2 human monocyte-derived macrophages: bioenergetic and secreted mediator profiles. PLoS One 18:e0279037

Hodge S, Hodge G, Ahern J, Jersmann H, Holmes M, Reynolds PN (2007) Smoking alters alveolar macrophage recognition and phagocytic ability: implications in chronic obstructive pulmonary disease. Am J Respir Cell Mol Biol 37:748–755

Hou F, Xiao K, Tang L, Xie L (2021) Diversity of macrophages in lung homeostasis and diseases. Front Immunol 12:753940

Hough KP, Curtiss ML, Blain TJ, Liu R-M, Trevor J, Deshane JS, Thannickal VJ (2020) Airway remodeling in asthma. Front Med 7

Huang X, Cao M, Xiao Y (2023) Alveolar macrophages in pulmonary alveolar proteinosis: origin, function, and therapeutic strategies. Front Immunol 14:1195988

Huynh ML, Malcolm KC, Kotaru C, Tilstra JA, Westcott JY, Fadok VA, Wenzel SE (2005) Defective apoptotic cell phagocytosis attenuates prostaglandin E2 and 15-hydroxyeicosatetraenoic acid in severe asthma alveolar macrophages. Am J Respir Crit Care Med 172:972–979

Ito K, Barnes PJ (2009) COPD as a disease of accelerated lung aging. Chest 135:173–180

Jasper AE, Sapey E, Thickett DR, Scott A (2021) Understanding potential mechanisms of harm: the drivers of electronic cigarette-induced changes in alveolar macrophages, neutrophils, and lung epithelial cells. Am J Physiol Lung Cell Mol Physiol 321:L336–l348

Kulikauskaite J, Wack A (2020) Teaching Old Dogs New Tricks? The Plasticity of Lung Alveolar Macrophage Subsets. Trends Immunol 41:864–877

Lacey DC, Achuthan A, Fleetwood AJ, Dinh H, Roiniotis J, Scholz GM, Chang MW, Beckman SK, Cook AD, Hamilton JA (2012) Defining GM-CSF and macrophage-CSF-dependent macrophage responses by in vitro models. J Immunol 188:5752–5765

Lee KA, Flores RR, Jang IH, Saathoff A, Robbins PD (2022) Immune senescence, immunosenescence and aging. Front Aging 3:900028

Liang Z, Zhang Q, Thomas CM, Chana KK, Gibeon D, Barnes PJ, Chung KF, Bhavsar PK, Donnelly LE (2014) Impaired macrophage phagocytosis of bacteria in severe asthma. Respir Res 15:72

Liu Z, Gu Y, Chakarov S, Bleriot C, Kwok I, Chen X, Shin A, Huang W, Dress RJ, Dutertre CA, Schlitzer A, Chen J, Ng LG, Wang H, Liu Z, Su B, Ginhoux F (2019) Fate mapping via Ms4a3-expression history traces monocyte-derived cells. Cell 178:1509–1525.e19

Ma J, Kummarapurugu AB, Hawkridge A, Ghosh S, Zheng S, Voynow JA (2021) Neutrophil elastase-regulated macrophage sheddome/secretome and phagocytic failure. Am J Physiol Lung Cell Mol Physiol 321:L555–l565

Maddock SD, Cirulis MM, Callahan SJ, Keenan LM, Pirozzi CS, Raman SM, Aberegg SK (2019) Pulmonary lipid-laden macrophages and vaping. N Engl J Med 381:1488–1489

Mah J, Ritchie AI, Finney LJ (2024) Selected updates on chronic obstructive pulmonary disease. Curr Opin Pulm Med 30

Maher TM, Bendstrup E, Dron L, Langley J, Smith G, Khalid JM, Patel H, Kreuter M (2021) Global incidence and prevalence of idiopathic pulmonary fibrosis. Respir Res 22:197

Martin FP, Jacqueline C, Poschmann J, Roquilly A (2021) Alveolar macrophages: adaptation to their anatomic niche during and after inflammation. Cells 10

Mccubbrey AL, Curtis JL (2013) Efferocytosis and lung disease. Chest 143:1750–1757

Mercado N, Thimmulappa R, Thomas CM, Fenwick PS, Chana KK, Donnelly LE, Biswal S, Ito K, Barnes PJ (2011) Decreased histone deacetylase 2 impairs Nrf2 activation by oxidative stress. Biochem Biophys Res Commun 406:292–298

Mills EL, Ryan DG, Prag HA, Dikovskaya D, Menon D, Zaslona Z, Jedrychowski MP, Costa ASH, Higgins M, Hams E, Szpyt J, Runtsch MC, King MS, Mcgouran JF, Fischer R, Kessler BM, Mcgettrick AF, Hughes MM, Carroll RG, Booty LM, Knatko EV, Meakin PJ, Ashford MLJ, Modis LK, Brunori G, Sévin DC, Fallon PG, Caldwell ST, Kunji ERS, Chouchani ET, Frezza C, Dinkova-Kostova AT, Hartley RC, Murphy MP, O'neill LA (2018) Itaconate is an anti-inflammatory metabolite that activates Nrf2 via alkylation of KEAP1. Nature 556:113–117

Morales-Nebreda L, Misharin AV, Perlman H, Budinger GR (2015) The heterogeneity of lung macrophages in the susceptibility to disease. Eur Respir Rev 24:505–509

Morimoto K, Janssen WJ, Terada M (2012) Defective efferocytosis by alveolar macrophages in IPF patients. Respir Med 106:1800–1803

Morrell ED, Holton SE, Lawrance M, Orlov M, Franklin Z, Mitchem MA, Deberg H, Gersuk VH, Garay A, Barnes E, Liu T, Peltan ID, Rogers A, Ziegler S, Wurfel MM, Mikacenic C (2023) The transcriptional and phenotypic characteristics that define alveolar macrophage subsets in acute hypoxemic respiratory failure. Nat Commun 14:7443

Morse C, Tabib T, Sembrat J, Buschur KL, Bittar HT, Valenzi E, Jiang Y, Kass DJ, Gibson K, Chen W, Mora A, Benos PV, Rojas M, Lafyatis R (2019) Proliferating SPP1/MERTK-expressing macrophages in idiopathic pulmonary fibrosis. Eur Respir J 54:1802441

Murray MA, Chotirmall SH (2015) The impact of immunosenescence on pulmonary disease. Mediators Inflamm 2015:692546

Mutsaers SE, Miles T, Prêle CM, Hoyne GF (2023) Emerging role of immune cells as drivers of pulmonary fibrosis. Pharmacol Ther 252:108562

Novak CM, Sethuraman S, Luikart KL, Reader BF, Wheat JS, Whitson B, Ghadiali SN, Ballinger MN (2023) Alveolar macrophages drive lung fibroblast function in cocultures of IPF and normal patient samples. Am J Physiol Lung Cell Mol Physiol 324:L507–l520

Ogger PP, Albers GJ, Hewitt RJ, O'sullivan BJ, Powell JE, Calamita E, Ghai P, Walker SA, Mcerlean P, Saunders P, Kingston S, Molyneaux PL, Halket JM, Gray R, Chambers DC, Maher TM, Lloyd CM, Byrne AJ (2020) Itaconate controls the severity of pulmonary fibrosis. Sci Immunol 5:eabc1884

Perrot CY, Karampitsakos T, Herazo-Maya JD (2023) Monocytes and macrophages: emerging mechanisms and novel therapeutic targets in pulmonary fibrosis. Am J Physiol Cell Physiol 325:C1046–c1057

Pham TH, Damera G, Newbold P, Ranade K (2016) Reductions in eosinophil biomarkers by benralizumab in patients with asthma. Respir Med 111:21–29

Reyfman PA, Walter JM, Joshi N, Anekalla KR, Mcquattie-Pimentel AC, Chiu S, Fernandez R, Akbarpour M, Chen CI, Ren Z, Verma R, Abdala-Valencia H, Nam K, Chi M, Han S, Gonzalez-Gonzalez FJ, Soberanes S, Watanabe S, Williams KJN, Flozak AS, Nicholson TT, Morgan VK, Winter DR, Hinchcliff M, Hrusch CL, Guzy RD, Bonham CA, Sperling AI, Bag R, Hamanaka RB, Mutlu GM, Yeldandi AV, Marshall SA, Shilatifard A, Amaral LAN, Perlman H, Sznajder JI, Argento AC, Gillespie CT, Dematte J, Jain M, Singer BD, Ridge KM, Lam AP, Bharat A, Bhorade SM, Gottardi CJ, Budinger GRS, Misharin AV (2019) Single-cell transcriptomic analysis of human lung provides insights into the pathobiology of pulmonary fibrosis. Am J Respir Crit Care Med 199:1517–1536

Rogerio AP, Haworth O, Croze R, Oh SF, Uddin M, Carlo T, Pfeffer MA, Priluck R, Serhan CN, Levy BD (2012) Resolvin D1 and aspirin-triggered resolvin D1 promote resolution of allergic airways responses. J Immunol 189:1983–1991

Russell RJ, Boulet LP, Brightling CE, Pavord ID, Porsbjerg C, Dorscheid D, Sverrild A (2024) The airway epithelium: an orchestrator of inflammation, a key structural barrier and a therapeutic target in severe asthma. Eur Respir J

Ryan EM, Sadiku P, Coelho P, Watts ER, Zhang A, Howden AJM, Sanchez-Garcia MA, Bewley M, Cole J, Mchugh BJ, Vermaelen W, Ghesquiere B, Carmeliet P, Rodriguez Blanco G, Von

Kriegsheim A, Sanchez Y, Rumsey W, Callahan JF, Cooper G, Parkinson N, Baillie K, Cantrell DA, Mccafferty J, Choudhury G, Singh D, Dockrell DH, Whyte MKB, Walmsley SR (2023) NRF2 activation reprograms defects in oxidative metabolism to restore macrophage function in chronic obstructive pulmonary disease. Am J Respir Crit Care Med 207:998–1011

Saeed S, Quintin J, Kerstens HH, Rao NA, Aghajanirefah A, Matarese F, Cheng SC, Ratter J, Berentsen K, Van Der Ent MA, Sharifi N, Janssen-Megens EM, Ter Huurne M, Mandoli A, Van Schaik T, Ng A, Burden F, Downes K, Frontini M, Kumar V, Giamarellos-Bourboulis EJ, Ouwehand WH, Van Der Meer JW, Joosten LA, Wijmenga C, Martens JH, Xavier RJ, Logie C, Netea MG, Stunnenberg HG (2014) Epigenetic programming of monocyte-to-macrophage differentiation and trained innate immunity. Science 345:1251086

Sagiv A, Krizhanovsky V (2013) Immunosurveillance of senescent cells: the bright side of the senescence program. Biogerontology 14:617–628

Salvi S, Barnes PJ (2010) Is exposure to biomass smoke the biggest risk factor for COPD globally? Chest 138:3–6

Schuliga M, Read J, Knight DA (2021) Ageing mechanisms that contribute to tissue remodeling in lung disease. Ageing Res Rev 70:101405

Scott A, Lugg ST, Aldridge K, Lewis KE, Bowden A, Mahida RY, Grudzinska FS, Dosanjh D, Parekh D, Foronjy R, Sapey E, Naidu B, Thickett DR (2018) Pro-inflammatory effects of e-cigarette vapour condensate on human alveolar macrophages. Thorax 73:1161–1169

Serhan CN (2014) Pro-resolving lipid mediators are leads for resolution physiology. Nature 510:92–101

Sharif R (2017) Overview of idiopathic pulmonary fibrosis (IPF) and evidence-based guidelines. Am J Manag Care 23:S176–s182

Sharma R (2021) Perspectives on the dynamic implications of cellular senescence and immunosenescence on macrophage aging biology. Biogerontology 22:571–587

She Y, Xu X, Yu Q, Yang X, He J, Tang XX (2023) Elevated expression of macrophage MERTK exhibits profibrotic effects and results in defective regulation of efferocytosis function in pulmonary fibrosis. Respir Res 24:118

Simonin-Le Jeune K, Le Jeune A, Jouneau S, Belleguic C, Roux PF, Jaguin M, Dimanche-Boitre MT, Lecureur V, Leclercq C, Desrues B, Brinchault G, Gangneux JP, Martin-Chouly C (2013) Impaired functions of macrophage from cystic fibrosis patients: CD11b, TLR-5 decrease and sCD14, inflammatory cytokines increase. PLoS One 8:e75667

Simpson JL, Gibson PG, Yang IA, Upham J, James A, Reynolds PN, Hodge S, Group ASR (2013) Impaired macrophage phagocytosis in non-eosinophilic asthma. Clin Exp Allergy 43:29–35

Singh R, Belchamber KBR, Fenwick PS, Chana K, Donaldson G, Wedzicha JA, Barnes PJ, Donnelly LE, Consortium C (2021) Defective monocyte-derived macrophage phagocytosis is associated with exacerbation frequency in COPD. Respir Res 22:113

Slimmen LJM, Giacalone VD, Schofield C, Horati H, Manaï B, Estevão SC, Garratt LW, Peng L, Tirouvanziam R, Janssens HM, Unger WWJ (2023) Airway macrophages display decreased expression of receptors mediating and regulating scavenging in early cystic fibrosis lung disease. Front Immunol 14:1202009

Suzuki T, Arumugam P, Sakagami T, Lachmann N, Chalk C, Sallese A, Abe S, Trapnell C, Carey B, Moritz T, Malik P, Lutzko C, Wood RE, Trapnell BC (2014) Pulmonary macrophage transplantation therapy. Nature 514:450–454

Tan SY, Krasnow MA (2016) Developmental origin of lung macrophage diversity. Development 143:1318–1327

Taylor AE, Finney-Hayward TK, Quint JK, Thomas CM, Tudhope SJ, Wedzicha JA, Barnes PJ, Donnelly LE (2010) Defective macrophage phagocytosis of bacteria in COPD. Eur Respir J 35:1039–1047

Tomita K, Caramori G, Lim S, Ito K, Hanazawa T, Oates T, Chiselita I, Jazrawi E, Chung KF, Barnes PJ, Adcock IM (2002) Increased p21(CIP1/WAF1) and B cell lymphoma leukemia-x(L) expression and reduced apoptosis in alveolar macrophages from smokers. Am J Respir Crit Care Med 166:724–731

Townsend EA, Guadarrama A, Shi L, Roti Roti E, Denlinger LC (2023) P2X(7) signaling influences the production of pro-resolving and pro-inflammatory lipid mediators in alveolar macrophages derived from individuals with asthma. Am J Physiol Lung Cell Mol Physiol 325:L399–l410

Vandivier RW, Fadok VA, Hoffmann PR, Bratton DL, Penvari C, Brown KK, Brain JD, Accurso FJ, Henson PM (2002) Elastase-mediated phosphatidylserine receptor cleavage impairs apoptotic cell clearance in cystic fibrosis and bronchiectasis. J Clin Invest 109:661–670

Wedzicha JA, Donaldson GC (2003) Exacerbations of chronic obstructive pulmonary disease. Respir Care 48:1204–1213; discussion 1213-5

Wohnhaas CT, Baßler K, Watson CK, Shen Y, Leparc GG, Tilp C, Heinemann F, Kind D, Stierstorfer B, Delić D, Brunner T, Gantner F, Schultze JL, Viollet C, Baum P (2024) Monocyte-derived alveolar macrophages are key drivers of smoke-induced lung inflammation and tissue remodeling. Front Immunol 15:1325090

Woods PS, Kimmig LM, Meliton AY, Sun KA, Tian Y, O'leary EM, Gökalp GA, Hamanaka RB, Mutlu GM (2020) Tissue-resident alveolar macrophages do not rely on glycolysis for LPS-induced inflammation. Am J Respir Cell Mol Biol 62:243–255

Wrench CL, Baker JR, Monkley S, Fenwick PS, Murray L, Donnelly LE, Barnes PJ (2024) Small airway fibroblasts from patients with chronic obstructive pulmonary disease exhibit cellular senescence. Am J Physiol Lung Cell Mol Physiol 326:L266–l279

Zhang S, Shrestha CL, Kopp BT (2018) Cystic fibrosis transmembrane conductance regulator (CFTR) modulators have differential effects on cystic fibrosis macrophage function. Sci Rep 8:17066

Zhang S, Shrestha CL, Robledo-Avila F, Jaganathan D, Wisniewski BL, Brown N, Pham H, Carey K, Amer AO, Hall-Stoodley L, Mccoy KS, Bai S, Partida-Sanchez S, Kopp BT (2023) Cystic fibrosis macrophage function and clinical outcomes after elexacaftor/tezacaftor/ivacaftor. Eur Respir J 61

Zhang H, Xue K, Li W, Yang X, Gou Y, Su X, Qian F, Sun L (2024a) Cullin5 drives experimental asthma exacerbations by modulating alveolar macrophage antiviral immunity. Nat Commun 15:252

Zhang WX, Neupane AS, David BA, Ginhoux F, Vargas ESCF, Kubes P (2024b) A functional assessment of fetal liver and monocyte-derived macrophages in the lung alveolar environment. J Immunol 212:1012–1021

Chapter 10
Macrophages in Lung Repair and Fibrosis

Yago A. P. Jannini-Sá, Brecht Creyns, Cory M. Hogaboam, William C. Parks, and Miriam S. Hohmann

Abstract Macrophages are key regulators of tissue repair and fibrosis. Following injury, macrophages undergo marked phenotypic and functional changes to play crucial roles throughout the phases of tissue repair. Idiopathic Pulmonary Fibrosis, which is the most common fibrosing lung disease, has been described as an aberrant reparative response to repetitive alveolar epithelial injury in a genetically susceptible aging individual. The marked destruction of the lung architecture results from the excessive secretion of extracellular matrix by activated fibroblasts and myofibroblasts. Accumulating evidence suggests that macrophages have a pivotal regulatory role in pulmonary fibrosis. The origins and characteristics of macrophages in the lung and their role in regulating lung homeostasis, repair, and fibrosis are reviewed herein. We discuss recent studies that have employed single-cell RNA-sequencing to improve the identification and characterization of macrophage populations in the context of homeostatic and fibrotic conditions. We also discuss the current understanding of the macrophage-mediated mechanisms underlying the initiation and progression of pulmonary fibrosis, with a focus on the phenotypic and functional changes that aging macrophages acquire and how these changes ultimately contribute to age-related chronic lung diseases.

10.1 Introduction

Macrophages, large eaters from the Greek (macro=large, phage=eaters), are engulfment-specialized cells. As part of the innate immune system, these cells represent an important component of the first-line defense against pathogens and tumor

Y. A. P. Jannini-Sá · B. Creyns · C. M. Hogaboam · W. C. Parks · M. S. Hohmann (✉)
Women's Guild Lung Institute, Division of Pulmonary & Critical Care Medicine, Department of Medicine, Cedars-Sinai Medical Center, Los Angeles, CA, USA
e-mail: Miriam.hohmann@cshs.org

M. Kloc et al. (eds.), *Monocytes and Macrophages in Development, Regeneration, and Disease*, Results and Problems in Cell Differentiation 74,
https://doi.org/10.1007/978-3-031-65944-7_10

cells. Macrophages are also able to clear old, senescent, and dead cells, as well as noxious particles (e.g., silica, titanium, iron), fulfilling important functions in tissue homeostasis (Byrne et al. 2015). Macrophages have important regulating functions during tissue repair, regeneration, and fibrosis. Thus, disturbances in macrophage function can lead to aberrant mechanisms of repair and development of pathological fibrosis (Wynn and Vannella 2016). In this context, much focus has been directed at understanding their potential profibrotic role in fibrotic diseases (Bargagli et al. 2011; Byrne et al. 2016).

Macrophages are impacted by aging and their responses to injury and infections are greatly altered. Macrophages acquire features of senescence, altered immune responses, and function, as well as higher expression levels of pro-inflammatory and -fibrotic mediators, all of which seem to contribute to the development of pathological fibrosis (McQuattie-Pimentel et al. 2021; Mogilenko et al. 2021; Morimoto et al. 2012; Cho and Stout-Delgado 2020).

In this chapter, we discuss the origins, characteristics, and functions of macrophages in the lung. We discuss recent robust studies that have employed single-cell RNA-sequencing (scRNA-seq) to improve the identification and characterization of macrophage populations, as well as their role in homeostasis and fibrosis. Great focus is given to discussing the phenotypic and functional changes that aging macrophages acquire and how these changes ultimately contribute to chronic age-related lung diseases such as Idiopathic Pulmonary Fibrosis (IPF). Lastly, the recent advances in therapeutic candidates focused on modulating macrophages are reviewed.

10.2 Functional Heterogeneity of Macrophages

10.2.1 Origin, Distribution, and Heterogeneity

To better understand the origin of macrophages, we need to delve into embryology to highlight their development and distribution (Fig. 10.1). Most studies on macrophage development have tracked their origins in hematopoiesis during embryogenesis in small mammals, mostly mice. These studies have found that macrophages originate from three successive waves of hematopoiesis. The first wave, known as primitive hematopoiesis, starts in an extraembryonic structure known as the yolk sac at 2nd week post conception. This extends about to the 5th week post-fertilization, giving rise to early erythrocytes, endothelial cells and to the first CSF1R+ hematopoietic progenitor cells in humans (Holt and Jones 2000). In mice, the first wave starts 7 days post-conception. The first wave continues with the migration of these progenitor cells to regions known as para-aortic splanchnopleure (P-Sp) and aorta-gonad-mesonephros (AGM) areas, which are committed to the expansion and generation of macrophages (Huang and Auerbach 1993; Hoeffel et al. 2015; Lin et al. 2014).

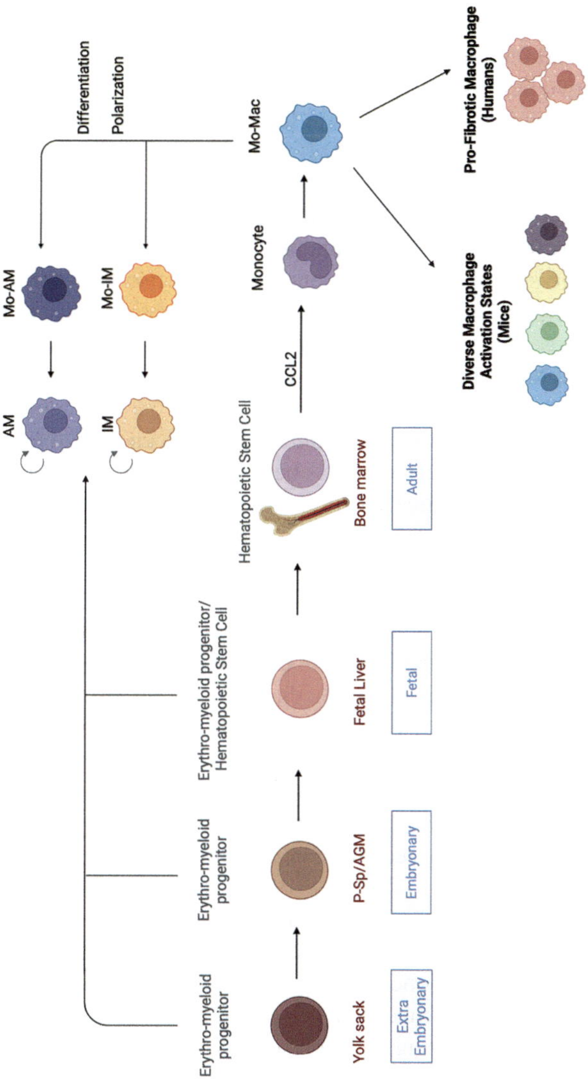

Fig. 10.1 Macrophage origin and fate in the lung

During early embryonic development, macrophages originate from the yolk sac (YS)—the initial site of their precursors. These YS progenitors then give rise to aorta-gonad-mesonephros (AGM) progenitors, intermediate cells originating from the para-aortic splanchnopleure (P-Sp) and AGM areas. As development progresses, these macrophage precursors further mature in the Fetal Liver (FL). In adulthood, FL progenitors migrate to the bone marrow, where they become the source of all blood cells. Erythro-myeloid progenitors (primitive macrophages), generated in both the yolk-sac and fetal liver, subsequently migrate to the lung. In the lung, they differentiate into two key populations: Alveolar Macrophages (AM), residing in the lung alveoli, and playing essential roles in lung defense, surfactant clearance, and tissue repair. Interstitial Macrophages (IM) are found in various tissues, including connective tissue, skin and organs, and contribute to tissue remodeling, wound healing, and immune responses. Monocytes (Mo), circulating immune cells originating from the bone marrow, exhibit remarkable plasticity and they can differentiate into various macrophage subsets. Notably, monocytes give rise to Monocyte-recruited Macrophages (Mo-Mac), which can acquire diverse activation states or become pro-fibrotic macrophages. The diverse activation states in mice have been described as M1, M2a, M2b, and M2c phenotypes. Mo-Macs also replenish alveolar and interstitial macrophage populations.

These cells migrate from the yolk sac and AGM between the 6th up to 18th week to the fetal liver. In mice, this period occurs between E10.5 and E13.5 (Hoeffel et al. 2015). In this phase, erythro-myeloid progenitor cells, commonly recognized by their expression of C-Myb(+), are described as the adult tissue-resident macrophage progenitors (Hoeffel et al. 2015; Guilliams et al. 2013).

By the 10th week, the embryonic mesoderm gives rise to the spongy and soft tissue inside the bones called the bone marrow. The cavities of the bone marrow are populated by migrating hematopoietic stem cells (HSC) establishing the conditions for the second wave and definitive HSC-mediated hematopoiesis, replacing erythrocytes, granulocytes, and monocytes, but not macrophages in most tissues (Gomez Perdiguero et al. 2015). In the adult lungs, monocytes, alveolar macrophages (AMs; found within the alveolus), and interstitial macrophages (IMs; found in the surrounding tissue) can be found; however, the origins of these subpopulations are still debated. By employing scRNA-seq, we can distinguish these populations by their gene expression.

Resident macrophages are named Langerhans cells in the skin, Küpffer cells in the liver, microglia in the central nervous system, red pulp macrophages in the spleen, and osteoclasts in the bone. Although these cells are mostly formed during embryogenesis, tissue-resident macrophages can be repopulated by auto-replication or by circulating monocytes during the adult stage (Guilliams et al. 2013; Hashimoto et al. 2013).

With next-generation sequencing pipelines, the "Immune progenitor" atlas was developed and the gaps in knowledge on the origin of the cells and their trajectory are being filled. A scRNA-seq analysis of 293095 human immune cells from 367 prenatal samples highlighted the presence of a pro-angiogenic macrophage (PraM) characterized by the expression of vascular endothelial growth factor A (VEGFA) and chemokine (C-X-C motif) ligand 8 (CXCL8) and the gene ontology analysis positioned these cells as active in different vascular and endothelial pathways. In the lung, the progenitor for PraM (pre-PraM) is the major population in the human lung between Carnegie Stage (CS) 19 and 23. After the 10th week post conception, the pre-PraM population starts to decline and the PraM becomes the major macrophage population in human fetal lung (Wang et al. 2023a). The correlation between the signatures observed in pre-PraM/PraM and AM/IM requires more study.

Alveolar macrophages originate mainly during the first and second wave of hematopoiesis, while monocytes are generated by the bone marrow. In *Rhesus macaque,* AM and IM constitute 70% of all immune cells in homeostasis. AMs are found only in the bronchoalveolar lavage fluid (BALF) and represent up to 90% of all leukocytes (Cai et al. 2014). Although the self-renewal of AMs is limited, its population can be replenished by recruited circulating monocytes or IMs (Cai et al. 2014). Granulocyte-macrophage colony-stimulating factor (GM-CSF) is known to modulate AM fate. Ablation of the production of GM-CSF by alveolar type II cells (ATII) in the pre-natal stage affects macrophage population and depletes this specific population from adult lungs (Guilliams et al. 2013).

There is still debate if interstitial macrophages are pre- or post-natal. Fate studies using $S100a4^{Cre}$-$Rosa^{EYFP}$ mice (bone morrow cells express EYFP constitutively) showed that almost 50% of the AM and IM express EYFP at birth but an increase in EYFP expression/detection was only found in the population of IM over time, indicating that bone marrow-derived monocytes can repopulate the IM niche in the lung. These monocyte-derived IM can be clustered in two distinct subpopulations, $MERTK^{+}CD64^{+}$ and $Lyve1^{lo}MHCII^{hi}$ or $Lyve1^{hi}MHCII^{lo}$. Each of these subpopulations distributes differentially through lung parenchyma and plays distinct roles in lung homeostasis (Chakarov et al. 2019).

10.2.2 *Lung Macrophages*

10.2.2.1 Resident Lung Macrophages: Gatekeepers of Pulmonary Health

Resident lung macrophages, predominantly found in the alveolar spaces, are critical components of the pulmonary immune system. As sentinels of the respiratory tract, these cells play a fundamental role in immune surveillance, tissue homeostasis, and response to environmental challenges. Here we describe the distinct characteristics, functions, and regulatory mechanisms of resident lung macrophages.

Originating from fetal liver monocytes are seeded in the first three days after birth. Resident macrophages exhibit self-renewal capacity, encompassing 90–95% of the cellular content in the steady state, thereby ensuring a stable population through local proliferation and maintenance (Guilliams et al. 2013; Guilliams and Svedberg 2021; Evren et al. 2022). Resident macrophages maintain tissue integrity and promote homeostasis through efficient phagocytosis, thereby clearing dying cells and preventing infections (Byrne et al. 2015). Based on their localization, resident macrophages are divided into two groups, AMs and IMs.

AMs are characterized by their unique localization within the lung alveoli, near the epithelial surface of the respiratory system, and the capillaries to sample and examine the air-borne and blood-borne material. Differentiation at birth and maintenance during the life of AMs are dependent on alveolar epithelial cell-derived GM-CSF and autocrine Tumor growth factor-β1 (TGF-β1) (Shi et al. 2021). These cells possess distinct morphological features that facilitate direct interaction with the alveolar epithelium (Wynn and Vannella 2016). As the total number of alveoli is well above the number of AMs in steady state, AMs are motile unlike other tissue-resident macrophages, and can cover approximately three alveoli through pores of Kohn, to ensure their role as vigilant guardians of the respiratory system for inhaled pathogens, particulate matter, and cellular debris (Westphalen et al. 2014; Neupane et al. 2020).

AMs are housekeeping cells of the lung. They are continuously phagocytosing debris in the alveolar lumen and are essential for catabolism of surfactant proteins produced by AT2 (Han and Mallampalli 2015; Hartl et al. 2018). By keeping the airways clean, AMs help maintain efficient gas exchange and prevent the triggering of an inflammatory reaction by inhaled particulates. Beyond their phagocytic capabilities, resident lung macrophages modulate immune responses by releasing extracellular vesicles and immunosuppressive factors (Branchett et al. 2021; Bissonnette et al. 2020). AM-derived extracellular vesicles contain suppressor of cytokine production proteins, which inhibit inflammatory STAT pathway in epithelial cells (Bissonnette et al. 2020). AM-derived TGF-β1 prevents self- activation and promotes the recruitment of regulatory T cells (Branchett et al. 2021). Nevertheless, the autocrine TGF-β1-positive feedback loop present in AMs and AT2-lineage cells is also critical in non-inflammatory lung fibrogenesis (Enomoto et al. 2023).

Upon injury or infection, Toll-like Receptors (TLRs) on AMs are activated by damage-associated molecular patterns (DAMPS) and pathogen-associated molecular patterns (PAMPS) resulting in the release of chemokines and either pro-inflammatory mediators like IL-1β, IL-6, and TNF-α, or anti-inflammatory mediators, including IL-10 and TGF-β1. Severe injury or infection, including Sars-Cov2, can result in the depletion of AMs, which will be replenished by recruited monocytes and by self-renewal (Bain et al. 2022). Additionally, AMs play a central role in tissue repair by releasing growth factors and participating in extracellular matrix (ECM) remodeling, crucial to restoring normal tissue architecture post-injury. Thus, dysregulation of these mechanisms can lead to alterations in macrophage function, contributing to the pathogenesis of lung diseases. By adopting a pro-inflammatory phenotype, can perpetuate inflammation and tissue damage and his series of events has been described in diseases such as chronic obstructive pulmonary disease (COPD) and some interstitial lung diseases (ILDs). Conversely, in fibrotic lung diseases, resident macrophages contribute to dysregulated repair processes by altering collagen synthesis and stimulating fibroblasts through the secretion of inflammatory mediators, such as CCL18, milk fat globule epidermal growth factor 8 (MFG-E8), fibroblast specific protein 1 (FSP-1), and matrix metalloproteinases (MMPs), which leads to the deposition of excess ECM (Dancer et al. 2011). Understanding the intricate interplay between resident lung macrophages and disease pathogenesis is essential for developing targeted therapeutic interventions.

In contrast, IMs are a more heterogeneous population derived from bone marrow and embryonic progenitor cells that can be replenished by both self-renewal and circulating monocytes (Chakarov et al. 2019; Evren et al. 2022). IMs are located in the parenchyma of the lung near bronchi, nerves, and blood vessels and are important for their homeostasis (Chakarov et al. 2019; Guilliams and Svedberg 2021; Kulle et al. 2022; Schyns et al. 2019). IMs are mainly distinguished from AM as smaller, shorter-lived cells that express CD36 in humans (Byrne et al. 2015, 2016; Leach et al. 2020). IMs are mainly immunoregulatory with a high capacity for antigen presentation (Chakarov et al. 2019; Misharin et al. 2013; Aran et al. 2019; Sajti et al. 2020).

While all IMs were found to share MERTK, CD64, CD11b, and CX3CR1 expression, scRNA-seq in mice identified Lyve1^{lo} MHCII^{hi} CX3CR1^{hi} IMs near nerves and Lyve1^{hi} MHCII^{lo}CX3CR1^{lo} IMs near blood vessels (Chakarov et al. 2019; Gibbings et al. 2015). Based on multiple scRNA-seq studies in the last decades, IMs are currently characterized as Lyve1^{hi} CD206^{hi} IMs and CX3CR1^{hi} MHCII^{hi} IMs based (Guilliams and Svedberg 2021). Furthermore, IMs are not in a defined state, since they are able to convert to AMs during pulmonary fibrosis (Misharin et al. 2013; Aran et al. 2019).

10.2.2.2 Infiltrating Lung Macrophages: Orchestrators of Pulmonary Immune Responses

Infiltrating lung macrophages are a diverse population of cells, including monocyte-derived macrophages and tissue-resident macrophages with migratory capabilities. These cells exhibit plasticity in response to local cues, adopting diverse phenotypes that contribute to both pro-inflammatory and anti-inflammatory responses. Unlike resident AMs, infiltrating macrophages are recruited from the circulation in response to various stimuli (Kulle et al. 2022; Herold et al. 2011a), such as infection, DAMPs, and more. The recruitment and differentiation of infiltrating macrophages to affected sites are tightly regulated by chemokines, cytokines, and adhesion molecules in the lung microenvironment (Kulle et al. 2022; Allard et al. 2018). Differentiated infiltrating lung macrophages phagocytose pathogens, dead cells, and debris and release pro-inflammatory cytokines and chemokines, promoting the recruitment of other immune cells and amplifying the immune response. Thereby infiltrating macrophages are essential for the elimination of pathogens and kick-starting an effective immune defense (Allard et al. 2018; Puttur et al. 2019).

10.2.3 Classification of Macrophages

For some time, the classification of human macrophages was limited to surface marker detection by flow cytometry and was defined as highly autofluorescent HLA-DR$^+$, CD43$^+$, CD88$^+$, CD169$^+$, CD206$^+$, CD10$^+$ (MME), CD11c$^+$, CD36$^+$, CD141$^+$, CD64$^+$, MARCO$^+$, CD14low AMs (Misharin et al. 2013; Bharat et al. 2016; Snelgrove et al. 2008). However, scRNA-seq of healthy human AMs highlights the heterogeneity of what was previously thought to be a homogenous population of AMs (Mould et al. 2021). Heterogeneity of AMs was universally observed as AM subclusters could be confirmed in both COPD and COVID-19 (Li et al. 2022). At least four superclusters of AMs, each comprising 14 subclusters, with *IFI27* and *APOC2* identified as the genes defining differential expression. Heterogeneity seems to be largely influenced by the tightly regulated production of specific combinations of chemokines, metallothioneins, IFN-inducible genes, cholesterol-biosynthesis-related genes, and IGF1 by subsets of AMs.

Unlike that for AMs, scRNA-seq has been hampered by the need for tissue digestion and redundancy of markers between IMs, AMs, and monocytes. Multiple groups identified two different subsets of IMs based in the level of CD206 expression; CD206^{hi} and CD206^{int} IMs (Chakarov et al. 2019; Lim et al. 2018; Ural et al. 2020). More recently, three IM subtypes have been identified in the lung, and confirmed across multiple organs, based on the expression of *Timd4, Lyve1, Folr2, Ccr2*, and *H2-Ab* (Dick et al. 2022). While these markers may be able to distinguish IM subtypes in homeostasis, the picture is more complicated after injury or infection, as expression of most IM markers is induced in infiltrating macrophages.

10.2.3.1 Shortcomings of M1/M2 Macrophage Dichotomy

Since about 2000, macrophages have been classified as either M1 or M2 macrophage subtypes based on the T helper (Th)1 and Th2 concept (Mills et al. 2000; Mosmann et al. 1986). In a steady state, macrophages reside in nearly every organ, including the lung, independent of monocytes. However, in response to a trigger, macrophages adapt to the local microenvironment, thereby showing a high degree of flexibility and adaptability. However, while the M1/M2 paradigm is based on a simplistic division of stimulation (LPS vs Interleukin-4 (IL-4)), pathway (STAT1 vs STAT6) and resulting cytokine production (TNF-α, IL-1 and nitric oxide vs IL-10, TGF-β1, CCL17, and CCL22) *in vitro*, this is an oversimplification of the polarized extremes of these cells that neglects the spectrum of cells that encompasses these extremes (Spiller et al. 2016; Isshiki et al. 2023). Moreover, although function and gene expression are shared between human and murine macrophages, markers established to define murine M2 macrophages like Arg-1, YM-1, and Fizz1 are not expressed in humans. Comparative studies identified only TGM2 as a shared gene and protein marker in both human and murine resting M2 macrophages (Martinez et al. 2013). Furthermore, a paradigm based on the *in vitro* behavior of macrophages will never completely translate to the *in vivo* complexity of these cells. Although dividing macrophages into M1 vs. M2 classes oversimplifies the complex continuum of functional and reversible states that these immune cells exist *in vivo* (Murray et al. 2014; Kratz et al. 2014; Xue et al. 2014), in this chapter we occasionally use the M1/M2 nomenclature as a shortcut for macrophage activation into either pro-inflammatory or immunosuppressive states, respectively.

Transcriptome-Based Network Analysis of stimulated human macrophages by Xue J *et al* revealed a spectrum of macrophage activation states extending the M1/M2-polarization model. In COPD the authors confirmed a complete loss of the profound signature of enriched modules observed in smokers, while IFN-γ stimulated macrophage subsets characterized by loss of antigen processing and regulation of immune response were depleted in these patients (Xue et al. 2014; Natoli and Monticelli 2014).

While current single cell sequencing approaches are undeniably valuable to unraveling the macrophage spectrum, they are not unequivocal and an enormous effort will be required to integrate human and murine transcriptome datasets, thereby avoiding redundancy of subsets that are identified as macrophages between different disease settings. Identification of human-specific regulation of gene expression will require new methodologies to study gene regulation in an entirely human context and without the respective animal models. Currently, in IFP, macrophages can be divided into three main populations based on single-cell transcriptomic analysis: FABP4^{hi}, SPP1^{hi} or IPFe, and FCN1^{hi} macrophages (Aran et al. 2019; Adams et al. 2020a; Ayaub et al. 2021; Morse et al. 2019; Reyfman et al. 2019). Whereby SPP1^{hi} macrophages were identified as the main pro-fibrotic subset, which could also be detected in patients suffering from lung fibrosis after a long COVID infection but are absent in COPD (Adams et al. 2020a; Wendisch et al. 2021).

10.3 Macrophages in Health and Disease

10.3.1 Macrophages in Health

Maintenance of pulmonary homeostasis is reliant on the presence and functions of resident macrophages. These cells patrol the alveolar space, recognizing foreign bodies and self-structures, and are constantly interacting with the lung epithelium. AMs perform tissue surveillance, clearance of pathogens, cellular debris, and noxious particles, as well as the catabolism of surfactant proteins. In parallel, interstitial macrophages play a role in immunoregulation and monocyte-derived macrophages (Mo-Macs) maintain the macrophage population stable in the lungs. Mo-Macs are a source of new interstitial and alveolar macrophages when these cells die (Dong et al. 2020; Nagre et al. 2019; Bedoret et al. 2009).

The crosstalk between macrophages and other cells within the environment contributes to lung homeostasis. For example, AMs are in physical contact with alveolar type 1 (ATI) and type 2 (ATII) cells, and their interactions are critical to maintaining conditions for efficient gas exchange. The ablation of GM-CSF production by ATII cells in the lung of adult mice, for example, decreases the AM population which results in pulmonary alveolar proteinosis due to lack of surfactant turnover (Guilliams et al. 2013). When lung macrophages phagocytose apoptotic ATII cells, an immunosuppressive phenotype is induced, observed as increased expression of Arginase and TGF-β1 in mice (Kim et al. 2018), the generation of these immunosuppressive niches helps to protect the fragile alveolar structure from a harsh inflammatory process. Macrophages can also interact with collagen through the leukocyte-associated immunoglobulin-like receptor 1 (LAIR1). In the absence of this receptor, the lack of interaction between the macrophages and the ECM triggers aberrant proliferation and apoptosis of monocytes, impairing the heterogeneity of interstitial macrophages in the lung (Keerthivasan et al. 2021). Moreover, AM transit in and between the alveoli is an important feature of these cells. Indeed, the global knockout of LFA-1 (CD11a), an integrin important for macrophage alveolar transit, reduces their motility and consequently increases BALF turbidity and number of cellular debris, and disrupts phagocytosis of bacteria (Neupane et al. 2020).

10.3.2 Macrophages in Lung Repair

Macrophages actively participate in the resolution of inflammation and the promotion of tissue repair after injury or infection (Byrne et al. 2015; Watanabe et al. 2019). During sterile LPS-induced inflammation in mice, IM expansion is mediated by $CCR2^+$ monocyte influx, and transient loss of AMs is restored by local proliferation only (Sabatel et al. 2017; Liu et al. 2019). In contrast during severe inflammation, including infections like influenza and Sars-CoV-2, monocyte-derived myeloid cells repopulate the lung after the loss of AMs, thereby reshaping the airway environment (Liu et al. 2019; Liao et al. 2020; Sheng et al. 2015). Acute lung injury (ALI) can lead to acute respiratory distress syndrome (ARDS), which is the primary cause of death in critical patients (Matthay et al. 2019).

10.3.2.1 Phagocytosis of Debris and Pathogens

After an injury or infection in the lung, tissue-resident AMs are the first line of defense against pathogens. Macrophages are recruited to the site to phagocytose cellular debris, apoptotic cells, and pathogens (Arandjelovic and Ravichandran 2015). PAMPs or DAMPS are released from cells dying from pyroptosis and necroptosis, thereby recruiting and activating neutrophils, monocytes, and macrophages. This process helps clear the damaged or infected tissue, preparing the ground for repair. Macrophage phagocytic function is regulated by multiple pathways, including β-adrenergic receptor, α7 nicotinic acetylcholine receptor, apoptosis inhibitor of macrophage (AIM), and transient receptor potential vanilloid 4 (TRPV4). TRPV4, a non-selective mechanosensitive ion channel, can be triggered by exogenous ligands and changes in ECM stiffness, thereby increasing macrophage phagocytosis during resolution of inflammation and remodeling (Nguyen et al. 2021; Scheraga et al. 2016; Dutta et al. 2020). Furthermore, catecholamines can signal through the β2-adrenergic receptor, resulting in alternate activation with enhanced macrophage phagocytosis and reduction of inflammation (Grailer et al. 2014; Freire et al. 2022). Neutrophils, the first recruited immune cells to arrive at an infection site, are the predominant infected phagocytic cells. Apoptosis of neutrophils can be induced by macrophage-derived IL-1β, resulting in the expression of find me signals including lysophosphatidylcholine, ATP, fractalkine (CX3CL1), and eat me signals like phosphatidylserine. Apoptotic cells are absorbed by opsonins, such as Mfge8 and AnxA1, inducing binding of Protein S and GAS6 that are recognized by macrophage receptors G2A and the TAM receptor tyrosine kinase family (Tyro3, Axl, and Mer), that results in attenuated inflammatory signaling pathways in macrophages (Nishi et al. 2019; Park and Kim 2017; Zheng et al. 2021; A-Gonzalez et al. 2017). Clearance processes for damaged and dying neutrophils are commonly attributed to macrophages and dendritic cells, thereby halting inflammation, limiting tissue injury, and promoting resolution (Singhal and Kumar 2022; Herold et al. 2011b). AIM can prevent macrophages from phagocytosing apoptotic neutrophils, accompanied by exacerbated histopathological damage and inflammation in the lungs. These experimental results further indicate that AIM regulates the resolution of inflammation (Kimura et al. 2017; Greenlee-Wacker 2016). Aging, smoking, and severe systemic diseases, such as trauma and sepsis, impair the phagocytosis of TR-AMs, thereby reducing their ability to fight pathogens (O'Beirne et al. 2020; Roquilly et al. 2020).

Microvesicles or exosomes originating from recovering cell populations can induce restorative pathways via signaling molecules including SOCS2 or signaling microRNAs (Bourdonnay et al. 2015). Additionally, regulatory T cells directly (CD40/CD800) and indirectly (amphiregulin, IL-10, and TGF-β) interact with macrophages to activate anti-inflammatory signaling pathways in macrophages (Arpaia et al. 2015). Furthermore, cytokine secretion by eosinophils and group 2 innate lymphoid cells (ILC2s) can contribute to a macrophage phenotypic switch by IL-4, IL-13, and amphiregulin production. Hijacking of these pathways in allergic airway diseases or by *Pseudomonas aeruginosa* can lead to chronicity of inflammation leading to persistent anti-inflammatory macrophage activation (Roquilly et al. 2017; Monticelli et al. 2016). After systemic inflammation, both circulating monocytes and alveolar macrophages undergo epigenetic reprogramming that can last for months impairing their activity, partly regulated by SIRPa (Roquilly et al. 2020; de Vos and van der Poll 2020; Bordon 2020).

10.3.2.2 Anti-Inflammatory Cytokine Production

Macrophages produce anti-inflammatory cytokines, such as IL10 and TGF-β1, growth factors [VEGF, platelet-derived growth factor A (PDGF-A)], itaconate and osteopontin, several proteinases (e.g., MMP-8, -10, -28 and various cathepsins), (Wynn and Vannella 2016). Macrophages are a major source of pro-resolving lipids like resolvins, protectins, and maresins, long-chain fatty acid–derived lipid mediators that drive the resolution of tissue injury in a range of pathologies. These lipid mediators help to dampen the inflammatory response, crucial for preventing excessive inflammation and tissue damage. Additionally, *in vitro* efferocytosis promotes a switch to a glycolytic phenotype with the extracellular release of lactate, which signals to inhibit inflammatory responses. This response to corpse uptake occurred via the upregulation of solute carrier family 16 member 1 (SLC16A1), which is part of a family of transporters involved in the sensing, sampling, and ingestion of apoptotic material (Morioka et al. 2018). Additionally, AMs express anti-inflammatory receptors such as TGF-R2 and FPR2 and synthesize increased concentrations of immune regulatory intracellular messengers such as cAMP (Varga et al. 2016).

Metabolic control of macrophage function differs depending on the phase of injury. During active inflammation, there is a shift from oxidative phosphorylation to aerobic glycolysis, known as the Warburg effect (O'Neill and Grahame Hardie 2013). Glycolysis is the main source of energy in inflammatory macrophages, while disruption of the Krebs cycles results in the accumulation of citrate and succinate, two proinflammatory molecules, via prostaglandin and HIF1α generation, respectively (O'Neill 2015; Ganeshan and Chawla 2014; Mills et al. 2016; Infantino et al. 2011). In contrast, anti-inflammatory macrophages get their energy from fatty acid oxidation through the Krebs cycle, thereby producing itaconate, which in turn activates Kelch-like ECH-associated protein-1, a key repressor of the anti-inflammatory transcription factor Nrf2 in macrophages (Mills et al. 2018). The enzyme that generates itaconate, IRG1, is involved in the induction of a pro-resolving macrophage phenotype (Michelucci et al. 2013).

10.3.2.3 Repair of the Endothelial and Epithelial Cell Barrier

A balanced crosstalk between epithelial cells and AMs is essential in maintaining homeostasis. During viral infections, endothelial cell (EC) pattern recognition receptors (PRRs) induce the production of IL-1β, IL-6, IL-8, TNF, GM-CSF, MIP-1α, RANTES, and MCP-1, which recruit and activate lung macrophages. Furthermore, expression of immunomodulatory surface molecules on ECs, such as CD47, PDL-1, and CD200 modulate the function of AMs by binding to SIRPα, CD200R, and PD-1, thereby respectively downregulating phagocytosis and dampening AM cytokine secretion (Snelgrove et al. 2008; Liu et al. 2020; Logtenberg et al. 2020; Barclay and Van Den Berg 2014). The close interaction between AMs and ECs allows ECs to further regulate macrophage function through IL-10 and

TGF-β production. Disruption of this interaction either through pathogens or epithelial damage leads to the activation of AMs and inflammation. Macrophages are also important in the process of restoring the epithelial barrier. After lung injury, goblet cells are replenished by *Scgb1a1*+cells, and the differentiation of ATII to ATI is impaired in the absence of CCR2, indicating a role for macrophages in epithelial lung regeneration (Lechner et al. 2017).

10.3.2.4 Stem Cell Recruitment and Activation

Macrophages produce chemotactic signals, including stromal cell-derived factor-1 (SDF-1) and monocyte chemoattractant protein-1 (MCP-1), that attract circulating stem cells to the site of injury essential for efficient regeneration of damaged lung tissue (Seo et al. 2022; Xu et al. 2007). Once recruited, macrophages create a microenvironment for stem cell survival and differentiation by secreting TGF-β, VEGF, and interleukins. Additionally, macrophages release factors that stimulate angiogenesis, including VEGF-A, PDGF, and fibroblast growth factors (FGFs), promoting the formation of new blood vessels. This is crucial for supplying nutrients and oxygen to the repairing tissue (Seo et al. 2022; Xu et al. 2007).

10.3.2.5 Resolution of Edema

Lung macrophages play a crucial role in the complex interplay associated with pulmonary edema due to their close association with vessels (Lapenna et al. 2018). Their involvement is two-fold, contributing to both the initiation and resolution of edema. In the context of inflammation, macrophages release inflammatory mediators that disrupt the integrity of the alveolar barrier, leading to increased permeability and fluid accumulation. This pro-inflammatory function exacerbates edema formation. Conversely, during the resolution phase, macrophages actively participate in clearing edematous fluid and debris, promoting tissue repair and restoring barrier function (Bissinger et al. 2021). The delicate balance between the pro-inflammatory and pro-resolving actions of lung macrophages highlights their dual role in the pathogenesis and resolution of pulmonary edema. Understanding these dynamics provides insights into potential therapeutic strategies for managing lung diseases characterized by disrupted fluid homeostasis.

10.3.3 Macrophages and Lung Fibrosis

It is still not entirely clear if macrophages initiate the fibrotic process or the inflammatory process that precedes parenchyma remodeling. However, these cells are involved in the pathogenesis of several fibrotic diseases. Single-cell RNA-seq of cells from the lungs of IPF patients has provided valuable insights into macrophage behavior during the fibrotic process, moreover it has allowed the identification of a specific subset of macrophages often referred to as IPF-mac or pro-fibrotic macrophages. However, several questions remain unanswered: Are these pro-fibrotic macrophages a cause or consequence of fibrosis? What impact do they have on the development of pulmonary fibrosis?

Joshi et al. used lineage tracing models to compare a fibrogenic (Asbestos) and a non-fibrogenic model (TiO2) to understand the macrophage population dynamics in the fibrotic response and niches (Joshi et al. 2020). In both models, the number of AMs (SiglecF^{hi} - 14 days post-challenge) showed a similar increase in AM numbers, but asbestos induced the recruitment of Mo-Macs (Pro-Fibrotic macrophage), which was not observed with TiO2 (Joshi et al. 2020). Additionally, the authors observed that resident AMs did not contribute directly to the fibrotic niche, but monocyte-derived AMs did. These cells can sustain themselves by autocrine activation of the M-CSF (Csf1)/M-CSFR pathway, which was shown to be targetable and reduce Mo-Mac populations (Joshi et al. 2020). A similar strategy was employed to map macrophage fate in a model of bleomycin lung injury. Bleomycin challenge reduced the numbers of tissue-resident AMs and a new population was recruited, referred to as Mo-AMs. Interestingly, these cells remain as part of the lung macrophage population after the resolution phase, suggesting a main role for circulating monocytes in replacing resident cells after loss or death (Misharin et al. 2017). Corroborating with previous findings (Joshi et al. 2020), depletion of tissue-resident macrophages did not protect the lung from bleomycin injury, although the induction of necroptosis in the Mo-AM subpopulation attenuated the fibrotic response, highlighting the heterogeneity of these cells (Misharin et al. 2017).

Multiple studies have highlighted a consistent set of genes associated with these macrophages. Among them, *SPP1, FN1, CHI3L1, MMP9, MMP7*, and *MERTK* are commonly upregulated in this profibrotic subpopulation in humans (Morse et al. 2019; Lv et al. 2023; Adams et al. 2020b). To uncover a shared pro-fibrotic macrophage phenotype, researchers integrated scRNA-seq datasets from human lung and liver fibrosis. Surprisingly, both fibrotic conditions exhibited a subpopulation of macrophages characterized by upregulation of *SPP1, GPNMB, CD9, TREM2, FABP5*, and *CD63*. These findings predict a common pro-fibrotic macrophage signature across different fibrotic diseases (Fabre et al. 2023). *In vitro* stimulation of monocytes with IL-17, TGF-β1, or GM-CSF led to an increase in CD9$^+$TREM2$^+$ macrophages, mirroring the profibrotic phenotype observed *in vivo*. Furthermore, therapeutic strategies targeting these mediators (TGF-β1, IL-17, or GM-CSF) showed promising results. In the liver fibrosis model induced by CCl4, blocking any of these three mediators resulted in reduced pro-fibrotic macrophages. Similarly, neutralizing TGF-β1 in a lung fibrosis model had a similar inhibitory effect on CD9$^+$TREM2$^+$SPP1$^+$ macrophage numbers (Fabre et al. 2023). Interestingly, *GPNMB*, another gene associated with fibrosis, was found to be upregulated in lung fibrosis induced by silica exposure in mice. Notably, depleting macrophages using liposome-clodronate improved fibrotic outcomes and reduced GPNMB production (Wang et al. 2023b). The integration of scRNA-seq datasets is a powerful tool for understanding how the profibrotic niche shapes macrophage populations. By employing similar analysis pipelines, researchers have identified *MCMP1* as an upregulated gene in IPF macrophages and lungs. MCMP1 plays a role in monocyte and alveolar macrophage chemotaxis and can be directly stimulated by TGF-β (Perrot et al. 2024).

Despite the limitations of gene expression in fully capturing macrophage function, experimental evidence supports a profibrotic profile in certain macrophage subpopulations. For instance, interstitial lung macrophages collected from bleomycin-challenged mice, when cultured and their conditioned media transferred to lung fibroblasts, resulted in increased α-SMA and collagen production (Liborio-Ramos et al. 2023). Macrophages are an important source of mediators that are considered biomarkers in lung fibrosis. Elevated levels of CCL18 in the BALF from IPF patients, primarily produced by macrophages, are closely associated with an increased risk of acute exacerbations (Schupp et al. 2015).

In the context of pulmonary fibrosis, immune responses are impaired, affecting phagocytosis and efferocytosis. Silica particles can activate the RhoA/Rock pathway in human Mo-Macs, leading to impaired efferocytosis (Lescoat et al. 2020). A similar phenotype with impaired phagocytosis has been observed in macrophages from bleomycin-challenged mice (Warheit-Niemi et al. 2022). Notably, this phenotype is also present in human lung macrophages from IPF patients (Morimoto et al. 2012). The recovery of efferocytosis capacity in mice and human macrophages was achieved when the RhoA/Rock pathway was inhibited or MERTK was regulated.

10.4 Macrophage Aging

Aging of the immune system dramatically increases the susceptibility to numerous chronic diseases and infections (Mahrouf-Yorgov et al. 2011; Sharma et al. 2009; Franceschi et al. 2018). The hallmark of immune aging is chronic, sterile, low-grade inflammation—called inflammaging—and a signature feature of this process is the chronic activation of the innate immune system, in which macrophages have a central role (Franceschi et al. 2000). In fact, it has been proposed that aging is an outcome of persistent stress impinging on macrophages, one of the major target cells in this process (Franceschi et al. 2000). Indeed, aging results in a plethora of functional and phenotypic changes in macrophages, among them changes in cytokine production, immunity, and role in tissue homeostasis.

10.4.1 Cytokine Production by Aging Macrophages

Inflammaging is characterized by high systemic levels of inflammatory cytokines such as IL-6, IL-8, and TNF in the elderly and macrophages are thought to be a major, although not exclusive, cellular source of these mediators (Clark and Peterson 1994; Ershler et al. 1993; Hasegawa et al. 2000). Markedly increased pro-inflammatory cytokine levels and mice death upon systemic immune stimulation (high-dose anti-CD40/IL-2 or IL-2/IL-12) can be prevented by depleting macrophages in the aged mice (Bouchlaka et al. 2013). Increased production of TNF and IL-6 in aged human and mouse precursor-derived macrophages in response to LPS-mediated TLR activation has been reported (Bouchlaka et al. 2013). This elevation in inflammatory response has been linked to a systemic increase in the expression and activation of components of the NLRP3 inflammasome in mouse myeloid cells. Furthermore, aged macrophages exhibit impaired termination of TLR responses compared to controls, *i.e.*, peritoneal macrophages from aged and young mice stimulated with LPS, exhibited similar levels of *Tnf* transcript, but blocking TLR4 signaling led to prolonged elevation of *Tnf, Il-1β*, and *Il-6* transcripts in aged macrophages compared to young ones (Pattabiraman et al. 2017). It is noteworthy that there have been contradicting reports on the age-dependent changes in TLR responses of macrophages. For example, diminished pro-inflammatory cytokine responses to TLR2 and TLR4 ligands in aged splenic macrophages compared to young ones have been described (Boehmer et al. 2005) and these differential responses may be due to the variations in experimental conditions and origin of the cells.

Activated macrophages from aged mice also produce more prostaglandin (PG) E_2, attributed to enhanced expression of cyclooxygenase 2 activity, than their younger counterparts (Chen et al. 2013). PGE_2 suppresses IL-12 secretion and enhances IL-10 secretion, which contributes to the age-associated decline in T-cell function (Kalim and Groettrup 2013; van der Pouw Kraan et al. 1995). Indeed, IL10 production is enhanced in stimulated mouse splenic macrophages and BMDM from aged mice (Chelvarajan et al. 2006).

10.4.2 Functional and Phenotypic Changes in Aging Macrophages

Unlike the output of cells of lymphoid lineage by hematopoietic stem cells that decline because of aging, the generation of myeloid lineage cells remains steady or may even increase with age (Beerman et al. 2010). While an increase in the number of myeloid precursors in the bone marrow of elderly individuals has been documented (Beerman et al. 2010), this is accompanied by a decline in the functional capacity of both macrophages and neutrophils (Mancuso et al. 2001; Ding et al. 1994; De La Fuente 1985). As specialized phagocytes, impairment of the phagocytic capacity of macrophages has a major impact on the clearance of infections and inflammation. Although there have been contradicting reports (Mancuso et al. 2001), overall, phagocytosis by macrophages declines with aging (De La Fuente 1985). In both a murine model of corneal inflammation induced by *Pseudomonas aeruginosa* infection in aged mice (Hazlett et al. 1990) and in elderly humans with chronic bronchitis (Fietta et al. 1993), the phagocytic activity of macrophages is diminished. Kupffer cells from aged mice also exhibit impaired functional capacity characterized by reduced phagocytosis, as well as diminished endocytic and respiratory burst activity, and increased oxidative stress (Videla et al. 2001). One important means of pathogen clearance in macrophages is 1A/1B-light chain-3 (LC3)-associated phagocytosis (LAP), a process by which cells target phagocytosed extracellular particles for efficient degradation. BMDMs from aged mice not only lack LAP and bacterial killing but also produce higher levels of pro-inflammatory cytokines, resulting in susceptibility to pathogens and heightened inflammation (Inomata et al. 2020). Various other reports using murine models indicate there is also a decline in the expression and function of TLRs, adherence (De La Fuente 1985; De la Fuente et al. 2000), opsonization (De La Fuente 1985), and respiratory burst (Ding et al. 1994) by aged macrophages also explain the altered immune response and heightened vulnerability to infections that occur with aging (Renshaw et al. 2002).

Although the onset of an inflammatory reaction is similar between young and elderly individuals, the resolution phase is substantially impaired and persistent chronic inflammation is seen in elderly individuals (De Maeyer et al. 2020). Impaired resolution has been attributed to a reduction in T cell immunoglobulin mucin receptor-4 (TIM-4), a phosphatidylserine receptor expressed on macrophages that enables the engulfment (i.e., efferocytosis) of apoptotic bodies. Reduced TIM-4 expression in macrophages was caused by an elevation in p38 mitogen-activated protein kinase (MAPK) activity and inhibition of p38 rescued TIM-4 expression, cleared apoptotic bodies, and restored a macrophage resolution phenotype (De Maeyer et al. 2020). An additional mechanism of suppression of the ability of macrophages to efferocytose in a senescent environment is via SIRPα-CD47-SHP-1. Senescent cells augment the expression of the "don't eat me" signal CD47 which interacts with SIRPα on macrophages through direct cell-cell contact and mediates negative regulatory signals that suppress efferocytosis of apoptotic corpses (Schloesser et al. 2023).

Accumulation of macrophages with high expression of $p16^{INK4a}$ and β-galactosidase is seen in the tissues of aging $p16^{Ink4a/LUC}$ mice (Hall et al. 2016). Expression of both markers was also shown to be induced in mouse peritoneal macrophages when senescent cells were implanted. Although $p16^{INK4a}$ and β-galactosidase are regarded as markers of the senescent status of cells (Hall et al. 2016), in macrophages their expression seems to be associated with physiological reprogramming towards an M2-like polarization phenotype (Hall et al. 2017). The increased M2 polarization of tissue macrophages in aging is supported by the observations of increased expression of transglutaminase-2, a preserved M2 marker conserved with age in mice and humans (Martinez et al. 2013). Other aging-related cellular processes, including increased ER stress, decreased autophagy, and mitochondrial dysfunction, may contribute to M2-like polarization and pro-inflammatory activation of macrophages (Oh et al. 2012; van Beek et al. 2019).

10.4.3 Markers of Senescence in Macrophages

Establishing reliable markers of senescence in macrophages is essential for developing therapies directed at or that modulate macrophages and/or their secretory phenotype, however, this has been challenging and unfruitful to date. The enzyme β-galactosidase, commonly referred to as senescence-associated (SA) β-galactosidase, has been used as a marker of cell senescence for decades. However, this enzyme is highly expressed in the lysosomes of macrophages undergoing phagocytosis, regardless of the senescence status (Lorimore et al. 2001). Cell cycle inhibitor proteins $p21^{Cip1}$ and $p16^{Ink4a}$ have also been used as indications of a senescent phenotype in macrophages in numerous studies (Zhou et al. 2024; Su et al. 2021; Young et al. 2021), but there are contradicting reports. Studies showing that p21 deficiency increases LPS-induced NFκB activation and pro-inflammatory cytokine production in macrophages contradict the notion of p21 as a marker of senescence in macrophages, since these cells are a major source of inflammatory mediators (Trakala et al. 2009), as discussed herein (Clark and Peterson 1994; Ershler et al. 1993; Hasegawa et al. 2000). The expression of $p16^{INK4a}$ and β-galactosidase is modulable, *i.e.*, stimuli that are M1-polarizing [LPS, IFN-α, Poly(I:C)] or M2-polarizing (IL-4, IL-13) can reduce or increase, respectively, their expression (Hall et al. 2017). Therefore, these markers may be more accurately described as markers of physiological programs of polarization in macrophages in response to immunomodulatory stimuli that are reversible and independent of p53 (Hall et al. 2017). This is distinct from cellular senescence in which the expression of these biomarkers is constitutive following p53-dependent establishment of proliferation arrest (Hall et al. 2017). A study exploring the cellular heterogeneity within fibrotic adipose tissue in obese mice identified a higher number of $CD11c^+$ $CD9^+$ macrophages exhibiting a senescent phenotype (Rabhi et al. 2022). These cells also showed blunted phagocytic capacity and increased expression of *Il1b*, *Cxcl1,* and *Spp1*. However, if $CD9^+$ macrophages exhibiting features of senescence also accumulate with age in humans and/or remains to be determined. Thus, future studies addressing the need to find reliable markers of senescence in macrophages are warranted.

10.4.4 Macrophage Senescence in Chronic Lung Diseases

As individuals age, multiple cell types in the lung, including epithelial cells, interstitial cells, lung progenitor cells, and immune cells, become senescent and undergo important modifications (Cho and Stout-Delgado 2020; Parimon et al. 2021). These alterations contribute to the development and progression of various chronic lung diseases, among them progressive pulmonary fibrosis (*e.g.,* IPF, scleroderma-associated interstitial lung disease (SSc-ILD)) and COPD (Cho and Stout-Delgado 2020; Parimon et al. 2021; Yang et al. 2024). The impact of ATII cell and fibroblast senescence in driving these chronic lung diseases has been widely studied and great progress has been made in the field. However, less attention has been given to understanding the impact of aging on macrophages and how these cells in turn contribute to the disease pathogenesis and progression.

scRNA-seq analyses of lungs from aged and young mice revealed alterations in the subpopulations of tissue-resident myeloid cells in the lung. Increased proportions of IMs expressing *Fn1* and chemokine receptor genes *Ccr2* and *Cx3cr1,* but reduced proportions of AMs in aged mice were observed (Mogilenko et al. 2021). AMs from aged mice also have a higher expression of genes such as *Ccl3* and *Spp1* (which encodes osteopontin) that contribute to inflammation and fibrosis (Mogilenko et al. 2021; Gao et al. 2020).

Other significant alterations in aging lung resident macrophages have also been described. Transcriptomic analysis of AMs revealed 1298 differentially expressed genes between the aged and young AMs and Gene Ontology (GO) enrichment analysis showed that these genes that were differentially expressed were mainly enriched in pathways that included cell junction/adhesion/migration/motility, cell proliferation/activation/differentiation, cytokine secretion, and chemotaxis pathways (Chen et al. 2023). Similarly to what has been described for macrophages in older humans (Hall et al. 2016, 2017; Martinez et al. 2013; Oh et al. 2012), AMs from the aged mice also exhibited both M1 and M2 features, *i.e.,* enhanced expression levels of *Nos2, Ccl8,* and *Cxcl9* for M1 and *Arg1*, *Cxcl13*, and *Tgf-β1* for M2 (Chen et al. 2023). Thus, AMs from aged mice also acquire a pro-inflammatory M2-like phenotype.

Altered cell metabolism is a feature of senescent cells and plays a pivotal role in skewing immune responses in macrophages (Minhas et al. 2019, 2021). Reduced glucose flux and mitochondrial respiration, for example, drive maladaptive proinflammatory responses. Changes in fat/carbohydrate digestion and absorption, glycosaminoglycan biosynthesis, and amino acid metabolism have also been described in aged AMs (Chen et al. 2023). Moreover, suppression in cell-autonomous NAD^+ synthesis in aged macrophages has been implicated in impaired resolution of inflammation (Minhas et al. 2019).

More recently, Chen et al. (2023) revealed that in aged mice, a distinct cluster of AMs appears, and these cells are characterized by higher expression levels of genes that indicate enhanced cell survival and senescence (*Gstm1, Serpine1, Cybb, Pdk4,* and *Cd63).* This cluster accounted for more than 77% of the AMs from the aged mice and it displayed higher expression levels of the chemokine receptors *Cxcr1*, Cxcr2 and *Cxcr3* and higher expression levels of several pro-inflammatory and/or pro-fibrotic chemokines (*Ccl6, Ccl8, Cxcl3, Cxcl1*, and *Ccl9*) (Chen et al. 2023). Enhanced susceptibility of aged mice to pulmonary fibrosis has been attributed to increased levels of CCL6 generated by macrophage receptor with collagenous structure $(MARCO)^+$ AMs (Chen et al. 2023). CCL6 is also critical to the development of bleomycin-induced pulmonary fibrosis, as the neutralization of CCL6 attenuates fibrosis in mice (Belperio et al. 2002). Although CCL6 is not produced by human cells, murine CCL6 shares homology with human CCL23 and CCL15, which seem to exert similar functions in lung inflammation (Du et al. 2021). CCL8 on the other hand, has been implicated in tissue fibrosis in humans and mice. CCL8 levels are significantly higher in the BALF and lung tissues of IPF patients and increased CCL8 levels were associated with mortality (Lee et al. 2017). A role for CCL8 in chronic kidney disease and peritoneal fibrosis has also been shown. Blocking CCL8 alleviates inflammation and fibrosis in both these tissues (Lee et al. 2022, 2023). Thus, the aged macrophages exhibit features that amplify lung inflammation and fibrosis, hampering appropriate resolution and wound healing responses in age-related chronic lung diseases.

10.4.5 Pulmonary Fibrosis

IPF is a chronic, progressive lung disease characterized by progressive lung scarring and the histological picture of usual interstitial pneumonia (UIP) (Martinez et al. 2017). The aging process plays a significant role in IPF and lung macrophages show altered function in this disease (Morimoto et al. 2012; Allden et al. 2019; Creyns et al. 2024). Analysis of bronchioalveolar lavage samples from two independent IPF patient cohorts identified increased proportions of a population of AMs that lack transferrin receptor (CD71), had impaired phagocytosis, reduced markers of macrophages maturity, and higher expression of profibrotic genes (Allden et al. 2019). Myeloid cells, mainly macrophages, from explanted lungs from IPF and Post-acute sequelae of COVID fibrosis (PASC-F) patients exhibit delayed and impaired capacity to phagocytose *Staphylococcus aureus* bioparticles and to clear cellular debris *in vitro* (preprint (Creyns et al. 2024)). In line with these reports, macrophages from mice with bleomycin-induced lung fibrosis exhibited downregulation of antimicrobial effector functions, which was attributed to defective phagocytosis but not intracellular bacterial killing (Warheit-Niemi et al. 2022). Collectively, these findings explain why patients with IPF are uniquely susceptible to severe respiratory infections (Warheit-Niemi et al. 2022).

Interestingly, hindered lung epithelial repair, which is at the center of the development of pulmonary fibrosis, may be aggravated by altered macrophage functions. AMs that phagocytosed apoptotic neutrophils produced higher levels of hepatocyte growth factor (HGF) (Morimoto et al. 2001), which can promote the proliferation of lung epithelial cells after lung injury and help restore the integrity of the alveolar and bronchial epithelium (Yanagita et al. 1993; Yaekashiwa et al. 1997). More recently, it was demonstrated that macrophages are at the center of ineffectual ATII cell differentiation to ATI cells in pulmonary fibrosis (Wang et al. 2023c). During physiological regeneration, ATII cells assume a transitional state that upregulates multiple keratins and ultimately differentiate into ATI cells. However, in IPF, these transitional alveolar epithelial cells accumulate and do not effectively differentiate into ATI. $Krt8^{hi}$ transitional cells express SASP chemokines, among them CCL2, that recruit profibrotic macrophages. Macrophages secrete IL-1β and TGF-β1, which promote the accumulation of the alveolar epithelial cell transitional state at the expense of ATI cell differentiation. TGF-β1 activated by transitional alveolar epithelial cell integrin β6 also stimulates fibroblast activation and collagen deposition. The stiffening of the lung matrix in turn supports the alveolar epithelial cell transitional state, establishing a positive feedback loop that drives fibrosis (Wang et al. 2023c).

10.4.6 Chronic Obstructive Pulmonary Disease

COPD is a heterogeneous chronic inflammatory disease that is characterized by small airway fibrosis and remodeling, with concomitant destruction of the lung parenchyma. It is a common disorder in the elderly, with cigarette smoking being the greatest risk factor for COPD development in genetically susceptible individuals. *Cellular senescence is* an important underlying feature in COPD, and senescence of alveolar epithelial cells, fibroblasts, endothelial cells, and immune cells occurs in COPD (Cho and Stout-Delgado 2020; Brandsma et al. 2017). Indeed, Mo-macs from COPD patients display increased expression of p21 and p27, as well as SASP factors (Hassibi et al. 2022). Other commonly described alterations of senescent cells, such as altered lipid metabolism and mitochondrial dysfunction (increased proton leakage and levels of reactive oxygen species, have been reported for AMs from COPD (Bassler et al. 2022). Functionally, AMs from COPD show reduced antigen presentation capacity, chemotaxis, and phagocytosis (Bassler et al. 2022; Taylor et al. 2010). Impaired phagocytosis of apoptotic airway epithelial cells and clearance of cellular debris by airway macrophages result in persistent inflammation, which is a feature of COPD (Hodge et al. 2003). Proinflammatory macrophages are highly abundant in the BALF of COPD patients and increased levels of IL-6 and TNF-α is associated with heightened COPD obstruction and increased risk of developing COPD-related complications (Hodge et al. 2003; Grumelli et al. 2004).

Thus, the heightened vulnerability of elderly individuals to chronic lung diseases can be attributed to age-related changes in the ability of lung macrophages to clear pathogens, apoptotic cells, and debris in the lung, as well as profoundly alter the surrounding lung environment.

10.4.7 The Effect of Senescent Lung Microenvironment on Macrophages

Although macrophages greatly alter their environment, there is evidence that the senescent lung microenvironment also shapes macrophage phenotype and responses to injury and infections (McQuattie-Pimentel et al. 2021).

Ligand-receptor pair analysis identified that changes in the ECM, where hyaluronan, a glycosoaminoglycan, is increased in the aged lung, alters the proliferative response of BMDM to GM-CSF. Moreover, resident macrophages from aged mice adoptively transferred to the lungs of young mice acquire a transcriptomic profile that reflects the age of the young recipient (McQuattie-Pimentel et al. 2021). The SASP from senescent cells has an important role in altering macrophages, as does the direct contact with senescent cells. The SASP from senescent fibroblasts can inhibit TNF-α mediated apoptosis by macrophages, as well as engulfment and clearance of the apoptotic cell (Ogata et al. 2021). Direct contact with senescent cells (fibroblasts and epithelial cells) also mediates negative regulatory signals that suppress the ability of macrophages to phagocytose and clear apoptotic corpses (Schloesser et al. 2023). These mechanisms impair the homeostatic clearance of senescent cells in tissues and allow these cells to accumulate in tissue. The accumulation of senescent cells is an important driver of fibrotic lung disease (Hohmann et al. 2019; Schafer et al. 2017). Studies have shown that senescent cells wield potent effects on adjacent cells through their SASP, promoting lung deterioration (Schafer et al. 2017).

Collectively, the body of literature reveals that macrophages acquire numerous intrinsic changes in the aged lung microenvironment. Among these changes, features of senescence, altered immune responses and function, and higher expression levels of pro-inflammatory and -fibrotic mediators are highlighted herein as they appear to be important driving mechanisms of lung fibrosis (Fig. 10.2).

10.5 Macrophages as Therapeutic Targets and Future Directions

Although great focus has been directed at understanding the modulatory effect of the standard-of-care drugs for IPF, Esbriet (Pirfenidone) and OFEV (Nintedanib), on other cell types in the lung, there is evidence that these drugs also modulate important mechanisms in macrophages. Pirfenidone exerts anti-fibrotic effects by suppressing macrophage M2 polarization by downregulating nuclear factor kappa-B (NF-κB) p50 activity or JAK2/STAT3 pathway in rodents (Ying et al. 2021; Toda et al. 2018). The effect of nintedanib, on the other hand, is less clear. It's been shown to inhibit colony-stimulating factor 1 (CSF1) receptor in both human and mouse macrophages and promotes IL4-associated repair macrophages (Watson et al. 2023). Nintedanib also reduces the adherence of human macrophages, *CCL2* gene expression, the production of IL-1β and IL-8 in M1 macrophages and reduces the expression of M2 markers (Bellamri et al. 2019).

These drugs have been effective at slowing down the progression of IPF, however they do not halt or reverse lung fibrosis (Richeldi et al. 2014; Noble et al. 2011). Moreover, therapeutic trials for drug candidates that target and/or modulate

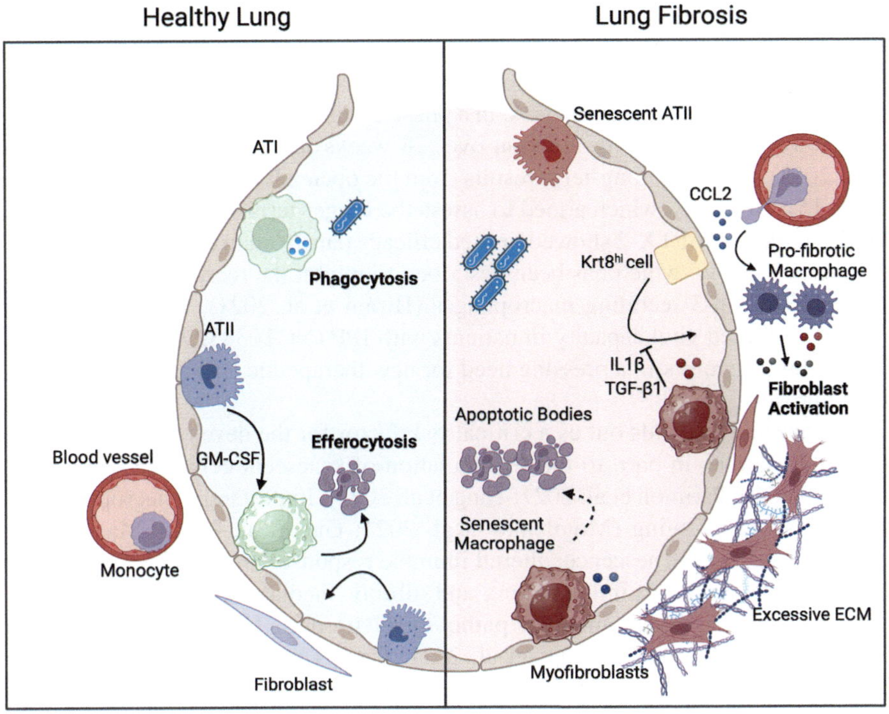

Fig. 10.2 Role of macrophage in lung homeostasis and fibrosis. Healthy Lung: Alveolar Type I (ATI) cells are responsible for efficient gas exchange, while Alveolar Type II (ATII) cells produce surfactants and differentiate into ATI after injury. Additionally, alveolar macrophages play crucial roles in maintaining alveolar homeostasis. These roles include surfactant clearance, phagocytosis of pathogens and particulates, and efferocytosis (removal of apoptotic cells). Lung Fibrosis: Pulmonary fibrosis has been described as an aberrant reparative response to repetitive alveolar epithelial injury in genetically susceptible aging individuals. The inflammatory responses that develop following tissue injury promote macrophage activation. Activated macrophages produce a series of growth factors, among them TGF-β1 which is a potent activator of fibroblasts. Activated fibroblasts and myofibroblasts secrete high levels of extracellular matrix (ECM) proteins causing scarring in the lung. Furthermore, impaired macrophage immune functions result in low clearance of apoptotic cells, debris, and pathogens result in non-resolving inflammation. Cellular senescence is an important underlying feature of pulmonary fibrosis. Senescence of ATII cells leads to decreased proliferation and differentiation, thereby suppressing alveolar repair. Moreover, there is an accumulation of keratin 8^{hi} transitional alveolar epithelial cells that do not effectively differentiate into ATI in response to IL-1β and TGF-β1 production by macrophages and the fibrotic environment. Importantly, in aged individuals, macrophages also acquire a senescent phenotype. Senescent macrophages are both pro-inflammatory and -fibrotic and they have altered immune responses and function (i.e., reduced phagocytosis and efferocytosis), which are important driving mechanisms of lung fibrosis

macrophage activation or skewing have resulted in modest or no impact in IPF progression. The recombinant human pentraxin-2 (rhPTX-2; previously known as PRM-151) inhibits the differentiation of monocytes into pro-inflammatory and pro-fibrotic macrophages and fibrocytes. In a phase 2 trial, PRM-151 vs placebo resulted in a slower decline in lung function over 28 weeks for patients with IPF (Raghu et al. 2018). However, long-term results from the open label extension period of the PRM-151-202 study, which aimed to assess the longer-term safety and efficacy (up to 128 weeks) of rhPTX-2 showed limited efficacy data (Raghu et al. 2022). GB0139 (formerly TD139), which has been described to inhibit the recruitment and expansion of Galectin-3-secreting macrophages (Hirani et al. 2021), failed to slow the decline of forced vital capacity in patients with IPF (NCT03832946) (unpublished data). Thus, there is still a pressing need for new therapeutic approaches for pulmonary fibrosis.

Advanced age stands out as a critical risk factor for the development of pulmonary fibrosis due, in part, to the accumulation of senescent cells (Cho and Stout-Delgado 2020; Parimon et al. 2021; Yang et al. 2024). Importantly, macrophages are also impacted by aging (Mogilenko et al. 2021; Gao et al. 2020). Macrophages acquire features of senescence, altered immune responses and function, and higher expression levels of pro-inflammatory and -fibrotic mediators, all of which seem to contribute to the development of pathological fibrosis (Cho and Stout-Delgado 2020; Chen et al. 2023; Minhas et al. 2021; Creyns et al. 2024). Thus, there's a growing realization that therapeutic strategies directed at senescent cells in the lung, could be beneficial and disease-modifying. However, studies with senolytic or senomorphic compounds have mainly focused on senescent epithelial cells and fibroblasts and for the most part left out the immune cells (Hohmann et al. 2019; Schafer et al. 2017). As cells that are highly responsive to the lung microenvironment and necessary to clear apoptotic cells and cellular debris, understanding the effects of these types of drugs on macrophages, in the context of an age-related fibrotic disease such as IPF, would be of extreme value.

Lastly, some outstanding questions remain. What are the mechanisms determining the balance of M1/M2 phenotype in IPF pathogenesis? How does the lung environment affect the ability of macrophages to be polarized in different directions? Related to macrophage aging, what are reliable markers to identify a senescent macrophage? How do macrophage polarization states relate to the expression of known senescence markers? Is there a dominant factor that leads macrophages to acquire a senescent phenotype? Are these intrinsic, or extrinsic (derived from aging tissue)? What are the effects of senolytic and senomorphic drugs on macrophages, their function, and phenotype? Future studies aimed at answering these questions are crucial for developing effective therapies against IPF and other fibrotic diseases.

Acknowledgements Our work on macrophage biology in the lung has been supported by NIH grants HL141078, HL108793, and HL159953.

Declaration of Competing Interest The authors declare that they have no competing interests related to this work.

Contributions All authors contributed equally to this work.

References

Adams TS, Schupp JC, Poli S, Ayaub EA, Neumark N, Ahangari F et al (2020a) Single-cell RNA-seq reveals ectopic and aberrant lung-resident cell populations in idiopathic pulmonary fibrosis

Adams TS, Schupp JC, Poli S, Ayaub EA, Neumark N, Ahangari F et al (2020b) Single-cell RNA-seq reveals ectopic and aberrant lung-resident cell populations in idiopathic pulmonary fibrosis. Sci Adv 6(28):eaba1983

A-Gonzalez N, Quintana JA, García-Silva S, Mazariegos M, de la Aleja AG, Nicolás-ávila JA et al (2017) Phagocytosis imprints heterogeneity in tissue-resident macrophages. J Exp Med 214(5):1281–1296

Allard B, Panariti A, Martin JG (2018) Alveolar macrophages in the resolution of inflammation, tissue repair, and tolerance to infection. Front Immunol 9:1777

Allden SJ, Ogger PP, Ghai P, McErlean P, Hewitt R, Toshner R et al (2019) The transferrin receptor CD71 delineates functionally distinct airway macrophage subsets during idiopathic pulmonary fibrosis. Am J Respir Crit Care Med 200(2):209–219

Aran D, Looney AP, Liu L, Wu E, Fong V, Hsu A et al (2019) Reference-based analysis of lung single-cell sequencing reveals a transitional profibrotic macrophage. Nat Immunol 20(2):163–172

Arandjelovic S, Ravichandran KS (2015) Phagocytosis of apoptotic cells in homeostasis. Nat Immunol 16(9):907–917

Arpaia N, Green JA, Moltedo B, Arvey A, Hemmers S, Yuan S et al (2015) A distinct function of regulatory T cells in tissue protection. Cell 162(5):1078–1089

Ayaub EA, Poli S, Ng J, Adams T, Schupp J, Quesada-Arias L et al (2021) Single cell RNA-seq and mass cytometry reveals a novel and a targetable population of macrophages in idiopathic pulmonary fibrosis. bioRxiv. 2021.01.04.425268-2021.01.04

Bain CC, Lucas CD, Rossi AG (2022) Pulmonary macrophages and SARS-Cov2 infection. Int Rev Cell Mol Biol 367:1

Barclay AN, Van Den Berg TK (2014) The interaction between signal regulatory protein alpha (SIRPα) and CD47: structure, function, and therapeutic target. Annu Rev Immunol 32:25–50

Bargagli E, Prasse A, Olivieri C, Muller-Quernheim J, Rottoli P (2011) Macrophage-derived biomarkers of idiopathic pulmonary fibrosis. Pulm Med 2011:717130

Bassler K, Fujii W, Kapellos TS, Dudkin E, Reusch N, Horne A et al (2022) Alveolar macrophages in early stage COPD show functional deviations with properties of impaired immune activation. Front Immunol 13:917232

Bedoret D, Wallemacq H, Marichal T, Desmet C, Quesada Calvo F, Henry E et al (2009) Lung interstitial macrophages alter dendritic cell functions to prevent airway allergy in mice. J Clin Invest 119(12):3723–3738

Beerman I, Bhattacharya D, Zandi S, Sigvardsson M, Weissman IL, Bryder D et al (2010) Functionally distinct hematopoietic stem cells modulate hematopoietic lineage potential during aging by a mechanism of clonal expansion. Proc Natl Acad Sci U S A 107(12):5465–5470

Bellamri N, Morzadec C, Joannes A, Lecureur V, Wollin L, Jouneau S et al (2019) Alteration of human macrophage phenotypes by the anti-fibrotic drug nintedanib. Int Immunopharmacol 72:112–123

Belperio JA, Dy M, Burdick MD, Xue YY, Li K, Elias JA et al (2002) Interaction of IL-13 and C10 in the pathogenesis of bleomycin-induced pulmonary fibrosis. Am J Respir Cell Mol Biol 27(4):419–427

Bharat A, Bhorade SM, Morales-Nebreda L, McQuattie-Pimentel AC, Soberanes S, Ridge K et al (2016) Flow cytometry reveals similarities between lung macrophages in humans and mice. Am J Respir Cell Mol Biol 54(1):147–149

Bissinger S, Hage C, Wagner V, Maser IP, Brand V, Schmittnaegel M et al (2021) Macrophage depletion induces edema through release of matrix-degrading proteases and proteoglycan deposition. Sci Transl Med 13(598):4550

Bissonnette EY, Lauzon-Joset JF, Debley JS, Ziegler SF (2020) Cross-talk between alveolar macrophages and lung epithelial cells is essential to maintain lung homeostasis. Front Immunol 11:583042

Boehmer ED, Meehan MJ, Cutro BT, Kovacs EJ (2005) Aging negatively skews macrophage TLR2- and TLR4-mediated pro-inflammatory responses without affecting the IL-2-stimulated pathway. Mech Ageing Dev 126(12):1305–1313

Bordon Y (2020) Blocking SIRPα could restore macrophage appetite and mitigate pneumonia. Nat Rev Immunol 20(6):349

Bouchlaka MN, Sckisel GD, Chen M, Mirsoian A, Zamora AE, Maverakis E et al (2013) Aging predisposes to acute inflammatory induced pathology after tumor immunotherapy. J Exp Med 210(11):2223–2237

Bourdonnay E, Zasłona Z, Penke LRK, Speth JM, Schneider DJ, Przybranowski S et al (2015) Transcellular delivery of vesicular SOCS proteins from macrophages to epithelial cells blunts inflammatory signaling. J Exp Med 212(5):729–742

Branchett WJ, Cook J, Oliver RA, Bruno N, Walker SA, Stölting H et al (2021) Airway macrophage-intrinsic TGF-β1 regulates pulmonary immunity during early-life allergen exposure. J Allergy Clin Immunol 147(5):1892–1906

Brandsma CA, de Vries M, Costa R, Woldhuis RR, Konigshoff M, Timens W (2017) Lung ageing and COPD: is there a role for ageing in abnormal tissue repair? Eur Respir Rev 26(146)

Byrne AJ, Mathie SA, Gregory LG, Lloyd CM (2015) Pulmonary macrophages: key players in the innate defence of the airways. Thorax 70(12):1189–1196

Byrne AJ, Maher TM, Lloyd CM (2016) Pulmonary macrophages: a new therapeutic pathway in fibrosing lung disease? Trends Mol Med 22(4):303–316

Cai Y, Sugimoto C, Arainga M, Alvarez X, Didier ES, Kuroda MJ (2014) In vivo characterization of alveolar and interstitial lung macrophages in rhesus macaques: implications for understanding lung disease in humans. J Immunol 192(6):2821–2829

Chakarov S, Lim HY, Tan L, Lim SY, See P, Lum J et al (2019) Two distinct interstitial macrophage populations coexist across tissues in specific subtissular niches. Science 363(6432)

Chelvarajan RL, Liu Y, Popa D, Getchell ML, Getchell TV, Stromberg AJ et al (2006) Molecular basis of age-associated cytokine dysregulation in LPS-stimulated macrophages. J Leukoc Biol 79(6):1314–1327

Chen H, Ma F, Hu X, Jin T, Xiong C, Teng X (2013) Elevated COX2 expression and PGE2 production by downregulation of RXRalpha in senescent macrophages. Biochem Biophys Res Commun 440(1):157–162

Chen Y, Hao X, Li M, Tian Z, Cheng M (2023) UGRP1-modulated MARCO(+) alveolar macrophages contribute to age-related lung fibrosis. Immun Ageing 20(1):14

Cho SJ, Stout-Delgado HW (2020) Aging and lung disease. Annu Rev Physiol 82:433–459

Clark JA, Peterson TC (1994) Cytokine production and aging: overproduction of IL-8 in elderly males in response to lipopolysaccharide. Mech Ageing Dev 77(2):127–139

Creyns B, MacKenzie B, Sa Y, Coelho AL, Christensen D, Parimon T, Windsor B, Hogaboam CM (2024) Caveolin scaffolding domain (CSD) peptide LTI-2355 modulates the phagocytic and synthetic activity of lung derived myeloid cells in Idiopathic Pulmonary Fibrosis (IPF) and post-acute sequelae of COVID-fibrosis (PASC-F). bioRxiv

Dancer RCA, Wood AM, Thickett DR (2011) Metalloproteinases in idiopathic pulmonary fibrosis. Eur Respir J 38(6):1461–1467

De La Fuente M (1985) Changes in the macrophage function with aging. Comp Biochem Physiol A Comp Physiol 81(4):935–938

De la Fuente M, Medina S, Del Rio M, Ferrandez MD, Hernanz A (2000) Effect of aging on the modulation of macrophage functions by neuropeptides. Life Sci 67(17):2125–2135

De Maeyer RPH, van de Merwe RC, Louie R, Bracken OV, Devine OP, Goldstein DR et al (2020) Blocking elevated p38 MAPK restores efferocytosis and inflammatory resolution in the elderly. Nat Immunol 21(6):615–625

de Vos AF, van der Poll T (2020) SIRP-α instructs alveolar macrophages to stop eating after pneumonia. Nat Immunol 21(6):601–603

Dick SA, Wong A, Hamidzada H, Nejat S, Nechanitzky R, Vohra S et al (2022) Three tissue resident macrophage subsets coexist across organs with conserved origins and life cycles. Sci Immunol 7(67)

Ding A, Hwang S, Schwab R (1994) Effect of aging on murine macrophages. Diminished response to IFN-gamma for enhanced oxidative metabolism. J Immunol 153(5):2146–2152

Dong Y, Arif AA, Guo J, Ha Z, Lee-Sayer SSM, Poon GFT et al (2020) CD44 loss disrupts lung lipid surfactant homeostasis and exacerbates oxidized lipid-induced lung inflammation. Front Immunol 11:29

Du X, Li F, Zhang C, Li N, Huang H, Shao Z et al (2021) Eosinophil-derived chemokine (hCCL15/23, mCCL6) interacts with CCR1 to promote eosinophilic airway inflammation. Signal Transduct Target Ther 6(1):91

Dutta B, Goswami R, Rahaman SO (2020) TRPV4 plays a role in matrix stiffness-induced macrophage polarization. Front Immunol 11

Enomoto Y, Katsura H, Fujimura T, Ogata A, Baba S, Yamaoka A et al (2023) Autocrine TGF-β-positive feedback in profibrotic AT2-lineage cells plays a crucial role in non-inflammatory lung fibrogenesis. Nat Commun 14(1):1–17

Ershler WB, Sun WH, Binkley N, Gravenstein S, Volk MJ, Kamoske G et al (1993) Interleukin-6 and aging: blood levels and mononuclear cell production increase with advancing age and in vitro production is modifiable by dietary restriction. Lymphokine Cytokine Res 12(4):225–230

Evren E, Ringqvist E, Doisne JM, Thaller A, Sleiers N, Flavell RA et al (2022) CD116+ fetal precursors migrate to the perinatal lung and give rise to human alveolar macrophages. J Exp Med 219(2)

Fabre T, Barron AMS, Christensen SM, Asano S, Bound K, Lech MP et al (2023) Identification of a broadly fibrogenic macrophage subset induced by type 3 inflammation. Sci Immunol 8(82):eadd8945

Fietta A, Merlini C, De Bernardi PM, Gandola L, Piccioni PD, Grassi C (1993) Non specific immunity in aged healthy subjects and in patients with chronic bronchitis. Aging (Milano) 5(5):357–361

Franceschi C, Bonafe M, Valensin S, Olivieri F, De Luca M, Ottaviani E et al (2000) Inflamm-aging. An evolutionary perspective on immunosenescence. Ann N Y Acad Sci 908:244–254

Franceschi C, Garagnani P, Parini P, Giuliani C, Santoro A (2018) Inflammaging: a new immune-metabolic viewpoint for age-related diseases. Nat Rev Endocrinol 14(10):576–590

Freire BM, de Melo FM, Basso AS (2022) Adrenergic signaling regulation of macrophage function: do we understand it yet? Immunother Adv 2(1):ltac010

Ganeshan K, Chawla A (2014) Metabolic regulation of immune responses. 32:609–634. https://doi.org/101146/annurev-immunol-032713-120236

Gao X, Jia G, Guttman A, DePianto DJ, Morshead KB, Sun KH et al (2020) Osteopontin links myeloid activation and disease progression in systemic sclerosis. Cell Rep Med 1(8):100140

Gibbings SL, Goyal R, Desch AN, Leach SM, Prabagar M, Atif SM et al (2015) Transcriptome analysis highlights the conserved difference between embryonic and postnatal-derived alveolar macrophages. Blood 126(11):1357–1366

Gomez Perdiguero E, Klapproth K, Schulz C, Busch K, Azzoni E, Crozet L et al (2015) Tissue-resident macrophages originate from yolk-sac-derived erythro-myeloid progenitors. Nature 518(7540):547–551

Grailer JJ, Haggadone MD, Sarma JV, Zetoune FS, Ward PA (2014) Induction of M2 regulatory macrophages through the β2-adrenergic receptor with protection during endotoxemia and acute lung injury. J Innate Immun 6(5):607–618

Greenlee-Wacker MC (2016) Clearance of apoptotic neutrophils and resolution of inflammation. Immunol Rev 273(1):357–370

Grumelli S, Corry DB, Song LZ, Song L, Green L, Huh J et al (2004) An immune basis for lung parenchymal destruction in chronic obstructive pulmonary disease and emphysema. PLoS Med 1(1):e8

Guilliams M, Svedberg FR (2021) Does tissue imprinting restrict macrophage plasticity? Nat Immunol 22(2):118–127

Guilliams M, De Kleer I, Henri S, Post S, Vanhoutte L, De Prijck S et al (2013) Alveolar macrophages develop from fetal monocytes that differentiate into long-lived cells in the first week of life via GM-CSF. J Exp Med 210(10):1977–1992

Hall BM, Balan V, Gleiberman AS, Strom E, Krasnov P, Virtuoso LP et al (2016) Aging of mice is associated with p16(Ink4a)- and beta-galactosidase-positive macrophage accumulation that can be induced in young mice by senescent cells. Aging (Albany NY) 8(7):1294–1315

Hall BM, Balan V, Gleiberman AS, Strom E, Krasnov P, Virtuoso LP et al (2017) p16(Ink4a) and senescence-associated beta-galactosidase can be induced in macrophages as part of a reversible response to physiological stimuli. Aging (Albany NY) 9(8):1867–1884

Han SH, Mallampalli RK (2015) The role of surfactant in lung disease and host defense against pulmonary infections. Ann Am Thorac Soc 12(5):765–774

Hartl D, Tirouvanziam R, Laval J, Greene CM, Habiel D, Sharma L et al (2018) Innate immunity of the lung: from basic mechanisms to translational medicine. J Innate Immun 10(5–6):487–501

Hasegawa Y, Sawada M, Ozaki N, Inagaki T, Suzumura A (2000) Increased soluble tumor necrosis factor receptor levels in the serum of elderly people. Gerontology 46(4):185–188

Hashimoto D, Chow A, Noizat C, Teo P, Beasley MB, Leboeuf M et al (2013) Tissue-resident macrophages self-maintain locally throughout adult life with minimal contribution from circulating monocytes. Immunity 38(4):792–804

Hassibi S, Baker J, Barnes P, Donnelly L (2022) COPD monocyte-derived macrophages display hallmarks of senescence. ERJ Open Res 8

Hazlett LD, Kreindler FB, Berk RS, Barrett R (1990) Aging alters the phagocytic capability of inflammatory cells induced into cornea. Curr Eye Res 9(2):129–138

Herold S, Mayer K, Lohmeyer J (2011a) Acute lung injury: how macrophages orchestrate resolution of inflammation and tissue repair. Front Immunol 2:65

Herold S, Tabar TS, Janßen H, Hoegner K, Cabanski M, Lewe-Schlosser P et al (2011b) Exudate macrophages attenuate lung injury by the release of IL-1 receptor antagonist in gram-negative pneumonia. Am J Respir Crit Care Med 183(10):1380–1390

Hirani N, MacKinnon AC, Nicol L, Ford P, Schambye H, Pedersen A et al (2021) Target inhibition of galectin-3 by inhaled TD139 in patients with idiopathic pulmonary fibrosis. Eur Respir J 57(5)

Hodge S, Hodge G, Scicchitano R, Reynolds PN, Holmes M (2003) Alveolar macrophages from subjects with chronic obstructive pulmonary disease are deficient in their ability to phagocytose apoptotic airway epithelial cells. Immunol Cell Biol 81(4):289–296

Hoeffel G, Chen J, Lavin Y, Low D, Almeida FF, See P et al (2015) C-Myb(+) erythro-myeloid progenitor-derived fetal monocytes give rise to adult tissue-resident macrophages. Immunity 42(4):665–678

Hohmann MS, Habiel DM, Coelho AL, Verri WA Jr, Hogaboam CM (2019) Quercetin enhances ligand-induced apoptosis in senescent idiopathic pulmonary fibrosis fibroblasts and reduces lung fibrosis in vivo. Am J Respir Cell Mol Biol 60(1):28–40

Holt PG, Jones CA (2000) The development of the immune system during pregnancy and early life. Allergy 55(8):688–697

Huang H, Auerbach R (1993) Identification and characterization of hematopoietic stem cells from the yolk sac of the early mouse embryo. Proc Natl Acad Sci U S A 90(21):10110–10114

Infantino V, Convertini P, Cucci L, Panaro MA, Di Noia MA, Calvello R et al (2011) The mitochondrial citrate carrier: a new player in inflammation. Biochem J 438(3):433–436

Inomata M, Xu S, Chandra P, Meydani SN, Takemura G, Philips JA et al (2020) Macrophage LC3-associated phagocytosis is an immune defense against Streptococcus pneumoniae that diminishes with host aging. Proc Natl Acad Sci U S A 117(52):33561–33569

Isshiki T, Vierhout M, Naiel S, Ali P, Yazdanshenas P, Kumaran V et al (2023) Therapeutic strategies targeting pro-fibrotic macrophages in interstitial lung disease. Biochem Pharmacol 211:115501

Joshi N, Watanabe S, Verma R, Jablonski RP, Chen CI, Cheresh P et al (2020) A spatially restricted fibrotic niche in pulmonary fibrosis is sustained by M-CSF/M-CSFR signalling in monocyte-derived alveolar macrophages. Eur Respir J 55(1)

Kalim KW, Groettrup M (2013) Prostaglandin E2 inhibits IL-23 and IL-12 production by human monocytes through down-regulation of their common p40 subunit. Mol Immunol 53(3):274–282

Keerthivasan S, Senbabaoglu Y, Martinez-Martin N, Husain B, Verschueren E, Wong A et al (2021) Homeostatic functions of monocytes and interstitial lung macrophages are regulated via collagen domain-binding receptor LAIR1. Immunity 54(7):1511–26 e8

Kim KK, Dotson MR, Agarwal M, Yang J, Bradley PB, Subbotina N et al (2018) Efferocytosis of apoptotic alveolar epithelial cells is sufficient to initiate lung fibrosis. Cell Death Dis 9(11):1056

Kimura H, Suzuki M, Konno S, Shindou H, Shimizu T, Nagase T et al (2017) Orchestrating role of apoptosis inhibitor of macrophage in the resolution of acute lung injury. J Immunol (Baltimore, Md: 1950) 199(11):3870–3882

Kratz M, Coats BR, Hisert KB, Hagman D, Mutskov V, Peris E et al (2014) Metabolic dysfunction drives a mechanistically distinct proinflammatory phenotype in adipose tissue macrophages. Cell Metab 20(4):614–625

Kulle A, Thanabalasuriar A, Cohen TS, Szydlowska M (2022) Resident macrophages of the lung and liver: the guardians of our tissues. Front Immunol 13:1029085

Lapenna A, De Palma M, Lewis CE (2018) Perivascular macrophages in health and disease. Nat Rev Immunol 18(11):689–702

Leach SM, Gibbings SL, Tewari AD, Atif SM, Vestal B, Danhorn T et al (2020) Human and mouse transcriptome profiling identifies cross-species homology in pulmonary and lymph node mononuclear phagocytes. Cell Rep 33(5)

Lechner AJ, Driver IH, Lee J, Conroy CM, Nagle A, Locksley RM et al (2017) Recruited monocytes and Type 2 immunity promote lung regeneration following pneumonectomy. Cell Stem Cell 21(1):120–34.e7

Lee JU, Cheong HS, Shim EY, Bae DJ, Chang HS, Uh ST et al (2017) Gene profile of fibroblasts identify relation of CCL8 with idiopathic pulmonary fibrosis. Respir Res 18(1):3

Lee J, Lee Y, Kim KH, Kim DK, Joo KW, Shin SJ et al (2022) Chemokine (C-C Motif) Ligand 8 and tubulo-interstitial injury in chronic kidney disease. Cells 11(4)

Lee Y, Lee J, Park M, Seo A, Kim KH, Kim S et al (2023) Inflammatory chemokine (C-C motif) ligand 8 inhibition ameliorates peritoneal fibrosis. FASEB J 37(1):e22632

Lescoat A, Ballerie A, Lelong M, Augagneur Y, Morzadec C, Jouneau S et al (2020) Crystalline silica impairs efferocytosis abilities of human and mouse macrophages: implication for silica-associated systemic sclerosis. Front Immunol 11:219

Li X, Kolling FW, Aridgides D, Mellinger D, Ashare A, Jakubzick CV (2022) ScRNA-seq expression of IFI27 and APOC2 identifies four alveolar macrophage superclusters in healthy BALF. Life Sci Alliance 5(11)

Liao M, Liu Y, Yuan J, Wen Y, Xu G, Zhao J et al (2020) Single-cell landscape of bronchoalveolar immune cells in patients with COVID-19. Nat Med 26(6):842–844

Liborio-Ramos S, Barbosa-Matos C, Fernandes R, Borges-Pereira C, Costa S (2023) Interstitial macrophages lead early stages of bleomycin-induced lung fibrosis and induce fibroblasts activation. Cells 12(3)

Lim HY, Lim SY, Tan CK, Thiam CH, Goh CC, Carbajo D et al (2018) Hyaluronan receptor LYVE-1-expressing macrophages maintain arterial tone through hyaluronan-mediated regulation of smooth muscle cell collagen. Immunity 49(6):1191

Lin Y, Yoder MC, Yoshimoto M (2014) Lymphoid progenitor emergence in the murine embryo and yolk sac precedes stem cell detection. Stem Cells Dev 23(11):1168–1177

Liu Z, Gu Y, Chakarov S, Bleriot C, Kwok I, Chen X et al (2019) Fate mapping via Ms4a3-expression history traces monocyte-derived cells. Cell 178(6):1509–25.e19

Liu JQ, Hu A, Zhu J, Yu J, Talebian F, Bai XF (2020) CD200-CD200R pathway in the regulation of tumor immune microenvironment and immunotherapy. Adv Exp Med Biol 1223:155–165

Logtenberg MEW, Scheeren FA, Schumacher TN (2020) The CD47-SIRPα Immune Checkpoint

Lorimore SA, Coates PJ, Scobie GE, Milne G, Wright EG (2001) Inflammatory-type responses after exposure to ionizing radiation in vivo: a mechanism for radiation-induced bystander effects? Oncogene 20(48):7085–7095

Lv J, Gao H, Ma J, Liu J, Tian Y, Yang C et al (2023) Dynamic atlas of immune cells reveals multiple functional features of macrophages associated with progression of pulmonary fibrosis. Front Immunol 14:1230266

Mahrouf-Yorgov M, Collin de l'Hortet A, Cosson C, Slama A, Abdoun E, Guidotti JE et al (2011) Increased susceptibility to liver fibrosis with age is correlated with an altered inflammatory response. Rejuvenation Res 14(4):353–363

Mancuso P, McNish RW, Peters-Golden M, Brock TG (2001) Evaluation of phagocytosis and arachidonate metabolism by alveolar macrophages and recruited neutrophils from F344xBN rats of different ages. Mech Ageing Dev 122(15):1899–1913

Martinez FO, Helming L, Milde R, Varin A, Melgert BN, Draijer C et al (2013) Genetic programs expressed in resting and IL-4 alternatively activated mouse and human macrophages: similarities and differences. Blood 121(9):e57–e69

Martinez FJ, Collard HR, Pardo A, Raghu G, Richeldi L, Selman M et al (2017) Idiopathic pulmonary fibrosis. Nat Rev Dis Primers 3:17074

Matthay MA, Zemans RL, Zimmerman GA, Arabi YM, Beitler JR, Mercat A et al (2019) Acute respiratory distress syndrome. Nat Rev Dis Primers 5(1):1–22

McQuattie-Pimentel AC, Ren Z, Joshi N, Watanabe S, Stoeger T, Chi M et al (2021) The lung microenvironment shapes a dysfunctional response of alveolar macrophages in aging. J Clin Invest 131(4)

Michelucci A, Cordes T, Ghelfi J, Pailot A, Reiling N, Goldmann O et al (2013) Immune-responsive gene 1 protein links metabolism to immunity by catalyzing itaconic acid production. Proc Natl Acad Sci U S A 110(19):7820–7825

Mills CD, Kincaid K, Alt JM, Heilman MJ, Hill AM (2000) M-1/M-2 macrophages and the Th1/Th2 paradigm. J Immunol 164(12):6166–6173

Mills EL, Kelly B, Logan A, Costa ASH, Varma M, Bryant CE et al (2016) Succinate dehydrogenase supports metabolic repurposing of mitochondria to drive inflammatory macrophages. Cell 167(2):457–70.e13

Mills EL, Ryan DG, Prag HA, Dikovskaya D, Menon D, Zaslona Z et al (2018) Itaconate is an anti-inflammatory metabolite that activates Nrf2 via alkylation of KEAP1. Nature 556(7699):113–117

Minhas PS, Liu L, Moon PK, Joshi AU, Dove C, Mhatre S et al (2019) Macrophage de novo NAD(+) synthesis specifies immune function in aging and inflammation. Nat Immunol 20(1):50–63

Minhas PS, Latif-Hernandez A, McReynolds MR, Durairaj AS, Wang Q, Rubin A et al (2021) Restoring metabolism of myeloid cells reverses cognitive decline in ageing. Nature 590(7844):122–128

Misharin AV, Morales-Nebreda L, Mutlu GM, Budinger GRS, Perlman H (2013) Flow cytometric analysis of macrophages and dendritic cell subsets in the mouse lung. Am J Respir Cell Mol Biol 49(4):503–510
Misharin AV, Morales-Nebreda L, Reyfman PA, Cuda CM, Walter JM, McQuattie-Pimentel AC et al (2017) Monocyte-derived alveolar macrophages drive lung fibrosis and persist in the lung over the life span. J Exp Med 214(8):2387–2404
Mogilenko DA, Shpynov O, Andhey PS, Arthur L, Swain A, Esaulova E et al (2021) Comprehensive profiling of an aging immune system reveals clonal GZMK(+) CD8(+) T cells as conserved hallmark of inflammaging. Immunity 54(1):99–115 e12
Monticelli LA, Buck MD, Flamar AL, Saenz SA, Wojno EDT, Yudanin NA et al (2016) Arginase 1 is an innate lymphoid-cell-intrinsic metabolic checkpoint controlling type 2 inflammation. Nat Immunol 17(6):656–665
Morimoto K, Amano H, Sonoda F, Baba M, Senba M, Yoshimine H et al (2001) Alveolar macrophages that phagocytose apoptotic neutrophils produce hepatocyte growth factor during bacterial pneumonia in mice. Am J Respir Cell Mol Biol 24(5):608–615
Morimoto K, Janssen WJ, Terada M (2012) Defective efferocytosis by alveolar macrophages in IPF patients. Respir Med 106(12):1800–1803
Morioka S, Perry JSA, Raymond MH, Medina CB, Zhu Y, Zhao L et al (2018) Efferocytosis induces a novel SLC program to promote glucose uptake and lactate release. Nature 563(7733):714–718
Morse C, Tabib T, Sembrat J, Buschur KL, Bittar HT, Valenzi E et al (2019) Proliferating SPP1/MERTK-expressing macrophages in idiopathic pulmonary fibrosis. Eur Respir J 54(2)
Mosmann TR, Cherwinski H, Bond MW, Giedlin MA, Coffman RL (1986) Two types of murine helper T cell clone. I. Definition according to profiles of lymphokine activities and secreted proteins. J Immunol 136(7):2348–2357
Mould KJ, Moore CM, McManus SA, McCubbrey AL, McClendon JD, Griesmer CL et al (2021) Airspace macrophages and monocytes exist in transcriptionally distinct subsets in healthy adults. Am J Respir Crit Care Med 203(8):946–956
Murray PJ, Allen JE, Biswas SK, Fisher EA, Gilroy DW, Goerdt S et al (2014) Macrophage activation and polarization: nomenclature and experimental guidelines. Immunity 41(1):14–20
Nagre N, Cong X, Pearson AC, Zhao X (2019) Alveolar macrophage phagocytosis and bacteria clearance in mice. J Vis Exp 145
Natoli G, Monticelli S (2014) Macrophage activation: glancing into diversity. Immunity 40(2):175–177
Neupane AS, Willson M, Chojnacki AK, Vargas E Silva Castanheira F, Morehouse C, Carestia A et al (2020) Patrolling alveolar macrophages conceal bacteria from the immune system to maintain homeostasis. Cell 183(1):110–25.e11
Nguyen TN, Siddiqui G, Veldhuis NA, Poole DP (2021) Diverse roles of TRPV4 in macrophages: a need for unbiased profiling. Front Immunol 12
Nishi C, Yanagihashi Y, Segawa K, Nagata S (2019) MERTK tyrosine kinase receptor together with TIM4 phosphatidylserine receptor mediates distinct signal transduction pathways for efferocytosis and cell proliferation. J Biol Chem 294(18):7221–7230
Noble PW, Albera C, Bradford WZ, Costabel U, Glassberg MK, Kardatzke D et al (2011) Pirfenidone in patients with idiopathic pulmonary fibrosis (CAPACITY): two randomised trials. Lancet 377(9779):1760–1769
O'Beirne SL, Kikkers SA, Oromendia C, Salit J, Rostmai MR, Ballman KV et al (2020) Alveolar macrophage immunometabolism and lung function impairment in smoking and chronic obstructive pulmonary disease. Am J Respir Crit Care Med 201(6):735–738
O'Neill LAJ (2015) A broken krebs cycle in macrophages. Immunity 42(3):393–394
O'Neill LAJ, Grahame Hardie D (2013) Metabolism of inflammation limited by AMPK and pseudo-starvation. Nature 493(7432):346–355
Ogata Y, Yamada T, Hasegawa S, Sanada A, Iwata Y, Arima M et al (2021) SASP-induced macrophage dysfunction may contribute to accelerated senescent fibroblast accumulation in the dermis. Exp Dermatol 30(1):84–91

Oh J, Riek AE, Weng S, Petty M, Kim D, Colonna M et al (2012) Endoplasmic reticulum stress controls M2 macrophage differentiation and foam cell formation. J Biol Chem 287(15):11629–11641
Parimon T, Hohmann MS, Yao C (2021) Cellular senescence: pathogenic mechanisms in lung fibrosis. Int J Mol Sci 22(12)
Park SY, Kim IS (2017) Engulfment signals and the phagocytic machinery for apoptotic cell clearance. Exp Mol Med 49(5):e331
Pattabiraman G, Palasiewicz K, Galvin JP, Ucker DS (2017) Aging-associated dysregulation of homeostatic immune response termination (and not initiation). Aging Cell 16(3):585–593
Perrot CY, Karampitsakos T, Unterman A, Adams T, Marlin K, Arsenault A et al (2024) MCEMP1 is expressed in classical monocytes and alveolar macrophages in IPF and regulates cell chemotaxis, adhesion, and migration in a TGFbeta dependent manner. Am J Physiol Cell Physiol
Puttur F, Gregory LG, Lloyd CM (2019) Airway macrophages as the guardians of tissue repair in the lung. Immunol Cell Biol 97(3):246–257
Rabhi N, Desevin K, Belkina AC, Tilston-Lunel A, Varelas X, Layne MD et al (2022) Obesity-induced senescent macrophages activate a fibrotic transcriptional program in adipocyte progenitors. Life Sci Alliance 5(5)
Raghu G, van den Blink B, Hamblin MJ, Brown AW, Golden JA, Ho LA et al (2018) Effect of recombinant human pentraxin 2 vs placebo on change in forced vital capacity in patients with idiopathic pulmonary fibrosis: a randomized clinical trial. JAMA 319(22):2299–2307
Raghu G, Hamblin MJ, Brown AW, Golden JA, Ho LA, Wijsenbeek MS et al (2022) Long-term evaluation of the safety and efficacy of recombinant human pentraxin-2 (rhPTX-2) in patients with idiopathic pulmonary fibrosis (IPF): an open-label extension study. Respir Res 23(1):129
Renshaw M, Rockwell J, Engleman C, Gewirtz A, Katz J, Sambhara S (2002) Cutting edge: impaired Toll-like receptor expression and function in aging. J Immunol 169(9):4697–4701
Reyfman PA, Walter JM, Joshi N, Anekalla KR, McQuattie-Pimentel AC, Chiu S et al (2019) Single-cell transcriptomic analysis of human lung provides insights into the pathobiology of pulmonary fibrosis. Am J Respir Crit Care Med 199(12):1517–1536
Richeldi L, du Bois RM, Raghu G, Azuma A, Brown KK, Costabel U et al (2014) Efficacy and safety of nintedanib in idiopathic pulmonary fibrosis. N Engl J Med 370(22):2071–2082
Roquilly A, McWilliam HEG, Jacqueline C, Tian Z, Cinotti R, Rimbert M et al (2017) Local modulation of antigen-presenting cell development after resolution of pneumonia induces long-term susceptibility to secondary infections. Immunity 47(1):135–47.e5
Roquilly A, Jacqueline C, Davieau M, Mollé A, Sadek A, Fourgeux C et al (2020) Alveolar macrophages are epigenetically altered after inflammation, leading to long-term lung immunoparalysis. Nat Immunol 21(6):636–648
Sabatel C, Radermecker C, Fievez L, Paulissen G, Chakarov S, Fernandes C et al (2017) Exposure to bacterial CpG DNA protects from airway allergic inflammation by expanding regulatory lung interstitial macrophages. Immunity 46(3):457–473
Sajti E, Link VM, Ouyang Z, Spann NJ, Westin E, Romanoski CE et al (2020) Transcriptomic and epigenetic mechanisms underlying myeloid diversity in the lung. Nat Immunol 21(2):221–231
Schafer MJ, White TA, Iijima K, Haak AJ, Ligresti G, Atkinson EJ et al (2017) Cellular senescence mediates fibrotic pulmonary disease. Nat Commun 8:14532
Scheraga RG, Abraham S, Niese KA, Southern BD, Grove LM, Hite RD et al (2016) TRPV4 mechanosensitive ion channel regulates lipopolysaccharide-stimulated macrophage phagocytosis. J Immunol (Baltimore, Md: 1950) 196(1):428–436
Schloesser D, Lindenthal L, Sauer J, Chung KJ, Chavakis T, Griesser E et al (2023) Senescent cells suppress macrophage-mediated corpse removal via upregulation of the CD47-QPCT/L axis. J Cell Biol 222(2)
Schupp JC, Binder H, Jager B, Cillis G, Zissel G, Muller-Quernheim J et al (2015) Macrophage activation in acute exacerbation of idiopathic pulmonary fibrosis. PLoS One 10(1):e0116775
Schyns J, Bai Q, Ruscitti C, Radermecker C, De Schepper S, Chakarov S et al (2019) Non-classical tissue monocytes and two functionally distinct populations of interstitial macrophages populate the mouse lung. Nat Commun 10(1)

Seo HR, Han HJ, Lee Y, Noh YW, Cho SJ, Kim JH (2022) Human pluripotent stem cell-derived alveolar organoid with macrophages. Int J Mol Sci 23(16)
Sharma G, Hanania NA, Shim YM (2009) The aging immune system and its relationship to the development of chronic obstructive pulmonary disease. Proc Am Thorac Soc 6(7):573–580
Sheng J, Ruedl C, Karjalainen K (2015) Most tissue-resident macrophages except microglia are derived from fetal hematopoietic stem cells. Immunity 43(2):382–393
Shi T, Denney L, An H, Ho LP, Zheng Y (2021) Alveolar and lung interstitial macrophages: definitions, functions, and roles in lung fibrosis. J Leukoc Biol 110(1):107–114
Singhal A, Kumar S (2022) Neutrophil and remnant clearance in immunity and inflammation. Immunology 165(1):22–43
Snelgrove RJ, Goulding J, Didierlaurent AM, Lyonga D, Vekaria S, Edwards L et al (2008) A critical function for CD200 in lung immune homeostasis and the severity of influenza infection. Nat Immunol 9(9):1074–1083
Spiller KL, Wrona EA, Romero-Torres S, Pallotta I, Graney PL, Witherel CE et al (2016) Differential gene expression in human, murine, and cell line-derived macrophages upon polarization. Exp Cell Res 347(1):1–13
Su L, Dong Y, Wang Y, Wang Y, Guan B, Lu Y et al (2021) Potential role of senescent macrophages in radiation-induced pulmonary fibrosis. Cell Death Dis 12(6):527
Taylor AE, Finney-Hayward TK, Quint JK, Thomas CM, Tudhope SJ, Wedzicha JA et al (2010) Defective macrophage phagocytosis of bacteria in COPD. Eur Respir J 35(5):1039–1047
Toda M, Mizuguchi S, Minamiyama Y, Yamamoto-Oka H, Aota T, Kubo S et al (2018) Pirfenidone suppresses polarization to M2 phenotype macrophages and the fibrogenic activity of rat lung fibroblasts. J Clin Biochem Nutr 63(1):58–65
Trakala M, Arias CF, Garcia MI, Moreno-Ortiz MC, Tsilingiri K, Fernandez PJ et al (2009) Regulation of macrophage activation and septic shock susceptibility via p21(WAF1/CIP1). Eur J Immunol 39(3):810–819
Ural BB, Yeung ST, Damani-Yokota P, Devlin JC, de Vries M, Vera-Licona P et al (2020) Identification of a nerve-associated, lung-resident interstitial macrophage subset with distinct localization and immunoregulatory properties. Sci Immunol 5(45)
van Beek AA, Van den Bossche J, Mastroberardino PG, de Winther MPJ, Leenen PJM (2019) Metabolic alterations in aging macrophages: ingredients for inflammaging? Trends Immunol 40(2):113–127
van der Pouw Kraan TC, Boeije LC, Smeenk RJ, Wijdenes J, Aarden LA (1995) Prostaglandin-E2 is a potent inhibitor of human interleukin 12 production. J Exp Med 181(2):775–779
Varga T, Mounier R, Horvath A, Cuvellier S, Dumont F, Poliska S et al (2016) Highly dynamic transcriptional signature of distinct macrophage subsets during sterile inflammation, resolution, and tissue repair. J Immunol 196(11):4771–4782
Videla LA, Tapia G, Fernandez V (2001) Influence of aging on Kupffer cell respiratory activity in relation to particle phagocytosis and oxidative stress parameters in mouse liver. Redox Rep 6(3):155–159
Wang Z, Wu Z, Wang H, Feng R, Wang G, Li M et al (2023a) An immune cell atlas reveals the dynamics of human macrophage specification during prenatal development. Cell 186(20):4454–71 e19
Wang J, Zhang X, Long M, Yuan M, Yin J, Luo W et al (2023b) Macrophage-derived GPNMB trapped by fibrotic extracellular matrix promotes pulmonary fibrosis. Commun Biol 6(1):136
Wang F, Ting C, Riemondy KA, Douglas M, Foster K, Patel N et al (2023c) Regulation of epithelial transitional states in murine and human pulmonary fibrosis. J Clin Invest 133(22)
Warheit-Niemi HI, Edwards SJ, SenGupta S, Parent CA, Zhou X, O'Dwyer DN et al (2022) Fibrotic lung disease inhibits immune responses to staphylococcal pneumonia via impaired neutrophil and macrophage function. JCI Insight 7(4)
Watanabe S, Alexander M, Misharin AV, Budinger GRS (2019) The role of macrophages in the resolution of inflammation. J Clin Invest 129(7):2619–2628
Watson CK, Schloesser D, Fundel-Clemens K, Lerner C, Gabler S, Baskaran P et al (2023) Antifibrotic drug nintedanib inhibits CSF1R to promote IL-4-associated tissue repair macrophages. Am J Respir Cell Mol Biol 68(4):366–380

Wendisch D, Dietrich O, Mari T, von Stillfried S, Ibarra IL, Mittermaier M et al (2021) SARS-CoV-2 infection triggers profibrotic macrophage responses and lung fibrosis. Cell 184(26):6243–61.e27

Westphalen K, Gusarova GA, Islam MN, Subramanian M, Cohen TS, Prince AS et al (2014) Sessile alveolar macrophages communicate with alveolar epithelium to modulate immunity. Nature 506(7489):503–506

Wynn TA, Vannella KM (2016) Macrophages in tissue repair, regeneration, and fibrosis. Immunity 44(3):450–462

Xu J, Mora A, Shim H, Stecenko A, Brigham KL, Rojas M (2007) Role of the SDF-1/CXCR4 axis in the pathogenesis of lung injury and fibrosis. Am J Respir Cell Mol Biol 37(3):291–299

Xue J, Schmidt SV, Sander J, Draffehn A, Krebs W, Quester I et al (2014) Transcriptome-based network analysis reveals a spectrum model of human macrophage activation. Immunity 40(2):274–288

Yaekashiwa M, Nakayama S, Ohnuma K, Sakai T, Abe T, Satoh K et al (1997) Simultaneous or delayed administration of hepatocyte growth factor equally represses the fibrotic changes in murine lung injury induced by bleomycin. A morphologic study. Am J Respir Crit Care Med 156(6):1937–1944

Yanagita K, Matsumoto K, Sekiguchi K, Ishibashi H, Niho Y, Nakamura T (1993) Hepatocyte growth factor may act as a pulmotrophic factor on lung regeneration after acute lung injury. J Biol Chem 268(28):21212–21217

Yang MM, Lee S, Neely J, Hinchcliff M, Wolters PJ, Sirota M (2024) Gene expression meta-analysis reveals aging and cellular senescence signatures in scleroderma-associated interstitial lung disease. Front Immunol 15:1326922

Ying H, Fang M, Hang QQ, Chen Y, Qian X, Chen M (2021) Pirfenidone modulates macrophage polarization and ameliorates radiation-induced lung fibrosis by inhibiting the TGF-beta1/Smad3 pathway. J Cell Mol Med 25(18):8662–8675

Young LV, Morrison W, Campbell C, Moore EC, Arsenault MG, Dial AG et al (2021) Loss of dystrophin expression in skeletal muscle is associated with senescence of macrophages and endothelial cells. Am J Physiol Cell Physiol 321(1):C94–C103

Zheng DJ, Taka MA, Heit B (2021) Role of apoptotic cell clearance in pneumonia and inflammatory lung disease. Pathogens 10(2):134

Zhou Z, Yao J, Wu D, Huang X, Wang Y, Li X et al (2024) Type 2 cytokine signaling in macrophages protects from cellular senescence and organismal aging. Immunity

Chapter 11
Macrophage Polarization and Its Impact on Osteoporosis

Junwei Yan, Chenyu Huang, Dongdong Jiang, Yan Xu, Zhuoli Zhang, Liming Wang, and Bin Liang

Abstract Owing to its reduced bone density and higher risk for fractures, osteoporosis remains an international public health crisis. Research highlights the essential role played by macrophage polarization in osteoporosis and indicates that the balance between pro-inflammatory M1 macrophages and anti-inflammatory M2 macrophages influences bone metabolism. This review examines how M1 and M2 macrophages contribute to the development of osteoporosis and evaluates existing therapeutic approaches aimed at controlling macrophage polarization. It also describes future study areas that will allow improved management and treatment of osteoporosis.

Junwei Yan and Chenyu Huang contributed equally with all other contributors.

J. Yan · D. Jiang · Y. Xu · L. Wang · B. Liang (✉)
Department of Orthopedic Surgery, Nanjing First Hospital, Nanjing Medical University, Nanjing, China

C. Huang
Department of Orthopedic Surgery, Nanjing First Hospital, Nanjing Medical University, Nanjing, China

Department of Biomedical Engineering, University of California, Irvine, CA, USA

Department of Medicine, University of California, Irvine, CA, USA

Z. Zhang
Department of Biomedical Engineering, University of California, Irvine, CA, USA

Department of Medicine, University of California, Irvine, CA, USA

M. Kloc et al. (eds.), *Monocytes and Macrophages in Development, Regeneration, and Disease*, Results and Problems in Cell Differentiation 74,
https://doi.org/10.1007/978-3-031-65944-7_11

11.1 Introduction

Osteoporosis is a condition affecting the entire skeletal system and characterized by decreased bone density and degeneration of bone structure, leading to fragile and more frequent fractures. Affecting millions worldwide, osteoporosis poses a serious public health concern due to illness, death, and healthcare expenses associated with its global prevalence (Hadji et al. 2013). While traditionally, osteoporosis development was thought to be related to factors like aging, hormonal fluctuations. or nutritional deficiencies, recent research has shed new light on macrophage polarization as an indirect contributor—opening fresh opportunities for understanding and treating this condition.

As a highly plastic and dynamic cell population, macrophages are capable of changing their phenotype and altering their function to adapt to specific biological functions and pathophysiological processes in reaction to different triggers from the environment, a process known as macrophage polarization (Razi et al. 2023).They can transform into two different types: pro-inflammatory (known as M1) and anti-inflammatory (known as M2), based on the signals they receive from their environment. M1 macrophages can initiate inflammation and damage tissue. In contrast, M2 macrophages aid in healing and regeneration, maintaining the body's balance (Murray et al. 2014; Yang et al. 2019). This equilibrium between M1 and M2 macrophages is crucial for overall health. Specifically, in osteoporosis, a disruption in the balance between M1 and M2 macrophages can shift the bone remodeling process towards destruction rather than formation. This imbalance accelerates bone loss, progressively exacerbating osteoporosis (Muñoz et al. 2020; Liu et al. 2015; Funes et al. 2018).

This review seeks to understand the intricate relationship between macrophage polarization and bone metabolism in osteoporosis. Here, we will investigate how M1 and M2 macrophages impact osteoclast activity and osteoblast activity; explore existing therapeutic approaches targeting macrophage polarization; and future research directions in this fascinating field that blends immunology with bone biology.

11.2 Macrophage Polarization and Osteoporosis

11.2.1 Macrophage Basics

Macrophages, key components of the immune system, possess the capacity to transform into either pro-inflammatory M1 or anti-inflammatory M2 phenotypes depending on signals they receive from their environment. Interferon-gamma (IFN-g) and lipopolysaccharide (LPS) stimulation typically induce pro-inflammatory M1 macrophages to produce pro-inflammatory cytokines such as TNF-a and IL-1b that facilitate immune responses against infections or tumor cells while M2 stimulation by IL-4/IL-13 stimulation results in producing anti-inflammatory macrophages producing IL-10 that promote tissue repair and wound healing (Orecchioni et al. 2019; Tian et al. 2023; He et al. 2022; Gharavi et al. 2022).

11.2.2 M1 Macrophages and Osteoporosis

M1 macrophages produce pro-inflammatory cytokines that promote bone loss by inducing osteoclast formation and bone tissue breakdown. TNF-a and IL-1b, two pro-inflammatory cytokines produced by M1 macrophages, increase osteoclast development and activity leading to disruption of homeostasis (Souza and Lerner 2013); TNF-a has an additional twofold effect of stimulating both osteoclast formation while simultaneously hindering osteoblast function, further worsening bone loss. With M1 macrophages present in osteoporosis cases, comes the possibility of targeting pro-inflammatory pathways as an approach towards treatment to decrease bone resorption (Yang and Yang 2019).

11.2.3 M2 Macrophage and Bone Health

M2 macrophages differ significantly from M1 in that they facilitate bone production and repair by secreting anti-inflammatory cytokines and growth factors such as IL-4 and IL-13 cytokines associated with M2 polarization (Porta et al. 2015); such cytokines have been demonstrated to suppress osteoclast development while increasing osteoblast activity, leading to bone accumulation and repair processes. Furthermore, M2 macrophages play an integral part in restoring normal bone remodeling processes (Azevedo et al. 2021); by activating M2 activity one may enhance regeneration or combat osteoporotic bone loss and improve bone regeneration as a whole.

11.2.4 M1/M2 Ratio and Its Clinical Implications

The ratio of M1 to M2 macrophages, commonly measured as M1/M2 ratio, plays an essential role in bone health (Dou et al. 2018). An excessive presence of M1 polarization has been linked with elevated bone resorption and osteoporosis while the predominance of M2 macrophages promotes production and healing (Vannella and Wynn 2017). By understanding what regulates this equilibrium and its effect on osteoporosis, insight could lead to innovative diagnostic and treatment approaches that focus on stimulating normal bone-rebuilding processes.

11.2.5 The Interplay of AD and Macrophage Polarization

The interplay between Alzheimer's disease (AD) and osteoporosis is increasingly recognized as interconnected through the process of macrophage polarization. The chronic systemic inflammation observed in AD not only affects neural pathways but also has profound effects on bone metabolism by influencing macrophage behavior. Specifically, AD-related inflammation tends to promote M1 polarization, which is associated with elevated osteoclast activity and bone resorption (Gharpure et al. 2024).

Research indicates that AD-driven cytokines such as IL-1β and TNF-α are elevated, leading to an increased M1/M2 ratio. This shift favors bone degradation over repair and contributes to the lower bone mineral density seen in AD patients. Furthermore, the presence of brain-derived factors like SPP1 and Cst7, implicated in AD, can exacerbate the imbalance in macrophage polarization, further promoting osteoclastic activity.

11.2.6 Therapeutic Insights

Pharmacological agents that target the polarization of macrophages have shown promise in treating osteoporosis. Estrogen has been found to promote an M2 phenotype, leading to decreased osteoclast activity and an increase in bone production (Manolagas 2000). Melatonin has also proven itself an effective therapy because of its capacity to influence macrophage polarization and its antioxidant properties that mitigate bone loss caused by oxidative stress (Mauriz et al. 2013).

11.2.7 Biomaterial Implication

Biomaterial developments for bone regeneration have focused on creating scaffolds that promote M2 macrophage polarization. Certain biomaterials have the ability to manipulate their environment in such a way as to foster more M2-dominated environments, which in turn boost bone formation processes and lead to superior outcomes with repair treatments. Osteoporosis therapy relies heavily on using materials that promote M2-dominated environments—bioactive compounds or engineered compounds releasing therapeutic agents being the focal points (Wu et al. 2021; Bai et al. 2020).

11.3 Future Direction and Challenges

The study of macrophage polarization and its effect on osteoporosis offers promising avenues of exploration and advancement. Future research should seek to clarify the exact molecular mechanisms through which macrophages affect bone cells. Furthermore, the investigation should include potential combination therapies targeting multiple pathways at once and developing biomaterials that better direct macrophage polarization towards bone-healing phenotypes. To advance in this sector it is also imperative that obstacles like the diversity of macrophage states be overcome as well as turning preclinical discoveries into practical treatments to make progress in this sector.

11.4 Conclusion

This review highlights the significance of macrophage polarization in osteoporosis development and its therapeutic potential by employing M2 macrophages' bone-regenerative abilities while decreasing bone-resorptive activities of M1 macrophages to create novel therapies that better address the root causes of osteoporosis while improving patient outcomes.

References

Azevedo MCS, Fonseca AC, Colavite PM et al (2021) Macrophage polarization and alveolar bone healing outcome: despite a significant M2 polarizing effect, VIP and PACAP treatments present a minor impact in alveolar bone healing in homeostatic conditions. Front Immunol 12:782566. https://doi.org/10.3389/fimmu.2021.782566

Bai J, Wang H, Chen H et al (2020) Biomimetic osteogenic peptide with mussel adhesion and osteoimmunomodulatory functions to ameliorate interfacial osseointegration under chronic inflammation. Biomaterials 255:120197. https://doi.org/10.1016/j.biomaterials.2020.120197

Dou C, Ding N, Zhao C et al (2018) Estrogen deficiency-mediated M2 macrophage osteoclastogenesis contributes to M1/M2 ratio alteration in ovariectomized osteoporotic mice. J Bone Miner Res 33(5):899–908. https://doi.org/10.1002/jbmr.3364

Funes SC, Rios M, Escobar-Vera J, Kalergis AM (2018) Implications of macrophage polarization in autoimmunity. Immunology 154(2):186–195. https://doi.org/10.1111/imm.12910

Gharavi AT, Hanjani NA, Movahed E, Doroudian M (2022) The role of macrophage subtypes and exosomes in immunomodulation. Cell Mol Biol Lett 27(1):83

Gharpure M, Vyavahare S, Ahluwalia P, Gupta SK, Lee TJ, Lohakare J, Kolhe R, Lei Y, Deak F, Lu XY, Isales CM, Fulzele S (2024) Alterations in Alzheimer's disease microglia transcriptome might be involved in bone pathophysiology. Neurobiol Dis 191:106404. https://doi.org/10.1016/j.nbd.2024.106404. Epub 2024 Jan 4. PMID: 38184014

Hadji P, Klein S, Gothe H, Haussler B, Kless T, Schmidt T, Steinle T, Verheyen F, Linder R (2013) The epidemiology of osteoporosis—bone evaluation study (best): an analysis of routine health insurance data. Dtsch Arztebl Int 110:52–57

He X, Tan S, Shao Z, Wang X (2022) Latitudinal and longitudinal regulation of tissue macrophages in inflammatory diseases. Genes Dis 9(5):1194–1207

Liu H, Wu X, Gang N et al (2015) Macrophage functional phenotype can be consecutively and reversibly shifted to adapt to microenvironmental changes. Int J Clin Exp Med 8(2):3044–3053

Manolagas SC (2000) Birth and death of bone cells: basic regulatory mechanisms and implications for the pathogenesis and treatment of osteoporosis. Endocr Rev 21(2):115–137. https://doi.org/10.1210/edrv.21.2.0395

Mauriz JL, Collado PS, Veneroso C, Reiter RJ, González-Gallego J (2013) A review of the molecular aspects of melatonin's anti-inflammatory actions: recent insights and new perspectives. J Pineal Res 54(1):1–14. https://doi.org/10.1111/j.1600-079X.2012.01014.x

Muñoz J, Akhavan NS, Mullins AP, Arjmandi BH (2020) Macrophage polarization and osteoporosis: a review. Nutrients 12(10):2999. https://doi.org/10.3390/nu12102999

Murray PJ, Allen JE, Biswas SK et al (2014) Macrophage activation and polarization: nomenclature and experimental guidelines. Immunity 41(1):14–20. https://doi.org/10.1016/j.immuni.2014.06.008

Orecchioni M, Ghosheh Y, Pramod AB, Ley K (2019) Macrophage polarization: different gene signatures in M1(LPS+) vs. classically and M2(LPS-) vs. alternatively activated macrophages. Front Immunol 10:1084

Porta C, Riboldi E, Ippolito A, Sica A (2015) Molecular and epigenetic basis of macrophage polarized activation. Semin Immunol 27(4):237–248. https://doi.org/10.1016/j.smim.2015.10.003

Razi S, Yaghmoorian Khojini J, Kargarijam F et al (2023) Macrophage efferocytosis in health and disease. Cell Biochem Funct 41(2):152–165. https://doi.org/10.1002/cbf.3780

Souza PP, Lerner UH (2013) The role of cytokines in inflammatory bone loss. Immunol Invest 42(7):555–622. https://doi.org/10.3109/08820139.2013.822766

Tian T, Wang Z, Chen L, Xu W, Wu B (2023) Photobiomodulation activates undifferentiated macrophages and promotes M1/M2 macrophage polarization via PI3K/AKT/mTOR signaling pathway. Laser Med Sci 38(1):86. https://doi.org/10.1007/s10753-022-01777-z

Vannella KM, Wynn TA (2017) Mechanisms of organ injury and repair by macrophages. Annu Rev Physiol 79(1):593–617. https://doi.org/10.1146/annurev-physiol-022516-034356

Wu PL, Liang YY, Sun GM (2021) Engineering immune-responsive biomaterials for skin regeneration. Biomater Transl 2(1):61–71. https://doi.org/10.3877/cma.j.issn.2096-112X.2021.01.008

Yang DH, Yang MY (2019) The role of macrophage in the pathogenesis of osteoporosis. Int J Mol Sci 20(9):2093. https://doi.org/10.3390/ijms20092093

Yang J, Park OJ, Kim J, Kwon Y, Yun CH, Han SH (2019) Modulation of macrophage subtypes by IRF5 determines osteoclastogenic potential. J Cell Physiol 234(12):23033–23042. https://doi.org/10.1002/jcp.28863

Chapter 12
Unlocking Macrophage Secrets: Histone Deacetylases in Chronic Transplant Rejection

Marta Halasa and Anna Wawruszak

Abstract Solid organ transplantation (SOT) offers life-saving therapy for patients with organ failure, yet chronic rejection remains a significant challenge despite advances in immunosuppression. Macrophages are central to chronic rejection, orchestrating fibrosis, and tissue damage. Since it became clear that histone deacetylases (HDACs), a family of epigenetic regulators, modulate macrophage function and polarization and eventually affect fibrosis progression, the HDACs modulation has gained great importance. This review explores the role of HDACs in chronic rejection, focusing on their impact on macrophage polarization and fibrosis. While some HDACs promote M2 polarization and fibrosis, others inhibit these processes, highlighting the complexity of HDAC function. Targeting HDACs holds promise as a therapeutic strategy for chronic rejection, offering a potential approach for intervention in transplant recipients. However, further research is needed to elucidate the specific roles of individual HDAC isoforms and their inhibition in chronic rejection.

12.1 Introduction

Solid organ transplantation (SOT) saves patients with organ failure. However, because of the limited number of donors, only some patients get a chance for transplantation. After transplantation, organ recipients are required to take immunosuppressive drugs to prevent transplant rejection for the rest of their lives (Black et al. 2018). Although the general mortality after SOT has decreased in recent years

M. Halasa (✉)
Houston Methodist Research Institute, Transplant Immunology, Houston, TX, USA

Houston Methodist Hospital, Department of Surgery, Houston, TX, USA
e-mail: mhalasa@houstonmethodist.org

A. Wawruszak
Department of Biochemistry and Molecular Biology, Medical University of Lublin, Lublin, Poland

M. Kloc et al. (eds.), *Monocytes and Macrophages in Development, Regeneration, and Disease*, Results and Problems in Cell Differentiation 74,
https://doi.org/10.1007/978-3-031-65944-7_12

(Søborg et al. 2022), there is still an urgent need to investigate signaling pathways involved in this phenomenon.

Transplant rejection is classified as hyperacute, acute, or chronic. Hyperacute rejection occurs immediately after transplantation and is caused by accidental AB0 blood type discordance between donor and recipient, which nowadays almost does not happen anymore. Acute rejection occurs within hours or days after transplantation and is observed in almost all transplants except between identical twins (Justiz Vaillant et al. 2023). Acute rejection is mostly driven by CD4+ T cells, which recruit other immune cells, including CD8+ T cells, natural killer (NK) cells, B cells, and macrophages. Chronic rejection develops within months or years and is characterized by the development of arteriosclerosis that results in thickening and stiffening of blood vessels over time. This phenomenon is often accompanied by graft tissue fibrosis resulting from macrophage infiltration into allograft. Following the infiltration, macrophages release a variety of inflammatory factors (which directly destroy graft tissue) or interplay with fibroblasts enabling their transition into myofibroblasts and eventually potentiating fibrosis. Simultaneously, they mediate other rejection-associated processes such as cell phagocytosis, damage of vascular endothelial cells, interaction with T cells, and release of exosomes (Yu and Lu 2022). Thus, macrophages are believed to be master regulators of chronic rejection (Vaillant et al. 2023; Vaillant and Mohseni 2023; Vitak 2014; Tai et al. 2021).

While acute rejection is manageable, chronic rejection cannot be efficiently treated with medications. Despite ongoing clinical trials related to long-term treatment against transplant rejection (ongoing clinical trials can be found here: www.clinicaltrials.gov), there is no therapy for chronic rejection. The search for new therapies leads to the conclusion that macrophages, due to their unique properties, may be a promising therapeutic target and should be deeply investigated.

The lack of success in anti-chronic rejection therapy could be also explained by the complexity of macrophage-mediated chronic rejection resulting from the involvement of various groups of proteins and signaling pathways. The family of proteins called histone deacetylases (HDACs) has gained a special interest since their role in macrophage polarization, proliferation and migration was confirmed. Thus, in this chapter, we will focus on HDACs and their role in chronic rejection in the context of macrophages as well as their pharmacological inhibition as potential targeted therapy against rejection.

12.2 Histone Deacetylases Are Involved in Physiological and Pathological Processes

Reversible histone acetylation is an epigenetic event that regulates chromatin modification, eventually affecting gene expression. Mechanistically, acetylation mediated by histone acetyltransferases (HATs) eliminates the positive charge on histones, leading to a reduction in the binding between their N-termini and the negatively charged phosphate groups of DNA. This results in the conversion of tightly packed chromatin into a more loosely organized structure, which is linked to an increase in gene transcription. The opposite process, called histone deacetylation, leads to the tightness and compaction of chromatin structure and blocks its accessibility for transcription factors silencing gene transcription. The regulatory phenomenon is mediated by histone deacetylases (HDACs), which often function as integral components within larger multiprotein complexes (Chen et al. 2015a). The balance between HDACs and HATs is critical for cellular homeostasis, and its dysregulation is often associated with disease development. Additionally, aberrant histone acetylation patterns can lead to inappropriate gene expression, also contributing to disease progression (Hałasa et al. 2019). The process of reversible histone acetylation is shown in Fig. 12.1.

The HDACs constitute an enzyme family highly conserved across all eukaryotes. They remove an acetyl group (O=C-CH3) from an ε-N-acetyl lysine from not only histones but also non-histone proteins. While histone deacetylation is mainly associated with gene regulation (Seto and Yoshida 2014), deacetylation of non-histone proteins mediates critical cellular processes, exerting a significant impact on cell fate. HDACs mediate various physiological events, including development, cellular homeostasis, and cellular prosperity (Shvedunova and Akhtar 2022). Simultaneously, they are involved in many pathological events such as cancer, diabetes, and neurodegenerative disorders (Ramaiah et al. 2021; Guo et al. 2018; Bassi et al. 2017; Liu et al. 2023; von Knethen et al. 2020).

The HDAC family consists of 18 members and is divided into 4 classes in terms of localization and sequence similarity. Class I, II, and IV require zinc ions for their catalytic activity, unlike class III (called sirtuins), which is activated by nicotinamide adenine dinucleotide (NAD+). Class I has 4 members: HDAC1, HDAC2,

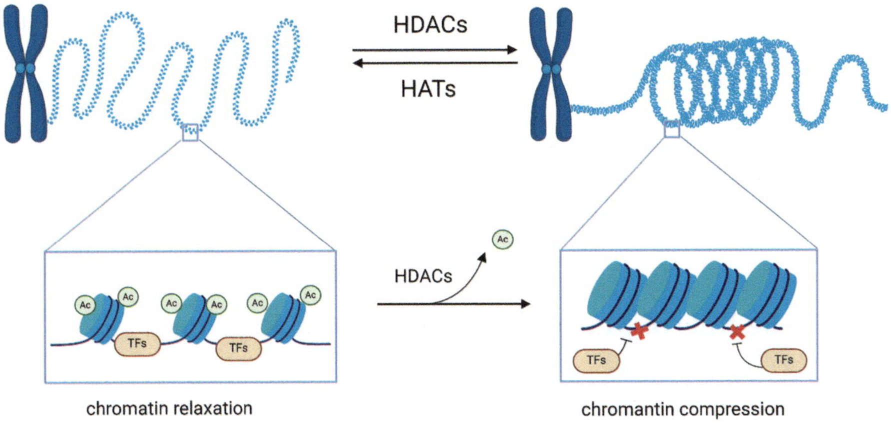

Fig. 12.1 Reversible histone deacetylation. Acetylation of histone lysine residues (Ac) mediated by histone transferases (HATs) causes relaxation of chromatin structure and provides easier access for transcription factors (TF), increasing gene transcription. Removal of acetyl-groups by histone deacetylases (HDACs) leads to the silencing of gene transcription through chromatin compaction and its inactivation. The figure was created with BioRender.com

HDAC3, and HDAC8. These proteins are ubiquitously expressed in many different tissues and are predominantly located in the nucleus exerting tremendous impact on histone deacetylation. They often recruit to the multi-protein complexes, which control transcription (Nakayama and Hayakawa 2011). Class II constitutes a heterogenic group of nuclear and cytosolic proteins. Due to their unique amino acid sequences, they are able to shuttle between the nucleus and cytoplasm, affecting gene expression and regulating cellular processes. Class II is divided into two subclasses: IIa (HDAC4, HDAC5, HDAC7, HDAC9) and IIb (HDAC6 and HDAC10) based on their structural similarity and is expressed in a more tissue-specific manner (Yang and Grégoire 2005). In the nucleus, HDACs from class II always form the multi-protein complexes as, in contrast to class I, they do not have catalytic activity alone. Class IV consists of only one member: HDAC11, whose role remains enigmatic (Morris and Monteggia 2013). The Sirtuins family (class III which consists of 7 members, from SIRT1 to SIRT7) is widely expressed and deacetylases not only histones but also many transcription factors and cytosolic proteins, influencing critical cellular processes such as aging (Grabowska et al. 2017).

Considering the complexity of HDAC function in cells, the designing of pharmacological modulators (inhibitors and activators) has gained great importance among scientists and attracted clinical attention (Hałasa et al. 2019; Wawruszak et al. 2019). Small molecules with the ability to decrease HDAC activity (HDAC inhibitors, HDACi) are of particular interest. They are divided into 4 main groups based on their chemical nature: hydroxamates (vorinostat, panobinostat, belinostat), benzamides (entinostat), short-chain fatty acids (sodium butyrate, valproic acid), and cyclic peptides (Shen and Kozikowski 2016; Sun et al. 2021). Largely, FDA-approved HDACi are not specific to a particular HDAC isoform (due to catalytic core similarities among HDACs) and work as pan-inhibitors (Bondarev et al. 2021). Pan-inhibitors are used to treat lymphomas (Vorinostat, Belinostat) and leukemia (Panabinostat). Additionally, just recently, a selective HDAC6 inhibitor (Ricolinostat) has been approved for multiple myeloma treatment (Vogl et al. 2017). In clinics, the majority of HDACi are used in anti-cancer therapy in combination with other drugs: alkylating agents, proteasome inhibitors, radiotherapy, or immunotherapy (Hontecillas-Prieto et al. 2020). However, due to the great potential of HDACi, some of them are tested in pre-clinical trials for neurodegenerative disorders and heart failure (Shukla and Tekwani 2020; Gillette 2021).

12.3 The Role of Macrophages in Chronic Rejection

Macrophages are widely distributed throughout the body, playing a pivotal role in immune defense. They exhibit remarkable phenotypic diversity and originate from either monocytes or the early hematopoietic progenitors located within the yolk sac or fetal liver. The latter constitutes a specialized population of resident macrophages with tissue-specific functions (Shukla and Tekwani 2020). In solid organ transplants, macrophages' behavior can vary depending on their activation state. These cells are recognized for their involvement in allograft injury and tissue remodeling, while also exhibiting immunoregulatory effects (Salehi and Reed 2015). Histological sections of the kidney cortex from chronically rejecting rats revealed the presence of infiltrated macrophages, indicating their potential as early markers for rejection before observable transplant failure (Beckmann et al. 2003). In humans, macrophages (and neutrophils) affect an immune cascade by which they potentiate early peri-transplant ischemia-reperfusion injury (IRI) resulting in tissue damage and eventually leading to organ loss and mortality (Nakamura et al. 2019). Their dynamic role in the IRI was demonstrated in kidney and liver transplant models (Kopecky et al. 2020).

Macrophages show remarkable adaptability in responding to diverse chemotactic factors and transitioning through various activation stages, spanning from the pro-inflammatory M1 to the anti-inflammatory M2. The process of macrophage polarization into M1 or M2 phenotypes is tightly regulated, encompassing various levels of control, including signaling pathways, transcriptional factors, epigenetic modifications, and post-transcriptional networks (Kapellos and Iqbal 2016; Komal et al. 2023; Sowers et al. 2022). It is crucial to note that the classification of macrophage activation stages is complex, and M1 and M2 are considered as the endpoints of the activation spectrum (Jayasingam et al. 2020). For simplicity, this chapter adopts the widely used classification, encompassing M0 (not activated), and the two extremes, M1 and M2.

The M1 phenotype is primarily induced by toll-like receptor ligands (bacterial lipopolysaccharides; LPS) or Th1 cytokines (tumor necrosis factor-alpha; TNFα), resulting in macrophages with heightened antigen-presenting capacity. When M1-activated, macrophages secrete pro-inflammatory factors (such as interleukin-6, interleukin-12, and interleukin-23) (Chen et al. 2023). M2 macrophage phenotype is shaped by Th2 cytokines (interleukin-4, interleukin-10, interleukin-13, and tumor growth factor beta; TGFβ). They are responsible for tissue regeneration as well as wound healing process (Kloc et al. 2019). Noteworthy, graft-infiltrated M2-activated macrophages are prevalent in chronic rejection allografts (Tan et al. 2022). In renal allograft biopsy samples from human subjects (taken one year after transplantation), M2-activated macrophages constituted 92% of all infiltrated macrophages (Toki et al. 2014). Similarly, in blood samples taken from patients after lung transplantation who have been using immunosuppressive medication for multiple years, monocyte-delivered macrophages presented an increased frequency of M2 phenotype compared to healthy control (Schreurs et al. 2020). Finally, in the murine cardiac allografts model, M1-activated macrophages rose at two weeks but declined by six weeks. Conversely, M2 macrophage markers increased by six weeks (Kaul et al. 2015). This observation was further confirmed in blood samples taken from allografts-rejected patients after one year of heart transplantation. The data showed a significantly higher level of M2-activated macrophages than M1-activated macrophages. Moreover, transplant rejection was also significantly associated with fibrosis (van den Bosch et al. 2017). Overall, macrophages play a crucial role in shaping transplant outcomes, and understanding their diverse functions is essential for advancing transplantation research. Thus, the appropriate approaches directing the polarization of macrophages as well as manipulation of their motility properties could be beneficial for patients after transplantation and serve as long-term anti-rejection treatment.

12.4 M2 Macrophages as Key Architects in Chronic Rejection

Chronic rejection is characterized by the gradual development of fibrosis. This fibrotic transformation often leads to organ failure and can result in fatality (Antar et al. 2023). This phenomenon is highly associated with fibroblast proliferation as well as excessive deposition of collagen and other extracellular matrix (ECM) components. ECM acts as a structural scaffold by forming fibrils providing a stable framework for tissue. While ECM interacts with cells through signal transmission (integrin and non-integrin receptors) affecting cellular phenotype and behavior, cells affect ECM by producing and secreting factors involved in its remodeling (Yue 2014). It is widely recognized that fibroblasts/myofibroblasts play a significant role as a primary source of ECM components potentiating fibrosis. Macrophages, however, also contribute substantially by providing proteins for the generation of active ECM fragments, such as Tissue Inhibitors of Metalloproteinases (TIMPs), cytokines, growth factors, fibronectin and laminin. Recently, transcriptomic analysis revealed elevated expression of ECM-related genes in both M1 and M2 macrophages, potentially contributing to fibrosis progression (Etich et al. 2019).

M2 macrophages are categorized as 'pro-fibrotic' due to their extensive production and secretion of various fibrosis-related factors. They release TGF-β, a master regulator, which induces fibroblast proliferation and facilitates their transformation into myofibroblasts. Importantly, this process is bilateral, as TGF-β plays a role in stimulating macrophages toward M2 polarization, thereby enhancing an anti-inflammatory phenotype (He et al. 2021). Macrophages also secrete Galectin-3, which regulates the activation of myofibroblasts, leading to hepatic fibrosis potentiation (Henderson et al. 2006). Moreover, they express and secrete a group of ECM-related proteins called metalloproteinases. Some of them tremendously impact fibrosis. For example, matrix metalloproteinase 12 (MMP12) mediates macrophage-to-myofibroblast transition (MMT) favoring fibrosis, while matrix metalloproteinase 28 (MMP28) sustains M2 macrophages phenotype, simultaneously reducing recruitment of repair cells (Yi et al. 2022; Gharib et al. 2014).

M2 macrophages promote collagen (type 1 and 3) production and expression in airway fibroblast in patients suffering from laryngotracheal stenosis (LTS) where pathologic scar tissue formation is observed (Motz et al. 2021). Reversibly, collagen monomers (type 1) produced by other immune cells can shift M0 macrophages into the M2 stage through Protein Kinase B (Akt) signaling, eventually inducing fibrosis (Stahl et al. 2013). Although the role of M2 macrophages in fibrosis is very complex and needs to be better understood, the newest findings focused on signaling pathways involved in this phenomenon give the impression that M2 macrophages could be a valuable target for anti-fibrosis treatment (Wang et al. 2024). Since fibrosis is associated with chronic transplant rejection, inhibition of M2 macrophage polarization could benefit transplant recipients.

12.5 Overview of HDACs Expression Pattern in Macrophages

HDACs are broadly expressed in human tissues and organs (Krusche et al. 2007; Zhao et al. 2020; Lucio-Eterovic et al. 2008). The single-cell transcriptomic map from The Human Protein Atlas database (https://www.proteinatlas.org/) illustrated the average level of HDACs in various organs and different types of blood and immune cells in healthy human tissues (Karlsson et al. 2021). In macrophages, the highest average expression was observed for *Hdac1*, *Hdac2*, *Hdac3*, and *Hdac9*, while *Hdac11* was barely expressed. Among Sirtuin family members, *Sirt1*, *Sirt2*, and *Sirt7* exhibited the highest expression levels, while *Sirt4* had marginal expression (Fig. 12.2a, b). HDACs' expression pattern is in line with the observation that HDAC1 and HDAC2 are ubiquitously expressed in many different cells, and they often occur in pairs, creating multi-protein co-repressor complexes that eventually impact cell prosperity. Their role, however, is not limited to chromatin modification, as they interplay with many non-histone proteins (Kelly and Cowley 2013; Delcuve et al. 2012). Moreover, they can act independently, as demonstrated in another immune cell type, CD4+ T cells, where differences in the distribution of HDAC1 and HDAC2 along regulatory and coding regions were observed (Wang et al. 2009). Moreover, the I class of HDACs is involved in the proliferation and differentiation of bone marrow-derived monocytes into macrophages and dendric cells (Wang et al. 2020). For example, HDAC3 is critical for lung alveolar macrophages (AM) development and homeostasis through direct binding to *Peroxisome Proliferator Activated Receptor Gamma (PParg)* gene enhancers and regulating PPAR-γ signaling (Yao et al. 2020). PPAR-γ is well known as a critical nuclear receptor for the differentiation of AM (Schneider et al. 2014). A decrease in HDACs activity was observed in AM isolated from patients suffering from chronic obstructive pulmonary disease (COPD) characterized by chronic inflammation (Shanmugam et al. 2022). In contrast, some HDACs, such as HDAC6, are upregulated in the lungs of patients with COPD, which is associated with fibrosis progression (Su et al. 2021). Indeed, the dysregulation of HDACs affects the profile of macrophages, directing them into M2 polarization. This phenomenon can be used for future COPD therapies (da Silva et al. 2023).

Consistently, the expression pattern of Sirtuins (particularly *Sirt1* and *Sirt2*) is unsurprising given their increasingly recognized role in controlling metabolic pathways, proliferation, and differentiation of cells involved in immune response (Fortuny and Sebastián 2021; Elesela et al. 2020; Chen et al. 2015b).

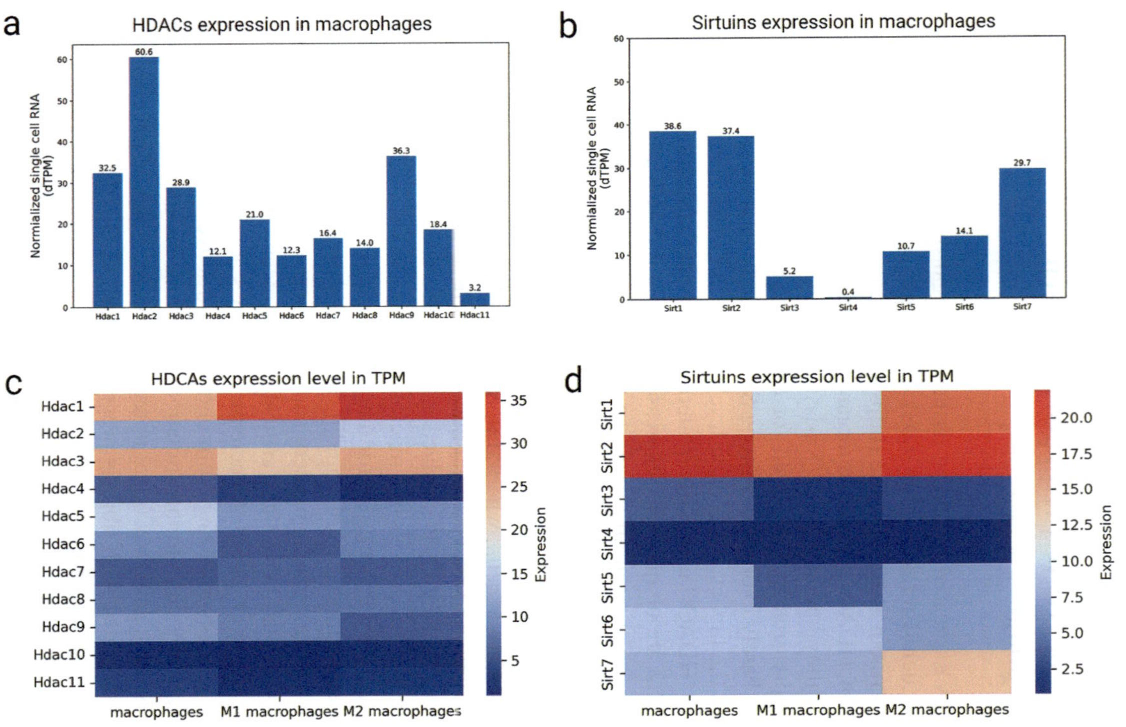

Fig. 12.2 The *Hdacs* and *Sirtuins* expression analysis. The gene expression analysis for all members of the HDACs family was performed based on data provided by The Human Protein Atlas database (**a**, **b**). The raw expression data (from single-cell RNA sequencing; scRNA-seq) of *Hdacs* from each tissue sample were grouped into clusters and pooled to generate normalized transcript per million (nTPM) for each gene and cell type. The Expression Atlas database was employed to assess gene expression for all members of the HDACs family in different activated macrophage stages (**c**, **d**). Macrophages from all stages (M0, M1, and M2) were isolated from venous blood from healthy individuals, and RNA-seq analysis was performed, with the data presented as transcripts per million (TPM). Expression levels are categorized as low (0–10 TPM), medium (11–100 TPM), and high (101–1000 TPM)

To access the HDACs expression pattern in activated macrophages, we used the Expression Atlas database (https://www.ebi.ac.uk/gxa/home) (Moreno et al. 2022). We analyzed mRNA levels of both *Hdacs* and *Sirtuins* across different stages of macrophage polarization, including M0, M1, and M2 macrophages. Transcriptomic analysis revealed a consistent expression pattern among non-activated macrophages (M0), aligning with the general macrophage expression profile from The Human Protein Atlas database. However, notable differences were observed in *Hdac2* expression between the datasets. This discrepancy may, at least partly, be explained by the fact that the first analysis was the average expression of macrophages isolated from different organs, while the second analysis had data from macrophages isolated exclusively from venous blood. Overall, the *Hdacs* expression is higher in M2 macrophages compared to M1 macrophages. For example, a slight upregulation is observed in the I class of HDACs (excluding *Hdac8*) compared to M0 macrophages, whereas this pattern is not maintained for M1 macrophages. Among Sirtuins, the most pronounced changes are observed for *Sirt1*, *Sirt2*, and *Sirt7*. However, a specific pattern is not discernible. It should be noted that the general levels of *Hdacs* and *Sirtuins* expression in macrophages were relatively moderate or low, irrespective of the activation stage.

12.6 HDACs Involvement in M2 Polarization

HDACs show higher activity in M2 macrophages than M1 macrophages, and some HDACi efficiently blocks M2 polarization without affecting cell proliferation (He et al. 2021). Inhibition of HDAC1 and HDAC2 activity promotes M2 polarization through an increase in histone 3 (H3) and histone 4 (H4) acetylation (Han et al. 2018). This highlights the complex role of HDACs in regulating macrophage polarization, where their activity levels and isoform-specific effects play crucial roles in determining the balance between M1 and M2 phenotypes.

The use of vorinostat, an inhibitor of class I and II HDACs, in mice with induced myocardial infarction elevated the concentration of M2 macrophages in hearts 1 day after surgery, which suggests that decreased HDAC expression is associated with early and potent recruitment of healing-related macrophages. Additionally, M2-polarization gene markers activation was observed in commercially available RAW264.7 macrophage cell line after vorinostat treatment (Kimbrough et al. 2018). This insight is particularly relevant considering findings from a COPD murine model study, where a myeloid-specific deficiency of HDAC2 amplified adverse effects under particle pollution from fine particulates (PM2.5) exposure. The absence of HDAC2 led to enhanced M2 polarization of AM, alongside significant upregulation of pro-fibrotic and tissue remodeling factors such as TGF-β, MMP-9, and MMP-12 (Jiang et al. 2020). Similarly, HDAC9 knock-out enhanced the M2 phenotype of macrophages upregulating M2-associated genes: *mannose receptor C-type 1* (*CD206*), *Pparg* in mice, and *Cd206* and *Cd209 molecule* (*Cd209*) in humans (Liu et al. 2021). Moreover, in HDAC9-deficient bone marrow macrophages with concurrent deletion of Low-Density Lipoprotein Receptor (LDLr), a noticeable surge in M2 markers and a decrease in M1 inflammatory genes was confirmed, signifying a shift towards an M2-like phenotype. This shift was correlated with heightened histone acetylation, specifically at *Pparg* promoters, leading to an upregulation of *Pparg* expression. Given the established association of LDLr with atherosclerosis and the observed effects of HDAC9 knockout on cholesterol efflux and M2 polarization in mice, HDAC9 emerges as a promising target for anti-atherosclerosis therapy (Cao et al. 2014).

For some HDACs, the data related to M2 polarization are contradictory. HDAC3 is an important regulator of M2 activation through deacetylation of histones at enhancer regions for M2-associated genes, silencing these genes and eventually suppressing the M2 polarization in mice (Mullican et al. 2011). In contrast, in humans, pharmacological inhibition of HDAC3 did not affect M2-associated cytokines secretions suggesting its minor role in macrophage polarization (Ghiboub et al. 2020).

Some HDACs appear to potentate M2 polarization. For example, in asthmatic patients and mice, macrophages highly express HDAC10 eventually promoting M2 polarization. The deficiency of HDAC10 notably reduced the level of transcription factor, Signal transducer and activator of transcription 3 (STAT3) and diminished M2 macrophage development after exposure to allergens. Regarding mechanism, HDAC10 binds directly to STAT3 in macrophages, promoting its deacetylation. This interaction boosted STAT3 expression and triggered the activation of the M2 macrophage program (Zhong et al. 2023). Recent findings showed a direct connection between pharmacological inhibition of HDAC6 and fibrosis inhibition by suppressing M2 macrophage polarization. The use of tubastatin (TA), selective HDAC6i, effectively halted peritoneal fibrosis progression by reducing ECM protein deposition and MMP-2/MMP-9 expression, as well as increased acetylation of H3 and α-tubulin. HDAC6 blockade hindered M2 macrophage polarization via TGF-β/Smad3, Phosphoinositide 3-kinase (PI3K)/Akt, STAT3, and STAT6 pathways. In cell injury models, TA dose-dependently impeded M2 macrophage polarization through TGF-β/Smad3, IL4/STAT6, and PI3K/Akt signaling. This study suggests that targeting HDAC6 holds promise for treating peritoneal fibrosis by inhibiting M2 macrophage polarization (Shi et al. 2022). The schematic representation of HDAC involvement in M2 polarization is presented in Fig. 12.3.

The intricate relationship between HDACs and M2 macrophages underscores the potential of targeted interventions to modulate immune responses, as evidenced by studies employing HDAC inhibitors and genetic manipulations to influence M2 polarization. Contradictory roles of specific HDACs in M2 macrophage polarization highlight the complexity of epigenetic regulation in immune responses, emphasizing the need for nuanced approaches in therapeutic targeting to achieve desired outcomes. The identification of HDACs as regulators of M2 polarization opens avenues for further research and therapeutic exploration, offering potential targets for addressing conditions involving chronic rejection and tissue remodeling.

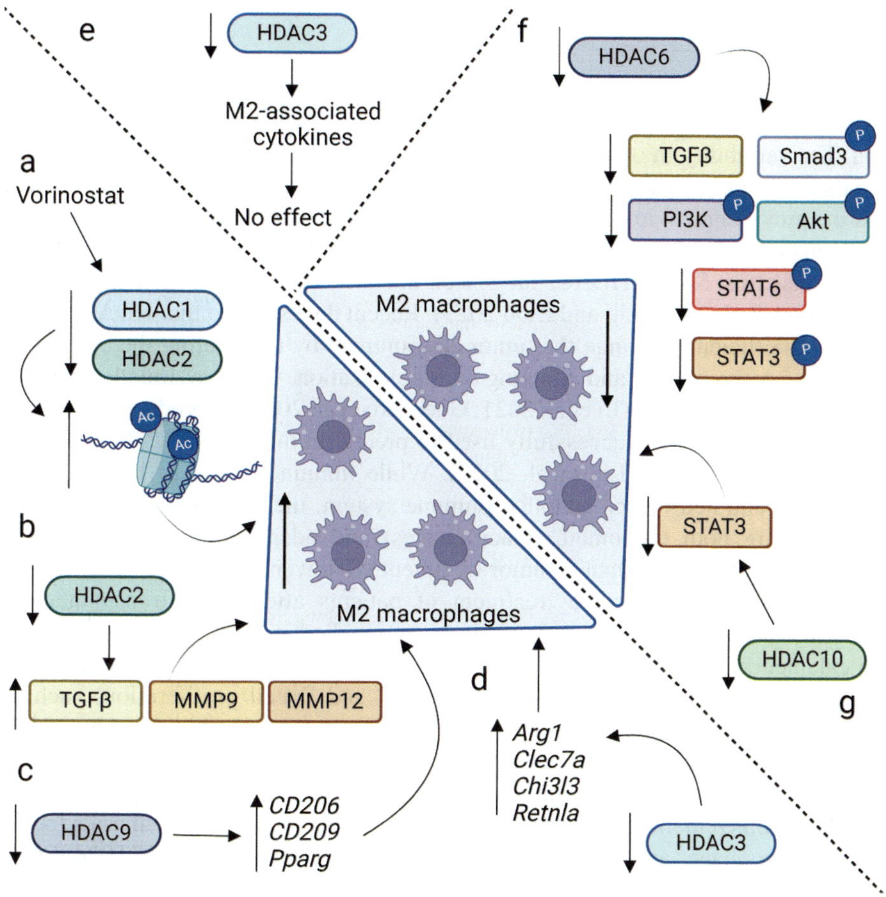

Fig. 12.3 HDACs involvement in M2 polarization of macrophages. Downregulation of HDAC1 and HDAC2 by vorinostat leads to upregulated acetylation of histones and eventually potentiates M2 polarization (**a**). Downregulation of HDAC2 is associated with upregulation of fibrosis-related proteins: TGFβ, MMP9 and MMP12 and, ultimately, potentiates M2 phenotype polarization (**b**). HDAC9 depletion enhances the transcription of M2-related genes: *CD206*, *CD209*, and *Pparg*, favoring the M2 phenotype (**c**). Downregulation of HDAC3 augments the expression of M2 gene markers *Agr1*, *Clec7a*, *Chi3l3*, and *Retnla* in mice or does not impact the production of M2-associated cytokines in humans (**d**, **e**). Pharmacological inhibition of HDAC6 abolishes IL-4 mediated M2 polarization through downregulating TGFβ expression and phosphorylation levels of Smad3, PI3K, Akt, STAT6, and STAT3 (**f**). Downregulation of HDAC10 inhibits the expression of STAT3 and eventually diminishes M2 macrophage development. The figure was created with BioRender.com. ↑: upregulation, ↓: downregulation, Ac: acetylated, P: phosphorylated, Akt: Protein kinase B, Arg1: Arginase 1, CD206: Mannose receptor C type 1, CD209: CD209 molecule, Chi3l3: Chitinase 3-like-3, Clec7a: C-type lectin domain containing 7A, MMP9: Matrix metalloproteinase-9, MMP12: Matrix metalloproteinase-12, PI3K: SMAD family member 3, Pprg: Peroxisome Proliferator-Activated Receptor Gamma, Rentla: Resistin-like molecule alpha, Smad3: SMAD family member 3, STAT3: signal transducer and activator of transcription 3, STAT6: signal transducer and activator of transcription 6, TGFβ: Transforming growth factor beta

12.7 Concluding Remarks

While HDACs within each class share some degree of structural similarity, they also exhibit differences that contribute to their specific functions and cellular localization. The variability in outcomes showing the context-dependent nature of HDACs and their inhibition differ markedly based on the disease model, the specific HDACs, and the prevailing environmental conditions. Understanding the specific molecular mechanism of HDACs inhibition appears beneficial in the context of immune-related diseases. So far, HDACs have been intensely investigated in many cancers (Mehmood et al. 2023; Liu and Liou 2023). Recent data showed that HDAC inhibition could efficiently change the tumor environment by modulating the activity of infiltrating macrophages and affecting their polarization, which eventually provides tumor-grow suppression (Li et al. 2021; Guerriero et al. 2017). Moreover, the HDAC expression pattern was successfully used to predict immunotherapy efficiency for gastric cancer patients (Lin et al. 2023). While immuno-oncology has predominantly concentrated on the adaptive immune system, the newest findings indicate that targeting both the innate (macrophages) and adaptive components of the immune system could enhance tumor treatment effectiveness. Similar observations might be considered for the treatment of patients after organ transplantation. Currently, the first line of immunosuppressive medications in transplant rejection are calcineurin inhibitors (cyclosporine, tacrolimus), mTOR inhibitors, corticosteroids, and antiproliferative agents (reducing T and B cell proliferation) such as mycophenolate mofetil or azathioprine (Barbarino et al. 2013; Alasfar et al. 2023). These drugs primarily target the adaptive immune system, although they may also have some effects on innate immunity. Yet, therapies based on targeting innate immunity cells remain left behind. The ongoing fibrosis-related clinical trials focusing on macrophages are usually in the early stages (NCT05976217, NCT02485886, NCT01777737). Considering the prominent role of macrophages in chronic rejection, these cells become a reasonable candidate for therapeutic intervention. Manipulation of macrophage polarization can be one of the suggested approaches. Among all molecular signaling pathways enabling such manipulation, HDACs involvement plays an important role. Thus, the use of HDACi can prevent M2 polarization and eventually reduce the progression of fibrosis. Despite the current gaps in understanding the molecular mechanisms of HDAC involvement in M2 polarization and contradictory data (for some HDACs), their inhibition remains a compelling therapeutic approach worth further investigation in the context of chronic rejection.

Although in this chapter we have focused on the role of HDACs in M2 polarization of macrophages, the HDACs expression pattern and HDACi also influence many other functions of these cells that ultimately contribute to chronic rejection and general inflammatory response, such as infiltration and motility. For example, HDAC6 regulates the macrophage's infiltration during inflammation, and HDAC6 knockout impairs macrophage motility (Yan et al. 2018). Trichostatin A (TSA), a potent pan-HDAC inhibitor, downregulates MMP-9 expression, which can also affect cell motility (Kim et al. 2015; Webb et al. 2017). It is worth noting that some

HDAC inhibitors also target other proteins. Sodium butyrate, a pan-HDAC inhibitor, reduces LPS-mediated macrophage migration through downregulation of migration-related proteins: Src tyrosine kinase and focal adhesion kinase (FAK) (Maa et al. 2010). Many other important macrophage properties are regulated by HDACs, which could eventually affect patients' outcomes after transplantation. HDACs and their inhibitors affect cytokines production, cells proliferation and can modulate cell fate (Choi et al. 2008; Ali et al. 2018; Kankaanranta et al. 2010).

The multi-level role of HDACs in macrophages in cellular function extends far beyond their involvement in the M2 polarization of macrophages. M2 polarization, however, remains critical for fibrosis progression and poor transplant patients outcomes. Considering these insights, continued research into the multifaceted roles of HDACs and their inhibitors in macrophage biology and immune-related diseases is warranted, offering prospects for novel therapeutic tools with the potential to enhance patient care and outcomes in transplant medicine and beyond.

References

Alasfar S, Kodali L, Schinstock CA (2023) Current therapies in kidney transplant rejection. J Clin Med 12(15). https://doi.org/10.3390/JCM12154927

Ali MN et al (2018) The HDAC inhibitor, SAHA, prevents colonic inflammation by suppressing pro-inflammatory cytokines and chemokines in DSS-induced colitis. Acta Histochem Cytochem 51(1):33–40. https://doi.org/10.1267/AHC.17033

Antar SA, Ashour NA, Marawan ME, Al-Karmalawy AA (2023) Fibrosis: types, effects, markers, mechanisms for disease progression, and its relation with oxidative stress, immunity, and inflammation. Int J Mol Sci 24(4). https://doi.org/10.3390/IJMS24044004

Barbarino JM, Staatz CE, Venkataramanan R, Klein TE, Altman RB (2013) PharmGKB summary: cyclosporine and tacrolimus pathways. Pharmacogenet Genomics 23(10):563. https://doi.org/10.1097/FPC.0B013E328364DB84

Bassi S, Tripathi T, Monziani A, Di Leva F, Biagioli M (2017) Epigenetics of Huntington's disease. Adv Exp Med Biol 978:277–299. https://doi.org/10.1007/978-3-319-53889-1_15

Beckmann N et al (2003) Macrophage labeling by SPIO as an early marker of allograft chronic rejection in a rat model of kidney transplantation. Magn Reson Med 49(3):459–467. https://doi.org/10.1002/MRM.10387

Black CK, Termanini KM, Aguirre O, Hawksworth JS, Sosin M (2018) Solid organ transplantation in the 21st century. Ann Transl Med 6(20):409–409. https://doi.org/10.21037/ATM.2018.09.68

Bondarev AD, Attwood MM, Jonsson J, Chubarev VN, Tarasov VV, Schiöth HB (2021) Recent developments of HDAC inhibitors: emerging indications and novel molecules. Br J Clin Pharmacol 87(12):4577–4597. https://doi.org/10.1111/BCP.14889

Cao Q, Rong S, Repa JJ, Clair RS, Parks JS, Mishra N (2014) HDAC9 represses cholesterol efflux and generation of alternatively activated macrophages in atherosclerosis development. Arterioscler Thromb Vasc Biol 34(9):1871. https://doi.org/10.1161/ATVBAHA.114.303393

Chen HP, Zhao YT, Zhao TC (2015a) Histone deacetylases and mechanisms of regulation of gene expression (histone deacetylases in cancer). Crit Rev Oncog 20(1–2):35. https://doi.org/10.1615/CRITREVONCOG.2015012997

Chen X, Lu Y, Zhang Z, Wang J, Yang H, Liu G (2015b) Intercellular interplay between Sirt1 signalling and cell metabolism in immune cell biology. Immunology 145(4):455. https://doi.org/10.1111/IMM.12473

Chen S et al (2023) Macrophages in immunoregulation and therapeutics. Signal Transduct Target Ther 8(1):1–35. https://doi.org/10.1038/s41392-023-01452-1

Choi Y et al (2008) Histone deacetylase inhibitor KBH-A42 inhibits cytokine production in RAW 264.7 macrophage cells and in vivo endotoxemia model. Exp Mol Med 40(5):574–581. https://doi.org/10.3858/emm.2008.40.5.574

da Silva CO et al (2023) COPD patients exhibit distinct gene expression, accelerated cellular aging, and bias to M2 macrophages. Int J Mol Sci 24(12). https://doi.org/10.3390/IJMS24129913

Delcuve GP, Khan DH, Davie JR (2012) Roles of histone deacetylases in epigenetic regulation: emerging paradigms from studies with inhibitors. Clin Epigenetics 4(1):1–13. https://doi.org/10.1186/1868-7083-4-5/FIGURES/2

Elesela S, Morris SB, Narayanan S, Kumar S, Lombard DB, Lukacs NW (2020) Sirtuin 1 regulates mitochondrial function and immune homeostasis in respiratory syncytial virus infected dendritic cells. PLoS Pathog 16(2):e1008319. https://doi.org/10.1371/JOURNAL.PPAT.1008319

Etich J, Koch M, Wagener R, Zaucke F, Fabri M, Brachvogel B (2019) Gene expression profiling of the extracellular matrix signature in macrophages of different activation status: relevance for skin wound healing. Int J Mol Sci 20(20). https://doi.org/10.3390/IJMS20205086

Fortuny L, Sebastián C (2021) Sirtuins as metabolic regulators of immune cells phenotype and function. Genes (Basel) 12(11). https://doi.org/10.3390/GENES12111698

Gharib SA et al (2014) MMP28 promotes macrophage polarization toward M2 cells and augments pulmonary fibrosis. J Leukoc Biol 95(1):9–18. https://doi.org/10.1189/JLB.1112587

Ghiboub M et al (2020) HDAC3 mediates the inflammatory response and LPS tolerance in human monocytes and macrophages. Front Immunol 11. https://doi.org/10.3389/FIMMU.2020.550769/FULL

Gillette TG (2021) HDAC inhibition in the heart: erasing hidden fibrosis. Circulation 143(19):1891–1893. https://doi.org/10.1161/CIRCULATIONAHA.121.054262

Grabowska W, Sikora E, Bielak-Zmijewska A (2017) Sirtuins, a promising target in slowing down the ageing process. Biogerontology 18(4):447. https://doi.org/10.1007/S10522-017-9685-9

Guerriero JL et al (2017) Class IIa HDAC inhibition reduces breast tumours and metastases through anti-tumour macrophages. Nature 543(7645):428–432. https://doi.org/10.1038/NATURE21409

Guo P, Chen W, Li H, Li M, Li L (2018) The histone acetylation modifications of breast cancer and their therapeutic implications. Pathol Oncol Res 24(4):807–813. https://doi.org/10.1007/S12253-018-0433-5

Hałasa M et al (2019) H3K18Ac as a marker of cancer progression and potential target of anticancer therapy. Cells 8(5). https://doi.org/10.3390/CELLS8050485

Han Z et al (2018) TOPK promotes microglia/macrophage polarization towards M2 phenotype via inhibition of HDAC1 and HDAC2 activity after transient cerebral ischemia. Aging Dis 9(2):235. https://doi.org/10.14336/AD.2017.0328

He L et al (2021) Global characterization of macrophage polarization mechanisms and identification of M2-type polarization inhibitors. Cell Rep 37(5):109955. https://doi.org/10.1016/J.CELREP.2021.109955

Henderson NC et al (2006) Galectin-3 regulates myofibroblast activation and hepatic fibrosis. Proc Natl Acad Sci U S A 103(13):5060–5065. https://doi.org/10.1073/PNAS.0511167103/SUPPL_FILE/05-11167FIG7.PDF

Hontecillas-Prieto L, Flores-Campos R, Silver A, de Álava E, Hajji N, García-Domínguez DJ (2020) Synergistic enhancement of cancer therapy using HDAC inhibitors: opportunity for clinical trials. Front Genet 11:578011. https://doi.org/10.3389/FGENE.2020.578011/BIBTEX

Jayasingam SD, Citartan M, Thang TH, Mat Zin AA, Ang KC, Ch'ng ES (2020) Evaluating the polarization of tumor-associated macrophages into M1 and M2 phenotypes in human cancer tissue: technicalities and challenges in routine clinical practice. Front Oncol 9:494426. https://doi.org/10.3389/FONC.2019.01512/BIBTEX

Jiang Y, Zhao Y, Wang Q, Chen H, Zhou X (2020) Fine particulate matter exposure promotes M2 macrophage polarization through inhibiting histone deacetylase 2 in the pathogenesis of chronic obstructive pulmonary disease. Ann Transl Med 8(20):1303–1303. https://doi.org/10.21037/ATM-20-6653

Justiz Vaillant AA, Misra S, Fitzgerald BM (2023) Acute transplantation rejection. StatPearls. Accessed 03 Jan 2024. [Online]. Available: https://pubmed.ncbi.nlm.nih.gov/30571031/

Kankaanranta H et al (2010) Histone deacetylase inhibitors induce apoptosis in human eosinophils and neutrophils. J Inflamm 7(1):1–15. https://doi.org/10.1186/1476-9255-7-9/FIGURES/12

Kapellos TS, Iqbal AJ (2016) Epigenetic control of macrophage polarisation and soluble mediator gene expression during inflammation. Mediators Inflamm 2016. https://doi.org/10.1155/2016/6591703

Karlsson M et al (2021) A single–cell type transcriptomics map of human tissues. Sci Adv 7(31). https://doi.org/10.1126/SCIADV.ABH2169/SUPPL_FILE/SCIADV.ABH2169_SM.PDF

Kaul AMK et al (2015) Acute and chronic rejection: compartmentalization and kinetics of counterbalancing signals in cardiac transplants. Am J Transplant 15(2):333–345. https://doi.org/10.1111/AJT.13014

Kelly RDW, Cowley SM (2013) The physiological roles of histone deacetylase (HDAC) 1 and 2: complex co-stars with multiple leading parts. Biochem Soc Trans 41(3):741–749. https://doi.org/10.1042/BST20130010

Kim S, Bong N, Kim OS, Jin JY, Kim DE, Lee DK (2015) Lithium chloride suppresses LPS-mediated matrix metalloproteinase-9 expression in macrophages through phosphorylation of GSK-3β. Cell Biol Int 39(2):177–184. https://doi.org/10.1002/CBIN.10352

Kimbrough D et al (2018) HDAC inhibition helps post-MI healing by modulating macrophage polarization. J Mol Cell Cardiol 119:51. https://doi.org/10.1016/J.YJMCC.2018.04.011

Kloc M, Ghobrial RM, Wosik J, Lewicka A, Lewicki S, Kubiak JZ (2019) Macrophage functions in wound healing. J Tissue Eng Regen Med 13(1):99–109. https://doi.org/10.1002/TERM.2772

Komal S et al (2023) Epigenetic regulation of macrophage polarization in cardiovascular diseases. Pharmaceuticals 16(2). https://doi.org/10.3390/PH16020141

Kopecky BJ, Frye C, Terada Y, Balsara KR, Kreisel D, Lavine KJ (2020) Role of donor macrophages after heart and lung transplantation. Am J Transplant 20(5):1225. https://doi.org/10.1111/AJT.15751

Krusche CA, Vloet AJ, Classen-Linke I, Von Rango U, Beier HM, Alfer J (2007) Class I histone deacetylase expression in the human cyclic endometrium and endometrial adenocarcinomas. Hum Reprod 22(11):2956–2966. https://doi.org/10.1093/HUMREP/DEM241

Li X et al (2021) HDAC inhibition potentiates anti-tumor activity of macrophages and enhances anti-PD-L1-mediated tumor suppression. Oncogene 40(10):1836–1850. https://doi.org/10.1038/s41388-020-01636-x

Lin Y et al (2023) Histone deacetylase-mediated tumor microenvironment characteristics and synergistic immunotherapy in gastric cancer. Theranostics 13(13):4574–4600. https://doi.org/10.7150/THNO.86928

Liu YM, Liou JP (2023) An updated patent review of histone deacetylase (HDAC) inhibitors in cancer (2020 – present). Expert Opin Ther Pat 33(5):349–369. https://doi.org/10.1080/13543776.2023.2219393

Liu Y, Du M, Lin HY (2021) Histone deacetylase 9 deficiency exaggerates uterine M2 macrophage polarization. J Cell Mol Med 25(16):7690. https://doi.org/10.1111/JCMM.16616

Liu R, Li L, Wang Z, Zhu J, Ji Y (2023) Acetylated histone modifications: intersection of diabetes and atherosclerosis. J Cardiovasc Pharmacol. https://doi.org/10.1097/FJC.0000000000001516

Lucio-Eterovic AKB et al (2008) Differential expression of 12 histone deacetylase (HDAC) genes in astrocytomas and normal brain tissue: Class II and IV are hypoexpressed in glioblastomas. BMC Cancer 8(1):1–10. https://doi.org/10.1186/1471-2407-8-243/FIGURES/3

Maa MC et al (2010) Butyrate reduced lipopolysaccharide-mediated macrophage migration by suppression of Src enhancement and focal adhesion kinase activity. J Nutr Biochem 21(12):1186–1192. https://doi.org/10.1016/J.JNUTBIO.2009.10.004

Mehmood SA et al (2023) Recent advancement of HDAC inhibitors against breast cancer. Med Oncol 40(7). https://doi.org/10.1007/S12032-023-02058-X

Moreno P et al (2022) Expression Atlas update: gene and protein expression in multiple species. Nucleic Acids Res 50(D1):D129–D140. https://doi.org/10.1093/NAR/GKAB1030

Morris MJ, Monteggia LM (2013) Unique functional roles for class I and class II histone deacetylases in central nervous system development and function. Int J Dev Neurosci 31(6):370. https://doi.org/10.1016/J.IJDEVNEU.2013.02.005

Motz K et al (2021) M2 macrophages promote collagen expression and synthesis in laryngotracheal stenosis fibroblasts. Laryngoscope 131(2):E346. https://doi.org/10.1002/LARY.28980

Mullican SE et al (2011) Histone deacetylase 3 is an epigenomic brake in macrophage alternative activation. Genes Dev 25(23):2480. https://doi.org/10.1101/GAD.175950.111

Nakamura K, Kageyama S, Kupiec-Weglinski JW (2019) Innate immunity in ischemia-reperfusion injury and graft rejection. Curr Opin Organ Transplant 24(6):687. https://doi.org/10.1097/MOT.0000000000000709

Nakayama JI, Hayakawa T (2011) Physiological roles of Class I HDAC complex and histone demethylase. J Biomed Biotechnol 2011. https://doi.org/10.1155/2011/129383

Ramaiah MJ, Tangutur AD, Manyam RR (2021) Epigenetic modulation and understanding of HDAC inhibitors in cancer therapy. Life Sci 277. https://doi.org/10.1016/J.LFS.2021.119504

Salehi S, Reed EF (2015) The divergent roles of macrophages in solid organ transplantation. Curr Opin Organ Transplant 20(4):446. https://doi.org/10.1097/MOT.0000000000000209

Schneider C, Nobs SP, Kurrer M, Rehrauer H, Thiele C, Kopf M (2014) Induction of the nuclear receptor PPAR-γ by the cytokine GM-CSF is critical for the differentiation of fetal monocytes into alveolar macrophages. Nat Immunol 15(11):1026–1037. https://doi.org/10.1038/NI.3005

Schreurs I, Meek B, van Moorsel CHM, van Kessel DA, Luijk HD, Grutters JC (2020) Monocyte derived macrophages from lung transplantation patients have an increased M2 profile. Transplant Rep 5(1):100038. https://doi.org/10.1016/J.TPR.2019.100038

Seto E, Yoshida M (2014) Erasers of histone acetylation: the histone deacetylase enzymes. Cold Spring Harb Perspect Biol 6(4). https://doi.org/10.1101/CSHPERSPECT.A018713

Shanmugam G, Rakshit S, Sarkar K (2022) HDAC inhibitors: targets for tumor therapy, immune modulation and lung diseases. Transl Oncol 16:101312. https://doi.org/10.1016/J.TRANON.2021.101312

Shen S, Kozikowski AP (2016) Why hydroxamates may not be the best histone deacetylase inhibitors—what some may have forgotten or would rather forget? ChemMedChem 11(1):15. https://doi.org/10.1002/CMDC.201500486

Shi Y et al (2022) Pharmacologic inhibition of histone deacetylase 6 prevents the progression of chlorhexidine gluconate-induced peritoneal fibrosis by blockade of M2 macrophage polarization. Front Immunol 13:899140. https://doi.org/10.3389/FIMMU.2022.899140/BIBTEX

Shukla S, Tekwani BL (2020) Histone deacetylases inhibitors in neurodegenerative diseases, neuroprotection and neuronal differentiation. Front Pharmacol 11:462398. https://doi.org/10.3389/FPHAR.2020.00537/BIBTEX

Shvedunova M, Akhtar A (2022) Modulation of cellular processes by histone and non-histone protein acetylation. Nat Rev Mol Cell Biol 23(5):329–349. https://doi.org/10.1038/S41580-021-00441-Y

Søborg A et al (2022) Trends in underlying causes of death in solid organ transplant recipients between 2010 and 2020: using the CLASS method for determining specific causes of death. PLoS One 17(7). https://doi.org/10.1371/JOURNAL.PONE.0263210

Sowers ML et al (2022) Multi-OMICs analysis reveals metabolic and epigenetic changes associated with macrophage polarization. J Biol Chem 298(10):102418. https://doi.org/10.1016/j.jbc.2022.102418

Stahl M et al (2013) Lung collagens perpetuate pulmonary fibrosis via CD204 and M2 macrophage activation. PLoS One 8(11):e81382. https://doi.org/10.1371/JOURNAL.PONE.0081382

Su Y, Han W, Kovacs-Kasa A, Verin AD, Kovacs L (2021) Hdac6 activates erk in airway and pulmonary vascular remodeling of chronic obstructive pulmonary disease. Am J Respir Cell Mol Biol 65(6):603–614. https://doi.org/10.1165/RCMB.2020-0520OC/SUPPL_FILE/DISCLOSURES.PDF

Sun H, Song J, Sun M, Shan C (2021) Histone deacetylase inhibitors (HDACi), the ongoing epigenetic agents to enhance chemotherapy sensitization. In: Epigenetic regulation in overcoming chemoresistance, pp 25–39. https://doi.org/10.1016/B978-0-12-823684-0.00014-1

Tai Y et al (2021) Myofibroblasts: function, formation, and scope of molecular therapies for skin fibrosis. Biomolecules 11(8):1095. https://doi.org/10.3390/BIOM11081095

Tan L, Guo Y, Feng C, Hou Y, Xie X, Zhao Y (2022) The dual regulatory roles of macrophages in acute allogeneic organ graft rejection. Engineering 10:21–29. https://doi.org/10.1016/J.ENG.2021.10.015

Toki D et al (2014) The role of macrophages in the development of human renal allograft fibrosis in the first year after transplantation. Am J Transplant 14(9):2126–2136. https://doi.org/10.1111/AJT.12803

Vaillant AAJ, Mohseni M (2023) Chronic transplantation rejection. StatPearls. Accessed 19 Dec 2023. [Online]. Available: https://www.ncbi.nlm.nih.gov/books/NBK535435/

Vaillant AAJ, Misra S, Fitzgerald BM (2023) Acute transplantation rejection. StatPearls. Accessed 19 Dec 2023. [Online]. Available: https://www.ncbi.nlm.nih.gov/books/NBK535410/

van den Bosch TPP et al (2017) CD16+ monocytes and skewed macrophage polarization toward M2 type hallmark heart transplant acute cellular rejection. Front Immunol 8(MAR). https://doi.org/10.3389/FIMMU.2017.00346/FULL

Vitak K (2014) Organ transplantation. In: Acute care handbook for physical therapists: fourth edition, pp 335–356. https://doi.org/10.1016/B978-1-4557-2896-1.00014-7

Vogl DT et al (2017) Ricolinostat, the first selective histone deacetylase 6 inhibitor, in combination with bortezomib and dexamethasone for relapsed or refractory multiple myeloma. Clin Cancer Res 23(13):3307–3315. https://doi.org/10.1158/1078-0432.CCR-16-2526

von Knethen A, Heinicke U, Weigert A, Zacharowski K, Brüne B (2020) Histone deacetylation inhibitors as modulators of regulatory T cells. Int J Mol Sci 21(7). https://doi.org/10.3390/IJMS21072356

Wang Z et al (2009) Genome-wide mapping of HATs and HDACs reveals distinct functions in active and inactive genes. Cell 138(5):1019–1031. https://doi.org/10.1016/J.CELL.2009.06.049

Wang P, Wang Z, Liu J (2020) Role of HDACs in normal and malignant hematopoiesis. Mol Cancer 19(1):1–21. https://doi.org/10.1186/S12943-019-1127-7

Wang Y et al (2024) M2 macrophages promote subconjunctival fibrosis through YAP/TAZ signalling. Ann Med 56(1):2313680. https://doi.org/10.1080/07853890.2024.2313680

Wawruszak A et al (2019) Histone deacetylase inhibitors and phenotypical transformation of cancer cells. Cancers (Basel) 11(2). https://doi.org/10.3390/CANCERS11020148

Webb AH et al (2017) Inhibition of MMP-2 and MMP-9 decreases cellular migration, and angiogenesis in in vitro models of retinoblastoma. BMC Cancer 17(1). https://doi.org/10.1186/S12885-017-3418-Y

Yan B et al (2018) Histone deacetylase 6 modulates macrophage infiltration during inflammation. Theranostics 8(11):2927. https://doi.org/10.7150/THNO.25317

Yang X-J, Grégoire S (2005) Class II histone deacetylases: from sequence to function, regulation, and clinical implication. Mol Cell Biol 25(8):2873. https://doi.org/10.1128/MCB.25.8.2873-2884.2005

Yao Y et al (2020) Histone deacetylase 3 controls lung alveolar macrophage development and homeostasis. Nat Commun 11(1):1–15. https://doi.org/10.1038/s41467-020-17630-6

Yi C et al (2022) Macrophage elastase (MMP12) critically contributes to the development of subretinal fibrosis. J Neuroinflammation 19(1):1–16. https://doi.org/10.1186/S12974-022-02433-X/FIGURES/8

Yu S, Lu J (2022) Macrophages in transplant rejection. Transpl Immunol 71:101536. https://doi.org/10.1016/J.TRIM.2022.101536

Yue B (2014) Biology of the extracellular matrix: an overview. J Glaucoma 23(8):S20. https://doi.org/10.1097/IJG.0000000000000108

Zhao TC, Wang Z, Zhao TY (2020) The important role of histone deacetylases in modulating vascular physiology and arteriosclerosis. Atherosclerosis 303:36. https://doi.org/10.1016/J.ATHEROSCLEROSIS.2020.04.020

Zhong Y et al (2023) The HDAC10 instructs macrophage M2 program via deacetylation of STAT3 and promotes allergic airway inflammation. Theranostics 13(11):3568. https://doi.org/10.7150/THNO.82535

Chapter 13
Monocytes/Macrophages in Helminth Infections: Key Players in Host Defence, Inflammation, and Tissue Repair

Anuradha Rajamanickam and Subash Babu

Abstract Monocytes/macrophages are pivotal in host defense, inflammation, and tissue repair. They are actively engaged during helminth infections, playing critical roles in trapping pathogens, eliminating them, repairing tissue damage, and mitigating type 2 inflammation. These cells are indispensable in preserving physiological equilibrium and overseeing pathogen resistance as well as metabolic processes. Furthermore, these immune cells are influenced by cellular metabolism, which adjusts in response to host-derived factors and environmental cues. They secrete effector molecules crucial for anti-helminthic immunity and healing tissues damaged by parasites. Helminth parasites manipulate the immune regulatory capabilities of monocytes/macrophages by secreting anti-inflammatory mediators to dodge host defenses. Infections, especially with helminths, induce metabolic adaptations involving monocytes/macrophages that can lead to enhanced insulin sensitivity. This review provides a synthesis of the activation and diversity of monocytes/macrophages, their involvement in inflammation, and the latest insights into the strategies of monocyte/macrophage-mediated host defense against helminth infections. It also sheds light on recent discoveries concerning the immune regulatory interactions between monocytes/macrophages and helminth parasites.

A. Rajamanickam
National Institute of Health-National Institute of Allergy and Infectious Diseases International Center for Excellence in Research, Chennai, India

S. Babu (✉)
National Institute of Health-National Institute of Allergy and Infectious Diseases International Center for Excellence in Research, Chennai, India

Laboratory of Parasitic Diseases, National Institute of Allergy and Infectious Diseases, National Institutes of Health, Bethesda, MD, USA
e-mail: sbabu@icerindia.org

M. Kloc et al. (eds.), *Monocytes and Macrophages in Development, Regeneration, and Disease*, Results and Problems in Cell Differentiation 74,
https://doi.org/10.1007/978-3-031-65944-7_13

13.1 Introduction

Macrophages and monocytes are vital cells in the innate immune system, playing crucial roles in maintaining homeostasis, defending against pathogens, facilitating tissue repair, and modulating inflammation (Murray and Wynn 2011). They also promote wound healing, manage inflammation, maintain tissue equilibrium, and aid in the evolution of adaptive immune responses (Gordon and Taylor 2005; Passos et al. 2017). The remarkable plasticity of these cells enables them to function across diverse microenvironments (Biswas and Mantovani 2012). Despite the challenge posed by helminths—large extracellular pathogens impervious to phagocytosis—macrophages exhibit extraordinary adaptive capabilities, devising a spectrum of defense strategies to counteract helminth infections and mitigate the associated tissue damage (Lechner et al. 2021; Maizels and Yazdanbakhsh 2003).

13.2 Heterogeneity of Monocytes and Macrophages

Monocytes play a pivotal role in maintaining tissue equilibrium and are key players in a range of infectious and inflammatory conditions (Shapouri-Moghaddam et al. 2018). The remarkable functional diversity seen in monocytes stems from the existence of distinct cellular subpopulations, a fact unveiled through various techniques such as fate mapping, unsupervised multi-dimensional analysis, and transcriptional and proteomic profiling (Ancuta et al. 2009; Ingersoll et al. 2010). In humans, monocytes are divided into 'classical' and 'non-classical' categories based on the presence of CD14 and CD16 antigens. While classical monocytes are known for their phagocytic and pro-inflammatory roles, non-classical monocytes specialize in patrolling and recognizing viruses (Narasimhan et al. 2019; Passlick et al. 1989). Additionally, an intermediate subset with distinct surface markers and characteristics has been recognized, further illustrating the heterogeneity and variability among these cells (Sampath et al. 2018; Kapellos et al. 2019).

In mice, monocytes are identified through the expression of CD115 and CD11b and further classified by the Ly6C surface marker (Ziegler-Heitbrock et al. 2010). These cells are integral to a variety of functions including whole-body metabolism regulation, arterial tissue renewal, liver expansion, replenishing intestinal cells, and differentiating into Kupffer cells after depletion (Qu et al. 2004; Scott et al. 2016).

During inflammation and infection, circulating monocytes are crucial for the expansion of the macrophage population (Watanabe et al. 2019). Macrophages themselves are marked by significant heterogeneity, influenced by their specific location and the physiological demands of their environment (Okabe and Medzhitov 2014; Lica et al. 2023). Tissue macrophages in adults originate from fetal liver and embryonic yolk sac precursors, evolving into specialized cells such as osteoclasts, Kupffer cells, perivascular macrophages, and alveolar macrophages (Mass et al. 2023). These cells demonstrate remarkable adaptability, shaped by exposure to a myriad of stimuli, signaling molecules, and metabolic substrates (Lica et al. 2023; Sakai et al. 2019).

The metabolism of macrophages in adipose tissue, for instance, varies with the availability of free fatty acids, lipoproteins, and carbohydrates (Yao et al. 2022). In lean conditions, these macrophages typically express CD206 and CD301, whereas, in obesity, they exhibit a pro-inflammatory phenotype, characterized by the expression of inducible nitric oxide synthase and tumor necrosis factor-alpha (TNF-α) (Lumeng et al. 2008; Guilliams et al. 2013). The lung presents another example of macrophage heterogeneity, with alveolar macrophages of embryonic origin primarily located in airways, contrasting with interstitial macrophages that derive from circulating monocytes (Gibbings et al. 2017).

The influence of the localized tissue microenvironment on macrophage behavior has been demonstrated in murine studies, showing alveolar macrophages' reduced responsiveness to interleukin-4 (IL-4) under steady-state conditions (Mass et al. 2023). However, this altered responsiveness can be reversed once the macrophages are isolated from lung tissues, suggesting the presence of lung-specific endogenous factors that regulate the unique characteristics of these niche-specific cells.

13.3 Mechanisms Underlying the Polarization of Macrophages

Macrophages are crucial in the innate immune system due to their phagocytic properties, which remove intruders and present antigens, respond to inflammatory and viral events, and maintain tissue integrity (Gordon and Taylor 2005; Passos et al. 2017; Shapouri-Moghaddam et al. 2018; Uribe-Querol and Rosales 2020; Rahman et al. 2017). Activation of macrophages occurs upon encountering pathogens or during phagocytosis, sparking inflammation and leading to the release of cytokines and other factors (Fu and Harrison 2021). Distinctly, macrophages are classified into M1 (inflammatory) and M2 (anti-inflammatory) subtypes, each playing significant roles in disease dynamics and paralleling the Th1 and Th2 cell response profiles, respectively (Lica et al. 2023; Schmall et al. 2015; Murray 2017; Parisi et al. 2018; Mills et al. 2015) (Fig. 13.1).

M1 macrophages polarize under the influence of Th1 cytokines such as IFN-γ and TNF-α, or upon exposure to inflammatory molecules like lipopolysaccharides (LPS), through mechanisms involving the JAK/STAT, NF-κB signaling pathways, and cytokine receptors (Lica et al. 2023). These macrophages are crucial early responders to infectious challenges and lesion formation, promoting inflammation and secreting pro-inflammatory cytokines. Characterized by high expression levels of iNOS, MHC II, CD16/32, CD80, CD86, and various chemokines, M1 macrophages possess antimicrobial and antitumor activities, primarily through NADPH oxidase activation and reactive oxygen species (ROS) production (Shapouri-Moghaddam et al. 2018; Yunna et al. 2020; Orecchioni et al. 2019; Murray et al. 2014; Almeida et al. 2023).

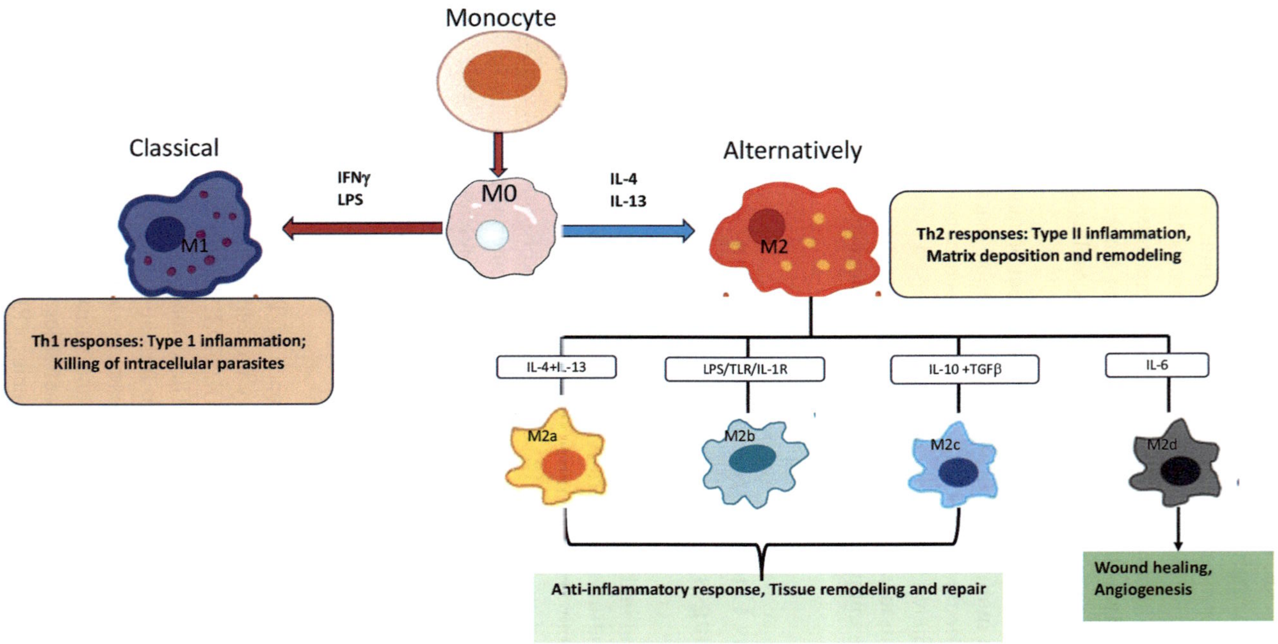

Fig. 13.1 Macrophage differentiation and polarization in helminth infections. Differential macrophage polarization is defined by macrophage microenvironment cues that either cause conventional activation (M1) or alternative activation (M2). When pathogen-derived LPS is administered alone or in conjunction with IFN-γ, M1 macrophages are classically activated, resulting in enhanced microbicidal activity and proinflammatory mediator production. Four subpopulations of alternative macrophage activation might be distinguished based on the host-parasite microenvironment. IL-4 or IL-3 promotes M2a differentiation, which is linked to the Th2 response, internalization, and the destruction of parasites. M2b are associated with TLR or IL-1R agonists, immune complexes, and the enhancement of immunoregulation. Differentiation into M2c is induced by glucocorticoids and IL-10 production; this process also triggers immunoregulation, tissue remodeling, and repair. Differentiation into M2d is induced by IL-6 and this process triggers wound healing and angiogenesis. LPS, or lipopolysaccharide; TLR, or toll-like receptor; IFN-γ, or interferon γ

Conversely, M2 macrophages emerge in response to Th2 cytokines like IL-4 and IL-13, with STAT6 activation playing a pivotal role (Chen et al. 2023). This subgroup, which IL-4 specifically pushes towards an M2 phenotype while inhibiting M1 characteristics, is crucial for anti-inflammatory actions and supports tissue repair and healing (Mamilos et al. 2023). The M2 phenotype is marked by a distinct cytokine profile, producing less IL-1, IL-6, and TNF-α, but more IL-10 and TGF-β (Chen et al. 2023; Perez and Rius-Perez 2022). Further subdivision into M2a, M2b, M2c, and M2d subtypes reveals a spectrum of functions ranging from promoting type 2 immune responses to facilitating tissue regeneration and modulating inflammation (He et al. 2021; Yao et al. 2023; Arora et al. 2018; Viola et al. 2019; Ross et al. 2021; Strizova et al. 2023).

The balance between M1 and M2 macrophages is essential for maintaining health and tissue homeostasis. An imbalance can lead to inflammatory disorders or immunosuppression, highlighting the importance of their regulated activity (Lechner et al. 2021; Sica et al. 2015; Luque-Campos et al. 2021). The adaptability of macrophages, especially in chronic helminth infections, exemplifies their role in preventing excessive tissue damage when healing from initial injuries proves challenging.

The simplistic dichotomy of macrophages into M1 and M2 categories, based on Th1/Th2 profiles, falls short of capturing their diverse and dynamic nature (Wei et al. 2021). In light of this, Mosser and Edwards have developed a more complex classification scheme that acknowledges the diverse functions of macrophages: regulatory macrophages that are essential for immunoregulation, classically activated macrophages oriented toward host defense, and alternatively activated macrophages focused on mending tissue damage (Passos et al. 2017).

13.4 Regulatory Macrophages

Regulatory macrophages, unlike their alternatively activated counterparts, do not actively participate in wound healing and lack markers associated with wound healing macrophages (Martinez and Gordon 2014). Additionally, regulatory macrophages exhibit distinct signaling pathways and chemokine receptor repertoires, setting them apart from alternatively activated macrophages (Mantovani et al. 2007). This distinction, observed in both murine and human macrophages, underscores the specialized function of regulatory macrophages in suppressing immunopathology and tempering inflammatory responses to maintain tissue homeostasis (Murray et al. 2014).

Macrophages produce various molecules for maintaining homeostasis. These molecules promote inflammatory responses and regulate inflammation to limit tissue destruction. Endogenous regulators like adenosine, PGE2, resolvins, and lipoxins contribute to inflammation resolution (Mosser et al. 2021). These endogenous mediators not only mitigate inflammation but also reprogram macrophages to support tissue repair, evidenced by the clearance of debris and apoptotic cells from infection sites (Passos et al. 2017). The interaction between macrophages and exogenous sources of these molecules from various tissue constituents underscores their central role in promoting tissue homeostasis and resolving inflammation (Martinez and Gordon 2014).

Macrophages can change their phenotype in response to environmental cues, either activating to kill microbes or assuming regulatory properties to modulate immune responses (Fleming and Mosser 2011). Regulatory macrophages are characterized by the expression of distinctive biochemical markers, indicating their potential to target specific tissues and serve as promising candidates for anti-inflammatory therapeutics (Mosser et al. 2021). A comprehensive understanding of macrophage phenotypes holds significant therapeutic potential, offering insights that could lead to the development of novel treatment strategies for various inflammatory conditions.

13.5 Fibrogenic Macrophages/Tissue Injury and Repair Macrophages

Macrophages play pivotal roles in tissue repair and fibrosis, displaying diverse functions across different stages of the process. Initially recognized as scavengers, recent studies highlight their complex involvement in repair and regeneration (Mosser and Edwards 2008). They orchestrate the initial inflammatory response and transition to promoting tissue repair through growth factor secretion and fibroblast activation (Wynn and Barron 2010; Wynn and Vannella 2016). Later, they adopt an anti-inflammatory phenotype to dampen inflammation and facilitate resolution (Khalil et al. 1989).

Tissue repair and regeneration are vital processes initiated in response to injury or infection, involving an inflammatory cascade initiated by damage-associated molecular patterns (DAMPs) and pathogen-associated molecular patterns (PAMPs), leading to the recruitment and activation of various cell types (Zhang et al. 2010). This cellular response involves hematopoietic and non-hematopoietic cells such as neutrophils, macrophages, innate lymphoid cells (ILCs), NK cells, B cells, T cells, fibroblasts, epithelial cells, endothelial cells, stem cells, orchestrating tissue repair (Wynn 2008; Fabre et al. 2023).

Effective wound healing involves well-organized and controlled inflammatory responses, leading to the restoration of normal tissue architecture. However, chronic or dysregulated wound healing can result in pathological fibrosis or scarring, impairing tissue function and leading to organ failure and death (Wynn and Ramalingam 2012; Gieseck et al. 2018). Thus, tightly regulated wound-healing responses are crucial.

Disturbances in macrophage function can lead to aberrant repair, emphasizing their significance as therapeutic targets (Duffield et al. 2005). Studies in various injury models, such as liver and cardiac injuries, delineate distinct roles for macrophage subsets in injury and recovery (Epelman et al. 2014; Vannella et al. 2014).

Additionally, macrophages contribute to fibrosis development through pro-inflammatory activities and direct interactions with myofibroblasts (Pradere et al. 2013). They produce matrix metalloproteinases and regulate collagen assembly, influencing tissue structure remodeling (Madala et al. 2010; Knipper et al. 2015).

Anti-fibrotic macrophages, induced by IL-10 and other mediators, play crucial roles in resolving inflammation and suppressing fibrosis (Shouval et al. 2014; De Nardo et al. 2014). They communicate immunosuppressive signals to epithelial cells, regulate cytokine production, and inhibit T-cell responses, exerting anti-fibrotic effects (Fabre et al. 2023; Westphalen et al. 2014; Bourdonnay et al. 2015).

In summary, macrophages exhibit dynamic functions in tissue repair and fibrosis, regulated by a complex interplay of cytokines and signaling pathways. Understanding their diverse roles and mechanisms of action is crucial for developing targeted therapies for chronic inflammatory and fibrotic conditions.

13.6 MDSC: Origin, Phenotype, and Function

Myeloid-derived suppressor cells (MDSCs) arise from myeloid progenitors, sharing lineage with mononuclear phagocytes, dendritic cells, and polymorphonuclear cells like neutrophils, eosinophils, basophils, and mast cells (Gabrilovich 2017). MDSCs exhibit heterogeneity and typically expresses CD11b and Gr1 markers. They comprise two main subtypes: granulocytic or polymorphonuclear MDSCs (PMN-MDSC) and monocytic MDSCs (M-MDSC) (Veglia et al. 2021). Identifying MDSCs has been challenging due to overlapping markers with conventional myeloid cells.

In mice, PMN-MDSC is identified as CD11b^{+}Ly6G^{+}Ly6Clo, while M-MDSC is CD11b^{+}Ly6G^{-}Ly6Chi. In humans, distinguishing between neutrophils and PMN-MDSC is intricate due to shared surface markers. However, human PMN-MDSCs can be isolated using low-density gradients, with lectin-type oxidized LDL receptor 1 (LOX1) emerging as a specific marker (Condamine et al. 2016). Conversely, M-MDSCs exhibit low or absent HLA-DR expression compared to monocytes. Additionally, CD84 has been identified as a marker expressed by both PMN-MDSC and M-MDSC in cancer patients (Alshetaiwi et al. 2020; Lewinsky et al. 2021).

Advancements in single-cell transcriptomics have shed light on MDSC heterogeneity and gene expression profiles. Studies have revealed the upregulation of Il1b and Arginase-2 (Arg-2) in MDSCs, emphasizing their role in immunosuppression (Alshetaiwi et al. 2020). Metabolically, MDSC exhibits increased lipid accumulation, fatty acid oxidation, and glycolysis, with hypoxia inducing a shift from oxidative phosphorylation to glycolysis via hypoxia-inducible factor 1 subunit alpha (HIF1A) activation (Tannahill et al. 2013; LaGory and Giaccia 2016; Stevenson et al. 2022).

Functionally, MDSC suppresses various immune cells, including T cells, B cells, and NK cells (Veglia et al. 2021). PMN-MDSC and M-MDSC employ common mechanisms such as STAT3 upregulation, ER stress induction, and expression of arginase-1 (Arg-1) and S100A8/A9. PMN-MDSC utilize ROS, peroxynitrite, and (Arg-1), while M-MDSC produces NO, IL-10, and TGF-β, and upregulates immunoregulatory molecules like PD-L1 (Stevenson et al. 2022).

Understanding MDSC diversity and mechanisms is crucial for developing targeted therapies to modulate their immunosuppressive effects and enhance antitumor or antimicrobial responses.

13.7 Helminth Parasites: An Introduction

Over 1.5 billion people worldwide grapple with helminth (worm) infections, posing a significant public health challenge due to their capacity to cause substantial morbidity and socioeconomic burden. Although seldom fatal, these infections elicit a range of pathological responses leading to considerable morbidity among affected individuals (Al Amin and Wadhwa 2024; Martviset et al. 2023). The immune reaction to helminths typically involves Type 2 responses, including Th2, Type 2 innate lymphoid cell, eosinophil, and antibody responses, along with regulatory responses featuring regulatory cytokines and cellular subsets (Babu and Nutman 2019).

Helminths boast complex lifecycles comprising multiple developmental stages. Human transmission occurs through ingestion of eggs or larvae, penetration of intact skin by larvae, or inoculation by biting insects. Consequently, hosts encounter various parasite lifecycle stages, each possessing a shared and unique antigenic repertoire. For instance, *Schistosoma mansoni* infection initiates with penetration of the skin by free-swimming cercariae, developing into tissue-dwelling schistosomula These eventually mature into sexually dimorphic adult worms in the liver and mesenteric veins, culminating in egg production and release into the intestine or bladder lumen (https://www.who.int/news-room/fact-sheets/detail/schistosomiasis; Sutrave and Richter 2023). Likewise, lymphatic filarial infection exposes hosts to infective-stage larvae in the skin, lymph nodes, and lymphatics, adult worms in lymph nodes and lymphatics, and microfilariae in the peripheral circulation (https://www.who.int/news-room/fact-sheets/detail/lymphatic-filariasis, 2023). Consequently, host–helminth interaction complexity arises from both the parasite's multiple lifecycle stages and their tissue tropism.

Antigenic disparities among lifecycle stages prompt distinct immune responses that evolve differentially throughout the infection course. Responses may be compartmentalized, such as in intestinal nematode infections (intestinal mucosa and draining lymph nodes) or onchocerciasis (skin/subcutaneous tissue and draining lymph nodes), or systemic, as observed in lymphatic filariasis or schistosomiasis (Mayer et al. 2020; Allen and Wynn 2011; Pearce et al. 2004). Furthermore, parasite migration patterns may induce varied cutaneous, pulmonary, and intestinal inflammatory pathologies, exemplified by *Ascaris* or *Strongyloides* infections during their migratory phase (Mati et al. 2014; Loukas et al. 2021; Yeh et al. 2023). Additionally, hosts often face simultaneous exposure to multiple parasite lifecycle stages, complicating immune responses as chronic infection patients with lymphatic filariasis may contend with insect bites transmitting infective-stage parasites. Consequently, ensuing immune responses reflect reactions to the invading organism alongside imprints of prior and concurrent exposures (Babu and Nutman 2014).

The canonical immune response mounted by hosts against helminth parasites predominantly follows a Th2 pattern, characterized by the secretion of cytokines such as interleukin (IL)-4, IL-5, IL-9, IL-10, and IL-13, along with the production of immunoglobulin G1 (IgG1), IgG4, and IgE antibodies (Babu and Nutman 2014; Allen and Sutherland 2014; Babu and Nutman 2016). This response also involves the proliferation of eosinophils, basophils, mast cells, type 2 innate lymphoid cells, and alternatively activated macrophages. However, recent research highlights the substantial regulatory component inherent in this response, encompassing both regulatory cytokines and cells (Kim et al. 2024).

The Th2 response initiated by helminth parasites follows a well-defined pattern, yet its initiation, progression, and culmination necessitate interactions with various cell types, notably: (i) Epithelial/stromal cells, (ii) Innate lymphoid cells (ILCs), (iii) Dendritic cells (DCs) and macrophages, (iv) T cells, (v) B cells, (vi) Eosinophils, (vii) Mast cells/basophils, and (viii) Neutrophils (Nutman 2015; Gause et al. 2020). The Th2 response serves dual roles in resistance and tolerance phenotypes against helminth infection. Resistance mechanisms aim to combat helminth infections, while tolerance mechanisms, such as wound healing processes, mitigate the impact of helminth infections on host immunity without causing pathological harm (Lekki-Jozwiak and Baska 2023; McSorley and Maizels 2012). Moreover, host–helminth interactions can prompt a spectrum of modulated immune responses, predominantly mediated by the induction of regulatory T cells (Tregs) and alternatively activated macrophages (AAMs). This intricate interplay underscores the complexity of host responses to helminth infections and underscores the need for comprehensive understanding to devise effective therapeutic interventions (White et al. 2020; Bhengu et al. 2022).

13.8 Helminth Infection and Macrophage Activation

The defense against helminth parasites primarily involves alternatively activated macrophages (AAMs). These macrophages help expel adult parasites and target helminth larvae through the production of growth factors, chemokines, and components for collagen synthesis. This not only facilitates tissue repair but also reduces damage from helminth-related injuries (Esser-von Bieren et al. 2013; Chen et al. 2014; Minutti et al. 2017) (Fig. 13.2).

Unlike classical macrophage activation triggered by interferon-γ and TNFα, which promotes microbicidal and tumoricidal activities through inducible nitric oxide synthase (iNOS) and arginine metabolism, the alternative activation pathway utilizes interleukins 4 and 13 (IL-4, IL-13). This pathway shifts macrophages towards enhancing tissue repair via increased arginase activity (Italiani and Boraschi 2014). Additionally, a subset of regulatory macrophages, which arise in response to

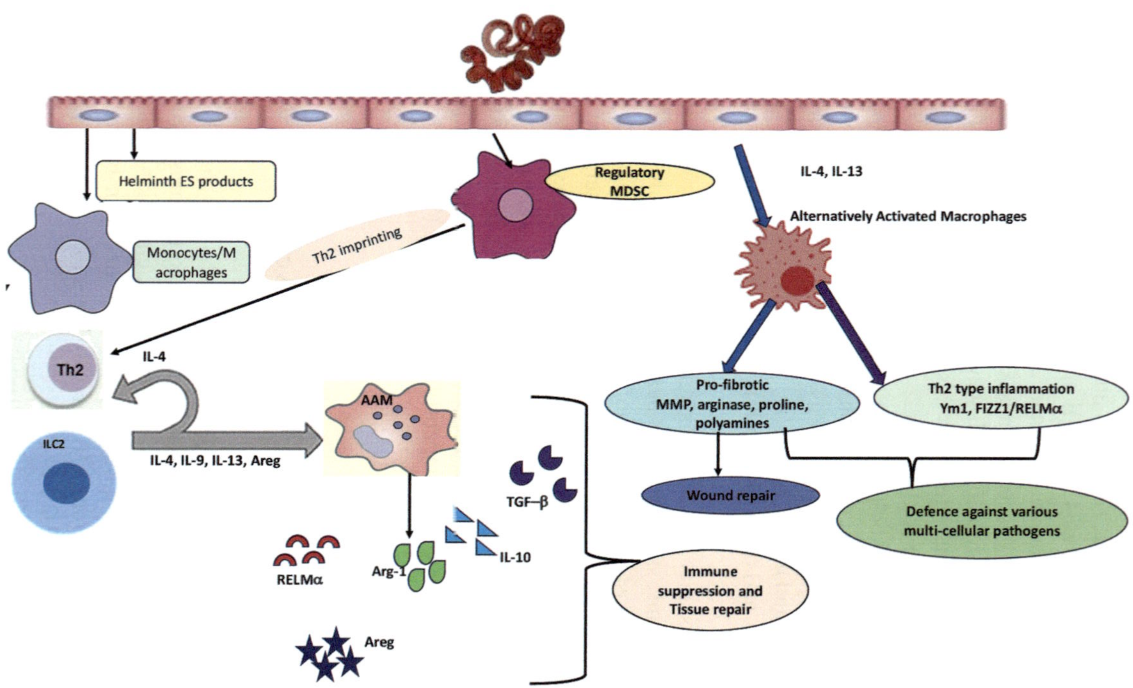

Fig. 13.2 Monocyte/macrophage response pathways following helminth infection. The host immune system polarizes into a Th2 response with parasitic worm infection, which is shown by Th2 cytokines. Products derived from helminth ES can induce macrophage development towards the M2 phenotype, which will elicit a Th2 immune response. Moreover, ES products can stimulate the synthesis of immunoregulatory molecules like TGFβ and IL-10 by dendritic cells while inhibiting their synthesis of pro-inflammatory cytokines. As immunoregulators, myeloid-derived suppressor cells (MDSC) produce reactive oxygen and nitrogen molecules that impede T cell function. Helminths cause alternatively activated macrophages to be activated, which in turn upregulates arginase and may be involved in several host defense mechanisms, such as parasite resistance and wound healing, which is necessary to repair the harm inflicted by these enormous, invasive, multicellular parasites. Areg, amphiregulin; Arg-1, arginase 1; RELM, resistin-like molecule

IL-10 and other factors, secrete IL-10, a critical anti-inflammatory cytokine, contributing to the anti-inflammatory processes (Orecchioni et al. 2019; Chen et al. 2023).

Helminth interactions prompt macrophages to predominantly express arginase, driven by IL-4 and IL-13-mediated upregulation of arginase-1, shifting away from nitric oxide production (Lang and Schick 2017). These alternatively activated macrophages (AAMs) also produce retinoic acid, facilitating the expansion of thymus-derived and inducible regulatory T cells at infection sites. Characteristically, AAMs increase levels of arginase-1, chitinase 3-like proteins (Ym1 and Ym2), and Resistin-like molecule alpha (RELM-α), playing crucial roles in tissue repair and recovery from helminth-induced tissue damage (Murray 2017; Faz-Lopez et al. 2016). Notably, AAMs are differentiated into two types: those originating from circulating blood with immune regulatory functions and tissue-resident macrophages that contribute to fibrosis in chronic helminth infections (Gundra et al. 2014). Various tissue-specific signals, such as surfactant protein A in the lungs and complement component C1q in the peritoneal cavity and liver, amplify AAM proliferation. These cells express regulatory molecules including IL-10, TGF-β, and programmed cell death 1 ligand 2 (PDL-2), predominantly acting in a regulatory capacity in helminth infections (Pastva et al. 2007; Bohlson et al. 2014). Through mechanisms involving Arg-1, PD-L2, triggering receptors expressed on myeloid cells 2 (TREM2), and RELM-α, these macrophages suppress conventional inflammatory responses and T-cell activity (Kieler et al. 2021). Additionally, RELM-α-expressing macrophages can lessen parasite burdens and prevent lung damage during infections (Wynn and Vannella 2016; Van den Bossche et al. 2015). In response to helminth infection, macrophage-derived human resistin is also triggered, heightening inflammatory responses and vulnerability (Coakley and Harris 2020). Helminth parasites skillfully modify immune reactions, eliciting type 2 immunity and activating AAMs through various immunoregulatory molecules. This modulation not only sparks type 2 immune responses but also potentially moderates harmful type 2 reactions, such as those seen in allergic asthma, suggesting therapeutic possibilities through the activation of regulatory macrophages that release anti-inflammatory mediators like IL-10 or PGE2 (Lechner et al. 2021; de Los Reyes Jimenez et al. 2020).

Despite macrophages' pivotal role in controlling effector functions during helminth infection to avoid pathological type 2 responses, the intricate details of how macrophages orchestrate host defense, tissue repair, and immune regulation remain only partially unraveled. The enduring impact of helminth infections on macrophage activation and their effector mechanisms calls for more in-depth investigation, particularly within human contexts where our understanding of macrophage functions is still evolving.

Soil-transmitted helminths (STH) are renowned for their ability to modulate the immune system, with infections by parasites such as *Ascaris lumbricoides*, *Ancylostoma duodenale*, *Necator americanus*, *Strongyloides stercoralis*, and *Trichuris sp.* eliciting robust Th2 responses (Lebu et al. 2023; Anuradha et al. 2016). Studies in both human populations and mouse models have shown that STH infections correlate with heightened insulin sensitivity and improved metabolic indices (de Gier et al. 2018; Wiria et al. 2015; Rajamanickam et al. 2019, 2020a). These effects on metabolic health are partly attributed to the regulation of metabolic homeostasis by IL-4-competent eosinophils in adipose tissue, as well as the dampening of Th1 inflammation, which is partially mediated by the modulation of fatty acid metabolism by STH (Cortes-Selva and Fairfax 2021).

Many STH species produce antigens that induce AAMs, which is crucial for expelling intestinal parasites and promoting wound repair (Cortes-Selva and Fairfax 2021; Su et al. 2018). For instance, *Trichuris suis* excretory/secretory products prompt the expression of *Arg-1* and *nitric oxide synthase 2 (NOS2)* in bone marrow-derived macrophages (BMDM), underscoring the phenotypic diversity of macrophages in vivo and suggesting an immunosuppressive state rather than a typical M2 phenotype (Leroux et al. 2018).

Changes in the activation state of macrophages in response to STH infections also correlate with the modulation of metabolic diseases. *Heligmosomoides polygyrus bakeri* (*Hbp*) infection, for instance, polarizes macrophages to an M2 phenotype, which, upon transfer to uninfected recipients on a high-fat diet, induces tissue browning and elevates levels of uncoupling protein 1 in adipose tissue. This phenomenon is accompanied by improved glucose, fat, and cholesterol levels, as well as reduced levels of leptin and TNF-α (Cortes-Selva and Fairfax 2021; Su et al. 2018).

The maintenance of the M2 phenotype necessitates the presence of eosinophils and innate lymphoid type 2 cells (ILC2), which produce IL-4 and IL-13 and are localized in perigonadal and other visceral adipose tissue areas (Cortes-Selva and Fairfax 2021; Houlden et al. 2015). The expansion of IL-4- and IL-13-expressing ILC2 following helminth infection underscores these parasites' ability to orchestrate a systemic response that shapes various tissue compartments and critical cellular components, leading to systemic metabolic regulation (Cortes-Selva and Fairfax 2021; Duarte et al. 2016).

13.9 Helminths and Monocytes

Monocytes play a vital role in the immune system, adapting to diverse roles during helminth infections, as highlighted by their phenotypic plasticity and functional versatility (Wong et al. 2012). These cells, traditionally classified by CD14 expression, are now also distinguished by CD16, enhancing their categorization and understanding of their immunological functions (Turner et al. 2014; Ziegler-Heitbrock and Hofer 2013). In human filarial infection, monocytes exhibit an alternatively activated immunoregulatory phenotype. Monocytes from asymptomatic filarial infection patients show decreased expression of NOS2 and increased expression of ARG1, along with elevated expression of resistin, mannose receptor C type 1 (MRC1), macrophage galactose type C lectin (MGL), and chemokine ligand 18 (CCL18) genes. Additionally, upon exposure to parasite antigen, purified monocytes from infected individuals demonstrate reduced expression of IL-12 and IL-18 but increased expression of transforming growth factor beta, IL-10, and suppressor of cytokine signaling 1 mRNA. Inhibition of Arg-1 leads to decreased expression of resistin, MRC1, MGL, and CCL18 genes, while enhancing the expression of NOS2, IL-12, and IL-18 genes (Babu et al. 2009; Roszer 2015; Tolouei Semnani et al. 2014). During filarial infections, there is a pronounced shift in monocyte subset distribution, with an increase in nonclassical monocytes, and changes in gene expression such as decreased *NOS2* and increased *ARG1*, altering cytokine profiles toward an anti-inflammatory state with higher TGF-β and IL-10 levels (Tolouei Semnani et al. 2014; O'Regan et al. 2014).

Moreover, immune dysregulation involves monocytes and T-cell populations. Monocytes in infected individuals show lower baseline expression of TLRs compared to uninfected individuals. Upon stimulation with filarial antigen, monocytes from infected individuals exhibit reduced TLR up-regulation. Stimulation with TLR ligands leads to decreased monocyte activation and cytokine production, suggesting immune tolerance. This dysregulation correlates with diminished $CD4^+$ T cell production of IFN-γ and IL-5, possibly due to chronic antigen stimulation (Babu et al. 2005; Babu and Nutman 2012; Nutman 2013; Rajasekaram et al. 2017). When interacting with *Brugia malayi* microfilariae, monocytes show increased PD-L1 and IL-10 levels, which suppress CD4+ T-cell activity, a reaction seen even in asymptomatic individuals from non-endemic areas (O'Regan et al. 2014). The engagement with Wolbachia heat shock protein 60 (Wmhsp60) highlights further complexities, as it provokes a pro-inflammatory response in THP-1 cells, impairs phagocytosis, and promotes lymphocyte apoptosis, which may lead to monocyte dysfunction in filarial diseases (Kamalakannan et al. 2012). These dynamics illustrate the complex interactions between monocytes and disease states, underlining their essential role in driving immune responses and influencing pathological outcomes in infectious and metabolic disorders.

The complex relationship between helminth infections and type 2 diabetes (T2D) may be significantly mediated by monocytes and their immunological activities. Studies suggest that individuals with T2D exhibit a lower prevalence of helminth infections, potentially indicating a protective effect of diabetes against these parasites (Chen et al. 2013; Hays et al. 2015). This protection may arise from helminth-driven modulation of monocyte responses, which could also help prevent the development or exacerbation of T2D (Silas et al. 2021).

Infections with helminths like *S. stercoralis* and hookworms influence metabolic parameters linked to T2D, and these effects are closely related to monocyte and macrophage activation. Infected individuals display alterations in glycemic control, hormone levels, and cytokine profiles, which are reversible with anthelmintic treatment (Rajamanickam et al. 2019; Silas et al. 2021; Yingklang et al. 2022; Lenz et al. 2023). Monocytes may play a crucial role in these processes by affecting insulin and glucagon secretion, thereby influencing glucose metabolism. Additionally, helminth infections can alter adipose tissue inflammation, which is integral to T2D, through monocyte-mediated mechanisms involving adiponectin and T-cell responses (Rajamanickam et al. 2019, 2020a; Dasan et al. 2023). Changes in various hormones and inflammatory markers like ghrelin, GLP-1, GIP, adiponectin, resistin, visfatin, leptin, and PAI-1 observed in helminth infections further underscore the central role of monocytes in modulating insulin sensitivity and glucose metabolism (Rajamanickam et al. 2020a, b).

Helminth infections also influence vascular health indicators in individuals with T2D, potentially affecting the vascular complications associated with the disease. For instance, changes in angiogenic factors and soluble receptors have been noted, which may be reversible with anthelmintic treatment (Rajamanickam et al. 2021). Additionally, individuals with T2D show increased levels of NOD1 and NOD2 mRNA, linked to poor glycemic control and enhanced monocyte activation markers. This heightened immune response is marked by elevated pro-inflammatory cytokines, such as TNF-α and IL-6, suggesting a connection between monocyte activation and the metabolic disruptions seen in diabetes (Shiny et al. 2013).

Helminth infections appear to stabilize metabolic health by promoting Th2 cell and eosinophil activity in white adipose tissue (WAT) and liver, as well as encouraging M2 macrophage polarization within WAT. This immune cell configuration, especially the balance between M1 and M2 macrophages and the presence of eosinophils in WAT, is critical for maintaining adipocyte insulin sensitivity and overall metabolic balance. Thus, the interplay between helminth infections, T2D, and monocyte and macrophage activation encompasses a complex array of metabolic, hormonal, and immunological factors. This relationship underscores the need for further research to fully understand these mechanisms and their potential therapeutic implications.

13.10 Helminth Infection and Intestinal Macrophages

The gastrointestinal tract serves as a habitat for soil-transmitted helminths, especially intestinal nematodes, which inflict considerable tissue damage, necessitating a type 2 cytokine-driven healing process (Allen and Wynn 2011). This regenerative response plays a vital role in safeguarding the gut from damage during parasite colonization and expulsion, as well as diminishing the parasitic load in afflicted hosts (Summers et al. 2005). There's evidence to suggest that helminth infections could aid in quelling inflammation in conditions like inflammatory bowel disease (IBD) by fostering an environment conducive to inflammation resolution (Summers et al. 2005). Helminth-triggered type 2 immunity can enhance goblet cell functionality and mucus layer regeneration, alongside stimulating the gut's alternatively-activated M2 macrophages to facilitate tissue repair and mitigate inflammation (Hine and Loke 2019).

Studies in mice have highlighted macrophages as key players in dampening intestinal inflammation. The detection of helminths by the nervous system can trigger ILC2s to release cytokines essential for the AAMs (Bouchery et al. 2015). These

M2 macrophages, in turn, can alter the gut microbiome towards a more anti-inflammatory state while suppressing pro-inflammatory bacteria (Ramanan et al. 2016). Consequently, directing type 2 cytokines to the gut emerges as a promising strategy to encourage inflammation resolution through the action of alternatively activated M2 macrophages and the restoration of the mucosal barrier (Hine and Loke 2019).

13.11 Helminth Infections and MDSC

Schistosomiasis, caused by *Schistosomes*, is characterized by liver and spleen enlargement due to granuloma formation. Studies in mice infected with *Schistosoma species* have shown an increase in myelopoiesis during acute infection, particularly in the liver and spleen (Dutra et al. 1998). Phenotypic analysis of spleen cells from chronically infected mice revealed a significant expansion of CD11b+Gr1+MHC-II-CD16+F4/80dull cells, resembling M-MDSC, capable of suppressing T cell proliferation (Marshall et al. 2001).

Infections with *Taenia crassiceps* larvae induce the accumulation of $CD11b^{+}Gr1^{+}$ cells in mice, which suppress T cell proliferation via NO-dependent mechanisms initially and later through ROS production, akin to PMN-MDSC (Gomez-Garcia et al. 2005; Brys et al. 2005). Similarly, infection with *Mesocitoides vogae* leads to an increase in $CD11b^{+}Gr1^{+}F4/80^{hi}MHC\text{-}II^{hi}$ cells early on, resembling M-MDSC, while later in infection, $CD11b^{+}Gr1^{+}F4/80^{low}MHC\ II^{low}$ cells, representing both M-MDSC and PMN-MDSC, become predominant. These cells suppress T cell proliferation and express immunosuppressive factors such as IL-10 and Arg-1 (Stevenson et al. 2022; Macak Kubaskova et al. 2021).

Infections with *B. malayi* induce the appearance of suppressive myeloid cells in mice, inhibiting antigen-specific $CD4^{+}$ T cell proliferation, possibly explaining the hyporesponsiveness observed in infected individuals (Piessens et al. 1980). Similarly, infections with *Nippostrongylus brasiliensis* and *Hbp* induce the expansion of MDSC (Saleem et al. 2012). In *Hbp* infections, MDSC, predominantly expressing Ly6G and Ly6C, suppresses Th2 responses crucial for worm expulsion, as evidenced by adoptive transfer experiments (Harris and Loke 2017). Gemcitabine depletion of MDSC in *N. brasiliensis*-infected mice resulted in higher worm burdens, suggesting a beneficial role of MDSC in some helminth infections (Saleem et al. 2012; Morales et al. 2014).

Overall, helminth infections induce the expansion of MDSC, contributing to immunosuppression and potentially influencing disease outcomes. Targeting MDSC may offer therapeutic opportunities in helminth infections, but further research is needed to elucidate their precise roles and mechanisms of action.

13.12 Unveiling the Epigenetic Remodeling of Macrophages in Helminth Immunity

Intriguing findings have emerged regarding the enduring effects of helminth-induced macrophage reprogramming on long-term immune regulation. *Fasciola hepatica* excretory-secretory products administered subcutaneously to mice induce sustained (>1.5 years) anti-inflammatory reprogramming of hematopoietic precursors. These precursors differentiate into hyporesponsive, anti-inflammatory monocytes through mTOR-mediated metabolic and epigenetic imprinting (Quinteros et al. 2023). This imprinting enables protection against autoimmune encephalitis, as observed in an experimental model of multiple sclerosis. Moreover, *Schistosoma mansoni* infection induces metabolically reprogrammed macrophages that are hyporesponsive to LPS stimulation and confer protection against high-fat diet-induced metabolic disease in Apoe−/− mice (Cortes-Selva and Fairfax 2021). Importantly, this protective effect is communicable via adoptive bone marrow transfer and persists for up to 10 weeks. Notably, the protective effect is not reproduced by IL-4 complex injection, suggesting additional factors beyond stereotypical type 2 activation contribute to helminth-induced anti-inflammatory myeloid reprogramming (Passos et al. 2017). Human studies have also revealed monocytosis and increased CD206 and IL-10 expression in monocytes of individuals previously infected with *N. americanus*, suggesting similar myeloid reprogramming in humans (Passos et al. 2017). In individuals with latent tuberculosis infection (LTBI), coinfection with *S. stercoralis* influences monocyte responses. Compared to *S. stercoralis*-negative LTBI individuals, those with *S. stercoralis* infection exhibit lower frequencies of monocytes expressing CD54, CD80, and CD86 both at baseline and in response to mycobacterial antigen stimulation. Conversely, *S. stercoralis*-positive individuals show higher frequencies of M2-like monocytes, characterized by CD206 and CD163 expression. These monocytes also display reduced phagocytic and respiratory burst activity upon mycobacterial antigen or LPS stimulation and decreased expression of pro-inflammatory cytokines (IL-1β, TNF-α, IL-6, and IL-12). Treatment of *S. stercoralis* infection leads to a significant reversal of these alterations in monocyte function after six months. Additionally, T cells from *S. stercoralis*-positive individuals exhibit lower activation levels both at baseline and following mycobacterial antigen stimulation. Overall, these findings highlight dampened monocyte activation, enhanced M2 polarization, and impaired monocyte function in helminth-LTBI coinfection, indicating a distinct mechanism by which helminth infection modulates immune function in LTBI (Rajamanickam et al. 2020a).

These findings suggest that helminth infections profoundly impact hematopoiesis, leading to enduring anti-inflammatory reprogramming that may extend to macrophage progenitors (Jenkins et al. 2011; Jackson-Jones et al. 2016). This reprogramming is likely to vary depending on the type, site, and duration of helminth infection, as well as the origin of macrophages (bone marrow-derived, resident, or recruited populations). The specific epigenetic mechanisms and associated chromatin landscapes underlying helminth-induced regulatory macrophage reprogramming remain to be fully elucidated. For instance, histone demethylase JMJD3 (KDM6B) has been implicated in mediating alternative activation of macrophages during *N. brasiliensis* infection by facilitating the expression of AAM-related genes through demethylation of H2K27 (Satoh et al. 2010). Additionally, HDAC3 has been implicated in negatively regulating AAM polarization, highlighting the central role of histone modifications in macrophage polarization and function during helminth infection (Lechner et al. 2021; Hoeksema et al. 2021).

Understanding the mechanisms and consequences of long-term helminth-induced epigenetic reprogramming of macrophages is crucial, particularly in the context of inflammatory diseases and susceptibility to subsequent infections with diverse pathogens.

13.13 Macrophage/Monocyte Regulation by Helminth Parasite Molecules

Parasites employ intricate strategies to manipulate host immune responses, impacting diseases such as chronic helminth infections and various inflammatory conditions like asthma and diabetes (Lechner et al. 2021). While chronic helminth infections often present as clinically silent diseases, they can significantly affect the host's immune system, leading to enhanced susceptibility to other infections and alterations in vaccine responses (Nouatin et al. 2021; Gent et al. 2019). However, the effects of helminth infections are diverse, with some parasites demonstrating the ability to enhance anti-viral immunity or exacerbate chronic inflammation (Briggs et al. 2016). Therefore, understanding the individual immunomodulatory molecules employed by helminths is crucial for their potential therapeutic application across different infection and inflammation scenarios.

One mechanism by which helminths exert their immunomodulatory effects is through the release of extracellular vesicles containing immune regulatory microRNAs (miRNAs) (Coakley et al. 2017). These miRNAs target macrophages, altering their gene expression profiles and dampening type 2 immune responses. Additionally, helminthic cystatins, such as those derived from *A. viteae* and *S. japonicum*, interfere with antigen processing in macrophages, thereby inhibiting pro-inflammatory responses and promoting the production of anti-inflammatory cytokines like IL-10 (Wang et al. 2015). Furthermore, helminths often mimic host immune regulators like TGF-β and macrophage inhibitory factor (MIF), influencing macrophage responses and promoting a type 2 suppressive phenotype (Johnston et al. 2017). For example, some helminth-derived molecules induce the production of regulatory T cells (Tregs) and modulate macrophage functions to create an immunosuppressive environment conducive to parasite survival (Lechner et al. 2021).

Moreover, certain helminth-derived enzymes, such as glutamate dehydrogenase from parasites like *Hbp*, suppress type 2 immune responses by inducing the production of immunosuppressive factors like IL-10 and PGE2 in macrophages (de Los Reyes Jimenez et al. 2020). Additionally, helminths have evolved mechanisms to neutralize danger signals such as damage-associated molecular patterns (DAMPs) and pathogen-associated molecular patterns (PAMPs), thereby modulating host immune responses and promoting parasite survival (Lechner et al. 2021). In summary, macrophages emerge as pivotal players in the host-parasite interaction, serving as both targets and mediators of helminth-induced immune regulation. Understanding the intricate mechanisms underlying helminth-driven macrophage regulation is essential for harnessing their therapeutic potential across various disease contexts, offering novel avenues for intervention and treatment strategies.

13.14 Conclusion and Future Directions

Studies on macrophage effector and regulatory functions in helminth infection have primarily focused on mice. However, human anti-helminth immunity often fails or develops slower due to the efficient immune evasion of parasites. It is crucial to translate mouse models into human settings to identify conserved mechanisms of macrophage-mediated chronicity, host defense, and repair. Macrophages initiate innate and adaptive immunity, activating granulocytes and T cells. They also regulate inflammation and aggressive tissue remodelling during helminth infection.

Future studies should aim to bridge the gap between mouse models and human settings in understanding macrophage effector and regulatory functions in helminth infection. It is crucial to identify conserved mechanisms of macrophage-mediated chronicity, host defense, and repair, considering the efficient immune evasion strategies employed by parasites in humans. Research endeavors should explore the functional diversity of macrophages in helminth infection, utilizing single-cell technologies and macrophage-specific deletions to elucidate their roles in host defense and immune regulation. Additionally, future investigations should focus on unraveling macrophage reprogramming during and after helminth infection through metabolomic and epigenetic analyses. This will provide insights into the mechanisms and duration of macrophage-mediated trained immunity in helminth infection, aiding in predicting and preventing pathological responses to infectious or inflammatory insults. Furthermore, there is a need to determine whether "trained type 2 immunity" is a common feature of protective and pathological type 2 immune responses, which could inform the design of new anti-helminthic treatments and therapeutics targeting detrimental functions of alternatively activated macrophages in other diseases.

Acknowledgement We would like to express our gratitude to Dr. Thomas B. Nutman of LPD, NIAID, NIH for his unwavering support. We also acknowledge the valuable contributions made to this effort by the members of the India ICER program.

References

Al Amin ASM, Wadhwa R (2024) Helminthiasis. In: StatPearls [Internet]. StatPearls, Treasure Island (FL)

Allen JE, Sutherland TE (2014) Host protective roles of type 2 immunity: parasite killing and tissue repair, flip sides of the same coin. Semin Immunol 26(4):329–340

Allen JE, Wynn TA (2011) Evolution of Th2 immunity: a rapid repair response to tissue destructive pathogens. PLoS Pathog 7(5):e1002003

Almeida FS, Vanderley SER, Comberlang FC, Andrade AG, Cavalcante-Silva LHA, Silva EDS et al (2023) Leishmaniasis: immune cells crosstalk in macrophage polarization. Trop Med Infect Dis 8(5)

Alshetaiwi H, Pervolarakis N, McIntyre LL, Ma D, Nguyen Q, Rath JA et al (2020) Defining the emergence of myeloid-derived suppressor cells in breast cancer using single-cell transcriptomics. Sci Immunol 5(44)

Ancuta P, Liu KY, Misra V, Wacleche VS, Gosselin A, Zhou X et al (2009) Transcriptional profiling reveals developmental relationship and distinct biological functions of CD16+ and CD16- monocyte subsets. BMC Genomics 10:403

Anuradha R, Munisankar S, Bhootra Y, Jagannathan J, Dolla C, Kumaran P et al (2016) Systemic cytokine profiles in Strongyloides stercoralis infection and alterations following treatment. Infect Immun 84(2):425–431

Arora S, Dev K, Agarwal B, Das P, Syed MA (2018) Macrophages: their role, activation and polarization in pulmonary diseases. Immunobiology 223(4–5):383–396

Babu S, Nutman TB (2012) Immunopathogenesis of lymphatic filarial disease. Semin Immunopathol 34(6):847–861

Babu S, Nutman TB (2014) Immunology of lymphatic filariasis. Parasite Immunol 36(8):338–346

Babu S, Nutman TB (2016) Helminth-tuberculosis co-infection: an immunologic perspective. Trends Immunol 37(9):597–607

Babu S, Nutman TB (2019) Immune responses to helminth infection. In: Clinical immunology, 5th edn. Elsevier, pp 437–447.e1

Babu S, Blauvelt CP, Kumaraswami V, Nutman TB (2005) Diminished expression and function of TLR in lymphatic filariasis: a novel mechanism of immune dysregulation. J Immunol 175(2):1170–1176

Babu S, Kumaraswami V, Nutman TB (2009) Alternatively activated and immunoregulatory monocytes in human filarial infections. J Infect Dis 199(12):1827–1837

Bhengu KN, Naidoo P, Singh R, Mpaka-Mbatha MN, Nembe N, Duma Z et al (2022) Immunological interactions between intestinal helminth infections and tuberculosis. Diagnostics (Basel) 12(11)

Biswas SK, Mantovani A (2012) Orchestration of metabolism by macrophages. Cell Metab 15(4):432–437

Bohlson SS, O'Conner SD, Hulsebus HJ, Ho MM, Fraser DA (2014) Complement, c1q, and c1q-related molecules regulate macrophage polarization. Front Immunol 5:402

Bouchery T, Kyle R, Camberis M, Shepherd A, Filbey K, Smith A et al (2015) ILC2s and T cells cooperate to ensure maintenance of M2 macrophages for lung immunity against hookworms. Nat Commun 6:6970

Bourdonnay E, Zaslona Z, Penke LR, Speth JM, Schneider DJ, Przybranowski S et al (2015) Transcellular delivery of vesicular SOCS proteins from macrophages to epithelial cells blunts inflammatory signaling. J Exp Med 212(5):729–742

Briggs N, Weatherhead J, Sastry KJ, Hotez PJ (2016) The hygiene hypothesis and its inconvenient truths about helminth infections. PLoS Negl Trop Dis 10(9):e0004944

Brys L, Beschin A, Raes G, Ghassabeh GH, Noel W, Brandt J et al (2005) Reactive oxygen species and 12/15-lipoxygenase contribute to the antiproliferative capacity of alternatively activated myeloid cells elicited during helminth infection. J Immunol 174(10):6095–6104

Chen Y, Lu J, Huang Y, Wang T, Xu Y, Xu M et al (2013) Association of previous schistosome infection with diabetes and metabolic syndrome: a cross-sectional study in rural China. J Clin Endocrinol Metab 98(2):E283–E287
Chen F, Wu W, Millman A, Craft JF, Chen E, Patel N et al (2014) Neutrophils prime a long-lived effector macrophage phenotype that mediates accelerated helminth expulsion. Nat Immunol 15(10):938–946
Chen S, Saeed A, Liu Q, Jiang Q, Xu H, Xiao GG et al (2023) Macrophages in immunoregulation and therapeutics. Signal Transduct Target Ther 8(1):207
Coakley G, Harris NL (2020) Interactions between macrophages and helminths. Parasite Immunol 42(7):e12717
Coakley G, McCaskill JL, Borger JG, Simbari F, Robertson E, Millar M et al (2017) Extracellular vesicles from a helminth parasite suppress macrophage activation and constitute an effective vaccine for protective immunity. Cell Rep 19(8):1545–1557
Condamine T, Dominguez GA, Youn JI, Kossenkov AV, Mony S, Alicea-Torres K et al (2016) Lectin-type oxidized LDL receptor-1 distinguishes population of human polymorphonuclear myeloid-derived suppressor cells in cancer patients. Sci Immunol 1(2)
Cortes-Selva D, Fairfax K (2021) Schistosome and intestinal helminth modulation of macrophage immunometabolism. Immunology 162(2):123–134
Dasan B, Rajamanickam A, Munisankar S, Menon PA, Ahamed SF, Nott S et al (2023) Hookworm infection induces glycometabolic modulation in South Indian individuals with type 2 diabetes. IJID Reg 9:18–24
de Gier B, Pita-Rodriguez GM, Campos-Ponce M, van de Bor M, Chamnan C, Junco-Diaz R et al (2018) Soil-transmitted helminth infections and intestinal and systemic inflammation in schoolchildren. Acta Trop 182:124–127
de Los Reyes Jimenez M, Lechner A, Alessandrini F, Bohnacker S, Schindela S, Trompette A et al (2020) An anti-inflammatory eicosanoid switch mediates the suppression of type-2 inflammation by helminth larval products. Sci Transl Med 12(540)
De Nardo D, Labzin LI, Kono H, Seki R, Schmidt SV, Beyer M et al (2014) High-density lipoprotein mediates anti-inflammatory reprogramming of macrophages via the transcriptional regulator ATF3. Nat Immunol 15(2):152–160
Duarte AM, Jenkins TP, Latrofa MS, Giannelli A, Papadopoulos E, de Carvalho LM et al (2016) Helminth infections and gut microbiota - a feline perspective. Parasit Vectors 9(1):625
Duffield JS, Forbes SJ, Constandinou CM, Clay S, Partolina M, Vuthoori S et al (2005) Selective depletion of macrophages reveals distinct, opposing roles during liver injury and repair. J Clin Invest 115(1):56–65
Dutra HS, El-Cheikh MC, Azevedo SP, Rossi MI, Borojevic R (1998) Murine schistosomiasis mansoni: experimental analysis of bone marrow and peripheral myelopoiesis. Parasitol Res 84(8):668–675
Epelman S, Lavine KJ, Randolph GJ (2014) Origin and functions of tissue macrophages. Immunity 41(1):21–35
Esser-von Bieren J, Mosconi I, Guiet R, Piersgilli A, Volpe B, Chen F et al (2013) Antibodies trap tissue migrating helminth larvae and prevent tissue damage by driving IL-4Ralpha-independent alternative differentiation of macrophages. PLoS Pathog 9(11):e1003771
Fabre T, Barron AMS, Christensen SM, Asano S, Bound K, Lech MP et al (2023) Identification of a broadly fibrogenic macrophage subset induced by type 3 inflammation. Sci Immunol 8(82):eadd8945
Faz-Lopez B, Morales-Montor J, Terrazas LI (2016) Role of macrophages in the repair process during the tissue migrating and resident helminth infections. Biomed Res Int 2016:8634603
Fleming BD, Mosser DM (2011) Regulatory macrophages: setting the threshold for therapy. Eur J Immunol 41(9):2498–2502
Fu YL, Harrison RE (2021) Microbial phagocytic receptors and their potential involvement in cytokine induction in macrophages. Front Immunol 12:662063
Gabrilovich DI (2017) Myeloid-derived suppressor cells. Cancer Immunol Res 5(1):3–8
Gause WC, Rothlin C, Loke P (2020) Heterogeneity in the initiation, development and function of type 2 immunity. Nat Rev Immunol 20(10):603–614

Gent V, Waihenya R, Kamau L, Nyakundi R, Ambala P, Kariuki T et al (2019) An investigation into the role of chronic Schistosoma mansoni infection on Human Papillomavirus (HPV) vaccine induced protective responses. PLoS Negl Trop Dis 13(8):e0007704

Gibbings SL, Thomas SM, Atif SM, McCubbrey AL, Desch AN, Danhorn T et al (2017) Three unique interstitial macrophages in the murine lung at steady state. Am J Respir Cell Mol Biol 57(1):66–76

Gieseck RL 3rd, Wilson MS, Wynn TA (2018) Type 2 immunity in tissue repair and fibrosis. Nat Rev Immunol 18(1):62–76

Gomez-Garcia L, Lopez-Marin LM, Saavedra R, Reyes JL, Rodriguez-Sosa M, Terrazas LI (2005) Intact glycans from cestode antigens are involved in innate activation of myeloid suppressor cells. Parasite Immunol 27(10–11):395–405

Gordon S, Taylor PR (2005) Monocyte and macrophage heterogeneity. Nat Rev Immunol 5(12):953–964

Guilliams M, De Kleer I, Henri S, Post S, Vanhoutte L, De Prijck S et al (2013) Alveolar macrophages develop from fetal monocytes that differentiate into long-lived cells in the first week of life via GM-CSF. J Exp Med 210(10):1977–1992

Gundra UM, Girgis NM, Ruckerl D, Jenkins S, Ward LN, Kurtz ZD et al (2014) Alternatively activated macrophages derived from monocytes and tissue macrophages are phenotypically and functionally distinct. Blood 123(20):e110–e122

Harris NL, Loke P (2017) Recent advances in type-2-cell-mediated immunity: insights from helminth infection. Immunity 47(6):1024–1036

Hays R, Esterman A, Giacomin P, Loukas A, McDermott R (2015) Does Strongyloides stercoralis infection protect against type 2 diabetes in humans? Evidence from Australian Aboriginal adults. Diabetes Res Clin Pract 107(3):355–361

He L, Jhong JH, Chen Q, Huang KY, Strittmatter K, Kreuzer J et al (2021) Global characterization of macrophage polarization mechanisms and identification of M2-type polarization inhibitors. Cell Rep 37(5):109955

Hine AM, Loke P (2019) Intestinal macrophages in resolving inflammation. J Immunol 203(3):593–599

Hoeksema MA, Shen Z, Holtman IR, Zheng A, Spann NJ, Cobo I et al (2021) Mechanisms underlying divergent responses of genetically distinct macrophages to IL-4. Sci Adv 7(25)

Houlden A, Hayes KS, Bancroft AJ, Worthington JJ, Wang P, Grencis RK et al (2015) Chronic Trichuris muris infection in C57BL/6 mice causes significant changes in host microbiota and metabolome: effects reversed by pathogen clearance. PLoS One 10(5):e0125945

Ingersoll MA, Spanbroek R, Lottaz C, Gautier EL, Frankenberger M, Hoffmann R et al (2010) Comparison of gene expression profiles between human and mouse monocyte subsets. Blood 115(3):e10–e19

Italiani P, Boraschi D (2014) From monocytes to M1/M2 macrophages: phenotypical vs. functional differentiation. Front Immunol 5:514

Jackson-Jones LH, Ruckerl D, Svedberg F, Duncan S, Maizels RM, Sutherland TE et al (2016) IL-33 delivery induces serous cavity macrophage proliferation independent of interleukin-4 receptor alpha. Eur J Immunol 46(10):2311–2321

Jenkins SJ, Ruckerl D, Cook PC, Jones LH, Finkelman FD, van Rooijen N et al (2011) Local macrophage proliferation, rather than recruitment from the blood, is a signature of TH2 inflammation. Science 332(6035):1284–1288

Johnston CJC, Smyth DJ, Kodali RB, White MPJ, Harcus Y, Filbey KJ et al (2017) A structurally distinct TGF-beta mimic from an intestinal helminth parasite potently induces regulatory T cells. Nat Commun 8(1):1741

Kamalakannan V, Kirthika S, Haripriya K, Babu S, Narayanan RB (2012) Wolbachia heat shock protein 60 induces pro-inflammatory cytokines and apoptosis in monocytes in vitro. Microbes Infect 14(7–8):610–618

Kapellos TS, Bonaguro L, Gemund I, Reusch N, Saglam A, Hinkley ER et al (2019) Human monocyte subsets and phenotypes in major chronic inflammatory diseases. Front Immunol 10:2035

Khalil N, Bereznay O, Sporn M, Greenberg AH (1989) Macrophage production of transforming growth factor beta and fibroblast collagen synthesis in chronic pulmonary inflammation. J Exp Med 170(3):727–737

Kieler M, Hofmann M, Schabbauer G (2021) More than just protein building blocks: how amino acids and related metabolic pathways fuel macrophage polarization. FEBS J 288(12):3694–3714

Kim HY, Jeong D, Kim JH, Chung DH (2024) Innate Type-2 cytokines: from immune regulation to therapeutic targets. Immune Netw 24(1):e6

Knipper JA, Willenborg S, Brinckmann J, Bloch W, Maass T, Wagener R et al (2015) Interleukin-4 receptor alpha signaling in myeloid cells controls collagen fibril assembly in skin repair. Immunity 43(4):803–816

LaGory EL, Giaccia AJ (2016) The ever-expanding role of HIF in tumour and stromal biology. Nat Cell Biol 18(4):356–365

Lang R, Schick J (2017) Review: impact of helminth infection on antimycobacterial immunity-a focus on the macrophage. Front Immunol 8:1864

Lebu S, Kibone W, Muoghalu CC, Ochaya S, Salzberg A, Bongomin F et al (2023) Soil-transmitted helminths: a critical review of the impact of co-infections and implications for control and elimination. PLoS Negl Trop Dis 17(8):e0011496

Lechner A, Bohnacker S, Esser-von Bieren J (2021) Macrophage regulation & function in helminth infection. Semin Immunol 53:101526

Lekki-Jozwiak J, Baska P (2023) The roles of various immune cell populations in immune response against helminths. Int J Mol Sci 25(1)

Lenz B, Andrew BA, Ritter M, Karunakaran I, Gandjui NVT, Nchang LC et al (2023) The design and development of a study protocol to investigate Onchocerca volvulus, Loa loa and Mansonella perstans-mediated modulation of the metabolic and immunological profile in lean and obese individuals in Cameroon. PLoS One 18(6):e0285689

Leroux LP, Nasr M, Valanparambil R, Tam M, Rosa BA, Siciliani E et al (2018) Analysis of the Trichuris suis excretory/secretory proteins as a function of life cycle stage and their immunomodulatory properties. Sci Rep 8(1):15921

Lewinsky H, Gunes EG, David K, Radomir L, Kramer MP, Pellegrino B et al (2021) CD84 is a regulator of the immunosuppressive microenvironment in multiple myeloma. JCI Insight 6(4)

Lica ICL, Frazao G, Nogueira RA, Lira MGS, Dos Santos VAF, Rodrigues JGM et al (2023) Immunological mechanisms involved in macrophage activation and polarization in schistosomiasis. Parasitology 150(5):401–415

Loukas A, Maizels RM, Hotez PJ (2021) The yin and yang of human soil-transmitted helminth infections. Int J Parasitol 51(13–14):1243–1253

Lumeng CN, DelProposto JB, Westcott DJ, Saltiel AR (2008) Phenotypic switching of adipose tissue macrophages with obesity is generated by spatiotemporal differences in macrophage subtypes. Diabetes 57(12):3239–3246

Luque-Campos N, Bustamante-Barrientos FA, Pradenas C, Garcia C, Araya MJ, Bohaud C et al (2021) The macrophage response is driven by mesenchymal stem cell-mediated metabolic reprogramming. Front Immunol 12:624746

Macak Kubaskova T, Mudronova D, Vargova M, Reiterova K, Hrckova G (2021) Cellular and humoral peritoneal immunity to Mesocestoides vogae metacestode infection in mice. Parasit Vectors 14(1):54

Madala SK, Pesce JT, Ramalingam TR, Wilson MS, Minnicozzi S, Cheever AW et al (2010) Matrix metalloproteinase 12-deficiency augments extracellular matrix degrading metalloproteinases and attenuates IL-13-dependent fibrosis. J Immunol 184(7):3955–3963

Maizels RM, Yazdanbakhsh M (2003) Immune regulation by helminth parasites: cellular and molecular mechanisms. Nat Rev Immunol 3(9):733–744

Mamilos A, Winter L, Schmitt VH, Barsch F, Grevenstein D, Wagner W et al (2023) Macrophages: from simple phagocyte to an integrative regulatory cell for inflammation and tissue regeneration-a review of the literature. Cells 12(2)

Mantovani A, Sica A, Locati M (2007) New vistas on macrophage differentiation and activation. Eur J Immunol 37(1):14–16

Marshall MA, Jankovic D, Maher VE, Sher A, Berzofsky JA (2001) Mice infected with Schistosoma mansoni develop a novel non-T-lymphocyte suppressor population which inhibits virus-specific CTL induction via a soluble factor. Microbes Infect 3(13):1051–1061

Martinez FO, Gordon S (2014) The M1 and M2 paradigm of macrophage activation: time for reassessment. F1000Prime Rep 6:13

Martviset P, Phadungsil W, Na-Bangchang K, Sungkhabut W, Panupornpong T, Prathaphan P et al (2023) Current prevalence and geographic distribution of helminth infections in the parasitic endemic areas of rural Northeastern Thailand. BMC Public Health 23(1):448

Mass E, Nimmerjahn F, Kierdorf K, Schlitzer A (2023) Tissue-specific macrophages: how they develop and choreograph tissue biology. Nat Rev Immunol 23(9):563–579

Mati VL, Raso P, de Melo AL (2014) Strongyloides stercoralis infection in marmosets: replication of complicated and uncomplicated human disease and parasite biology. Parasit Vectors 7:579

Mayer JU, Brown SL, MacDonald AS, Milling SW (2020) Defined intestinal regions are drained by specific lymph nodes that mount distinct Th1 and Th2 responses against Schistosoma mansoni eggs. Front Immunol 11:592325

McSorley HJ, Maizels RM (2012) Helminth infections and host immune regulation. Clin Microbiol Rev 25(4):585–608

Mills CD, Lenz LL, Ley K (2015) Macrophages at the fork in the road to health or disease. Front Immunol 6:59

Minutti CM, Jackson-Jones LH, Garcia-Fojeda B, Knipper JA, Sutherland TE, Logan N et al (2017) Local amplifiers of IL-4Ralpha-mediated macrophage activation promote repair in lung and liver. Science 356(6342):1076–1080

Morales JK, Saleem SJ, Martin RK, Saunders BL, Barnstein BO, Faber TW et al (2014) Myeloid-derived suppressor cells enhance IgE-mediated mast cell responses. J Leukoc Biol 95(4):643–650

Mosser DM, Edwards JP (2008) Exploring the full spectrum of macrophage activation. Nat Rev Immunol 8(12):958–969

Mosser DM, Hamidzadeh K, Goncalves R (2021) Macrophages and the maintenance of homeostasis. Cell Mol Immunol 18(3):579–587

Murray PJ (2017) Macrophage polarization. Annu Rev Physiol 79:541–566

Murray PJ, Wynn TA (2011) Protective and pathogenic functions of macrophage subsets. Nat Rev Immunol 11(11):723–737

Murray PJ, Allen JE, Biswas SK, Fisher EA, Gilroy DW, Goerdt S et al (2014) Macrophage activation and polarization: nomenclature and experimental guidelines. Immunity 41(1):14–20

Narasimhan PB, Marcovecchio P, Hamers AAJ, Hedrick CC (2019) Nonclassical monocytes in health and disease. Annu Rev Immunol 37:439–456

Nouatin O, Mengue JB, Dejon-Agobe JC, Fendel R, Ibanez J, Ngoa UA et al (2021) Exploratory analysis of the effect of helminth infection on the immunogenicity and efficacy of the asexual blood-stage malaria vaccine candidate GMZ2. PLoS Negl Trop Dis 15(6):e0009361

Nutman TB (2013) Insights into the pathogenesis of disease in human lymphatic filariasis. Lymphat Res Biol 11(3):144–148

Nutman TB (2015) Looking beyond the induction of Th2 responses to explain immunomodulation by helminths. Parasite Immunol 37(6):304–313

O'Regan NL, Steinfelder S, Venugopal G, Rao GB, Lucius R, Srikantam A et al (2014) Brugia malayi microfilariae induce a regulatory monocyte/macrophage phenotype that suppresses innate and adaptive immune responses. PLoS Negl Trop Dis 8(10):e3206

Okabe Y, Medzhitov R (2014) Tissue-specific signals control reversible program of localization and functional polarization of macrophages. Cell 157(4):832–844

Orecchioni M, Ghosheh Y, Pramod AB, Ley K (2019) Macrophage polarization: different gene signatures in M1(LPS+) vs. classically and M2(LPS-) vs. alternatively activated macrophages. Front Immunol 10:1084

Parisi L, Gini E, Baci D, Tremolati M, Fanuli M, Bassani B et al (2018) Macrophage polarization in chronic inflammatory diseases: killers or builders? J Immunol Res 2018:8917804

Passlick B, Flieger D, Ziegler-Heitbrock HW (1989) Identification and characterization of a novel monocyte subpopulation in human peripheral blood. Blood 74(7):2527–2534

Passos LS, Gazzinelli-Guimaraes PH, Oliveira Mendes TA, Guimaraes AC, Silveira Lemos DD, Ricci ND et al (2017) Regulatory monocytes in helminth infections: insights from the modulation during human hookworm infection. BMC Infect Dis 17(1):253

Pastva AM, Wright JR, Williams KL (2007) Immunomodulatory roles of surfactant proteins A and D: implications in lung disease. Proc Am Thorac Soc 4(3):252–257

Pearce EJ, Kane CM, Sun J, Taylor JJ, AS MK, Cervi L (2004) Th2 response polarization during infection with the helminth parasite Schistosoma mansoni. Immunol Rev 201:117–126

Perez S, Rius-Perez S (2022) Macrophage polarization and reprogramming in acute inflammation: a redox perspective. Antioxidants (Basel) 11(7)

Piessens WF, McGreevy PB, Piessens PW, McGreevy M, Koiman I, Saroso JS et al (1980) Immune responses in human infections with Brugia malayi: specific cellular unresponsiveness to filarial antigens. J Clin Invest 65(1):172–179

Pradere JP, Kluwe J, De Minicis S, Jiao JJ, Gwak GY, Dapito DH et al (2013) Hepatic macrophages but not dendritic cells contribute to liver fibrosis by promoting the survival of activated hepatic stellate cells in mice. Hepatology 58(4):1461–1473

Qu C, Edwards EW, Tacke F, Angeli V, Llodra J, Sanchez-Schmitz G et al (2004) Role of CCR8 and other chemokine pathways in the migration of monocyte-derived dendritic cells to lymph nodes. J Exp Med 200(10):1231–1241

Quinteros SL, von Krusenstiern E, Snyder NW, Tanaka A, O'Brien B, Donnelly S (2023) The helminth derived peptide FhHDM-1 redirects macrophage metabolism towards glutaminolysis to regulate the pro-inflammatory response. Front Immunol 14:1018076

Rahman K, Vengrenyuk Y, Ramsey SA, Vila NR, Girgis NM, Liu J et al (2017) Inflammatory Ly6Chi monocytes and their conversion to M2 macrophages drive atherosclerosis regression. J Clin Invest 127(8):2904–2915

Rajamanickam A, Munisankar S, Bhootra Y, Dolla C, Thiruvengadam K, Nutman TB et al (2019) Metabolic consequences of concomitant strongyloides stercoralis infection in patients with type 2 diabetes mellitus. Clin Infect Dis 69(4):697–704

Rajamanickam A, Munisankar S, Dolla C, Menon PA, Thiruvengadam K, Nutman TB et al (2020a) Helminth infection modulates systemic pro-inflammatory cytokines and chemokines implicated in type 2 diabetes mellitus pathogenesis. PLoS Negl Trop Dis 14(3):e0008101

Rajamanickam A, Munisankar S, Thiruvengadam K, Menon PA, Dolla C, Nutman TB et al (2020b) Impact of helminth infection on metabolic and immune homeostasis in non-diabetic obesity. Front Immunol 11:2195

Rajamanickam A, Munisankar S, Menon PA, Nutman TB, Babu S (2021) Diminished circulating levels of angiogenic factors and rage ligands in helminth-diabetes comorbidity and reversal following anthelmintic treatment. J Infect Dis 224(9):1614–1622

Rajasekaram S, Anuradha R, Manokaran G, Bethunaickan R (2017) An overview of lymphatic filariasis lymphedema. Lymphology 50(4):164–182

Ramanan D, Bowcutt R, Lee SC, et al (2016) Helminth infection promotes colonization resistance via type 2 immunity. Science 352(6285):608–612. https://doi.org/10.1126/science.aaf3229

Ross EA, Devitt A, Johnson JR (2021) Macrophages: the good, the bad, and the gluttony. Front Immunol 12:708186

Roszer T (2015) Understanding the mysterious M2 macrophage through activation markers and effector mechanisms. Mediators Inflamm 2015:816460

Sakai M, Troutman TD, Seidman JS, Ouyang Z, Spann NJ, Abe Y et al (2019) Liver-derived signals sequentially reprogram myeloid enhancers to initiate and maintain Kupffer cell identity. Immunity 51(4):655–70 e8

Saleem SJ, Martin RK, Morales JK, Sturgill JL, Gibb DR, Graham L et al (2012) Cutting edge: mast cells critically augment myeloid-derived suppressor cell activity. J Immunol 189(2):511–515

Sampath P, Moideen K, Ranganathan UD, Bethunaickan R (2018) Monocyte subsets: phenotypes and function in tuberculosis infection. Front Immunol 9:1726

Satoh T, Takeuchi O, Vandenbon A, Yasuda K, Tanaka Y, Kumagai Y et al (2010) The Jmjd3-Irf4 axis regulates M2 macrophage polarization and host responses against helminth infection. Nat Immunol 11(10):936–944

Schmall A, Al-Tamari HM, Herold S, Kampschulte M, Weigert A, Wietelmann A et al (2015) Macrophage and cancer cell cross-talk via CCR2 and CX3CR1 is a fundamental mechanism driving lung cancer. Am J Respir Crit Care Med 191(4):437–447

Scott CL, Zheng F, De Baetselier P, Martens L, Saeys Y, De Prijck S et al (2016) Bone marrow-derived monocytes give rise to self-renewing and fully differentiated Kupffer cells. Nat Commun 7:10321

Shapouri-Moghaddam A, Mohammadian S, Vazini H, Taghadosi M, Esmaeili SA, Mardani F et al (2018) Macrophage plasticity, polarization, and function in health and disease. J Cell Physiol 233(9):6425–6440

Shiny A, Regin B, Balachandar V, Gokulakrishnan K, Mohan V, Babu S et al (2013) Convergence of innate immunity and insulin resistance as evidenced by increased nucleotide oligomerization domain (NOD) expression and signaling in monocytes from patients with type 2 diabetes. Cytokine 64(2):564–570

Shouval DS, Biswas A, Goettel JA, McCann K, Conaway E, Redhu NS et al (2014) Interleukin-10 receptor signaling in innate immune cells regulates mucosal immune tolerance and anti-inflammatory macrophage function. Immunity 40(5):706–719

Sica A, Erreni M, Allavena P, Porta C (2015) Macrophage polarization in pathology. Cell Mol Life Sci 72(21):4111–4126

Silas E, Ndlovu S, Tshilwane SI, Mukaratirwa S (2021) Immunological and pathophysiological outcomes of helminth infections and type 2 diabetes comorbidity studies in humans and experimental animals—a scoping review. Appl Sci 11(17):8079

Stevenson MM, Valanparambil RM, Tam M (2022) Myeloid-derived suppressor cells: the expanding world of helminth modulation of the immune system. Front Immunol 13:874308

Strizova Z, Benesova I, Bartolini R, Novysedlak R, Cecrdlova E, Foley LK et al (2023) M1/M2 macrophages and their overlaps - myth or reality? Clin Sci (Lond) 137(15):1067–1093

Su CW, Chen CY, Li Y, Long SR, Massey W, Kumar DV et al (2018) Helminth infection protects against high fat diet-induced obesity via induction of alternatively activated macrophages. Sci Rep 8(1):4607

Summers RW, Elliott DE, Urban JF Jr, Thompson RA, Weinstock JV (2005) Trichuris suis therapy for active ulcerative colitis: a randomized controlled trial. Gastroenterology 128(4):825–832

Sutrave S, Richter MH (2023) The truman show for human helminthic parasites: a review of recent advances in in vitro cultivation platforms. Microorganisms 11(7)

Tannahill GM, Curtis AM, Adamik J, Palsson-McDermott EM, McGettrick AF, Goel G et al (2013) Succinate is an inflammatory signal that induces IL-1beta through HIF-1alpha. Nature 496(7444):238–242

Tolouei Semnani R, Moore V, Bennuru S, McDonald-Fleming R, Ganesan S, Cotton R et al (2014) Human monocyte subsets at homeostasis and their perturbation in numbers and function in filarial infection. Infect Immun 82(11):4438–4446

Turner JD, Bourke CD, Meurs L, Mbow M, Dieye TN, Mboup S et al (2014) Circulating CD14brightCD16+ ‘intermediate’ monocytes exhibit enhanced parasite pattern recognition in human helminth infection. PLoS Negl Trop Dis 8(4):e2817

Uribe-Querol E, Rosales C (2020) Phagocytosis: our current understanding of a universal biological process. Front Immunol 11:1066

Van den Bossche J, Laoui D, Naessens T, Smits HH, Hokke CH, Stijlemans B et al (2015) E-cadherin expression in macrophages dampens their inflammatory responsiveness in vitro, but does not modulate M2-regulated pathologies in vivo. Sci Rep 5:12599

Vannella KM, Barron L, Borthwick LA, Kindrachuk KN, Narasimhan PB, Hart KM et al (2014) Incomplete deletion of IL-4Ralpha by LysM(Cre) reveals distinct subsets of M2 macrophages controlling inflammation and fibrosis in chronic schistosomiasis. PLoS Pathog 10(9):e1004372

Veglia F, Sanseviero E, Gabrilovich DI (2021) Myeloid-derived suppressor cells in the era of increasing myeloid cell diversity. Nat Rev Immunol 21(8):485–498

Viola A, Munari F, Sanchez-Rodriguez R, Scolaro T, Castegna A (2019) The metabolic signature of macrophage responses. Front Immunol 10:1462

Wang L, Li Z, Shen J, Liu Z, Liang J, Wu X et al (2015) Exosome-like vesicles derived by Schistosoma japonicum adult worms mediates M1 type immune- activity of macrophage. Parasitol Res 114(5):1865–1873

Watanabe S, Alexander M, Misharin AV, Budinger GRS (2019) The role of macrophages in the resolution of inflammation. J Clin Invest 129(7):2619–2628

Wei Y, Wang M, Ma Y, Que Z, Yao D (2021) Classical dichotomy of macrophages and alternative activation models proposed with technological progress. Biomed Res Int 2021:9910596

Westphalen K, Gusarova GA, Islam MN, Subramanian M, Cohen TS, Prince AS et al (2014) Sessile alveolar macrophages communicate with alveolar epithelium to modulate immunity. Nature 506(7489):503–506

White MPJ, McManus CM, Maizels RM (2020) Regulatory T-cells in helminth infection: induction, function and therapeutic potential. Immunology 160(3):248–260

Wiedemann M, Voehringer D (2020) Immunomodulation and immune escape strategies of gastrointestinal helminths and schistosomes. Front Immunol 11:572865

Wiria AE, Hamid F, Wammes LJ, Prasetyani MA, Dekkers OM, May L et al (2015) Infection with soil-transmitted helminths is associated with increased insulin sensitivity. PLoS One 10(6):e0127746

Wong KL, Yeap WH, Tai JJ, Ong SM, Dang TM, Wong SC (2012) The three human monocyte subsets: implications for health and disease. Immunol Res 53(1–3):41–57

Wynn TA (2008) Cellular and molecular mechanisms of fibrosis. J Pathol 214(2):199–210

Wynn TA, Barron L (2010) Macrophages: master regulators of inflammation and fibrosis. Semin Liver Dis 30(3):245–257

Wynn TA, Ramalingam TR (2012) Mechanisms of fibrosis: therapeutic translation for fibrotic disease. Nat Med 18(7):1028–1040

Wynn TA, Vannella KM (2016) Macrophages in tissue repair, regeneration, and fibrosis. Immunity 44(3):450–462

Yao J, Wu D, Qiu Y (2022) Adipose tissue macrophage in obesity-associated metabolic diseases. Front Immunol 13:977485

Yao Z, Qi W, Zhang H, Zhang Z, Liu L, Shao Y et al (2023) Down-regulated GAS6 impairs synovial macrophage efferocytosis and promotes obesity-associated osteoarthritis. Elife 12

Yeh MY, Aggarwal S, Carrig M, Azeem A, Nguyen A, Devries S et al (2023) Strongyloides stercoralis infection in humans: a narrative review of the most neglected parasitic disease. Cureus 15(10):e46908

Yingklang M, Chaidee A, Dangtakot R, Jantawong C, Haonon O, Sitthirach C et al (2022) Association of Strongyloides stercoralis infection and type 2 diabetes mellitus in northeastern Thailand: impact on diabetic complication-related renal biochemical parameters. PLoS One 17(5):e0269080

Yunna C, Mengru H, Lei W, Weidong C (2020) Macrophage M1/M2 polarization. Eur J Pharmacol 877:173090

Zhang QZ, Su WR, Shi SH, Wilder-Smith P, Xiang AP, Wong A et al (2010) Human gingiva-derived mesenchymal stem cells elicit polarization of m2 macrophages and enhance cutaneous wound healing. Stem Cells 28(10):1856–1868

Ziegler-Heitbrock L, Hofer TP (2013) Toward a refined definition of monocyte subsets. Front Immunol 4:23

Ziegler-Heitbrock L, Ancuta P, Crowe S, Dalod M, Grau V, Hart DN et al (2010) Nomenclature of monocytes and dendritic cells in blood. Blood 116(16):e74–e80

Chapter 14
Macrophages, Metabolism, Mitochondria, Circadian Rhythmicity and the Pathogen: The Multidimensional Nature of Tuberculosis

Paula Guzmán-Téllez, Miguel Angel Rivero-Silva, María Maximina Bertha Moreno-Altamirano, and Francisco Javier Sánchez-García

Abstract *Mycobacterium tuberculosis,* the causative agent of tuberculosis (TB) was first identified in 1882 by Robert Koch, and it is estimated that this pathogen has been around for as long as 3 million years.

The World Health Organization (WHO) reported that in 2022 alone an estimated 10.6 million people developed TB worldwide, making TB the world's second leading cause of death from a single infectious agent, just after coronavirus disease (COVID-19), despite TB being a preventable and usually curable disease.

Moreover, epidemiological studies suggest that approximately a quarter of the global population has been infected with TB bacteria, of which 5–10% will eventually develop symptoms and TB disease. Poverty, obesity, diabetes, and alcohol use contribute to the burden of TB.

Alveolar macrophages play a pivotal role in the clearance of airborne pathogenic microorganisms and are the primary target of *M. tuberculosis.*

Macrophage activity depend on metabolism and circadian rhythmicity, and mitochondria are a central hub that coordinates the communication between metabolism, circadian rhythmicity, and the immune system.

Recent evidence has thrown light on how *M. tuberculosis* metabolism may regulate macrophage activity and the overall host responses to *M. tuberculosis* infection.

P. Guzmán-Téllez · M. A. Rivero-Silva · F. J. Sánchez-García (✉)
Laboratorio de Inmunorregulación, Departamento de Inmunología, Escuela Nacional de Ciencias Biológicas, Instituto Politécnico Nacional, Mexico City, Mexico
e-mail: fjsanchez@ipn.mx

M. M. B. Moreno-Altamirano
Laboratorio de Inmunoparasitología, Departamento de Inmunología, Escuela Nacional de Ciencias Biológicas, Instituto Politécnico Nacional, Mexico City, Mexico

M. Kloc et al. (eds.), *Monocytes and Macrophages in Development, Regeneration, and Disease*, Results and Problems in Cell Differentiation 74,
https://doi.org/10.1007/978-3-031-65944-7_14

This chapter explores how all these biological domains relate to each other, highlighting the multidimensional nature of TB, and positioning macrophages at center stage.

14.1 Tuberculosis

In 2024 tuberculosis (TB) remains the leading cause of human death caused by pathogenic bacteria, despite of global efforts to overcome it (Floyd et al. 2018; WHO 2023). Globally, in the year 2022 alone, an estimated 1.3 million people died as a consequence of TB, whereas 7.5 million people were newly diagnosed with TB. Likewise, it is estimated that about 25% of the world population has been infected with *M. tuberculosis* although being asymptomatic (WHO 2023). Of note, and fortunately, over 90% of individuals infected with *M. tuberculosis* spontaneously control the infection (Cambier et al. 2014).

In March 2023, the WHO Director-General, Dr. Tedros Adhanom Ghebreyesus, launched a new 5-year (2023–2027) flagship initiative to accelerate progress towards ending TB. This initiative includes a call for the licensing of at least one new vaccine by the year 2027 (WHO 2023).

During *M. tuberculosis* infection the host immune cells, including macrophages, undergo significant metabolic alterations (Shi et al. 2019); concomitantly, *M. tuberculosis* may engage in one of several host-pathogen outcomes, it can cause active disease, be kept within the host in a latency state for prolonged time or leave the latency state for re-entry into a bacterial replicative cycle that may cause active disease and allow for transmission to new hosts. In doing so, *M. tuberculosis* must adapt to a set of metabolic pressures (Ehrt et al. 2018), amongst these, hypoxia (Eoh et al. 2017).

M. tuberculosis can survive within infected resident alveolar macrophages and dendritic cells, and the phagocytosis of the bacteria triggers specific cell signaling that culminates in the secretion of cytokines such as tumor necrosis factor α (TNF-α), the IL-12 family, IL-6, IL-1α/β and IL-10, and chemokines involved in the recruitment of polymorphonuclear leukocytes. Increasing anti-microbial activity and the accumulation of mononuclear cells around the infected areas leads to the formation of granulomas. Macrophages are essential for granuloma, and different types of macrophages contribute to this unique structure, including multinucleated giant cells, epithelioid cells and, foamy macrophages (Krishnan et al. 2023, Russell et al. 2009).

Polarization of macrophages plays an important role in the establishment of *M. tuberculosis* infection. In general terms, the capacity of intracellular bacteria to modulate macrophage polarization constitutes an important pathogenesis mechanism. Thus, M1 macrophages are related to intracellular microbicidal activity due in part to the efficient production of reactive oxygen species (ROS) and nitric oxide (NO), pro-inflammatory cytokines and chemokines, and the recruitment of other immune cells. In particular, M1 macrophages promote the TB-related granuloma formation. On the other hand, the M2 macrophages are found in chronic infections, where those cells promote tissue repair and resolution of inflammation in necrotic and no necrotic granulomas (BoseDasgupta and Pieters 2018; Thiriot et al. 2020).

Tuberculosis is a complex disease. Its symptoms vary depending on the patient's age, nutritional status, site of infection (pulmonary or extra pulmonary TB), co-morbidities and, quite importantly, the host's immunological status (Natarajan et al. 2020). In pulmonary TB (primary TB), the lung parenchyma and the tracheobronchial tree are compromised, and the most common clinical manifestations include chronic cough, sputum production, appetite loss, weight loss, fever, night sweats, and hemoptysis. In case of complications, pulmonary TB also causes lymph node enlargement that leads to bronchial obstruction. During TB, there is the possibility that the patient develops extra pulmonary dissemination to secondary infection sites, through hematogenous spread (secondary TB). The most common secondary TB infections include tubercular lymphadenitis, pleural TB, abdominal TB, central nervous system TB, bone and joint TB, genito-urinary TB, and miliary TB (Loddenkemper et al. 2015).

Rapid and accurate diagnosis tests are indispensable. They must demonstrate the presence of the *M. tuberculosis* bacilli. Microscopy identification in pulmonary and extrapulmonary samples through acid-fast staining (including Ziehl-Neelsen, Kinyoun and Auramine fluorochrome stains) is very useful. Other diagnosis methods include culture using selective and specific media from which specific biochemical tests are performed, molecular methods (*in situ* hybridization, PCR, line probe assay (LPA), DNA sequencing, among others), and imaging methods that allow to determine the severity of the affectations in the infected organs (radiographs, tomography, etc.) (Natarajan et al. 2020; Loddenkemper et al. 2015).

Usually, the treatment for drug-sensitive *M. tuberculosis* infections consist of a daily regimen with isoniazid, rifampicin, pyrazinamide and ethambutol for at least 16 weeks. The presence of multidrug-resistant strains has complicated treatment; in these cases, the duration of the treatment extends to 6–9 months with kanamycin, levofloxacin, ethionamide, cycloserine, pyrazinamide, ethambutol, isoniazid, and 18 months with levofloxacin, ethionamide, cycloserine, ethambutol, isoniazid on a daily basis (Natarajan et al. 2020).

14.2 Macrophages

Because macrophages are the main cell target for *M. tuberculosis*, any attempt to understand TB must consider macrophage functional complexity.

Macrophages have phagocytic and endocytic ability and a complex embryonic origin. Most tissue-resident populations derive from erythro-myeloid (EM) progenitors and fetal liver precursors, and after birth, the circulating monocytes derive from myeloid progenitors in the bone marrow that are also related to the development of dendritic cells and neutrophils. Tissue macrophages are part of the mononuclear phagocyte cellular system involved in the physiological responses to infection and host protection, metabolic processes, and tissue repair (Gordon and

Pluddemann 2017; Gordon and Taylor 2005). All these functions are orchestrated by the specialized cell membrane and intracellular proteins, integrated into specific signaling pathways whose activation results in phenotypically heterogeneous cells. The recruitment to the site of infection, and the differentiation processes that the peripheral monocytes undergo upon infection are modulated by metabolic requirements, tissue microenvironment and the, coordinated action of innate or adaptive immune responses (Makuch et al. 2024; Mosser and Edwards 2008).

Considering the phenotypic diversity, the range of functions in which macrophages are involved, the differential mechanism of activation and their role in promoting or resolving inflammation, the macrophages have been phenotypically and functionally classified as polarized cells, resembling the Th lymphocytes classification. The most widely used classification of macrophages considers M1 (classically activated) macrophages at one pole, and M2 (alternatively activated) macrophages at the other pole. Notably, in humans, it is possible to identify different subsets within the M2 population (Mosser and Edwards 2008).

The concept of macrophage polarization was proposed by Mills (Mills et al. 2000) and originates from the observation of the differential response to immune stimuli (cytokines and microbial products) and the innate or adaptive response that is established (Thapa and Lee 2019).

M1 or classically activated macrophages were described for the first time by Mackaness (Mackaness 1962), who characterized macrophage activation as the enhanced anti-microbial activity induced by the action of interferon-γ (IFN-γ), TNF and lipopolysaccharide (LPS) over macrophages. IFN-γ is produced by the Th1 lymphocytes, natural killer cells, and macrophages, and induces expression of cytokine receptors (CSF2RB, IL2RA or IL6R), cell activation markers (CD3, CD38, CD69, and CD97) and adhesion molecules (ICAM1, ITGAL, ITGA4) in M1 macrophages. This differential state of macrophages is also associated with anti-tumoral capacity, secretion of high levels of pro-inflammatory cytokines (TNF-α, IL-1 β, IL-6, IL-12, and IL-18), other soluble mediators (ROS, nitrogen intermediates, and antimicrobial peptides), and enhanced antigen presentation (Martínez and Gordon 2014; Mosser and Edwards 2008). It is worth mentioning that the pro-inflammatory cytokines produced by M1 macrophages are key components of the host defense but, at the same time, they may cause a severe damage (Martínez and Gordon 2014).

The alternatively activated macrophages (M2) participate in wound healing and immune regulation. Th2 cytokines (mainly IL-4 and IL-13) promote the polarization of macrophages to the M2 phenotype characterized by the secretion of anti-inflammatory molecules such as IL-10, TGF-β, IL-1RA, and glucocorticoids and the upregulation of the mannose receptor. M2 macrophages are less efficient than M1 macrophages to eliminate intracellular pathogens or to induce a pro-inflammatory milieu (Martínez and Gordon 2014; Mosser and Edwards 2008; Thapa and Lee 2019).

Macrophages are polarized into the wound healing phenotype by IL-4, which stimulates the arginase activity in these cells, leading to the conversion of arginine to ornithine, and eventually to the synthesis of polyamines and collagen, thus contributing to the regeneration of the extracellular matrix (Mantovani et al. 2004). M2 macrophages also produce vascular endothelial growth factor (VEGF), epidermal growth factor (EGF), and platelet-derived growth factor (PDGF), thus contributing to tissue repair (Liu et al. 2021). M2 regulatory macrophages result from the inhibition of cytokine gene transcription by IL-10 and glucocorticoids (Xie et al. 2019). These cells do not contribute to extracellular matrix production, but instead, they harbor high levels of co-stimulatory molecules (CD80 and CD86), which allow them to present efficiently antigens to T lymphocytes. An important feature of M2 regulatory macrophages is the production of IL-10 and the downregulation of IL-12 (Mantovani et al. 2004; Martínez and Gordon 2014).

An extended classification of M2 macrophages includes M2a to M2d macrophages. The M2a includes the macrophages that were polarized by IL-4 or IL-13 and contribute to tissue repair; the M2b subgroup contains macrophages that were activated by TLR agonists and by immune complexes; M2c macrophages respond to IL-10/TGF-β, IL-21 or glucocorticoids and also participate in tissue repair; the M2d are those that were co-stimulated through TLRs and adenosine receptor, and produce IL-10 and VEGF (Thiriot et al. 2020).

A more recently described subpopulation of polarized macrophages has been dubbed as M3 or "switching phenotype". The name comes from the observation that in response to reprogramming-inducing factors that favor the production of anti-inflammatory cytokines (RF-M2), which would normally lead to M2 macrophage polarization, an M1 phenotype results instead. Conversely, in the presence of RF-M1, which would lead to a pro-inflammatory response, M2 macrophages arise (Ge et al. 2021).

Recent studies show that macrophages exposed to murine mesothelioma microenvironment secrete IL-10 and TNF-α and express a mixed M1 and M2 phenotype ($CD11b^{+}$ $F4/80^{+}$ $Ly6C^{hi}$ $CX3CR1^{hi}$). The authors concluded that M3 macrophages are induced and expanded within a tumor microenvironment (Jackaman et al. 2016).

M3 macrophages with a switching phenotype may be reprogrammed by the simultaneous activation with RF-M1, and the concomitant inhibition of RF-M2-mediated signalling, and thus it is likely that M3 macrophages represent a relevant component of macrophage plasticity that provides multifunctional activation and an efficient anti-inflammatory response (Kalish et al. 2017).

On the other hand, the M4 macrophages correspond to a subpopulation of cells characterized by the expression of CD206, CD68, matrix metalloproteinase 7 (MMP 7) and myeloid-related protein 8 (MRP8; S100A8). These are stimulated by CXCL4 and produce IL-6 and TNF-α, MMP7, and MMP12. The presence of M4 macrophages has been described in atherosclerosis, cardiac remodeling, and leprosy (Nakai 2021).

In leprosy, the M4 macrophage phenotype has been found in patients with lepromatous leprosy, thus suggesting that M4 macrophages are inefficient in the elimination of *M. leprae* and thus favor the multibacillary form of the disease (de Sousa et al. 2018).

The role of M3 and M4 macrophages in tuberculosis has not as yet been addressed in detail (Ge et al. 2021). Quite clearly, this topic deserves further investigation.

14.3 The Role of Macrophages in Fighting *M. tuberculosis*

Lung-resident alveolar macrophages play a pivotal role in the clearance of airborne pathogenic microorganisms, including bacteria, fungi, and viruses. These cells are endowed with a highly phagocytic activity related to their high expression of bacteria and fungi recognizing receptors (Mitsi et al. 2018).

The phenotypical and functional polarization of macrophages determines the course of *M. tuberculosis* infection. The alveolar macrophages interact with the bacteria in the lungs, the bacteria are phagocytosed, and the recruitment of inflammatory cells begins (Fig. 14.1). This process will eventually lead to the formation of granulomas (Khan et al. 2019). Other specialized macrophages present in the lung are the interstitial macrophages (Fig. 14.1), which include the intra-epithelial macrophages found in larger airways, and a population found in the pleural cavity (Marakalala et al. 2018).

In TB patients, granulomas may present different stages and types and thus, it is possible to find macrophages with diverse immune phenotypes compartmentalized across distinct granuloma regions, and differentiation of macrophages may generate foam cells, epithelioid cells, and multinucleated giant cells (Fig. 14.1). Modern high resolution mass-spectrometry-based approaches have shown that the center of lung granulomas obtained from TB patients, contain pro-inflammatory and antibacterial molecules such as cathelicidins, activators of ROS, eicosanoids precursors, and anti-microbial effectors downstream of TNF and IFN signaling pathways, whereas the periphery of the granulomas contain metabolically active and intact cells, proteins related with arachidonic acid (ALOX5, ALOX5AP and LTA4H), and anti-inflammatory and tissue-preserving mediators (Marakalala et al. 2016, 2018). Epithelioid cells are another phenotype of macrophages found in granulomas. These cells form junctions containing adherent components associated with epithelial cells (desmoplakin, desmoglein, and desmocollin), adherent junction proteins (E-cadherin, plakoglobin, α-, β-, and δ-catenin), and tight junction proteins (ZO-1 and ZO-2) (Fig. 14.1). All of which suggest that the development of TB granuloma and the fate of infection are dependent on the macrophage differentiation (Cronan et al. 2016). On the other hand, foam cells and the caseation are important components in the pathogenesis of TB. The caseation originates from the cell death that follows the interaction mycobacteria/host immune response, while the foam macrophages result from altered lipid metabolism encompassing mobilization from lipid stores, endocytosis, and deposition in the cytoplasm, enhanced synthesis of lipids either locally or systemically, and failure to degrade accumulated lipids (Fig. 14.1) (Marakalala et al. 2018; Neyen and Gordon 2014). Lipid metabolism in macrophages is dependent on the transcription factor Peroxisome proliferator-activated receptor γ (PPAR-γ), cytokines, and adipokines, so, in the course of TB infection,

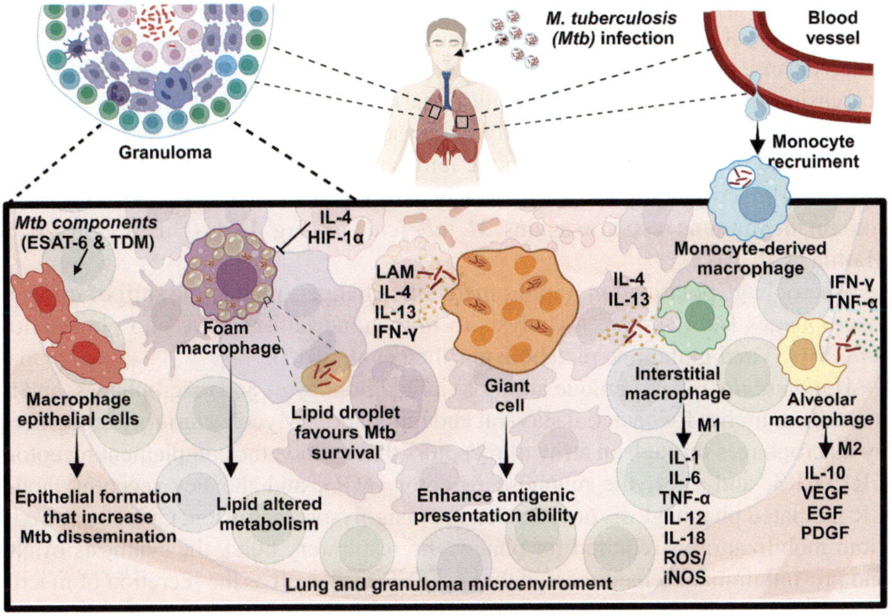

Fig. 14.1 Major macrophage subpopulations implicated in the immunopathogenesis of tuberculosis. Following the inhalation of Mtb contained in the microdroplets expelled by a tuberculosis-infected person, the mycobacteria establish in the parenchyma of the lung. The strong inflammatory response in the lung induces the recruitment of circulating monocytes, which then form granulomas. The recruited monocytes and resident macrophages (alveolar and interstitial) adopt either an M1 or M2 polarization profile, dependent on microenvironment factors. Other macrophage phenotypes that play a relevant role in the granuloma structure are macrophage epithelial cells, foam macrophages and giant cells. The signalling pathways that induce them remain poorly understood, and their function in the immunopathogenesis of tuberculosis requires further investigation. Abbreviations: *Mtb Mycobacterium tuberculosis*, *ROS* Radical Oxygen Species, *iNOS* inducible Nitric Oxygen Synthase, *VEGF* Vascular Endothelial Growth Factor, *EGF* Epidermal Growth Factor, *PDGF* Platelet-Derived Growth Factor, *ESAT-6* Early Secretory Antigenic Target 6, *TDM* Trehalose 6,6′-DiMycolate, *HIF-1α* Hypoxia-Inducible Factor-1α, *LAM* LipoArabinoMannan

an interaction between bacteria, macrophages, and adipocytes takes place leading to the infection of adipocytes and the development of giant Langhans cells (Fig. 14.1) (Huang et al. 2018). *M. tuberculosis* takes advantage of fatty acid and cholesterol uptake from the host cells to survive intracellularly (Daniel et al. 2011; Huang et al. 2018; Marakalala et al. 2018).

M1 macrophages are predominant at the early stages of TB infection, concomitant with the production of ROS, NO, CXCL11, and TNF-α, all of which contribute to the intracellular elimination of the pathogen. Meanwhile, the pro-inflammatory milieu drives granuloma formation (Fig. 14.1) (Huang et al. 2015).

The polarization of macrophages TB is a highly dynamic process, and the presence of both M1 and M2 phenotypes has been detected in non-necrotic granulomas of TB patients; the expression of both M1 and M2 markers (CXCL11, CCL17,

TNF-α, and IL-10) may persist during the granuloma formation i.e., the pro-inflammatory (M1) and the anti-inflammatory (M2) macrophages may coexist. Later during infection, M2 macrophage polarization takes over, starting with a decreasing expression of M1-associated markers, while the M2 markers (CD206, CCL17, IL-4, and IL-10) gradually increase. M2 macrophages are present in necrotic and non-necrotic lesions in the lungs, and their biological activity is opposite to M1 macrophages; M2 macrophages provide one of the main mechanisms to contain the immune response against *M. tuberculosis* (Fig. 14.1) (Huang et al. 2015; Mattila et al. 2013).

Macrophage polarization implies metabolic changes, mainly the shift of infected macrophages from aerobic glycolysis to mitochondrial oxidative phosphorylation (OXPHOS) and glutamine metabolism (Shen et al. 2017).

M. tuberculosis has a unique ability to infect monocytes and macrophages, and to make them niches for bacterial survival and replication. Mycobacteria are recognized by macrophages through an array of receptors that include the complement receptors CR1, CR3, and CR4, the mannose receptor (MR), and the Fcγ receptor FcγR; CR-mediated phagocytosis does not lead to phagolysosome fusion. Cytoplasmic calcium mobilization is required for phagocytic respiratory burst, the synthesis of NO and pro-inflammatory mediators including TNF-α, as well as the secretion of macrophage granule components (Malik et al. 2000). Scavenger receptors such as MARCO (macrophage receptor with collagenous structure) activate macrophages through NF-κB, in response to *M. tuberculosis* infection (Bowdish et al. 2009). CD14 binds *M. tuberculosis* lipoarabinomannan, facilitating the phagocytosis of non-opsonized mycobacteria. This engagement, along with cell activation through other cell membrane receptors activates inflammatory responses (Shams et al. 2003).

Also, it has been described that the pattern recognition receptor Dectin-1 is recognized by an unknown mycobacterial ligand which triggers a spleen tyrosine kinase (Syk), protein kinase C-δ (PKC-δ), and caspase recruitment domain family member 9 (CARD9) signaling pathway that helps control TB infection (Wagener et al. 2018).

The IRAK-M pathway is as a negative regulator triggered by the PAMP-TLR interaction in macrophages. This kinase plays a key role in the intracellular survival of *M. tuberculosis* on infected cells and in the M2 polarization acting through hipoxya-inducible factor-1α (HIF-1α) and MAPK signaling (Khan et al. 2019; Shen et al. 2017). A role for microRNAs in the macrophage polarization during TB has been recently described. The microRNA-26a (miR-26a) has been related to M2 polarization (Thiriot et al. 2020).

It is worth mentioning that an important variable that determines the outcome of the granuloma is the strain of *M. tuberculosis* involved (Manabe et al. 2003). Within the TB granulomas, macrophages undergoing apoptosis are degraded by neighboring macrophages via efferocytosis (Martin et al. 2012). Also, macrophages that undergo autophagy inhibit the replication of *M. tuberculosis* (Gutierrez et al. 2004). An inducer of autophagy in macrophages is IFN-γ, and thus, the Th1 immune response is a key component for the control of the infection (Lo et al. 2021).

As part of the adaptative immune response, *M. tuberculosis*-specific antibodies are produced. These are recognized by Fc Receptors (FcR). Macrophage activation through FcγR enhances phagolysosomal fusion increasing the bactericidal function of macrophages (Maglione et al. 2008).

The innate immune training allows macrophages a rapid and robust response when they face a second homologous or heterologous challenge (Chakraborty et al. 2023). This phenomenon is evident in BCG-vaccinated individuals who became protected against *M. tuberculosis* but also against other pathogens, such as *Listeria monocytogenes, Salmonella typhimurium, Staphylococcus aureus, Candida albicans,* and *Schistosoma mansoni* (Kleinnijenhuis et al. 2012; Clark et al. 1976).

14.4 Macrophages, Mitochondria, and Tuberculosis

Macrophages phagocytose and destroy *M. tuberculosis,* but mycobacteria may also establish a niche within alveolar macrophages, and several macrophage subtypes act at different stages of infection making the host-pathogen interaction dynamic and complex (Ahmad et al. 2022).

In macrophages infected with *M. tuberculosis*, the virulence of the strain may determine the response of mitochondria as shown by changes in mitochondrial shape (size and inner membrane ultrastructure) and activity (mitochondrial membrane potential and cytochrome c release), for instance (Abarca-Rojano et al. 2003).

One of the bactericidal mechanisms used by macrophages is the production of ROS. Although NADPH oxidase is the first source of ROS identified in macrophages and is considered fundamental for the elimination of invasive microorganisms (Canton et al. 2021). Macrophage mitochondria also produce ROS (mROS), which has long been regarded as a harmful byproduct of the electron transport chain. More recent evidence highlights the signaling properties of mROS as well as its important role in cellular adaptation and stress resistance, all of which are essential to keep cells and whole multicellular organisms alive (Palma et al. 2024).

Mitochondria play a central role in innate and adaptive immune responses in addition to their more widely acknowledged role as energy-producing organelles (Patil et al. 2022). Because of the bacterial origin of mitochondria (endosymbiotic origin), these organelles contain molecules that function as DAMPs (mitochondrial DNA, formylated peptides, ATP, cardiolipin, etc.), which may stimulate the innate immune response (Nakahira et al. 2015).

Mitochondrial ROS is also a multifunctional mitochondrial DAMP, linked to pro- and anti-microbial immune outcomes and a particularly important mediator of pathogenesis during infection with *M. tuberculosis* (Ellzey et al. 2023).

Since mROS may be detrimental to *M. tuberculosis*, this bacterial species has evolved mechanisms to regulate its effects, such as decreasing the mROS levels in cells by utilizing the major *M. tuberculosis* antigen, ESAT-6 secretion complex -1, (ESX-1 or type VII secretion system) (Yabaji et al. 2020; Ellzey et al. 2023).

Mitochondria homeostasis requires continuous and timely fusion and fission of mitochondria (mitochondrial dynamics) (Tilokani et al. 2018) and, amongst other biological roles, mitochondrial fission allows the removal of damaged mitochondria by a form of autophagy called mitophagy (Twig et al. 2008). Interestingly, damaged mitochondria activate the expression of cytosolic nucleic acid sensing-induced type I IFN, inflammasome activation, and mitophagy i.e., the same innate immune pathways that evolved to sense and respond to intracellular pathogens (Patrick and Watson 2021).

The above suggests that *M. tuberculosis* control over mitochondrial dynamics would have important consequences for the host-pathogen interaction. In this regard, it has been shown that *M. tuberculosis* virulent factor PE_PGRS33 colocalizes with the mitochondrial network in transfected cells (Cadieux et al. 2011), and that treatment of monocyte-derived macrophages with recombinant LprG (an *M. tuberculosis* virulence factor) induces mitochondrial fission and decreases the basal respiratory rate, while the addition of PE_PGRS33 induces mitochondrial fusion (Aguilar-López et al. 2019; Patrick and Watson 2021). On the other hand, there is evidence that mitochondrial fusion, dependent on mitofusin 1 (MFN1), is essential for the mycobactericidal activity of macrophages, through the regulation of ATP-dependent autophagy (Ning et al. 2021).

Macrophages that have been infected with *M. tuberculosis* undergo metabolic reprogramming, and there is increasing evidence that mitochondria play a key part in that process. Cellular metabolism is controlled by energy sensors, including a family of seven NAD^+-dependent protein deacetylases, known as sirtuins (SIRT) (Schwer and Verdin 2008; Smulan et al. 2021); one of them, called SIRT3, and the major mitochondrial sirtuin, enhances anti-*M. tuberculosis* activity in macrophages by mediating autophagy, preserving mitochondrial function, and limiting ROS accumulation (Kim et al. 2019; Smulan et al. 2021). Recently, Smulan et al. (2021) showed that *M. tuberculosis* infection of macrophages downregulates SIRT3 in a TLR2-dependent manner, reprogramming tricarboxylic acid (TCA) cycle, electron transport chain, and glycolytic pathways, and leading to increased mROS, and cell death. The same authors propose that SIRT3 modulation by *M. tuberculosis* is a key upstream event leading to metabolic reprogramming and macrophage dysfunction in TB.

14.5 Macrophage Function and Circadian Rhythmicity

The word 'circadian' derives from the Latin 'Circa' and 'Diem', meaning 'about a day'. Circadian rhythms refer to biological rhythms with cycles of approximately 24-hours, corresponding to the Earth's 24-hour rotation (Halberg et al. 1959). From the turn of this century, it has increasingly been clear that both innate and adaptive immune responses, from leukocyte blood counts, immune cell trafficking, chemotaxis, cytokine synthesis, and T cell activation, are closely related to the time of the day (Scheiermann et al. 2013; Haspel et al. 2020).

Macrophages display rhythmicity in their metabolism and effector functions such as pathogen sensing, phagocytosis, migration, and the production of inflammatory mediators (Timmons et al. 2020; Shirato and Sato 2022). The role of circadian rhythmicity is best illustrated by the finding that BCG vaccination in the morning induces stronger trained immunity and adaptive responses than the evening vaccination (de Bree et al. 2020).

Earlier work showed that in the lungs of endotoxemic mice, inflammation produces a complex re-organization of cellular and molecular circadian rhythms relevant to early events in lung injury (Haspel et al. 2014). Likewise, bacterial products such as LPS disrupt circadian rhythms in macrophages by a mechanism dependent on ROS (Wang et al. 2016), and the macrophage phagocytic and bactericidal activity is dependent on the mitochondrial dynamics and circadian rhythmicity crosstalk (Oliva-Ramírez et al. 2014).

Circadian disruption and sleep disorders are detrimental to the immune system and chronic insomnia may exacerbate chronic infections, including TB (Vadakkan Devassy et al. 2022).

The relationship between circadian misalignment and TB infection and the role of *M. tuberculosis*-infected alveolar macrophages in the context of circadian rhythmicity deserves further investigation.

14.6 Systemic Metabolism and Cell Metabolism in the Context of Tuberculosis

Diabetes mellitus (DM) is a severe and chronic metabolic disorder characterized by high glucose concentration in serum due to the inability to generate good-quality insulin (Kumar et al. 2024).

Monocytes/macrophages exposed to a high concentration of glucose for prolonged periods have diminished phagocytosis and NO and ROS production, simultaneously with enhanced production of IL-1β and TNF-α (Pavlou et al. 2018). Macrophages obtained from two different murine models of type 1 diabetes exhibit a pro-inflammatory profile associated with the higher expression of ACSL1 (long-chain acyl-CoA synthetase 1), the enzyme that catalyzes the thioesterification of fatty acids (Kanter et al. 2012). In addition, hyperlipidemia and hyperglycemia in type 2 diabetes leads to epigenetic modifications that favor a pro-inflammatory profile in macrophages (Ahmed et al. 2017).

Epidemiological studies have shown that DM is a risk factor for the development of complicated forms of TB (Chai et al. 2018). People with DM have about three times more probability of developing active TB and have more severe symptoms than non-diabetic TB patients (Jeon and Murray 2008). Studies on the Th1 and Th17 profiles in DM patients concomitantly coursing with active TB show an elevated response of $CD4^+$ T lymphocytes and increased Th1 and Th17-derived cytokines in plasma (Kumar et al. 2013). In contrast, patients with DM or pre-DM coursing with latent *M. tuberculosis* infection show a diminished Th1 and Th17 profile, thus suggesting an increased risk of developing active TB (Kumar et al. 2014).

Peritoneal macrophages from diabetic mice (high-fat diet plus streptozotocin), and hyperglycemic THP-1-derived macrophages show impaired ROS and NO production in response to mycobacteria and the activation (TLR-4) and differentiation (CD11b and CD11c) markers are also affected, in addition to a higher production of anti-inflammatory cytokines (Chaubey et al. 2024).

Because obesity is a risk factor for type 2 diabetes (Narayan et al. 2007), it would follow that obesity is also a risk factor for TB. However, there is evidence that overweight and obese individuals are somehow protected against TB, perhaps due to immunomodulatory functions of adipose tissue during *M. tuberculosis* infection and the influence of the nutritional status that highlights the complex relationship between DM, obesity, and TB (Badawi et al. 2020; Chai et al. 2018; Yen et al. 2017).

Cellular metabolism particularly that of macrophages, is an essential factor to consider when discussing TB. *M. tuberculosis* tends to subvert host immune defenses by interfering with the metabolic system thus, any attempt to understand TB has to take macrophage metabolism into account. In this regard, macrophage polarization is a biochemical process that implies metabolic reprogramming (Cheng et al. 2022). According to recent studies, polarized macrophages employ different metabolic pathways. Of note, M1 macrophages preferentially engage in glycolysis and have an interrupted Krebs cycle. In contrast, M2 macrophages privilege fatty acid oxidation and OXPHOS (O'Neill and Pearce 2016; Stunault et al. 2018).

In addition, metabolic substrate transporters such as glucose transporters (GLUT family), fatty acids transporters (FATP family), and amino acids transporters, amongst others, help to drive macrophage polarization. Higher expression of Glut transporters (GLUT-1 and GLUT-3) leads to an increase in the glycolytic rate and the pentose phosphate pathway that, together, have a direct impact on the induction of pro-inflammatory mediators (TNF-α and IL-6) and increased ROS production, characteristic of M1 macrophages (Table 14.1) (Stunault et al. 2018; Cheng et al. 2022).

M1 macrophages accumulate iso-citrate, which results in increased synthesis of itaconate. Itaconate has antibacterial properties, and it is also an inhibitor of succinate dehydrogenase. The action of itaconate drives the accumulation of malate and the induction of the aspartate argininosuccinate shunt, necessary for the synthesis of NO (Jha et al. 2015; Sheedy and Divangahi 2021; O'Neill and Pearce 2016). Moreover, succinate accumulation stabilizes HIF-1α which results in the binding of HIF-1 to the promoters and the consequent production of IL-1β (Table 14.1) (Tannahill et al. 2013).

At the other side of the M1/M2 macrophage metabolite profile is glutaminolysis, in which glutamine is converted into glutamate and this to α-ketoglutarate (α-KG) which integrates into the Krebs cycle. The α-KG functions as an anti-inflammatory metabolite and it is related to M2 macrophage polarization, through the α-KG-Jumonji domain-containing 3 (Jmjd3)- pathway (Stunault et al. 2018; Yang et al. 2024).

Table 14.1 Key characteristics of M1, M2 and Foam macrophages during tuberculosis infection

Macrophage Phenotype	Stimulus	Cytokines	Metabolism	Role in Tuberculosis	Reference
Interstitial Macrophage (M1 phenotype)	IFN-γ TNF-α	IL-12 TNF-α IL1-β IL-8 GM-CSF IL1-α IL-17 IL-18 NO ROS CXCL11 CCL2 CCL13 CCL18 CCL22	↑ Glycolysis Warburg Effect- like ↑ Itaconate ↑ NO production ↑ ROS production Succinate release → induction IL1-β Citrate-derived fatty acid synthesis ↑ Pentose-phosphate pathway	Recognition of bacteria Pro-inflammatory activity Granuloma promotion Elimination of bacteria Antigen presentation	Khan et al. (2022), Sheedy and Divangahi (2021), Jha et al. (2015), O'Neill and Pearce (2016), Stunault et al. (2018)
Alveolar Macrophage (M2 phenotype)	IL-4 IL-1β IL-8 IL-10 IL-13 Glucocorticoids TLR-ligands	IL-4 IL-6 IL-10 TGF-β	↑ OXPHOS ↑ Fatty acids and β oxidation ↑Glutamine metabolism ↑ ARG1 →↓ NO production	Anti-inflammatory and immunomodulatory activity Reservoir of bacterial proliferation Wound healing	Khan et al. (2022), Stunault et al. (2018), Yang et al. (2024), Huang et al. (2014), Zhu et al. (2015)
Foam Macrophage	Mtb specific envelope compounds (oxygenated mycolic acids)	–	Altered lipid metabolism Enhanced synthesis of lipids	Mtb persistence in a dormant state Nutrient source for the bacteria	Silva Miranda et al. (2012), Sheedy and Divangahi (2021).

The macrophages are the main players in tuberculosis. It is widely described the multiple roles of these cells during the immunopathogenesis of the disease. Their metabolism and profile of cytokine secretion varies according with their phenotype

The M2 macrophages preferentially use the OXPHOS-dependent metabolism, without modification in the Krebs cycle. In addition, the HIF-2α-induced activation leads to the expression of the enzyme arginase 1 (ARG1) which inhibits the synthesis of NO and estimulates the production of polyamines; these promote the production of collagen and tissue remodelling. The metabolism of lipids (fatty acids) in M2 macrophages also has a role in the phagocytosis mechanism, since lipids function as the energetic pool of these cells and determine the cell membrane fluidity (Table 14.1) (Huang et al. 2014; Zhu et al. 2015). An important subpopulation of M2 macrophages is the foam macrophages, characteristic of TB granulomas, these cells undergo evident changes in their lipidic profile that is responsible for their foamy appearance. The foam macrophages are enriched in triglycerides and contain intracellular lipid droplets in which *M. tuberculosis* resides (Sheedy and Divangahi 2021).

In *in vitro* models, alveolar and interstitial macrophages undergo different metabolic programs in the presence of *M. tuberculosis*. M1 lung macrophages are glycolytically active, nutritional restrictive, and do not allow efficient intracellular growth of the pathogen (Table 14.1). In contrast, the M2 lung macrophages are characterized by an up-regulation of fatty acids and β oxidation pathways, providing an environment rich in triacylglycerols, cholesterol, and fatty acids that favor bacterial growth (Table 14.1) (Huang et al. 2018). The anti-*M. tuberculosis* activity of macrophages is related to HIF-1α, which coordinates the metabolic shift to increased glycolysis and NO, IL-1β, and PGE2 production in IFN-γ activated macrophages (Braverman et al. 2016) (Fig. 14.2.).

In vivo M. tuberculosis infection metabolically reprograms the infected macrophages. In this regard, interstitial macrophages down-regulate glycolysis, which depends on bacterial virulence and the induction of the miR-21. The expression of miR-21 correlates with an anti-inflammatory state that includes the regulation of phosphofructokinase (PFK-M) activity and limiting IL-1β production, all of which allows for *M. tuberculosis* immune evasion (Hackett et al. 2020) (Fig. 14.2.).

The M1 macrophages initiate the granuloma formation during the acute phase of infection, whereas in the chronic phase, the M2 macrophages undergo a metabolic shift towards mitochondrial OXPHOS and glutamine metabolism, with an upregulation of COX-1 (Ahmad et al. 2022). In addition, *M. tuberculosis*-infected M2 macrophages produce higher amounts of NAD^+, creatine, creatine phosphate, and glutathione, which limits the production of cytokines (Vrieling et al. 2020).

Another metabolite relevant for macrophage polarization is lactate. This metabolite constitutes an inflammatory signal related to the M1 macrophage phenotype, since lactate, in coordination with TLR-4-mediated signaling, increases the expression of inflammatory genes. In contrast, lactate acts as a modulator of inflammation for M2 macrophages, negatively regulating glycolysis and inducing the expression of genes related to wound healing (Cheng et al. 2022) (Fig. 14.2.).

Equally relevant to macrophage polarization is the metabolism of fatty acids. The IL-4-mediated differentiation towards the M2 macrophage phenotype involves the activity of the transcription factor PPAR-γ (peroxisome proliferator-activated receptor-γ), which together with its coactivator PGC1β results in a substantial increase of fatty acid oxidation and mitochondrial biogenesis (Cheng et al. 2022).

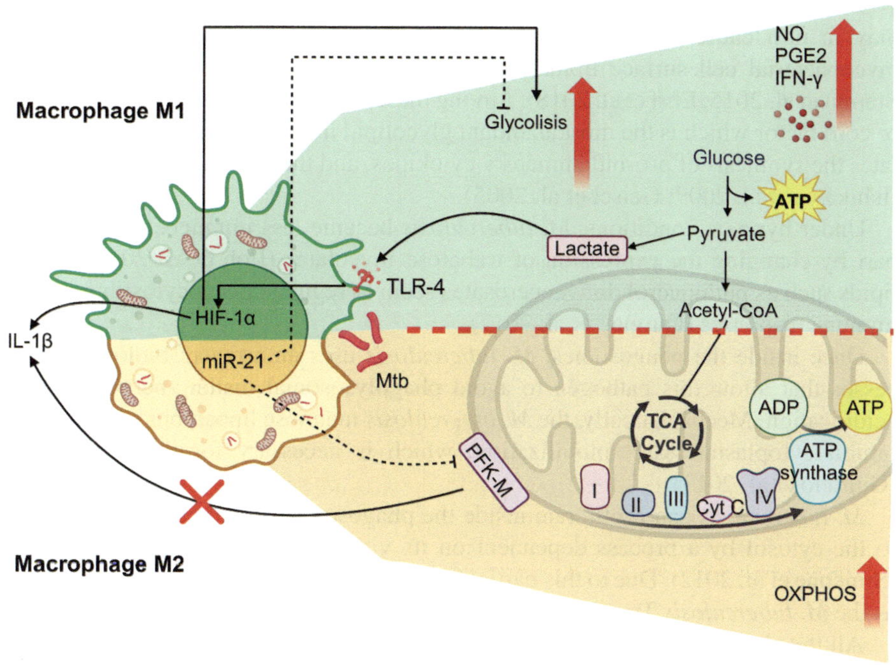

Fig. 14.2. Macrophages M1 and M2 metabolism in tuberculosis context. The infection of macrophages by Mtb induces their polarization to either the M1 or M2 profile. M1 macrophages primarily utilize a glycolytic metabolism. The presence of lactate, the final product of glycolysis, stimulates the transcription factor HIF-1α through TLR-4 signalling. As a result, a pro-inflammatory profile is established (IL-1β, NO, PGE2, IFN-γ). Conversely, M2 macrophages prioritize an oxidative phosphorylation metabolism. The presence of Mtb induces the expression of miR-21, which inhibits glycolysis and PFK-M activity, thereby reducing IL-1β concentration. Abbreviations: *Mtb Mycobacterium tuberculosis*, *HIF-1α* Hypoxia-Inducible Factor α, *NO* Nitric Oxide, *IL-1β* Interleukin 1β, *PGE2* Prostaglandin E2, *IFN-γ* Interferon γ, *miR-21* microRNA 21, *PFK-M* 6-phosphofructokinase, muscle type

14.7 *M. tuberculosis* Metabolism

The most abundant components in *M. tuberculosis* are complex lipids, which constitute approximately 60% of their total mass (Singh et al. 2018). *M. tuberculosis* can adapt its metabolism to the host cell microenvironment thus, because lipids are predominant constituents, it is not surprising that lipid metabolism is fundamental for bacterial survival and pathogenicity. Eoh and colleagues have pointed out that *M. tuberculosis* adaptation to its host cell includes changes in the catabolism of mycobacterial trehalose mycolates, which facilitate the initiation of *de novo* synthesis of peptidoglycan. Of note, the inhibition of peptidoglycan synthesis is essential for bacterial cell cycle re-entry. This highlights the existence of an *M. tuberculosis* metabolism-dependent regulatory response for mycobacterial cell cycle exit and re-entry (Eoh et al. 2017).

On the other hand, *M. tuberculosis* can induce an inflammatory response that may in turn cause tissue necrosis, facilitating its transmission. In this process, mycobacterial cell surface lipids play a determinant role (Ishikawa et al. 2009; Stamm et al. 2015; Ehrt et al. 2018). Among these, trehalose 6,6′-dimycolate (TDM) or cord factor which is the most abundant glycolipid in the bacteria cell wall, stimulates the synthesis of pro-inflammatory cytokines, and the formation of granulomas (Ishikawa et al. 2009; Geisel et al. 2005).

Under hypoxia conditions, *M. tuberculosis* became less immunostimulatory, in part by changing the expression of trehalose mycolates (Eoh et al. 2017). Other lipids such as phthiocerol dimycocerosates seem to regulate phagocytosis and other immune responses (Quigley et al. 2017).

Once inside the phagosomes, *M. tuberculosis* uses an array of biological processes that allow this pathogen to avoid phagolysosomal fusion and phagosome acidification. Mechanistically, the *M. tuberculosis* mannose lipoarabinomannan can inhibit cytoplasmic Ca^{2+} mobilization, which is necessary for these processes (Guirado et al. 2013).

M. tuberculosis can proliferate inside the phagosomes and to escape from them to the cytosol by a process dependent on its virulence factor ESAT-6 and RD-1 (Simeone et al. 2012). Due to this particularity, the macrophages have been regarded as the *M. tuberculosis* Trojan horses (Nguyen and Pieters 2005).

All the above offers a glimpse of the *M. tuberculosis* metabolism and its role in host-pathogen interaction. For a detailed account of the bacterial lipid metabolism and its role in pathogenicity and immune response, several comprehensive reviews are available; of note that of Eoh et al. (2017).

A main challenge in the effort to overcome TB is the existence of multidrug-resistance *M. tuberculosis* strains. Therefore, the discovery of biomarkers that would allow their identification is mandatory. In this regard, gas chromatography/mass spectrometry (GC/MS)-based total metabolome and fatty acid metabolome was performed on two rpoB mutant *M. tuberculosis* drug-resistant strains (S522L and S531L) and on a drug-susceptible wild-type strain. The results showed that drug resistant bacteria use alkanes and fatty acids as a carbon/energy source and synthesize antioxidant compounds such as ascorbic acid (Loots 2014). This could help the rapid identification of multidrug-resistant *M. tuberculosis* strains.

14.8 An Integrated View on Tuberculosis, Macrophages, Metabolism, Mitochondria, Circadian Rhythmicity and the Pathogen

This chapter has attempted to provide an integrated view of the central role that macrophages play in TB, how in the host-pathogen interaction metabolism is at play at the cellular and the systemic levels, and how the metabolism of both *M. tuberculosis* and the host metabolism interact to each other, making the metabolic reprograming a driving force for pathogenicity on the one hand, and infection control on the other. Since timing is also a primordial variable, circadian rhythmicity must be considered when addressing TB. Finally, we have tried to convey the view that mitochondria constitute a central hub that coordinates diverse mechanism at play during the *M. tuberculosis* infection of macrophages (Aguilar-López et al. 2020).

14.9 What Is Ahead?

The need to make diagnosis and treatment of TB more effective and timely has met the OMICS (genomics, transcriptomics, proteomics, metabolomics, lipidomics, glycomics, and glycoproteomics) technological revolution, and these approaches are currently helping to elucidate the mechanisms of *M. tuberculosis* pathogenicity, and the identification of new pharmacological targets and biomarkers (Ahamad et al. 2022). The discovery of TB biomarkers using metabolomics is currently a fast-moving field (Hu et al. 2022), and a variety of biological samples have been used, such as sputum, plasma, serum, urine, and breath (Preez et al. 2017).

Plasma samples from newly diagnosed TB patients and their asymptomatic contacts have been analyzed using liquid chromatography high-resolution mass spectrometry (LC-MS), in search of differential metabolic profiles. In this way, 61 metabolites were found to be significantly distinctive between TB patients and contacts, including specific D-series resolvins, glutamate, and *Mycobacterium tuberculosis*-derived trehalose-6-mycolate; these metabolites are potential biomarkers for a prompt and accurate diagnosis (Frediani et al. 2014).

In another study, the metabolomic profiles of plasma from 27 TB patients were compared to those of the plasma from 37 smear-negative pulmonary tuberculosis patients and the respective controls. A distinctive metabolic profile (mainly fatty acids and amino acid composition) was observed in TB patients as compared to healthy individuals (Hu et al. 2022).

The use of an automated thermal desorption gas chromatography/mass spectrometry (GC–MS), and breath as the biological samples showed a differential pattern in at least 130 constituents (mostly alkanes, naphthalene, and benzene derivatives) between TB patients and negative controls, with a sensitivity of 84% and specificity of 64.7% (Phillips et al. 2010).

Nuclear Magnetic Resonance techniques performed on urine samples, have allowed the identification of metabolic differences between pediatric TB patients and healthy individuals (Comella-del-Barrio et al. 2021). Metabolic differences in the urine from active TB and latent TB patients have been observed, specifically in the content of glutathione and histamine. These molecules could be used as biomarkers to assess the course of TB infection (Deng et al. 2021).

High-resolution metabolomics, lipidomics and cytokine profiling on plasma samples have proved to be very useful approaches to characterize tuberculosis caused by multidrug or susceptible strains of *M. tuberculoisis*. Of note, differences in TCA cycle intermediates related to the synthesis of IL-1β correlate with pathogen persistence (Collins et al. 2021).

Hopefully, the target set by the STOP TB initiative (stoptb.org) (UNOPS 2024) will be reached and humans will be free of TB by 2030, and the tremendous success of that initiative will promote global partnerships and health policies, and scientific breakthroughs to fight other infectious diseases. By then, even if only four of the 16 vaccine candidates under clinical trials in August 2023 (four in Phase I, eight in Phase II and four in Phase III) (WHO 2023) are released to the public, they will pave the way to global anti-TB protection in the years to come. In parallel, new antibiotics will solve the problem of multidrug-resistant *M. tuberculosis* strains.

By then, even if TB does no longer pose a threat, there will be a much better understanding of the complex *M. tuberculosis*-host interaction.

References

Abarca-Rojano E, Rosas-Medina P, Zamudio-Cortéz P, Mondragón-Flores R, Sánchez-García FJ (2003) *Mycobacterium tuberculosis* virulence correlates with mitochondrial cytochrome c release in infected macrophages. Scand J Immunol 58:419–427

Aguilar-López BA, Correa F, Moreno-Altamirano MMB, Espitia C, Hernández-Longoria R, Oliva-Ramírez J, Padierna-Olivos J, Sánchez-García FJ (2019) LprG and PE_PGRS33 *Mycobacterium tuberculosis* virulence factors induce differential mitochondrial dynamics in macrophages. Scand J Immunol 89:e12728

Aguilar-López BA, Moreno-Altamirano MMB, Dockrell HM, Duchen MR, Sánchez-García FJ (2020) Mitochondria: an integrative hub coordinating circadian rhythms, metabolism, the microbiome, and immunity. Front Cell Dev Biol 8:51

Ahamad N, Gupta S, Parashar D (2022) Using omics to study leprosy, tuberculosis, and other mycobacterial diseases. Front Cell Infect Microbiol 12:792617

Ahmad F, Rani A, Alam A, Zarin S, Pandey S, Singh H, Hasnain SE, Ehtesham NZ (2022) Macrophage: a cell with many faces and functions in tuberculosis. Front Immunol 13:747799

Ahmed M, De Winther MPJ, van den Bossche J (2017) Epigenetic mechanisms of macrophage activation in type 2 diabetes. Immunobiology 222:937–943

Badawi A, Gregg B, Vasileva D (2020) Systematic analysis for the relationship between obesity and tuberculosis. Public Health 186:246–256

BoseDasgupta S, Pieters J (2018) Macrophage-microbe interaction: lessons learned from the pathogen *Mycobacterium tuberculosis*. Semin Immunopathol 40:577–591

Bowdish DM, Sakamoto K, Kim MJ, Kroos M, Mukhopadhyay S, Leifer CA, Tryggvason K, Gordon S, Russell DG (2009) MARCO, TLR2, and CD14 are required for macrophage cytokine responses to mycobacterial trehalose dimycolate and *Mycobacterium tuberculosis*. PLoS Pathog 5:e1000474

Braverman J, Sogi KM, Benjamin D, Nomura DK, Stanley SA (2016) HIF-1alpha is an essential mediator of IFN-gamma-dependent immunity to *Mycobacterium tuberculosis*. J Immunol 197:1287–1297

Cadieux N, Parra M, Cohen H, Maric D, Morris SL, Brennan MJ (2011) Induction of cell death after localization to the host cell mitochondria by the *Mycobacterium tuberculosis* PE_PGRS33 protein. Microbiology (Reading) 157:793–804

Cambier CJ, Falkow S, Ramakrishnan L (2014) Host evasion and exploitation schemes of *Mycobacterium tuberculosis*. Cell 159:1497–1509

Canton M, Sánchez-Rodríguez R, Spera I, Venegas FC, Favia M, Viola A, Castegna A (2021) Reactive oxygen species in macrophages: sources and targets. Front Immunol 12:734229

Chai Q, Zhang Y, Liu CH (2018) *Mycobacterium tuberculosis:* an adaptable pathogen associated with multiple human diseases. Front Cell Infect Microbiol 8:158

Chakraborty S, Singh A, Wang L, Wang X, Sanborn MA, Ye Z, Maienschein-Cline M, Mukhopadhyay A, Ganesh BB, Malik AB, Rehman J (2023) Trained immunity of alveolar macrophages enhances injury resolution via KLF4-MERTK-mediated efferocytosis. J Exp Med 220

Chaubey GK, Modanwal R, Dilawari R, Talukdar S, Dhiman A, Chaudhary S, Patidar A, Raje CI, Raje M (2024) Chronic hyperglycemia impairs anti-microbial function of macrophages in response to *Mycobacterium tuberculosis* infection. Immunol Res

Cheng J, Cai W, Zong S, Yu Y, Wei F (2022) Metabolite transporters as regulators of macrophage polarization. Naunyn Schmiedebergs Arch Pharmacol 395:13–25

Clark IA, Allison AC, Cox FE (1976) Protection of mice against *Babesia* and *Plasmodium* with BCG. Nature 259:309–311

Collins JM, Jones DP, Sharma A, Khadka M, Liu KH, Kempker RR, Prideaux B, Maner-Smith K, Tukvadze N, Shah NS, Brust JCM, Sekaly RP, Gandhi NR, Blumberg HM, Ortlund EA, Ziegler TR (2021) TCA cycle remodeling drives proinflammatory signaling in humans with pulmonary tuberculosis. PLoS Pathog 17:e1009941

Comella-del-Barrio P, Izquierdo-García JL, Gautier J, Doresca MJC, Campos-Olivas R, Santiveri CM, Muriel-Moreno B, Prat-Aymerich C, Abellana R, Pérez-Porcuna TM, Cuevas LE, Ruíz-Cabello J, Domínguez J (2021) Urine NMR-based TB metabolic fingerprinting for the diagnosis of TB in children. Sci Rep 11:12006

Cronan MR, Beerman RW, Rosenberg AF, Saelens JW, Johnson MG, Oehlers SH, Sisk DM, Jurcic Smith KL, Medvitz NA, Miller SE, Trinh LA, Fraser SE, Madden JF, Turner J, Stout JE, Lee S, Tobin DM (2016) Macrophage epithelial reprogramming underlies mycobacterial granuloma formation and promotes infection. Immunity 45:861–876

Daniel J, Maamar H, Deb C, Sirakova TD, Kolattukudy PE (2011) *Mycobacterium tuberculosis* uses host triacylglycerol to accumulate lipid droplets and acquires a dormancy-like phenotype in lipid-loaded macrophages. PLoS Pathog 7:e1002093

de Bree LCJ, Mourits VP, Koeken VA, Moorlag SJ, Janssen R, Folkman L, Barreca D, Krausgruber T, Fife-Gernedl V, Novakovic B, Arts RJ, Dijkstra H, Lemmers H, Bock C, Joosten LA, Van Crevel R, Benn CS, Netea MG (2020) Circadian rhythm influences induction of trained immunity by BCG vaccination. J Clin Invest 130:5603–5617

de Sousa JR, Lucena Neto FD, Sotto MN, Quaresma JAS (2018) Immunohistochemical characterization of the M4 macrophage population in leprosy skin lesions. BMC Infect Dis 18:576

Deng J, Liu L, Yang Q, Wei C, Zhang H, Xin H, Pan S, Liu Z, Wang D, Liu B, Gao L, Liu R, Pang Y, Chen X, Zheng J, Jin Q (2021) Urinary metabolomic analysis to identify potential markers for the diagnosis of tuberculosis and latent tuberculosis. Arch Biochem Biophys 704:108876

Ehrt S, Schnappinger D, Rhee KY (2018) Metabolic principles of persistence and pathogenicity in *Mycobacterium tuberculosis*. Nat Rev Microbiol 16:496–507

Ellzey LM, Patrick KL, Watson RO (2023) Mitochondrial reactive oxygen species: double agents in *Mycobacterium tuberculosis* infection. Curr Opin Immunol 84:102366

Eoh H, Wang Z, Layre E, Rath P, Morris R, Branch Moody D, Rhee KY (2017) Metabolic anticipation in *Mycobacterium tuberculosis*. Nat Microbiol 2:17084

Floyd K, Glaziou P, Zumla A, Raviglione M (2018) The global tuberculosis epidemic and progress in care, prevention, and research: an overview in year 3 of the End TB era. Lancet Respir Med 6:299–314

Frediani JK, Jones DP, Tukvadze N, Uppal K, Sanikidze E, Kipiani M, Tran VT, Hebbar G, Walker DI, Kempker RR, Kurani SS, Colas RA, Dalli J, Tangpricha V, Serhan CN, Blumberg HM, Ziegler TR (2014) Plasma metabolomics in human pulmonary tuberculosis disease: a pilot study. PLoS One 9:e108854

Ge G, Jiang H, Xiong J, Zhang W, Shi Y, Tao C, Wang H (2021) Progress of the art of macrophage polarization and different subtypes in mycobacterial infection. Front Immunol 12:752657

Geisel RE, Sakamoto K, Russell DG, Rhoades ER (2005) *In vivo* activity of released cell wall lipids of *Mycobacterium bovis* bacillus Calmette-Guerin is due principally to trehalose mycolates. J Immunol 174:5007–5015

Gordon S, Pluddemann A (2017) Tissue macrophages: heterogeneity and functions. BMC Biol 15:53

Gordon S, Taylor PR (2005) Monocyte and macrophage heterogeneity. Nat Rev Immunol 5:953–964

Guirado E, Schlesinger LS, Kaplan G (2013) Macrophages in tuberculosis: friend or foe. Semin Immunopathol 35:563–583

Gutierrez MG, Master SS, Singh SB, Taylor GA, Colombo MI, Deretic V (2004) Autophagy is a defense mechanism inhibiting BCG and *Mycobacterium tuberculosis* survival in infected macrophages. Cell 119:753–766

Hackett EE, Charles-Messance H, O'leary SM, Gleeson LE, Muñoz-Wolf N, Case S, Wedderburn A, Johnston DGW, Williams MA, Smyth A, Ouimet M, Moore KJ, Lavelle EC, Corr SC, Gordon SV, Keane J, Sheedy FJ (2020) *Mycobacterium tuberculosis* limits host glycolysis and IL-1 beta by restriction of PFK-M via MicroRNA-21. Cell Rep 30(124–136):e4

Halberg F, Halberg E, Barnum CP, Bittner J (1959) Physiologic 24-hour periodicity in human beings and mice, the lighting regimen and daily routine. In: Withrow RB (ed) Photoperiodism and related phenomena in plants and animals. AAAS, Washington, DC

Haspel JA, Chettimada S, Shaik RS, Chu JH, Raby BA, Cernadas M, Carey V, Process V, Hunninghake GM, Ifedigbo E, Lederer JA, Englert J, Pelton A, Coronata A, Fredenburgh LE, Choi AM (2014) Circadian rhythm reprogramming during lung inflammation. Nat Commun 5:4753

Haspel JA, Anafi R, Brown MK, Cermakian N, Depner C, Desplats P, Gelman AE, Haack M, Jelic S, Kim BS, Laposky AD, Lee YC, Mongodin E, Prather AA, Prendergast BJ, Reardon C, Shaw AC, Sengupta S, Szentirmai E, Thakkar M, Walker WE, Solt LA (2020) Perfect timing: circadian rhythms, sleep, and immunity - an NIH workshop summary. JCI Insight 5

Hu X, Wang J, Ju Y, Zhang X, Qimanguli W, Li C, Yue L, Tuohetaerbaike B, Li Y, Wen H, Zhang W, Chen C, Yang Y, Wang J, Chen F (2022) Combining metabolome and clinical indicators with machine learning provides some promising diagnostic markers to precisely detect smear-positive/negative pulmonary tuberculosis. BMC Infect Dis 22:707

Huang L, Nazarova EV, Tan S, Liu Y, Russell DG (2018) Growth of *Mycobacterium tuberculosis in vivo* segregates with host macrophage metabolism and ontogeny. J Exp Med 215:1135–1152

Huang SC, Everts B, Ivanova Y, O'sullivan D, Nascimento M, Smith AM, Beatty W, Love-Gregory L, Lam WY, O'Neill CM, Yan C, Du H, Abumrad NA, Urban JF Jr, Artyomov MN, Pearce EL, Pearce EJ (2014) Cell-intrinsic lysosomal lipolysis is essential for alternative activation of macrophages. Nat Immunol 15:846–855

Huang Z, Luo Q, Guo Y, Chen J, Xiong G, Peng Y, Ye J, Li J (2015) *Mycobacterium tuberculosis*-induced polarization of human macrophage orchestrates the formation and development of tuberculous granulomas *In Vitro*. PLoS One 10:e0129744

Ishikawa E, Ishikawa T, Morita YS, Toyonaga K, Yamada H, Takeuchi O, Kinoshita T, Akira S, Yoshikai Y, Yamasaki S (2009) Direct recognition of the mycobacterial glycolipid, trehalose dimycolate, by C-type lectin Mincle. J Exp Med 206:2879–2888

Jackaman C, Yeoh TL, Acuil ML, Gardner JK, Nelson DJ (2016) Murine mesothelioma induces locally-proliferating IL-10(+)TNF-alpha(+)CD206(-)CX3CR1(+) M3 macrophages that can be selectively depleted by chemotherapy or immunotherapy. Oncoimmunology 5:e1173299

Jeon CY, Murray MB (2008) Diabetes mellitus increases the risk of active tuberculosis: a systematic review of 13 observational studies. PLoS Med 5:e152

Jha AK, Huang SC, Sergushichev A, Lampropoulou V, Ivanova Y, Loginicheva E, Chmielewski K, Stewart KM, Ashall J, Everts B, Pearce EJ, Driggers EM, Artyomov MN (2015) Network integration of parallel metabolic and transcriptional data reveals metabolic modules that regulate macrophage polarization. Immunity 42:419–430

Kalish S, Lyamina S, Manukhina E, Malyshev Y, Raetskaya A, Malyshev I (2017) M3 macrophages stop division of tumor cells in vitro and extend survival of mice with ehrlich ascites carcinoma. Med Sci Monit Basic Res 23:8–19

Kanter JE, Kramer F, Barnhart S, Averill MM, Vivekanandan-Giri A, Vickery T, Li LO, Becker L, Yuan W, Chait A, Braun KR, Potter-Perigo S, Sanda S, Wight TN, Pennathur S, Serhan CN, Heinecke JW, Coleman RA, Bornfeldt KE (2012) Diabetes promotes an inflammatory macrophage phenotype and atherosclerosis through acyl-CoA synthetase 1. Proc Natl Acad Sci U S A 109:E715–E724

Khan A, Singh VK, Hunter RL, Jagannath C (2019) Macrophage heterogeneity and plasticity in tuberculosis. J Leukoc Biol 106:275–282

Khan A, Zhang K, Singh VK, Mishra A, Kachroo P, Bing T, Won JH, Mani A, Papanna R, Mann LK, Ledezma-Campos E, Aguillon-Duran G, Canaday DH, David SA, Restrepo BI, Viet NN, Phan H, Graviss EA, Musser JM, Kaushal D, Gauduin MC, Jagannath C (2022) Human M1 macrophages express unique innate immune response genes after mycobacterial infection to defend against tuberculosis. Commun Biol 5:480

Kim TS, Jin YB, Kim YS, Kim S, Kim JK, Lee HM, Suh HW, Choe JH, Kim YJ, Koo BS, Kim HN, Jung M, Lee SH, Kim DK, Chung C, Son JW, Min JJ, Kim JM, Deng CX, Kim HS, Lee SR, Jo EK (2019) SIRT3 promotes antimycobacterial defenses by coordinating mitochondrial and autophagic functions. Autophagy 15:1356–1375

Kleinnijenhuis J, Quintin J, Preijers F, Joosten LA, Ifrim DC, Saeed S, Jacobs C, Van Loenhout J, De Jong D, Stunnenberg HG, Xavier RJ, van der Meer JW, van Crevel R, Netea MG (2012) Bacille Calmette-Guerin induces NOD2-dependent nonspecific protection from reinfection via epigenetic reprogramming of monocytes. Proc Natl Acad Sci U S A 109:17537–17542

Krishnan V, Nath S, Nair P, Das B (2023) *Mycobacterium tuberculosis* and its clever approaches to escape the deadly macrophage. World J Microbiol Biotechnol 39:300

Kumar A, Gangwar R, Zargar AA, Kumar R, Sharma A (2024) Prevalence of diabetes in india: a review of IDF diabetes atlas 10th edition. Curr Diabetes Rev 20:e130423215752

Kumar NP, Sridhar R, Banurekha VV, Jawahar MS, Nutman TB, Babu S (2013) Expansion of pathogen-specific T-helper 1 and T-helper 17 cells in pulmonary tuberculosis with coincident type 2 diabetes mellitus. J Infect Dis 208:739–748

Kumar NP, George PJ, Kumaran P, Dolla CK, Nutman TB, Babu S (2014) Diminished systemic and antigen-specific type 1, type 17, and other proinflammatory cytokines in diabetic and prediabetic individuals with latent *Mycobacterium tuberculosis* infection. J Infect Dis 210:1670–1678

Liu J, Geng X, Hou J, Wu G (2021) New insights into M1/M2 macrophages: key modulators in cancer progression. Cancer Cell Int 21:389

Lo CY, Huang YC, Huang HY, Chung FT, Lin CW, Chung KF, Wang CH (2021) Increased Th1 cells with disease resolution of active pulmonary tuberculosis in non-atopic patients. Biomedicines 9

Loddenkemper R, Lipman M, Zumla A (2015) Clinical aspects of adult tuberculosis. Cold Spring Harb Perspect Med 6:a017848

Loots DT (2014) An altered *Mycobacterium tuberculosis* metabolome induced by katG mutations resulting in isoniazid resistance. Antimicrob Agents Chemother 58:2144–2149

Mackaness GB (1962) Cellular resistance to infection. J Exp Med 116:381–406

Maglione PJ, Xu J, Casadevall A, Chan J (2008) Fc gamma receptors regulate immune activation and susceptibility during *Mycobacterium tuberculosis* infection. J Immunol 180:3329–3338

Makuch M, Stepanechko M, Bzowska M (2024) The dance of macrophage death: the interplay between the inevitable and the microenvironment. Front Immunol 15:1330461

Malik ZA, Denning GM, Kusner DJ (2000) Inhibition of Ca(2+) signaling by *Mycobacterium tuberculosis* is associated with reduced phagosome-lysosome fusion and increased survival within human macrophages. J Exp Med 191:287–302

Manabe YC, Dannenberg AM Jr, Tyagi SK, Hatem CL, Yoder M, Woolwine SC, Zook BC, Pitt ML, Bishai WR (2003) Different strains of *Mycobacterium tuberculosis* cause various spectrums of disease in the rabbit model of tuberculosis. Infect Immun 71:6004–6011

Mantovani A, Sica A, Sozzani S, Allavena P, Vecchi A, Locati M (2004) The chemokine system in diverse forms of macrophage activation and polarization. Trends Immunol 25:677–686

Marakalala MJ, Raju RM, Sharma K, Zhang YJ, Eugenin EA, Prideaux B, Daudelin IB, Chen PY, Booty MG, Kim JH, Eum SY, Via LE, Behar SM, Barry CE 3rd, Mann M, Dartois V, Rubin EJ (2016) Inflammatory signaling in human tuberculosis granulomas is spatially organized. Nat Med 22:531–538

Marakalala MJ, Martínez FO, Pluddemann A, Gordon S (2018) Macrophage Heterogeneity in the Immunopathogenesis of Tuberculosis. Front Microbiol 9:1028

Martin CJ, Booty MG, Rosebrock TR, Nunes-Alves C, Desjardins DM, Keren I, Fortune SM, Remold HG, Behar SM (2012) Efferocytosis is an innate antibacterial mechanism. Cell Host Microbe 12:289–300

Martinez FO, Gordon S (2014) The M1 and M2 paradigm of macrophage activation: time for reassessment. F1000Prime Rep 6:13

Mattila JT, Ojo OO, Kepka-Lenhart D, Marino S, Kim JH, Eum SY, Via LE, Barry CE 3rd, Klein E, Kirschner DE, Morris SM Jr, Lin PL, Flynn JL (2013) Microenvironments in tuberculous granulomas are delineated by distinct populations of macrophage subsets and expression of nitric oxide synthase and arginase isoforms. J Immunol 191:773–784

Mills CD, Kincaid K, Alt JM, Heilman MJ, Hill AM (2000) M-1/M-2 macrophages and the Th1/Th2 paradigm. J Immunol 164:6166–6173

Mitsi E, Kamng'ona R, Rylance J, Solorzano C, Jesus Reine J, Mwandumba HC, Ferreira DM, Jambo KC (2018) Human alveolar macrophages predominately express combined classical M1 and M2 surface markers in steady state. Respir Res 19:66

Mosser DM, Edwards JP (2008) Exploring the full spectrum of macrophage activation. Nat Rev Immunol 8:958–969

Nakahira K, Hisata S, Choi AM (2015) The roles of mitochondrial damage-associated molecular patterns in diseases. Antioxid Redox Signal 23:1329–1350

Nakai K (2021) Multiple roles of macrophage in skin. J Dermatol Sci 104:2–10

Narayan KM, Boyle JP, Thompson TJ, Gregg EW, Williamson DF (2007) Effect of BMI on lifetime risk for diabetes in the U.S. Diabetes Care 30:1562–1566

Natarajan A, Beena PM, Devnikar AV, Mali S (2020) A systemic review on tuberculosis. Indian J Tuberc 67:295–311

Neyen CD, Gordon S (2014) Macrophages in lipid and immune homeostasis. In: John Wiley & Sons L (ed) Encyclopedia of life sciences

Nguyen L, Pieters J (2005) The Trojan horse: survival tactics of pathogenic mycobacteria in macrophages. Trends Cell Biol 15:269–276

Ning Y, Cai Y, Dai Y, Li F, Mo S, Werz O, Chen X (2021) Mitochondrial fusion mediated by Mitofusin 1 regulates macrophage mycobactericidal activity by enhancing autophagy. Infect Immun 89:e0030621

O'Neill LA, Pearce EJ (2016) Immunometabolism governs dendritic cell and macrophage function. J Exp Med 213:15–23

Oliva-Ramírez J, Moreno-Altamirano MM, Pineda-Olvera B, Cauich-Sánchez P, Sánchez-García FJ (2014) Crosstalk between circadian rhythmicity, mitochondrial dynamics and macrophage bactericidal activity. Immunology 143:490–497

Palma FR, Gantner BN, Sakiyama MJ, Kayzuka C, Shukla S, Lacchini R, Cunniff B, Bonini MG (2024) ROS production by mitochondria: function or dysfunction? Oncogene 43:295–303

Patil NK, Bohannon JK, Vachharajani V, Mccall CE (2022) Editorial: the roles of mitochondria in immunity. Front Immunol 13:914639

Patrick KL, Watson RO (2021) Mitochondria: powering the innate immune response to *Mycobacterium tuberculosis* infection. Infect Immun 89

Pavlou S, Lindsay J, Ingram R, Xu H, Chen M (2018) Sustained high glucose exposure sensitizes macrophage responses to cytokine stimuli but reduces their phagocytic activity. BMC Immunol 19:24

Phillips M, Basa-Dalay V, Bothamley G, Cataneo RN, Lam PK, Natividad MP, Schmitt P, Wai J (2010) Breath biomarkers of active pulmonary tuberculosis. Tuberculosis (Edinb) 90:145–151

Preez ID, Luies L, Loots DT (2017) Metabolomics biomarkers for tuberculosis diagnostics: current status and future objectives. Biomark Med 11:179–194

Quigley J, Hughitt VK, Velikovsky CA, Mariuzza RA, El-Sayed NM, Briken V (2017) The cell wall lipid PDIM contributes to phagosomal escape and host cell exit of *Mycobacterium tuberculosis*. mBio, 8

Russell DG, Cardona PJ, Kim MJ, Allain S, Altare F (2009) Foamy macrophages and the progression of the human tuberculosis granuloma. Nat Immunol 10:943–948

Scheiermann C, Kunisaki Y, Frenette PS (2013) Circadian control of the immune system. Nat Rev Immunol 13:190–198

Schwer B, Verdin E (2008) Conserved metabolic regulatory functions of sirtuins. Cell Metab 7:104–112

Shams H, Wizel B, Lakey DL, Samten B, Vankayalapati R, Valdivia RH, Kitchens RL, Griffith DE, Barnes PF (2003) The CD14 receptor does not mediate entry of *Mycobacterium tuberculosis* into human mononuclear phagocytes. FEMS Immunol Med Microbiol 36:63–69

Sheedy FJ, Divangahi M (2021) Targeting immunometabolism in host defence against *Mycobacterium tuberculosis*. Immunology 162:145–159

Shen P, Li Q, Ma J, Tian M, Hong F, Zhai X, Li J, Huang H, Shi C (2017) IRAK-M alters the polarity of macrophages to facilitate the survival of *Mycobacterium tuberculosis*. BMC Microbiol 17:185

Shi L, Jiang Q, Bushkin Y, Subbian S, Tyagi S (2019) Biphasic dynamics of macrophage immunometabolism during *Mycobacterium tuberculosi*s Infection. mBio, 10

Shirato K, Sato S (2022) Macrophage meets the circadian clock: implication of the circadian clock in the role of macrophages in acute lower respiratory tract infection. Front Cell Infect Microbiol 12:826738

Silva Miranda M, Breiman A, Allain S, Deknuydt F, Altare F (2012) The tuberculous granuloma: an unsuccessful host defence mechanism providing a safety shelter for the bacteria? Clin Dev Immunol 2012:139127

Simeone R, Bobard A, Lippmann J, Bitter W, Majlessi L, Brosch R, Enninga J (2012) Phagosomal rupture by *Mycobacterium tuberculosis* results in toxicity and host cell death. PLoS Pathog 8:e1002507

Singh P, Rameshwaram NR, Ghosh S, Mukhopadhyay S (2018) Cell envelope lipids in the pathophysiology of *Mycobacterium tuberculosis*. Future Microbiol 13:689–710

Smulan LJ, Martinez N, Kiritsy MC, Kativhu C, Cavallo K, Sassetti CM, Singhal A, Remold HG, Kornfeld H (2021) Sirtuin 3 downregulation in Mycobacterium tuberculosis-infected macrophages reprograms mitochondrial metabolism and promotes cell death. mBio, 12

Stamm CE, Collins AC, Shiloh MU (2015) Sensing of *Mycobacterium tuberculosi*s and consequences to both host and bacillus. Immunol Rev 264:204–219

Stunault MI, Bories G, Guinamard RR, Ivanov S (2018) Metabolism plays a key role during macrophage activation. Mediat Inflamm 2018:2426138

Tannahill GM, Curtis AM, Adamik J, Palsson-Mcdermott EM, Mcgettrick AF, Goel G, Frezza C, Bernard NJ, Kelly B, Foley NH, Zheng L, Gardet A, Tong Z, Jany SS, Corr SC, Haneklaus M, Caffrey BE, Pierce K, Walmsley S, Beasley FC, Cummins E, Nizet V, Whyte M, Taylor CT, Lin H, Masters SL, Gottlieb E, Kelly VP, Clish C, Auron PE, Xavier RJ, O'Neill LA (2013) Succinate is an inflammatory signal that induces IL-1beta through HIF-1alpha. Nature 496:238–242

Thapa B, Lee K (2019) Metabolic influence on macrophage polarization and pathogenesis. BMB Rep 52:360–372

Thiriot JD, Martinez-Martinez YB, Endsley JJ, Torres AG (2020) Hacking the host: exploitation of macrophage polarization by intracellular bacterial pathogens. Pathog Dis 78
Tilokani L, Nagashima S, Paupe V, Prudent J (2018) Mitochondrial dynamics: overview of molecular mechanisms. Essays Biochem 62:341–360
Timmons GA, O'siorain JR, Kennedy OD, Curtis AM, Early JO (2020) Innate rhythms: clocks at the center of monocyte and macrophage function. Front Immunol 11:1743
Twig G, Hyde B, Shirihai OS (2008) Mitochondrial fusion, fission and autophagy as a quality control axis: the bioenergetic view. Biochim Biophys Acta 1777:1092–1097
UNOPS (2024) Stop TB partnership [Online]. The Stop TB Partnership, Geneva, Switzerland. Available: https://www.stoptb.org/stop-tb-partnership/our-journey Accessed 5 May 5 2024
Vadakkan Devassy T, Ps N, Sharma D, Thomas AM (2022) Sleep disorders in elderly population suffering from TB and respiratory diseases. Indian J Tuberc 69(Suppl 2):S272–S279
Vrieling F, Kostidis S, Spaink HP, Haks MC, Mayboroda OA, Ottenhoff THM, Joosten SA (2020) Analyzing the impact of *Mycobacterium tuberculosis* infection on primary human macrophages by combined exploratory and targeted metabolomics. Sci Rep 10:7085
Wagener M, Hoving JC, Ndlovu H, Marakalala MJ (2018) Dectin-1-Syk-CARD9 signaling pathway in TB immunity. Front Immunol 9:225
Wang Y, Pati P, Xu Y, Chen F, Stepp DW, Huo Y, Rudic RD, Fulton DJ (2016) Endotoxin disrupts circadian rhythms in macrophages via reactive oxygen species. PLoS One 11:e0155075
WHO (2023) Global tuberculosis report 2023. World Health Organization, Geneva
Xie Y, Tolmeijer S, Oskam JM, Tonkens T, Meijer AH, Schaaf MJM (2019) Glucocorticoids inhibit macrophage differentiation towards a pro-inflammatory phenotype upon wounding without affecting their migration. Dis Model Mech 12
Yabaji SM, Dhamija E, Mishra AK, Srivastava KK (2020) ESAT-6 regulates autophagous response through SOD-2 and as a result induces intracellular survival of *Mycobacterium bovis* BCG. Biochim Biophys Acta Proteins Proteom 1868:140470
Yang Y, Cui BB, Li J, Shan JJ, Xu J, Zhang CY, Wei XT, Zhu RR, Wang JY (2024) Tricarboxylic acid cycle metabolites: new players in macrophage. Inflamm Res 73:531–539
Yen YF, Hu HY, Lee YL, Ku PW, Lin IF, Chu D, Lai YJ (2017) Obesity/overweight reduces the risk of active tuberculosis: a nationwide population-based cohort study in Taiwan. Int J Obes (Lond) 41:971–975
Zhu L, Zhao Q, Yang T, Ding W, zhao, Y. (2015) Cellular metabolism and macrophage functional polarization. Int Rev Immunol 34:82–100

Chapter 15
Role of Monocyte/Macrophages in the Pathogenesis of NeuroHIV

David Ajasin and Eliseo Eugenin

Abstract Monocyte/macrophages are cells of myeloid origin which play critical roles in innate and adaptive immune responses as well as surveillance and tissue repair. Only recently, the role of monocytes/macrophages in acute and chronic HIV Infection has become accepted. Here, we will focus on monocyte/macrophages on transmigration events and their role as viral reservoirs.

Abbreviation

AIDS	Acquired immune deficiency syndrome
ART	Anti-retroviral therapy
BBB	Blood-brain barrier
CCL2/MCP-1	Monocyte chemoattractant protein-1
CNS	Central nervous system
CSF	Cerebrospinal fluid
Glu	Glutamate
HIV	Human immunodeficiency virus-1
HAND	HIV-associated neurocognitive disorders
IL-1β	Interleukin-1β
PBMCs	peripheral blood mononuclear cells
SIV	Simian immunodeficiency virus
TNFα	Tumor necrosis factor-α
TGF-β	Transforming growth factor beta

D. Ajasin · E. Eugenin (✉)
Department of Neurobiology, The University of Texas Medical Branch (UTMB), Galveston, TX, USA
e-mail: eleugeni@UTMB.edu

M. Kloc et al. (eds.), *Monocytes and Macrophages in Development, Regeneration, and Disease*, Results and Problems in Cell Differentiation 74,
https://doi.org/10.1007/978-3-031-65944-7_15

15.1 Introduction

Approximately 38.7 million people worldwide are infected with HIV (www.unaids.org/en/resources/fact-sheet). In addition to immune compromise, HIV infection causes a broad spectrum of comorbidities including cardiovascular and brain-associated disorders (Lerner et al. 2020; Verheij et al. 2023). Several groups propose that most of this chronic HIV toxicity is mediated by the early generation of long-lived viral reservoirs in different tissues. A viral reservoir is a cell type or compartment where the replication-competent virus can quiescently persist longer than the main pool of actively replicating viruses (Churchill et al. 2016). Currently, the best-described viral reservoirs are several $CD4^{+}$ T lymphocyte subpopulations, mainly due to the easy access to biological material (blood), lymph tissue, and associated clinical data (Churchill et al. 2016). However, data from our laboratory and others indicate that viral reservoirs are localized in multiple tissues including the brain, and the cell type containing latent virus in these tissues correspond not only to $CD4^{+}$ T lymphocytes but also to myeloid cells such as microglia, macrophages, and potentially other cell types (Mitchell et al. 2019; Veenhuis et al. 2023). Here, we will discuss the role of monocytes/macrophages in acute and chronic HIV infection.

15.2 Monocyte/Macrophages: A Flexible Cell Type to Respond to Infection

Blood monocytes are made in the bone marrow and correspond to 5–10% of the total leukocyte population which circulates for 3 days prior to migrating into specific tissues (Sender et al. 2023). They exert a homeostatic function in the steady state and differentiate into tissue macrophages (Mitchell et al. 2019; Sender et al. 2023; Krombach et al. 1997). During an inflammatory process, monocytes are recruited to the site of inflammation and ultimately differentiate into tissue-associated macrophages, resting, inflammatory, or anti-inflammatory (Mitchell et al. 2019; Sender et al. 2023; Krombach et al. 1997). Normally, macrophages play a critical role in tissue development, remodeling, surveillance, and tissue homeostasis (Gentek et al. 2014; Shapouri-Moghaddam et al. 2018; Sheu and Hoffmann 2022). During adulthood, upon migration or homing into tissues, monocytes become macrophages with various differentiation and immune properties such as Kupffer's cells in the liver, osteoclast in the bone, Langerhans cells in the skin, alveolar macrophages in the lung and microglia in the brain (Locati et al. 2020; Wculek et al. 2022; Willenborg et al. 2022). In contrast, dermal, heart, and intestinal macrophages are first seeded by the embryonic liver-derived progenitors but quickly after birth are replaced by monocytes derived from hematopoietic stem cells (Dick et al. 2022; Epelman et al. 2014; Wang et al. 2023). However, new data indicate that long-lived tissue macrophages could self-renew independently of circulating monocytes (Epelman et al. 2014; Wang et al. 2021; Yap et al. 2023). These cells are generated from the yolk sack progenitors and fetal liver-derived monocytes and remain in tissues probably for the life of the individual (Röszer 2018). They can switch their phenotype from a basal state (homeostatic functions) to a pro-inflammatory state (eliminating pathogens and battling inflammation). During resolutive inflammation, macrophages can switch from a pro-inflammatory to an anti-inflammatory/pro-resolving phenotype, thereby promoting the re-establishment of homeostasis (Gentek et al. 2014; Epelman et al. 2014; Röszer 2018). Thus, depending on the different environmental signals or pathophysiological conditions, monocytes, and macrophages can undergo different forms of phenotypic polarization and thus acquire distinct functional phenotypes to recover from injury.

Monocytes/macrophages and dendritic cells, play a key role in linking the innate with the adaptive immune response (Gabrilovich et al. 2012). In HIV, monocytes and macrophages become infected with HIV (Burdo et al. 2013; Campbell et al. 2014; Coleman and Wu 2009; Crowe et al. 2003; Fantuzzi et al. 2019; Fischer-Smith et al. 2008; Herskovitz and Gendelman 2019). Studies have shown that the founder virus that initiates primary HIV infection is CCR5 tropic and early work from Dr. Gendelman's lab identified some of the viral isolates that use CD4 and CCR5 for entry, both receptors present on the surface of monocytes and macrophages (Swanstrom and Coffin 2012; Cotter et al. 2001; Ghorpade et al. 1998). Later experiments *in vitro* and *in vivo* further demonstrated that HIV targeted these cells compromising their immune response and associated inflammation (Collini et al. 2010). Also, inflammatory molecules that are induced by HIV infection through cytokines/chemokines/virus/viral proteins further activate monocytes and macrophages (Fantuzzi et al. 2003), and as we will describe below, these cells spread the virus *via* the "Trojan horse" mechanisms into most tissues where they can become viral reservoirs (Buckner et al. 2011; Eugenin et al. 2006; León-Rivera et al. 2021).

Overall, monocytes are classified according to the cell surface expression profile of specific differentiation markers, degree of activation, function, and gene expression profile (Geissmann et al. 2003; Ziegler-Heitbrock et al. 1991; Ziegler-Heitbrock 2007). Therefore, monocytes are divided into a classical phenotype ($CD14^{+}CD16^{-}$ or $CD14^{+}CD16^{+}$) that comprises 80–90% of the circulating monocytes (Geissmann et al. 2003; Ziegler-Heitbrock et al. 1991; Ziegler-Heitbrock 2007; Williams et al. 2014a). In addition, different degrees of expression for CD14 and CD16 can be found including intermediate ($CD14^{++}CD16^{+}$) and non-classical ($CD14^{+}CD16^{++}$) (Geissmann et al. 2003; Narasimhan et al. 2019). The main functions of classical monocytes are to promote phagocytosis, production of free radicals or ROS, and secretion of cytokines (Geissmann et al. 2003; Ziegler-Heitbrock 2007). Intermediate monocytes are ~2–8% of the circulating population and their functions include the generation of ROS, antigen presentation, T-cell proliferation, and angiogenesis in specific conditions (Merbecks et al. 2020; Ziegler-Heitbrock et al. 2010). This intermediate monocyte subpopulation is the main cell type with robust expression of CCR5, one of the critical receptors for HIV (Swanstrom and Coffin 2012; Merbecks et al. 2020; Ziegler-Heitbrock et al. 2010). In contrast, non-classical monocytes correspond to 2–10% of the circulating monocyte pool (Sender et al. 2023; Schyns et al. 2019). These are the main subpopulations of patrolling monocytes that respond rapidly to endothelium damage. Further, monocytes in this subgroup with the Ly6C phenotype have been classified as anti-inflammatory because they control vascular permeability and endothelial function (Ziegler-Heitbrock et al. 1991; Ziegler-Heitbrock 2007; Ziegler-Heitbrock et al. 2010; Zhao et al. 2018) (see Fig. 15.1).

As mentioned earlier, transmigrated monocytes into different tissue compartments become macrophages that upon activation and/or changes in their microenvironment, also can be cataloged into two main differentiation profiles, M1/pro-inflammatory or M2/anti-inflammatory (Shapouri-Moghaddam et al. 2018; Locati et al. 2020; Yunna et al. 2020). Their classification depends on the expression of surface receptors, secretion of cytokines and chemokines, and ROS. M1 macrophages are the first line of defense against pathogens including HIV (Yunna et al. 2020). The M1 phenotype is mostly associated with inflammation, expression, and activation of interferons and Toll-like receptor-related genes and signaling (Yunna et al. 2020). Most M1 phenotypes promote Th1 responses and secretion of pro-inflammatory cytokines such as TNF-α, IL1β, IL-6, NO, and chemokines as well as promote the expression of key surface activators such as CD40, and MHC class II (Yunna et al. 2020). For example, MHC- class II is expressed on the surface of

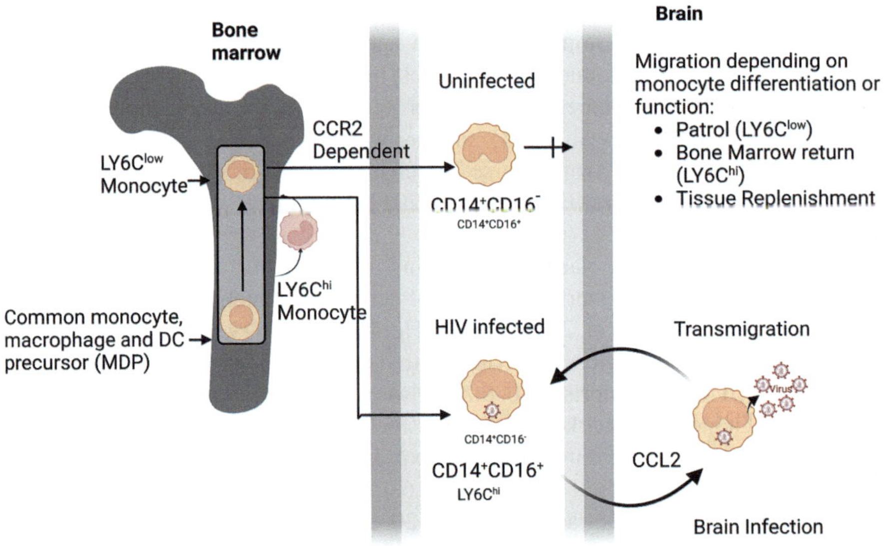

Fig. 15.1 Macrophages and dendritic cell precursors (MDPs) in the bone marrow generate LY6C^{hi} monocytes and release them into the circulation becoming LY6C^{low} monocytes. LY6C^{hi} monocytes egress the bone marrow in a CCR2-dependent manner and are recruited to different tissues in a homeostatic or inflammatory manner. As described in the text, LY6C^{low} monocytes patrol the blood vessels by direct association with the vascular endothelium. Under specific conditions, LY6C^{hi} monocytes can transmigrate and differentiate into microglia in a CX3CR1-dependent manner. In the absence of inflammation, LY6C^{hi} monocytes return to the bone marrow. Thus, has been proposed that LY6C^{hi} monocytes, and LY6^{low}, are recruited into the brain in HIV conditions also in a CCL2-dependent manner (image created with Biorender.com)

macrophages, B cells, and dendritic cells to present exogenous derived antigens from viruses, fungi, and bacteria to $CD4^+$ T lymphocytes to initiate an adaptive immune response (Collini et al. 2010). In comparison, M2 macrophages or repair macrophages promote Th2 responses to prevent or control inflammation and promote tissue repair (Yunna et al. 2020). M2 phenotype is associated with the secretion of IL-4 and IL-13, resulting in further secretion of IL-10 TGF-β, VEGF, EGF, and arginase-1 (Yunna et al. 2020). Therefore, the differentiation and polarization of monocyte/macrophages is associated with their degree of activation and function required.

15.3 Monocyte/Macrophages and HIV Infection

HIV infects an individual mainly *via* the mucosa or intravenously (Shaw and Hunter 2012; Salazar-Gonzalez et al. 2008). Here, HIV interacts with resident macrophage, dendritic cells, or circulating/patrolling T cells expressing CD4 and the chemokine receptor, CCR5 resulting in local infection and replication but also in apoptosis of a large number of T cells, especially memory T cells compromising an effective immune response (Shaw and Hunter 2012; Salazar-Gonzalez et al. 2008; Février et al. 2011). Early on after primary infection, the central nervous system becomes infected, thereby initiating a process called NeuroHIV. Monocytes ($CD14^+CD16^+$) are the main drivers of HIV invasion and spread within the brain (Williams et al. 2014a; Williams et al. 2013; Williams et al. 2012). Dr. Berman's laboratory identified that monocytes become infected with HIV triggering a migratory phenotype across the BBB dependent on CCR2, PECAM-1, ALCAM, and JAM-A (Williams et al. 2014a; Williams et al. 2013; Williams et al. 2012). After transmigration and BBB disruption in response to CCL2 gradients, circulating infected monocytes spread the virus into resident cells such as perivascular macrophages, microglia, and a small portion of astrocytes (Burdo et al. 2013; Churchill et al. 2015). Early in the infection, an inflammatory status is observed in the brains of HIV Infected individuals which later becomes chronic inflammation in association with elevated levels of TNF-α, IL1β, IL-6, IL-10, and Interferons (α and γ) resulting in minor cognitive decline to dementia (Abassi et al. 2017; Chaudhuri et al. 2008; Wei and Ho 2022). The introduction of ART resulted in controlled systemic viral replication but is not a cure due to the early generation of viral reservoirs. Most groups agreed that high levels of cognitive decline (currently around 50%, similar to the early epidemic rates) in the current HIV-infected population under effective ART is due to the early generation of viral reservoirs, residual replication, and bystander damage. Thus, understanding the mechanisms of invasion, viral reservoir formation and amplification of bystander damage is essential to prevent cognitive decline but also to design cure strategies.

15.4 Role of Monocytes and Macrophages in NeuroHIV

In the human brain, most groups accept that microglia/macrophages are the predominant cell types infected with HIV. HIV-infected astrocytes have also been detected, but most of these data are pre-ART or correspond to viral sequencing or reactivation experiments, which are not specific for cell type or provide spatial data (Marban et al. 2016; Gray et al. 2016; Li et al. 2016; Lutgen et al. 2020; Valdebenito et al. 2021). Thus, the real size, identity, and distribution of the viral reservoirs within the CNS were unknown. Recently, we and others have developed methods to identify, quantify, and characterize rare viral reservoirs in human blood and tissues, including human brain tissues obtained from HIV-infected individuals under effective ART for years (Lutgen et al. 2020; Valdebenito et al. 2021; Real et al. 2020; Prevedel et al. 2019; Ganor et al. 2019; Castellano et al. 2019; Castellano et al. 2017; Zhu et al. 2022). Using these methods, we already described rare viral reservoirs in human bone marrow and urethral tissue as well as lymph nodes and brains (Prevedel et al. 2019; Ganor et al. 2019; Castellano et al. 2017). In the human brain, we identify that the latent HIV-infected cells are microglia/macrophages and a small population of astrocytes. In contrast, in lymph nodes, the viral reservoirs correspond to a subpopulation of $CD4^+$ T lymphocytes and dendritic cells. We identified that brain reservoirs are localized in clusters of cells containing integrated HIV-DNA, some have viral mRNA transcript(s), and a subset of these cells express residual viral proteins even in samples obtained from individuals under effective ART (Castellano et al. 2017). Surprisingly, we observed the presence of viral proteins such as Nef, gp120, Vpr, and Tat in uninfected neighboring cells suggesting that these proteins are expressed and released into uninfected cells, probably to amplify HIV-CNS toxicity to vast regions of the brain that are beyond these clusters of cells containing integrated HIV-DNA (Castellano et al. 2017). These findings demonstrate that viral reservoirs within the CNS exist, and they are active, despite ART. Analysis of urethral and bone marrow also confirmed the presence and distribution of these myeloid viral reservoirs in close contact with mucosa and niche progenitor cells, respectively (Ganor et al. 2019; Zhu et al. 2022). Therefore, our group proposes that chronic inflammation in long-term infected individuals is due to the presence of viral reservoirs and bystander local damage in surrounding uninfected cells.

Conversely, several groups reported that long-term infection could induce anti-inflammatory cytokines such as IL-10 decreasing the overall immune and inflammatory response (Blackburn and Wherry 2007; Elrefaei et al. 2007; Liu et al. 2014). For instance, IL-10 has been demonstrated to be essential in viral persistence and preventing viral clearance in monocytes and T cells, supporting that IL-10 can prevent apoptosis (Elrefaei et al. 2007). Moreover, others have identified that weak IL-10 signaling can promote viral clearance and immune reconstitution and proliferation of CD4 and CD8 T lymphocytes (Wilson and Brooks 2011). This suggests that IL-10 participates in energy and exhaustion, but also in clearance (Blackburn and Wherry 2007; Liu et al. 2014; Wilson and Brooks 2011). Thus, IL-10 has a dual role in promoting infection or clearance.

A major reason for long-term brain and immune damage is the persistent residual expression of viral proteins and local inflammation as observed in our *in-situ* analysis (Hu 2016; Price et al. 2005; Ranki et al. 1995; Smith et al. 2014; Sviridov and Bukrinsky 2023). In agreement, although the HIV reservoir has been traditionally referred to as "latent or silenced", a small population of viral reservoirs still produce and release proteins and viruses (1–3 copies/ml in plasma) that can be detected with ultra-sensitive methods (Hu 2016; Price et al. 2005; Ranki et al. 1995; Smith et al. 2014; Sviridov and Bukrinsky 2023; Bertrand et al. 2018; Williams et al. 2020; Jacobs et al. 2019). In addition, latency is in fact "leaky" due to multiple observations that substantial levels of HIV transcriptional activity can be found in ART-suppressed individuals (Einkauf et al. 2022; Telwatte et al. 2018; Yukl et al. 2018). This leaky phenotype can be observed in individuals with uncontrolled viremia who show plasma viremia (>50 copies/ml) despite ART adherence and lack of drug resistance. Furthermore, it has been calculated that this phenotype is 1 out of 250 ART-suppressed individuals (Halvas et al. 2020; Katusiime et al. 2020; Siliciano and Siliciano 2020; White et al. 2023). Thus, we have to understand that latency does not mean total suppression and as we observed in post-mortem brain tissues, residual replication, viral protein production and secretion, and associated inflammation is still occurring and probably is the main contributor to long-term comorbidities in the HIV population (Gay et al. 2022).

A critical question during long-term infection and ART is whether the immune system is properly reconstituted. So far, the answer is inconclusive. As indicated above, during acute and chronic HIV infection, there is persistent immune activation (Sokoya et al. 2017). For example, during acute infection in T cells and astrocytes, studies have described that the upregulation of the enzyme telomerase is associated with senescence (Côté et al. 2012; Palmer et al. 1997). Senescence occurs when the functional ability of immune cells is altered by the exhaustion of T cells (Chou et al. 2013). Acute and chronic HIV infection accelerates the rate of turnover of new T cells leading to premature aging (Chou et al. 2013; Wallis and Williams 2022). These changes are evident by the accumulation of aging markers on T cells such as $CD28^{-}$ and $CD57^{+}$ (Lee et al. 2014). Similarly, in young HIV-positive individuals, the monocytes show the same changes in phenotype, function, and telomere length that are also seen in uninfected older adults (Deeks 2011; Hearps et al. 2012). More importantly, none of these aging-related changes are prevented by ART, again showing that the immune system is not fully restored due to the observed chronic/debilitating inflammation even in HIV-infected individuals with ART adherence.

Further, similar dysfunctions is seen in monocyte/macrophages where we observed in chronic HIV-infected individuals including increased numbers of the population $CD14^{+}CD16^{+}$ and enhanced expression of critical adhesion/junctional proteins required for migration or transmigration into tissues as well as changes in phagocytosis (Williams et al. 2014a; Williams et al. 2013; Williams et al. 2012; Espíndola et al. 2018). Clearly, alterations in the myeloid lineage will alter the CD4/

CD8 differentiation by compromising the crosstalk between the innate and adaptive immune response, enhancing susceptibility to tumorigenesis and other infections. Interestingly, these changes are further accelerated by drug abuse and other comorbidities, but the mechanism is poorly understood (Melvan et al. 2012; Siggins et al. 2014).

The amplified $CD14^+CD16^+$ monocyte population is preferentially susceptible to HIV infection with high levels of CCR5 (Ellery et al. 2007; Jaworowski et al. 2007) and CCR2 (Veenstra et al. 2019; Williams et al. 2014b). In addition, drug abuse in combination with HIV as well as buprenorphine treatment modulates the activation stage of these monocyte subpopulations (Carvallo et al. 2015; Jaureguiberry-Bravo et al. 2021; Murphy et al. 2022). The expansion of both intermediate and non-classical monocyte populations in the blood has been essential to explain cardiovascular disease and HIV-associated cognitive decline (Williams et al. 2014a; Murphy et al. 2022). These cells had an increased pro-inflammatory phenotype secreting multiple cytokines, and chemokines, and shedding adhesional/junctional proteins that compromise endothelial function (Williams et al. 2012; Williams et al. 2014b; Williams et al. 2016). Further, studies have observed that the phagocytic activity of monocytes and macrophages from HIV-infected individuals is significantly less compared to those from uninfected individuals. As a cell type that is critical for innate immunity and acts as an interface to the adaptive immune system, reduced phagocytic activity will affect the control of other pathogens and the ability of macrophages to present peptides to $CD4^+$ T cells for adaptive immune response in HIV-infected individuals, hence the emergence of opportunistic infection. For instance, *Mycobacterium tuberculosis* (MTb) co-infection is common in endemic regions of the world and some studies have shown that monocytes in HIV-MTb co-infection have different gene expression profiles compared to HIV only especially in patients with TB-associated immune reconstitution inflammatory syndrome, further suggesting that the initial HIV infection makes the individual vulnerable to MTb infection (Meintjes et al. 2008).

15.5 Monocyte/Macrophages and HIV Replication

Until recently, HIV has been categorized as either X4 tropic or R5 tropic virus. The X4 tropic viruses were generally seen as virions that preferentially infect T cells because the CXCR4 co-receptor facilitates viral entry of these virions in target cells, while the R5 tropic viruses were seen as virions that mostly infect monocyte/macrophages *via* CCR5 co-receptor (Swanstrom and Coffin 2012). This has been misleading and erroneously suggests that X4 tropic viruses cannot infect monocytes/macrophages and vice versa. However, a study clarified the distinction between the two viral strains based on their ability to infect cells with low or high CD4 receptor density (Arrildt et al. 2015; Beauparlant et al. 2017; Joseph and Swanstrom 2018). It has been shown that HIV's main receptor is CD4, and the co-receptors are CCR5 and CXCR4 (Kwong et al. 2002). The expression and density of these receptors differ between CD4⁺ T cells and monocytes/macrophages. CD4⁺ T cells are known to express a high level of both CD4 and CXCR4, but low levels of CCR5. Conversely, monocytes/macrophages express the same receptors, but at a different density. Studies have shown that monocytes/macrophages express lower CD4, CXCR4, and a high density of CCR5 on its surface. Therefore, the main distinction between monocytes/macrophage-tropic viruses is that they can infect cells with low CD4 density while the T cell trophic viruses depend on high CD4 density for infection (Arrildt et al. 2015; Joseph and Swanstrom 2018). This is the case with primary CD4⁺ T cells and monocytes/macrophages which suggest that two categories of primary cells can support both X4 and R5 viruses. However, when studies use CD4⁺ T cell lines or monocyte/macrophage cell lines, X4 tropic viruses infect the CD4⁺ T cell lines while the R5 tropic viruses preferentially infect monocyte/macrophage cell lines (Joseph and Swanstrom 2018). Another observation that has been suggested to distinguish these two viral strains is the level of intracellular dNTPs that can support replication. Studies have suggested that X4 tropic viruses replicate better in CD4⁺ T cells because of their high pool of intracellular dNTPs compared to monocytes/macrophages (Amie et al. 2013; Baldauf et al. 2012; Kim et al. 2012; Pan et al. 2013; Schüssler et al. 2023; Shepard et al. 2019). All these observations further show that monocytes/macrophages are more restrictive to HIV replication, when compared to CD4 T cells, and only a specific strain can replicate in this cell type.

15.6 Monocyte/Macrophages and HIV Egress

Generally, it has been observed that monocytes and macrophages are more resistant to HIV infection compared to the CD4⁺ T cells (Koppensteiner et al. 2012). This is due to various factors such as low expression or lack of expression of some HIV-facilitating host factors or the expression of restriction factors such as sterile motif and histidine aspartic domain and HD domain-containing protein 1 (SAMHD1) and the (apolipoprotein B mRNA editing enzyme) 3G family (APOBEC) (Nguyen et al. 2014; Peng et al. 2007). However, a subpopulation of monocytes that is susceptible to productive HIV infection is the intermediate ($CD14^{++}/CD16^{+}$) or non-classical ($CD14^{+}/CD16^{+}$) monocytes which are ~10% of the monocyte population (Williams et al. 2012). They exhibit more mature phenotypes which are similar to macrophages. Interestingly, these subsets of monocytes are significantly increased in HIV-infected individuals when compared to their uninfected peers (Williams et al. 2013; Williams et al. 2012). More importantly, this same subpopulation has been implicated in seeding the CNS (Williams et al. 2013; Williams et al. 2012). When these monocyte subpopulations traffic to different tissues or organs, they differentiate into macrophages which are more permissive of productive HIV infection. One of the features of macrophages is the large size of the cell compared to other cells of the immune system (Sender et al. 2023; Krombach et al. 1997). Therefore, the surface area of the cells allows for phagocytosis which requires a robust endocytic mechanism. As part of this is the presence of very active multivesicular bodies which not only participate in endocytosis and recycling of surface proteins but are hijacked by HIV for budding (Bennett et al. 2009; Costa et al. 2006; Ghossoub et al. 2014; Gousset et al. 2008; Joshi et al. 2009; Morita and Sundquist 2004; Tan and Sattentau 2013; Yasen et al. 2017). Typically, in CD4$^+$ T cells, HIV buds from the plasma membrane, and requires the activity of the endosomal sorting complex required for transport (ESCRT) machinery (Morita and Sundquist 2004). ESCRT is made up of four complexes (ESCRT 0,1,2,3) and other associated proteins referred to as adaptor proteins. Studies have shown that blocking ESCRT especially the ESCRT 1 and 3 leads to significant inhibition of HIV budding (Bieniasz 2009; Bleck et al. 2014; Martin-Serrano et al. 2003). More importantly, ESCRT is also critical for multivesicular body activities which is enhanced in macrophages. There has been some debate about if HIV only buds from the plasma membrane and few studies have shown through electron microscopy experiments that HIV budding in macrophages occurs both at the plasma membrane and in multivesicular (Bennett et al. 2009; Gousset et al. 2008; Joshi et al. 2009). The implication of such observations has not been fully explored. The question that comes up is, are these virions eventually released, and if yes, how are they released? We could speculate that some of the virions will be released via extracellular vesicles (EVs) or the unfolding of the plasma membrane invaginations (Bennett et al. 2009; Joshi et al. 2009). If these virions are released via an EV mechanism, how they are targeted to specific cell types will be an interesting finding (see Fig. 15.2). Other questions are, how the virions maturation, long-term accumulation/storage, and release occur as well as the signals involved.

15.7 The Monocyte/Macrophage Is a Latent Reservoir

As stated above most studies on HIV reservoirs have shown that well-characterized viral reservoirs are latently infected CD4$^+$ T lymphocytes (naive and memory), but the wholistic examination of the long-term viral reservoir probably is comprised of multiple cell types that are present in most tissues and microenvironment. It is calculated that this pool of CD4$^+$ T lymphocytes corresponds to around 0.05% of the circulating cells during effective ART (Sender et al. 2023). Monocytes and macrophages account for high levels of the viral reservoirs in several anatomical sites such as the lungs and brain despite ART, and they persist without immune detection due to the restricted access of the immune cells into these anatomical sites (Campbell et al. 2014; Cotter et al. 2001; Castellano et al. 2017; Koppensteiner et al. 2012). Our data on several post-mortem tissues such as the heart, brain, and lymph nodes support that each tissue has its own viral reservoir identity peculiar to its cellular composition, but overall, macrophages are also a major and distinct viral reservoir in most of these tissues (Ganor et al. 2019; Castellano et al. 2017; Zhu et al. 2022). Upon infection and replication, most cells undergo apoptosis, but a small population of cells survive infection and can silence the virus after ART (Churchill et al. 2016).

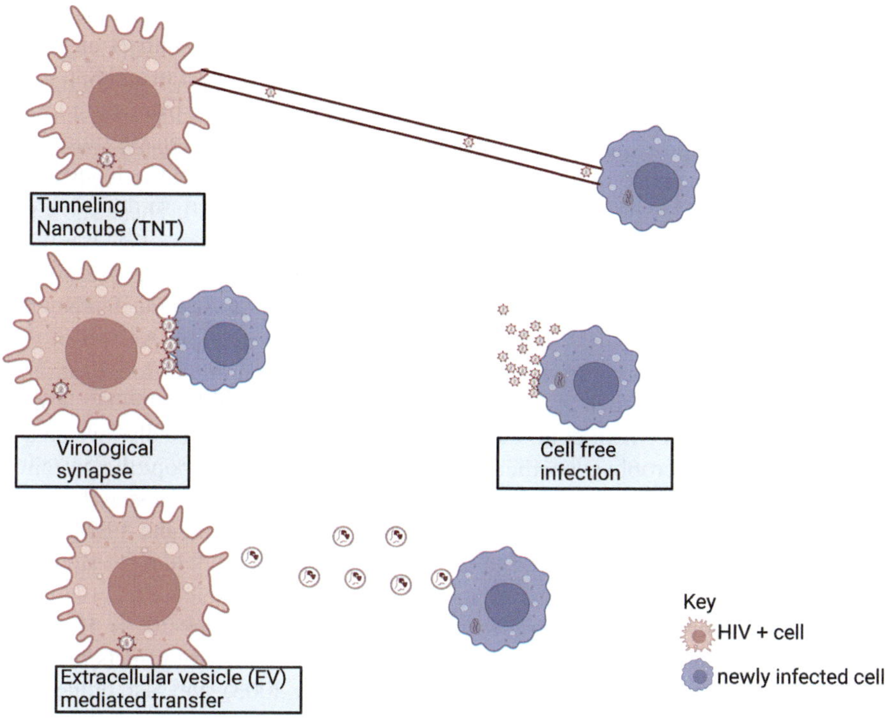

Fig. 15.2 There are various modes of viral transfer or infection from infected to uninfected target cells. As discussed, tunneling nanotube is a recently discovered mode of long-range viral transfer from infected to uninfected target cells through long actin-rich nanotubes (bridges). TNTs allow for the transfer of viral components sufficient for HIV replication. Importantly, this mode of viral transfer is resistant to antibody-mediated neutralization and is induced by acute infection and reactivation. The virological synapse-mediated infection occurs between an HIV-infected cell and an uninfected targeted cell *via* close cell-to-cell contact which is promoted by the interaction between viral protein (gp120) on the plasma membrane of the infected cell(s) and CD4/CCR5 or CXCR4 receptors on the surface of the yet-to-be-infected target cell(s). Cell-free (soluble virus) infection is the canonical mode of viral transfer where mature free virions enter new target cells *via* CD4 receptors and either CXCR4 or CCR5 and it is the best described mechanism of viral infection and spread. Extracellular vesicle-mediated viral transfer is still not fully understood but in theory, indicates that sufficient viral components are packaged in extracellular vesicles (EV). EV-containing viruses or viral components are released into the extracellular space from HIV-infected cells and taken up by uninfected cells. These mechanisms are poorly described but the proposal is that these packed viral components can establish infection or prime uninfected cells for infection (image created with Biorender.com)

Overall, it is widely accepted that resident cells such as monocyte/macrophage cell lineage are the first to be targeted by HIV and help in viral transmission into $CD4^+$ T lymphocytes, especially in mucosa (Koppensteiner et al. 2012). HIV infection kills most $CD4^+$ T lymphocytes, and HIV-infected macrophages survive for extended periods by maintaining virions in membrane invaginations that protect the viruses from ART and neutralizing antibodies (Bennett et al. 2009). Macrophages are terminally differentiated, nondividing cells and are present in all tissues (Koppensteiner et al. 2012). Their long life and wide distributions in all tissues and anatomical sites contribute to the spread of viral infection, and reactivation mechanisms (Valdebenito et al. 2023). As part of the uniqueness of monocytes/macrophages, a study demonstrated that macrophages, as well as differentiated monocytes ($CD14^+CD16^+$), can sustain not only HIV entry and infection but also viral "retention" on the surface or multivesicular bodies (Bennett et al. 2009). These show that monocytes and macrophages play a critical role in HIV pathogenesis, yet, compared to the $CD4^+$ T cells, their roles as viral reservoirs have not been extensively studied.

In addition, monocyte/macrophages have extensive cell survival mechanism that prevents the activation of HIV-induced apoptosis during infection. As part of this, several viral proteins have been shown to alter the expression and localization of pro-apoptotic proteins such as Bcl-2, Bax, cFLIP, and XIAP in $CD4^+$ T cells (Andersen et al. 2006; Kim et al. 2018; Wang et al. 2009). Conversely, our data in latently infected macrophages indicates that Bim, a critical pro-apoptotic protein is involved in the survival of HIV-infected macrophages, and it is sequestered into the mitochondria of macrophages (Chen et al. 2002). In astrocytes and PBMCs, we identified that nef protein can bind to the IP_3R1 and prevent calcium overload and apoptosis by preventing the proper organelle-organelle communication between the endoplasmic reticulum and the mitochondria (Ganta and Chaubey 2019). However, uninfected cells surrounding viral reservoirs lack these protective effects resulting in chronic damage and even apoptosis (Lum and Badley 2003). Nef also induces the phosphorylation and inactivation of Bad, resulting in extended survival (Salminen et al. 2013). Further, HIV-Tat can induce inflammation and upregulate host apoptotic proteins such as Bcl-2, FLIP, and XIAP (Chen et al. 2002). All of these data are different when compared to findings from studies in T cells, where the formation of the transition pore was shown to be affected by Bcl-2 mediated mechanisms resulting in controlled HIV replication, reactivation, and susceptibility to apoptosis (Jacotot et al. 2000; Kim et al. 2014).

Our results that Bim is a critical marker of myeloid viral reservoirs also opened the possibility that the protein contributes to the survival metabolism observed in viral reservoirs (Castellano et al. 2017). In agreement, myeloid viral reservoirs with high mitochondrial sequestration of Bim had fused mitochondria (Castellano et al. 2017). Bim has been shown to interact with creatine kinase and hexokinase supporting its metabolic role (Le Bras et al. 2006). Our data indicates that in latently infected macrophages, the metabolic steps in the tricarboxylic acid (TCA) cycle preceding the oxidative phosphorylation (OXPHOS) were compromised, thereby resulting in large lipid accumulation and the use of unusual carbon sources to produce energy such as amino acids (Castellano et al. 2019; Gorska and Eugenin 2020). More importantly, we identified that latently infected macrophages lack the capacity to shift between different carbon sources and rely heavily on glutamine/glutamate to survive (Castellano et al. 2019; Gorska and Eugenin 2020). Interestingly, blocking the use of glutamine and glutamate by viral reservoirs resulted in significant cell death for the latently infected macrophages even in the absence of viral reactivation (Castellano et al. 2019). In the CNS, HIV-infected microglia/

macrophages, especially the ones in the brain parenchyma in close contact with glutamatergic areas, use glutamine and glutamate as a major source of energy suggesting a potential adaptation of these reservoirs to this abundant source of energy (Castellano et al. 2019). Further, our *in situ* data in human brains indicates that viral reservoirs are comprised of both microglia/macrophages and astrocytes (Valdebenito et al. 2021; Castellano et al. 2017). Astrocytes are the main regulator of the glutamine/glutamate cycle further supporting the "symbiosis" between both cell types (Oya et al. 2023; Tardito et al. 2015). Thus, in the brain, the use of this alternative fuel source could be critical for the reservoir. For instance, glutamine can be used to produce nucleotides for DNA replication or signaling including ATP and GMP, which would potentially increase cell viability of the viral reservoirs compared to other cell types not using glutamine (Wise and Thompson 2010). In line with the above observation, we have identified that extracellular ATP signaling is upregulated in HIV-infected individuals and its concentration in the blood can predict cognitive decline in the HIV-infected population (Velasquez et al. 2020). In the last years, beautiful work from The CNS HIV Antiretroviral Therapy Effect Research (CHARTER), National NeuroAIDS Tissue Consortium (NNTC), Neuroimaging Consortium, and the Neurobiobank have demonstrated that in glioblastoma and HIV, there are local alterations in metabolites such as citrate, creatine, glutamine, glucose, inositol, glutamic acid, and CSF mtDNA in association with tissue dysfunction (Castellano et al. 2019). An additional metabolic adaptation was described for S100A8 controlling persistence in macrophages, further confirming a unique metabolic profile (Endoh et al. 2009; Real et al. 2022). This metabolic profile is extremely similar to the one observed in glioblastoma stem cells, suggesting a potential overlap between both diseases and a promising therapeutic approach to reach a cure (Deininger et al. 2001).

15.8 Macrophages as a Critical Cell Type to Maintain and Spread HIV During Acute and Reactivation Conditions

During the last few years, it has become evident that latently infected macrophages have unique mechanisms of silencing viral replication (Veenhuis et al. 2023; Koppensteiner et al. 2012; Cobos-Jiménez et al. 2011; Poli and Fauci 1992; Saphire et al. 2001; Sattentau and Stevenson 2016). One of the main questions is how HIV spreads efficiently during acute infection and reactivation in the absence of a significant soluble virus. A key mechanism that is inhibited by the limitations to viral reservoirs such as; efficiency of viral infection, reactivation, and avoidance of immune recognition or ART was the discovery of HIV transfer by the virological synapses (VS) (Dufloo et al. 2018; Duncan et al. 2014; Dupont and Sattentau 2020; Felts et al. 2010; Groot et al. 2008; Han et al. 2022; Real et al. 2018) (see Fig. 15.2). VS is efficient even in the absence of a soluble virus but requires close cell-to-cell interactions between the HIV-infected and uninfected cells, adding another level of complexity because HIV compromises cell migration as well. Further, we identified a novel mechanism of long-range cell-to-cell mediated infection by Tunneling Nanotubes (TNTs) that avoids most of the limitations described above for both the soluble and VS modes of infection (Valdebenito et al. 2023).

TNTs are long-range (30–500 μm) cell membrane projections that mediate the direct transfer of organelles and signaling molecules between connected cells (Dufrançais et al. 2021; Dupont et al. 2020; Hashimoto et al. 2016; Lotfi et al. 2020; Uhl et al. 2019) (see Fig. 15.2). TNTs are not expressed in healthy individuals, but upon disease, TNT proliferates (Dupont et al. 2020; Hashimoto et al. 2016; Lotfi et al. 2020) (see Fig. 15.2). We identified that primary HIV infection and reactivation induce TNT formation in macrophages and microglia (Okafo et al. 2017). TNTs transport HIV and/or viral components that result in the infection of target cells as demonstrated *in vitro* and *in vivo* (Lotfi et al. 2020; Okafo et al. 2017). Our data indicate that TNT-mediated HIV transmission is 100–1000-fold more efficient than a cell-free virus. TNTs, like VS, protect the virus from neutralizing antibodies, and ART and promote multiple strain infections (Valdebenito et al. 2023; Dufrançais et al. 2021; Hashimoto et al. 2016; Uhl et al. 2019; Okafo et al. 2017) (see Fig. 15.2). Thus, microglia/macrophages had specific mechanisms of viral reactivation to ensure the survival of viral reservoirs and efficient viral transfer into neighboring uninfected cells. Thus, new mechanisms of monocyte/macrophage/microglia survival, persistence, reactivation, and bystander damage have just been discovered to explain the pathogenesis of NeuroHIV, opening new potential therapies.

15.9 Conclusion

HIV research has demonstrated that myeloid cells are a critical viral reservoir in humans. An early ART initiation indicated that there is a two-phase decay in viremia in plasma, attributing the second slower phase to longer-lived cells such as macrophages. Also, macrophages sustain HIV replication even independently than T lymphocytes suggesting a different kind of reservoir. Overall, all these studies highlight the importance of understanding and targeting myeloid and other viral reservoirs to reach an HIV cure.

Acknowledgments This work was funded by The National Institute of Mental Health grants, MH128082 and MH134761, the National Institute of Neurological Disorders and Stroke, NS105584, and UTMB internal funding (to E.A.E).

Conflicts of Interest None.

Competing Financial Interest None.

Author Contributions D.A. and E.A.E. wrote the paper.

References

Abassi M et al (2017) Cerebrospinal fluid biomarkers and HIV-associated neurocognitive disorders in HIV-infected individuals in Rakai, *Uganda*. J Neurovirol 23(3):369–375

Amie SM, Noble E, Kim B (2013) Intracellular nucleotide levels and the control of retroviral infections. Virology 436(2):247–254

Andersen JL et al (2006) HIV-1 Vpr-induced apoptosis is cell cycle dependent and requires Bax but not ANT. PLoS Pathog 2(12):e127

Arrildt KT et al (2015) Phenotypic correlates of HIV-1 macrophage tropism. J Virol 89(22):11294–11311

Baldauf HM et al (2012) SAMHD1 restricts HIV-1 infection in resting CD4(+) T cells. Nat Med 18(11):1682–1687

Beauparlant D et al (2017) Delineating CD4 dependency of HIV-1: adaptation to infect low level CD4 expressing target cells widens cellular tropism but severely impacts on envelope functionality. PLoS Pathog 13(3):e1006255

Bennett AE et al (2009) Ion-abrasion scanning electron microscopy reveals surface-connected tubular conduits in HIV-infected macrophages. PLoS Pathog 5(9):e1000591

Bertrand SJ et al (2018) HIV-1 proteins dysregulate motivational processes and dopamine circuitry. Sci Rep 8(1):7869

Bieniasz PD (2009) The cell biology of HIV-1 virion genesis. Cell Host Microbe 5(6):550–558

Blackburn SD, Wherry EJ (2007) IL-10, T cell exhaustion and viral persistence. Trends Microbiol 15(4):143–146

Bleck M et al (2014) Temporal and spatial organization of ESCRT protein recruitment during HIV-1 budding. Proc Natl Acad Sci USA 111(33):12211–12216

Buckner CM et al (2011) Characterization of monocyte maturation/differentiation that facilitates their transmigration across the blood-brain barrier and infection by HIV: implications for NeuroAIDS. Cell Immunol 267(2):109–123

Burdo TH, Lackner A, Williams KC (2013) Monocyte/macrophages and their role in HIV neuropathogenesis. Immunol Rev 254(1):102–113

Campbell JH et al (2014) The importance of monocytes and macrophages in HIV pathogenesis, treatment, and cure. AIDS 28(15):2175–2187

Carvallo L et al (2015) Buprenorphine decreases the CCL2-mediated chemotactic response of monocytes. J Immunol 194(7):3246–3258

Castellano P, Prevedel L, Eugenin EA (2017) HIV-infected macrophages and microglia that survive acute infection become viral reservoirs by a mechanism involving Bim. Sci Rep 7(1):12866

Castellano P et al (2019) HIV infection and latency induce a unique metabolic signature in human macrophages. Sci Rep 9(1):3941

Chaudhuri A et al (2008) HIV-1 activates proinflammatory and interferon-inducible genes in human brain microvascular endothelial cells: putative mechanisms of blood-brain barrier dysfunction. J Cereb Blood Flow Metab 28(4):697–711

Chen D et al (2002) HIV-1 Tat targets microtubules to induce apoptosis, a process promoted by the pro-apoptotic Bcl-2 relative Bim. EMBO J 21(24):6801–6810

Chou JP et al (2013) Accelerated aging in HIV/AIDS: novel biomarkers of senescent human CD8+ T cells. PLoS One 8(5):e64702

Churchill MJ et al (2015) HIV-1 transcriptional regulation in the central nervous system and implications for HIV cure research. J Neurovirol 21(3):290–300

Churchill MJ et al (2016) HIV reservoirs: what, where and how to target them. Nat Rev Microbiol 14(1):55–60

Cobos-Jiménez V et al (2011) Macrophages and HIV-1. Curr Opin HIV AIDS 6(5):385–390

Coleman CM, Wu L (2009) HIV interactions with monocytes and dendritic cells: viral latency and reservoirs. Retrovirology 6:51

Collini P et al (2010) Monocyte and macrophage dysfunction as a cause of HIV-1 induced dysfunction of innate immunity. Curr Mol Med 10(8):727–740

Costa LJ et al (2006) Interactions between Nef and AIP1 proliferate multivesicular bodies and facilitate egress of HIV-1. Retrovirology 3:33
Côté HC et al (2012) Leukocyte telomere length in HIV-infected and HIV-exposed uninfected children: shorter telomeres with uncontrolled HIV viremia. PLoS One 7(7):e39266
Cotter RL et al (2001) Regulation of human immunodeficiency virus type 1 infection, beta-chemokine production, and CCR5 expression in CD40L-stimulated macrophages: immune control of viral entry. J Virol 75(9):4308–4320
Crowe S, Zhu T, Muller WA (2003) The contribution of monocyte infection and trafficking to viral persistence, and maintenance of the viral reservoir in HIV infection. J Leukoc Biol 74(5):635–641
Deeks SG (2011) HIV infection, inflammation, immunosenescence, and aging. Annu Rev Med 62:141–155
Deininger MH et al (2001) Macrophage/microglial cell subpopulations in glioblastoma multiforme relapses are differentially altered by radiochemotherapy. J Neuro-Oncol 55(3):141–147
Dick SA et al (2022) Three tissue-resident macrophage subsets coexist across organs with conserved origins and life cycles. Sci Immunol 7(67):eabf7777
Dufloo J, Bruel T, Schwartz O (2018) HIV-1 cell-to-cell transmission and broadly neutralizing antibodies. Retrovirology 15(1):51
Dufrançais O et al (2021) Cellular and molecular actors of myeloid cell fusion: podosomes and tunneling nanotubes call the tune. Cell Mol Life Sci 78(17–18):6087–6104
Duncan CJ et al (2014) High-multiplicity HIV-1 infection and neutralizing antibody evasion mediated by the macrophage-T cell virological synapse. J Virol 88(4):2025–2034
Dupont M, Sattentau QJ (2020) Macrophage cell-cell interactions promoting HIV-1 infection. Viruses 12(5)
Dupont M et al (2020) Tuberculosis-associated IFN-I induces Siglec-1 on tunneling nanotubes and favors HIV-1 spread in macrophages. elife:9
Einkauf KB et al (2022) Parallel analysis of transcription, integration, and sequence of single HIV-1 proviruses. Cell 185(2):266–282.e15
Ellery PJ et al (2007) The CD16+ monocyte subset is more permissive to infection and preferentially harbors HIV-1 in vivo. J Immunol 178(10):6581–6589
Elrefaei M et al (2007) HIV-specific IL-10-positive CD8+ T cells suppress cytolysis and IL-2 production by CD8+ T cells. J Immunol 178(5):3265–3271
Endoh Y et al (2009) IL-10-dependent S100A8 gene induction in monocytes/macrophages by double-stranded RNA. J Immunol 182(4):2258–2268
Epelman S et al (2014) Embryonic and adult-derived resident cardiac macrophages are maintained through distinct mechanisms at steady state and during inflammation. Immunity 40(1):91–104
Espíndola MS et al (2018) Epigenetic alterations are associated with monocyte immune dysfunctions in HIV-1 infection. Sci Rep 8(1):5505
Eugenin EA et al (2006) CCL2/monocyte chemoattractant protein-1 mediates enhanced transmigration of human immunodeficiency virus (HIV)-infected leukocytes across the blood-brain barrier: a potential mechanism of HIV-CNS invasion and NeuroAIDS. J Neurosci 26(4):1098–1106
Fantuzzi L, Belardelli F, Gessani S (2003) Monocyte/macrophage-derived CC chemokines and their modulation by HIV-1 and cytokines: a complex network of interactions influencing viral replication and AIDS pathogenesis. J Leukoc Biol 74(5):719–725
Fantuzzi L et al (2019) Dual CCR5/CCR2 targeting: opportunities for the cure of complex disorders. Cell Mol Life Sci 76(24):4869–4886
Felts RL et al (2010) 3D visualization of HIV transfer at the virological synapse between dendritic cells and T cells. Proc Natl Acad Sci USA 107(30):13336–13341
Février M, Dorgham K, Rebollo A (2011) CD4+ T cell depletion in human immunodeficiency virus (HIV) infection: role of apoptosis. Viruses 3(5):586–612

Fischer-Smith T et al (2008) Monocyte/macrophage trafficking in acquired immunodeficiency syndrome encephalitis: lessons from human and nonhuman primate studies. J Neurovirol 14(4):318–326
Gabrilovich DI, Ostrand-Rosenberg S, Bronte V (2012) Coordinated regulation of myeloid cells by tumours. Nat Rev Immunol 12(4):253–268
Ganor Y et al (2019) HIV-1 reservoirs in urethral macrophages of patients under suppressive antiretroviral therapy. Nat Microbiol 4(4):633–644
Ganta KK, Chaubey B (2019) Endoplasmic reticulum stress leads to mitochondria-mediated apoptosis in cells treated with anti-HIV protease inhibitor ritonavir. Cell Biol Toxicol 35(3):189–204
Gay CL et al (2022) Stable latent HIV infection and low-level Viremia despite treatment with the broadly neutralizing antibody VRC07-523LS and the latency reversal agent vorinostat. J Infect Dis 225(5):856–861
Geissmann F, Jung S, Littman DR (2003) Blood monocytes consist of two principal subsets with distinct migratory properties. Immunity 19(1):71–82
Gentek R, Molawi K, Sieweke MH (2014) Tissue macrophage identity and self-renewal. Immunol Rev 262(1):56–73
Ghorpade A et al (1998) Role of the beta-chemokine receptors CCR3 and CCR5 in human immunodeficiency virus type 1 infection of monocytes and microglia. J Virol 72(4):3351–3361
Ghossoub R et al (2014) Syntenin-ALIX exosome biogenesis and budding into multivesicular bodies are controlled by ARF6 and PLD2. Nat Commun 5:3477
Gorska AM, Eugenin EA (2020) The glutamate system as a crucial regulator of CNS toxicity and survival of HIV reservoirs. Front Cell Infect Microbiol 10:261
Gousset K et al (2008) Real-time visualization of HIV-1 GAG trafficking in infected macrophages. PLoS Pathog 4(3):e1000015
Gray LR, Brew BJ, Churchill MJ (2016) Strategies to target HIV-1 in the central nervous system. Curr Opin HIV AIDS 11(4):371–375
Groot F, Welsch S, Sattentau QJ (2008) Efficient HIV-1 transmission from macrophages to T cells across transient virological synapses. Blood 111(9):4660–4663
Halvas EK et al (2020) HIV-1 viremia not suppressible by antiretroviral therapy can originate from large T cell clones producing infectious virus. J Clin Invest 130(11):5847–5857
Han M et al (2022) Mechanisms of HIV-1 cell-to-cell transfer to myeloid cells. J Leukoc Biol 112(5):1261–1271
Hashimoto M et al (2016) Potential role of the formation of tunneling nanotubes in HIV-1 spread in macrophages. J Immunol 196(4):1832–1841
Hearps AC et al (2012) HIV infection induces age-related changes to monocytes and innate immune activation in young men that persist despite combination antiretroviral therapy. AIDS 26(7):843–853
Herskovitz J, Gendelman HE (2019) HIV and the macrophage: from cell reservoirs to drug delivery to viral eradication. J Neuroimmune Pharmacol 14(1):52–67
Hu XT (2016) HIV-1 tat-mediated calcium dysregulation and neuronal dysfunction in vulnerable brain regions. Curr Drug Targets 17(1):4–14
Jacobs IR et al (2019) Inhibitory control deficits associated with upregulation of CB(1)R in the HIV-1 tat transgenic mouse model of Hand. J Neuroimmune Pharmacol 14(4):661–678
Jacotot E et al (2000) The HIV-1 viral protein R induces apoptosis via a direct effect on the mitochondrial permeability transition pore. J Exp Med 191(1):33–46
Jaureguiberry-Bravo M et al (2021) Treatment with buprenorphine prior to EcoHIV infection of mice prevents the development of neurocognitive impairment. J Leukoc Biol 109(3):675–681
Jaworowski A et al (2007) CD16+ monocyte subset preferentially harbors HIV-1 and is expanded in pregnant Malawian women with plasmodium falciparum malaria and HIV-1 infection. J Infect Dis 196(1):38–42
Joseph SB, Swanstrom R (2018) The evolution of HIV-1 entry phenotypes as a guide to changing target cells. J Leukoc Biol 103(3):421–431

Joshi A et al (2009) Evidence that productive human immunodeficiency virus type 1 assembly can occur in an intracellular compartment. J Virol 83(11):5375–5387
Katusiime MG et al (2020) Intact HIV proviruses persist in children seven to nine years after initiation of antiretroviral therapy in the first year of life. J Virol 94(4)
Kim B et al (2012) Tight interplay among SAMHD1 protein level, cellular dNTP levels, and HIV-1 proviral DNA synthesis kinetics in human primary monocyte-derived macrophages. J Biol Chem 287(26):21570–21574
Kim M et al (2014) A primary CD4(+) T cell model of HIV-1 latency established after activation through the T cell receptor and subsequent return to quiescence. Nat Protoc 9(12):2755–2770
Kim Y, Anderson JL, Lewin SR (2018) Getting the "Kill" into "Shock and Kill": strategies to eliminate latent HIV. Cell Host Microbe 23(1):14–26
Koppensteiner H, Brack-Werner R, Schindler M (2012) Macrophages and their relevance in human immunodeficiency virus type I infection. Retrovirology 9:82
Krombach F et al (1997) Cell size of alveolar macrophages: an interspecies comparison. Environ Health Perspect **105**(Suppl 5):1261–1263
Kwong PD et al (2002) HIV-1 evades antibody-mediated neutralization through conformational masking of receptor-binding sites. Nature 420(6916):678–682
Le Bras M, Rouy I, Brenner C (2006) The modulation of inter-organelle cross-talk to control apoptosis. Med Chem 2(1):1–12
Lee SA et al (2014) Impact of HIV on CD8+ T cell CD57 expression is distinct from that of CMV and aging. PLoS One 9(2):e89444
León-Rivera R et al (2021) Central nervous system (CNS) viral seeding by mature monocytes and potential therapies to reduce CNS viral reservoirs in the cART era. MBio 12(2)
Lerner AM, Eisinger RW, Fauci AS (2020) Comorbidities in persons with HIV: the lingering challenge. JAMA 323(1):19–20
Li GH, Henderson L, Nath A (2016) Astrocytes as an HIV reservoir: mechanism of HIV infection. Curr HIV Res 14(5):373–381
Liu J et al (2014) IL-10-producing B cells are induced early in HIV-1 infection and suppress HIV-1-specific T cell responses. PLoS One 9(2):e89236
Locati M, Curtale G, Mantovani A (2020) Diversity, mechanisms, and significance of macrophage plasticity. Annu Rev Pathol 15:123–147
Lotfi S et al (2020) M-sec facilitates intercellular transmission of HIV-1 through multiple mechanisms. Retrovirology 17(1):20
Lum JJ, Badley AD (2003) Resistance to apoptosis: mechanism for the development of HIV reservoirs. Curr HIV Res 1(3):261–274
Lutgen V et al (2020) HIV infects astrocytes in vivo and egresses from the brain to the periphery. PLoS Pathog 16(6):e1008381
Marban C et al (2016) Targeting the brain reservoirs: toward an HIV cure. Front Immunol 7:397
Martin-Serrano J, Zang T, Bieniasz PD (2003) Role of ESCRT-I in retroviral budding. J Virol 77(8):4794–4804
Meintjes G et al (2008) Tuberculosis-associated immune reconstitution inflammatory syndrome: case definitions for use in resource-limited settings. Lancet Infect Dis 8(8):516–523
Melvan JN et al (2012) Alcohol impairs the myeloid proliferative response to bacteremia in mice by inhibiting the stem cell antigen-1/ERK pathway. J Immunol 188(4):1961–1969
Merbecks MB et al (2020) Intermediate monocytes exhibit higher levels of TLR2, TLR4 and CD64 early after congenital heart surgery. Cytokine 133:155153
Mitchell BI, Laws EI, Ndhlovu LC (2019) Impact of myeloid reservoirs in HIV cure trials. Curr HIV/AIDS Rep 16(2):129–140
Morita E, Sundquist WI (2004) Retrovirus budding. Annu Rev Cell Dev Biol 20:395–425
Murphy AJ et al (2022) Buprenorphine reverses neurocognitive impairment in EcoHIV infected mice: a potential therapy for HIV-NCI. Front Immunol 13:1004985
Narasimhan PB et al (2019) Nonclassical monocytes in health and disease. Annu Rev Immunol 37:439–456

Nguyen LA et al (2014) Host SAMHD1 protein promotes HIV-1 recombination in macrophages. J Biol Chem 289(5):2489–2496
Okafo G, Prevedel L, Eugenin E (2017) Tunneling nanotubes (TNT) mediate long-range gap junctional communication: implications for HIV cell to cell spread. Sci Rep 7(1):16660
Oya M et al (2023) Increased glutamate and glutamine levels and their relationship to astrocytes and dopaminergic transmissions in the brains of adults with autism. Sci Rep 13(1):11655
Palmer LD et al (1997) Telomere length, telomerase activity, and replicative potential in HIV infection: analysis of CD4+ and CD8+ T cells from HIV-discordant monozygotic twins. J Exp Med 185(7):1381–1386
Pan X et al (2013) Restrictions to HIV-1 replication in resting CD4+ T lymphocytes. Cell Res 23(7):876–885
Peng G et al (2007) Myeloid differentiation and susceptibility to HIV-1 are linked to APOBEC3 expression. Blood 110(1):393–400
Poli G, Fauci AS (1992) The role of monocyte/macrophages and cytokines in the pathogenesis of HIV infection. Pathobiology 60(4):246–251
Prevedel L et al (2019) Identification, localization, and quantification of HIV reservoirs using microscopy. Curr Protoc Cell Biol 82(1):e64
Price TO et al (2005) HIV-1 viral proteins gp120 and tat induce oxidative stress in brain endothelial cells. Brain Res 1045(1–2):57–63
Ranki A et al (1995) Abundant expression of HIV Nef and Rev proteins in brain astrocytes in vivo is associated with dementia. AIDS 9(9):1001–1008
Real F et al (2018) Live imaging of HIV-1 transfer across T cell Virological synapse to epithelial cells that promotes stromal macrophage infection. Cell Rep 23(6):1794–1805
Real F et al (2022) S100A8-mediated metabolic adaptation controls HIV-1 persistence in macrophages in vivo. Nat Commun 13(1):5956
Real F, Capron C, Sennepin A, Arrigucci R, Sannier G, Zhu A, Zheng J, Xu L, Massé J-M, Greffe S, Eugenin E, Gennaro ML, Rouveix E, Bordé EC, Bomsel M (2020) Platelets from cART-suppressed HIV-infected patients with poor CD4+T cell recovery carry infectious HIV. Sci Transl Med 18:12(535):eaat6263
Röszer T (2018) Understanding the biology of self-renewing macrophages. Cells 7(8)
Salazar-Gonzalez JF et al (2008) Deciphering human immunodeficiency virus type 1 transmission and early envelope diversification by single-genome amplification and sequencing. J Virol 82(8):3952–3970
Salminen A et al (2013) Impaired autophagy and APP processing in Alzheimer's disease: the potential role of Beclin 1 interactome. Prog Neurobiol 106-107:33–54
Saphire AC et al (2001) Syndecans serve as attachment receptors for human immunodeficiency virus type 1 on macrophages. J Virol 75(19):9187–9200
Sattentau QJ, Stevenson M (2016) Macrophages and HIV-1: an unhealthy constellation. Cell Host Microbe 19(3):304–310
Schüssler M et al (2023) Gene editing of SAMHD1 in macrophage-like cells reveals complex relationships between SAMHD1 phospho-regulation, HIV-1 restriction, and cellular dNTP levels. MBio 14(5):e0225223
Schyns J et al (2019) Non-classical tissue monocytes and two functionally distinct populations of interstitial macrophages populate the mouse lung. Nat Commun 10(1):3964
Sender R et al (2023) The total mass, number, and distribution of immune cells in the human body. Proc Natl Acad Sci USA 120(44):e2308511120
Shapouri-Moghaddam A et al (2018) Macrophage plasticity, polarization, and function in health and disease. J Cell Physiol 233(9):6425–6440
Shaw GM, Hunter E (2012) HIV transmission. Cold Spring Harb Perspect Med 2(11)
Shepard C et al (2019) Effect of induced dNTP pool imbalance on HIV-1 reverse transcription in macrophages. Retrovirology 16(1):29
Sheu KM, Hoffmann A (2022) Functional hallmarks of healthy macrophage responses: their regulatory basis and disease relevance. Annu Rev Immunol 40:295–321

Siggins RW et al (2014) Dysregulation of myelopoiesis by chronic alcohol administration during early SIV infection of rhesus macaques. Alcohol Clin Exp Res 38(7):1993–2000
Siliciano JD, Siliciano RF (2020) Nonsuppressible HIV-1 viremia: a reflection of how the reservoir persists. J Clin Invest 130(11):5665–5667
Smith DB, Simmonds P, Bell JE (2014) Brain viral burden, neuroinflammation and neurodegeneration in HAART-treated HIV positive injecting drug users. J Neurovirol 20(1):28–38
Sokoya T et al (2017) HIV as a cause of immune activation and immunosenescence. Mediat Inflamm 2017:6825493
Sviridov D, Bukrinsky M (2023) Neuro-HIV-New insights into pathogenesis and emerging therapeutic targets. FASEB J 37(12):e23301
Swanstrom R, Coffin J (2012) HIV-1 pathogenesis: the virus. Cold Spring Harb Perspect Med 2(12):a007443
Tan J, Sattentau QJ (2013) The HIV-1-containing macrophage compartment: a perfect cellular niche? Trends Microbiol 21(8):405–412
Tardito S et al (2015) Glutamine synthetase activity fuels nucleotide biosynthesis and supports growth of glutamine-restricted glioblastoma. Nat Cell Biol 17(12):1556–1568
Telwatte S et al (2018) Gut and blood differ in constitutive blocks to HIV transcription, suggesting tissue-specific differences in the mechanisms that govern HIV latency. PLoS Pathog 14(11):e1007357
Uhl J et al (2019) Myosin-X is essential to the intercellular spread of HIV-1 Nef through tunneling nanotubes. J Cell Commun Signal 13(2):209–224
Valdebenito S et al (2021) Astrocytes are HIV reservoirs in the brain: a cell type with poor HIV infectivity and replication but efficient cell-to-cell viral transfer. J Neurochem 158(2):429–443
Valdebenito S et al (2023) The role of tunneling nanotubes during early stages of HIV infection and reactivation: implications in HIV cure. NeuroImmune Pharm Ther 2(2):169–186
Veenhuis RT et al (2023) Monocyte-derived macrophages contain persistent latent HIV reservoirs. Nat Microbiol 8(5):833–844
Veenstra M et al (2019) CCR2 on peripheral blood CD14(+)CD16(+) monocytes correlates with neuronal damage, HIV-associated neurocognitive disorders, and peripheral HIV DNA: reseeding of CNS reservoirs? J Neuroimmune Pharmacol 14(1):120–133
Velasquez S et al (2020) Circulating levels of ATP is a biomarker of HIV cognitive impairment. EBioMedicine 51:102503
Verheij E et al (2023) Long-term evolution of comorbidities and their disease burden in individuals with and without HIV as they age: analysis of the prospective AGE(h)IV cohort study. Lancet HIV 10(3):e164–e174
Wallis ZK, Williams KC (2022) Monocytes in HIV and SIV infection and aging: implications for Inflamm-aging and accelerated aging. Viruses 14(2)
Wang M et al (2021) Two populations of self-maintaining monocyte-independent macrophages exist in adult epididymis and testis. Proc Natl Acad Sci USA 118(1)
Wang X et al (2009) c-FLIPL regulates PKC via AP-2 to inhibit Bax-mediated apoptosis induced by HIV-1 gp120 in Jurkat cells. Mol Cell Biochem 330(1–2):23–29
Wang Z et al (2023) An immune cell atlas reveals the dynamics of human macrophage specification during prenatal development. Cell 186(20):4454–4471.e19
Wculek SK et al (2022) Metabolism of tissue macrophages in homeostasis and pathology. Cell Mol Immunol 19(3):384–408
Wei Y, Ho YC (2022) Interferon opens up: HIV-induced inflammation reconfigures 3D chromatin conformation and affects where HIV integrates. Mol Cell 82(24):4585–4587
White JA et al (2023) Clonally expanded HIV-1 proviruses with 5′-leader defects can give rise to nonsuppressible residual viremia. J Clin Invest 133(6)
Willenborg, S., L. Injarabian, and S.A. Eming, Role of macrophages in wound healing. Cold Spring Harb Perspect Biol, 2022. 14(12)a041216

Williams B et al (2016) Short communication: SCD14 and SCD163 levels are correlated with VACS index scores: initial data from the blunted immune recovery in CORE patients with HIV (BIRCH) cohort. AIDS Res Hum Retrovir 32(2):144–147
Williams DW et al (2012) Monocyte maturation, HIV susceptibility, and transmigration across the blood brain barrier are critical in HIV neuropathogenesis. J Leukoc Biol 91(3):401–415
Williams DW et al (2013) Mechanisms of HIV entry into the CNS: increased sensitivity of HIV infected CD14+CD16+ monocytes to CCL2 and key roles of CCR2, JAM-A, and ALCAM in diapedesis. PLoS One 8(7):e69270
Williams DW et al (2014a) Monocytes mediate HIV neuropathogenesis: mechanisms that contribute to HIV associated neurocognitive disorders. Curr HIV Res 12(2):85–96
Williams DW et al (2014b) CCR2 on CD14(+)CD16(+) monocytes is a biomarker of HIV-associated neurocognitive disorders. Neurol Neuroimmunol Neuroinflamm 1(3):e36
Williams ME et al (2020) Signatures of HIV-1 subtype B and C Tat proteins and their effects in the neuropathogenesis of HIV-associated neurocognitive impairments. Neurobiol Dis 136:104701
Wilson EB, Brooks DG (2011) The role of IL-10 in regulating immunity to persistent viral infections. Curr Top Microbiol Immunol 350:39–65
Wise DR, Thompson CB (2010) Glutamine addiction: a new therapeutic target in cancer. Trends Biochem Sci 35(8):427–433
Yap J et al (2023) Macrophages in cardiac remodelling after myocardial infarction. Nat Rev Cardiol 20(6):373–385
Yasen A et al (2017) Release of HIV-1 sequestered in the vesicles of oral and genital mucosal epithelial cells by epithelial-lymphocyte interaction. PLoS Pathog 13(2):e1006247
Yukl SA et al (2018) HIV latency in isolated patient CD4(+) T cells may be due to blocks in HIV transcriptional elongation, completion, and splicing. Sci Transl Med 10(430)
Yunna C et al (2020) Macrophage M1/M2 polarization. Eur J Pharmacol 877:173090
Zhao Y et al (2018) The origins and homeostasis of monocytes and tissue-resident macrophages in physiological situation. J Cell Physiol 233(10):6425–6439
Zhu A et al (2022) Infection of lung megakaryocytes and platelets by SARS-CoV-2 anticipate fatal COVID-19. Cell Mol Life Sci 79(7):365
Ziegler-Heitbrock HW et al (1991) Small (CD14+/CD16+) monocytes and regular monocytes in human blood. Pathobiology 59(3):127–130
Ziegler-Heitbrock L (2007) The CD14+ CD16+ blood monocytes: their role in infection and inflammation. J Leukoc Biol 81(3):584–592
Ziegler-Heitbrock L et al (2010) Nomenclature of monocytes and dendritic cells in blood. Blood 116(16):e74–e80